Graduate Texts in Physics

Graduate Texts in Physics

Graduate Texts in Physics publishes core learning/teaching material for graduate- and advanced-level undergraduate courses on topics of current and emerging fields within physics, both pure and applied. These textbooks serve students at the MS- or PhD-level and their instructors as comprehensive sources of principles, definitions, derivations, experiments and applications (as relevant) for their mastery and teaching, respectively. International in scope and relevance, the textbooks correspond to course syllabi sufficiently to serve as required reading. Their didactic style, comprehensiveness and coverage of fundamental material also make them suitable as introductions or references for scientists entering, or requiring timely knowledge of, a research field.

More information about this series at http://www.springer.com/series/8431

Jorge Loureiro • Jayr Amorim

Kinetics and Spectroscopy of Low Temperature Plasmas

Springer

Jorge Loureiro
Instituto Superior Técnico
Universidade de Lisboa
Lisbon, Portugal

Jayr Amorim
Instituto Tecnológico de Aeronáutica
São José dos Campos
São Paulo, Brazil

The original version of this book was revised. An erratum to this book can be found at DOI 10.1007/978-3-319-09253-9_12

ISSN 1868-4513 ISSN 1868-4521 (electronic)
Graduate Texts in Physics
ISBN 978-3-319-79166-1 ISBN 978-3-319-09253-9 (eBook)
DOI 10.1007/978-3-319-09253-9

Softcover reprint of the hardcover 1st edition 2016

Printed on acid-free paper

This Springer imprint is published by Springer Nature
The registered company is Springer International Publishing AG Switzerland

Preface

This textbook is a general introduction to the electron kinetics and plasma spectroscopy in the context of electrical gas discharges. It was designed to be read by advanced undergraduate and graduate students that have followed before an introductory course on plasma physics, of the type of Francis F. Chen's book (*Introduction to Plasma Physics and Controlled Fusion*), and that want to pursue their studies by acquiring basic knowledge in electron kinetics and plasma spectroscopy. Both topics are presented assuming that no significant previous knowledge exists. The exposition is based on other textbooks and some journal references, but in this latter case, the choice of the references reflects only a pedagogical option of the authors. The journal references listed in the book do not intend to cover a given topic in a very exhaustive and updated way because, in authors' opinion, this is not relevant to whom that intends to establish a first contact with the area. The aim of this book is therefore to supply the students with a comprehensive tool in which the basic concepts and formulae are totally derived and to which they can easily return whenever they need. The book is divided in two parts.

The structure of the first part is as follows. Chapter 1 presents the fundamentals of electrical gas discharges. It describes in general trends as a gas discharge works from a microscopic point of view using in certain passages a qualitative description only. Chapter 2 is devoted to the transport Boltzmann equation, in a first stage remembering the conditions of application to a gas of neutral molecules and in a second stage with its further application to a gas of electrons in the case of electron-molecule collisions. Chapter 3 presents the analysis and the solutions of the electron Boltzmann equation in velocity space for the case of an applied direct-current electric field. Chapter 4 presents an extension of the previous chapter to the case of a time-varying electric field, both for the case of high-frequency fields, where no time modulation exists in the so-called electron energy distribution function, and for the case of radio-frequency fields, where large time modulation may exist. Chapter 5 treats the Boltzmann analysis in the space of positions by considering the electron transport and the electron diffusion. Since the diffusion constitutes a loss term for electrons, other source or sink terms need to be included as well for consistency, such as electron ionization and electron attachment. Finally, Chap. 6

treats different aspects associated with the presence of space-charge electric fields in the medium. This firstly includes the concept of ambipolar diffusion, but other aspects are analysed as well, such as the transition from ambipolar to free diffusion as the space-charge density decreases, either in a discharge or at the beginning of a post-discharge, the ambipolar diffusion with an external magnetic field, the Boltzmann equation in a glow discharge, in which a radial space-charge electric field exists, and some Boltzmann treatments such as the local-field approximation that cannot be used anymore.

The second part of the textbook is composed of five chapters with the aim of introducing the students to equilibria in plasmas, spectroscopy diagnostics of electrical discharges and finally an overview of current applications of these plasmas. Chapter 7 presents and discusses basic concepts of the most relevant collisional-radiative models usually found in low-temperature plasmas. The corona model is introduced followed by a description of the excitation-saturation balance and the partial local Saha equilibrium. The possibilities of optical emission spectroscopy and its limitations and the interpretation of spectra are shown in Chap. 8. Some notions of line radiation and reminders of atomic and molecular physics are given without being exhaustive about the subject. In subsequent sections, the authors present the notions of spontaneous emission, absorption and stimulated emission and a brief discussion of molecular bands. Applications of optical emission spectroscopy to infer some plasma parameters, such as electron density and gas and vibrational temperatures, are shown. Some experimental techniques such as actinometry and titration for determination of species concentrations are presented. In Chap. 9, the bases of the incoherent absorption are presented and discussed. A case study of metastable kinetics in the argon positive column is proposed with the aid of classical absorption spectroscopy. Chapter 10 is devoted to introducing the principles of laser spectroscopy followed by an explanation of many gas lasers, solid-state lasers and liquid lasers. Experiments with absorption of one photon are presented together with absolute density measurements. Multiphoton laser-induced fluorescence is studied, and the bases of multiphoton absorption are addressed. The purpose of Chap. 11 is not to be exhaustive but to present some important industrial and technological applications of discharge plasmas. The fields where low-temperature plasmas are being employed today are vast and rapidly growing and cannot be described in a single textbook chapter. Much will be left out; however, a special emphasis is given in new breakthrough applications of plasmas in health science, production of biofuels and agriculture.

The sequence of the chapters in the book was thought having in mind that either advanced undergraduate or graduate students should be introduced to the field of kinetics and spectroscopy of low-temperature plasmas in at pedagogical and self-contained format. The authors suggest to the instructors of a two-semester course, for advanced undergraduate students, the exclusion of the following sections and subsections in Part I: Sects. 3.4, 3.5 and 3.6 in Chap. 3, Sects. 4.2 and 4.3 in Chap. 4, Sect. 5.3 in Chap. 5 and Sects. 6.2.3 and 6.2.4 in Chap. 6. In Part II, the sections from 8.4 to 8.7 may be removed from Chap. 8, as well as the Sects. 10.2

to 10.4 in Chap. 10. In case of graduate students, all the contents are advised and may be covered in a two-semester course.

This textbook relies on the experience of more than 20 years of Jorge Loureiro in teaching the plasma kinetics of low-temperature plasmas at Instituto Superior Técnico (IST), Universidade de Lisboa, Portugal, and of Jayr Amorim in teaching the different techniques and diagnostics of plasma spectroscopy at Instituto Tecnológico de Aeronáutica (ITA), São José dos Campos, Brazil. The authors would like hence to thank both institutions for all the support received along the teaching of these matters, as well as for the encouragement received from many colleagues and the contributions from students to improve this book. One of the authors (J. Loureiro) would also like to thank the support received from Instituto de Plasmas e Fusão Nuclear, which is the centre where his research has been realized at IST.

Lisboa, Portugal
São José dos Campos, Brazil
June 2016

Jorge Loureiro
Jayr Amorim

Contents

Part II Plasma Spectroscopy

Part I
Electron Kinetics

Chapter 1
Fundamentals of Electrical Gas Discharges

This chapter is a brief, and in certain passages qualitative, description of how an electrical gas discharge works. The main purpose of Part I of this textbook is not to present a detailed description of the operation of the various types of discharges, but instead to analyse the various aspects of the electron kinetics that are in the origin of the electrical discharges operation from a microscopic point of view. However, the present chapter is not in line with this perspective, since its purpose is to bring the reader to the physical system under study, which will be studied from Chap. 2 throughout based on the electron Boltzmann transport equation.

1.1 Non-self-Sustained Discharges

1.1.1 Primary Discharge Characteristics

A gas becomes a conductor of electricity if by any mechanism primary free charges are produced. Usually the electrons and the positive ions are produced at the same rate, but due to the lighter mass of electrons, and consequently larger mobility, the electrons carry the majority of the current. In a gas there always exists a very small background conductivity produced by permanent external ionizing agents such as cosmic rays and natural radioactivity. They do not exist hence perfect electrical insulating gases. At the lower atmosphere layers of Earth, for instance, the electrical conductivity produced by the solar wind and cosmic rays is of the order of $\sigma_c \sim 10^{-14}\,\Omega^{-1}\,\text{m}^{-1}$, whereas in a good metallic conductor as copper this value is $\sim 6 \times 10^{7}\,\Omega^{-1}\,\text{m}^{-1}$ at 20 °C.

Let us consider first the conduction of an electrical current in a gas produced by an external ionizing agent strong enough to produce a non-self-sustained discharge. The discharge exists as long as the gas is under the effects of the external ionizing

J.M.A.H. Loureiro, J. de Amorim Filho, *Kinetics and Spectroscopy of Low Temperature Plasmas*, Graduate Texts in Physics,
DOI 10.1007/978-3-319-09253-9_1

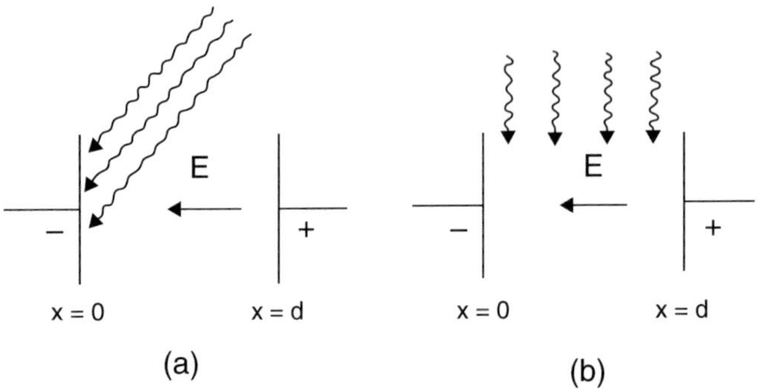

Fig. 1.1 (**a**) Photo-electron emission from the cathode with the surface rate γ_o; (**b**) Volume ionization originated by uniform radiation with the rate R_{ion} (Badareu and Popescu 1968)

agent and it vanishes when the ionizing source is removed. We assume that the current takes place between two electrodes localized in the gas, between which an electrical potential difference is applied, but with magnitude small enough in order the electrons in their movement do not gain energy larger than the ionizing threshold energy. Therefore, the ionization by multiplication does not occur. We also further assume that the space-charge density is negligibly small as compared with the charge in the electrode surfaces, so that the electric field established between the electrodes is not appreciably modified. The electrodes are geometrical slab, parallel, and large enough to assume a homogeneous electrical field in the gas. We will consider two situations in what concerns the external ionizing agent:

(i) The electron emission is produced from the cathode by an external radiating source characterized by the secondary emission coefficient γ_o, expressed in electrons $m^{-2}\,s^{-1}$ (see Fig. 1.1a);
(ii) The electron emission is produced by an external homogeneous radiating source acting upon the whole volume of the gas characterized by the photo-ionization rate R_{ion}, expressed in electron-ion pairs $m^{-3}\,s^{-1}$ (see Fig. 1.1b).

In the case *(i)* assuming the electric field directed towards the negative x axis, $\mathbf{E} = -E\,\mathbf{e_x}$, the electron current density on the cathode surface is given by the difference between the emission current density $\mathbf{J_{eo}} = -e\gamma_o\,\mathbf{e_x}$, with e denoting the absolute value of the electron charge, and the backward diffusion current due to the thermal agitation $\mathbf{J_{eb}} = en_e\langle v_e\rangle/4\,\mathbf{e_x}$ (see Appendix A.1.1), with n_e denoting the electron number density and $\langle v_e\rangle$ the average electron velocity,

$$\mathbf{J_e} = \left(-e\gamma_o + \frac{en_e\langle v_e\rangle}{4}\right)\mathbf{e_x}. \tag{1.1}$$

In a plan localized between the electrodes, the continuity of current $(\nabla.\,\mathbf{J_e}) = 0$, implies that

$$\mathbf{J_e} = en_e\mu_e\,\mathbf{E} = \text{const}, \tag{1.2}$$

being μ_e the electron mobility (here assumed >0) and $E = V/d$, with E, V and d denoting the field strength, the applied potential, and the interspacing electrode distance. Eliminating n_e between equations (1.1) and (1.2), we obtain

$$\mathbf{J_e} = \frac{e\gamma_o\mu_e}{\mu_e\,E + \langle v_e\rangle/4}\,\mathbf{E}, \tag{1.3}$$

At low electric fields such that $\mu_e\,E \ll \langle v_e\rangle/4$, the discharge current is linear with the field, $\mathbf{J_e} = \sigma_{ce}\,\mathbf{E}$, being $\sigma_{ce} = 4e\gamma_o\mu_e/\langle v_e\rangle$ the electron conductivity. At high enough fields we obtain the saturation current $\mathbf{J_e} = \mathbf{J_{eo}} = -\,e\gamma_o\,\mathbf{e_x}$, in which all emitted electrons from the cathode travel along the interelectrode spacing and are captured at the anode. Since $E = V/d$ the current-potential characteristic $J_e(V)$ is similar to the relation $J_e = J_e(E)$ given by equation (1.3).

Let us consider now the case *(ii)* in which the discharge gap between the large parallel plane electrodes is uniformly irradiated, for instance by a strong X-rays source, producing a uniform ionization throughout the gas volume at a rate R_{ion} of electron-ion pairs produced per volume and time units. For a relatively low electric field the electrons are mainly lost by electron-ion recombination in the gas volume, being negligible the number of electrons arrived at the anode. Then, the electron density is determined by the balance equation assuming a second order process for electron-ion recombination

$$\frac{dn_e}{dt} = R_{ion} - \alpha\,n_e\,n_i, \tag{1.4}$$

being α the electron-ion recombination coefficient in $m^{-3}s^{-1}$. Under steady-state conditions and assuming the quasineutrality of the medium, we simply obtain $n_e = \sqrt{R_{ion}/\alpha}$.

However, if the applied voltage is gradually increased, while the strength of the irradiation is kept constant, the current rises, at the beginning relatively steeply and later on more slowly, until it becomes constant at a relatively large voltage. This indicates that at weak fields only a small portion of the electrons and ions produced can reach the electrodes because most of them recombine in the gas before having reached the electrodes. On the contrary, at high fields most of the charged species are captured at the electrodes. In this latter case, the saturation electron particle current density, i.e. the number of electrons removed from the interspacing distance d per surface and time units, is $\Gamma_e = dR_{ion}$, so that the electron current density is $\mathbf{J_e} = -\,edR_{ion}\,\mathbf{e_x}$. In the range of medium field strengths, some electrons recombine in the volume gap while others are neutralized at the electrodes. Assuming the loss rates of two mechanisms in the same ratio throughout the whole volume, we may write

$$\begin{aligned}\frac{dn_e}{dt} &= R_{ion} - \left(\frac{dn_e}{dt}\right)_{rec} - \left(\frac{dn_e}{dt}\right)_{elec} \\ &= R_{ion} - \alpha\,n_e^2 - \frac{J_e}{ed},\end{aligned} \tag{1.5}$$

so that we obtain at stationary conditions

$$n_e = \sqrt{\frac{1}{\alpha}\left(R_{ion} - \frac{J_e}{ed}\right)}. \tag{1.6}$$

The total discharge current density ($\mathbf{J} = -\,J\,\mathbf{e_x}$) is

$$J = J_e + J_i = e\,(n_e\mu_e + n_i\mu_i)\,E \simeq en_e\mu_e\,E, \tag{1.7}$$

due to the large differences between the mobilities of electrons and ions, while for the two charged species densities we have $n_e \simeq n_i$. Substituting equation (1.6) in (1.7) and resolving the quadratic form for J_e, we obtain

$$J_e = e\,\delta\left(-1 + \sqrt{1 + (2d/\delta)\,R_{ion}}\right), \tag{1.8}$$

with $\delta = \mu_e^2 E^2/(2\alpha d)$. The characteristic $J_e = J_e(E)$ given by equation (1.8) can be represented in Fig. 1.2. The two asymptotes are for equation (1.7) with $n_e = \sqrt{R_{ion}/\alpha}$, in the case of low electric fields, and the saturation electron current density $J_{es} = edR_{ion}$, in the case of high electric fields.

1.1.2 Space-Charge Effects

So far it has been assumed that the charges move through the gas independently of each other and that the charge per volume unit is so small that the electric field at any point within the gap is constant (i.e. $(\nabla.\ \mathbf{E}) = \rho/\epsilon_0 \simeq 0$). The field is simply obtained from the applied potential on the plane parallel electrodes $E = V/d$. However, when the current density increases the field will also depend on the distribution of the charges in volume. The current density at which the field

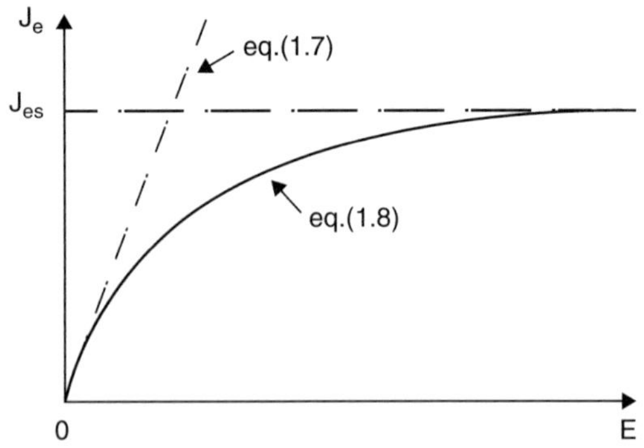

Fig. 1.2 Dependence of the electron current density with the electric field in a plane parallel condenser when the gas is uniformly irradiated. J_{es} is the saturation electron current density (von Engel 1965)

distortion by space-charge occurs can be estimated by comparing the surface charge density on the electrodes, $\sigma = \epsilon_0 E$, with that from charges in volume (whose field lines are assumed to be collected at the electrodes), $\sigma^* = en_e d$. Since $J_e = en_e \mu_e E$, we obtain for this latter $\sigma^* = J_e d/(\mu_e E)$. The distortion of the field occurs when $\sigma^* \geq \sigma$, i.e. at electron discharge currents larger than

$$J_e \geq \frac{\epsilon_0 \mu_e}{d} E^2. \tag{1.9}$$

Let us start by considering a discharge in a gas between two large plane electrodes in which the cathode emits electrons uniformly (see Fig. 1.1a), obtained in practice by irradiating or heating the cathode. The nature of the gas and the pressure will determine the mobility of the emitted electrons. The current density is now assumed to be large enough to make it necessary to consider field distortion by the space-charge. Under these circumstances the current-voltage discharge characteristic can be derived from the steady-state continuity equation $(\nabla . \mathbf{J_e}) = 0$, from which $\mathbf{J_e} = en_e\mu_e \mathbf{E} = \text{const}$, and Poisson's equation $(\nabla.\mathbf{E}) = -en_e/\epsilon_0$. Assuming $\mathbf{J_e} = -J_e \mathbf{e_x}$ and $\mathbf{E} = -E\,\mathbf{e_x}$, we may combine the two equations as

$$\frac{dE}{dx} = \frac{J_e}{\epsilon_0 \mu_e E}. \tag{1.10}$$

Assuming also $\mu_e = \text{const}$, this equation can be integrated from $(x = 0, E_0)$ to $(x, E(x))$, yielding to write

$$E^2(x) = E_0^2 + \frac{2J_e}{\epsilon_0 \mu_e} x. \tag{1.11}$$

Since the electrons are produced at nearly unlimited number, we may assume that the field at the cathode is very small (if $n_{e0} \to \infty$ then $E_0 \to 0$ because $J_e = \text{const}$) and we may write

$$E(x) = \sqrt{\frac{2J_e}{\epsilon_0 \mu_e}}\, x^{1/2}, \tag{1.12}$$

whereas for the electrical potential, we obtain from $E = dV/dx$ with $V_0 = 0$,

$$V(x) = \frac{2\sqrt{2}}{3}\sqrt{\frac{J_e}{\epsilon_0 \mu_e}}\, x^{3/2}, \tag{1.13}$$

while for the electron density, we get from $n_e = J_e/(e\mu_e E)$

$$n_e(x) = \frac{1}{e}\sqrt{\frac{\epsilon_0 J_e}{2\mu_e}}\, x^{-1/2}. \tag{1.14}$$

A plot of $E(x)$, $V(x)$ and $n_e(x)$ is shown in Fig. 1.3 for typical conditions.

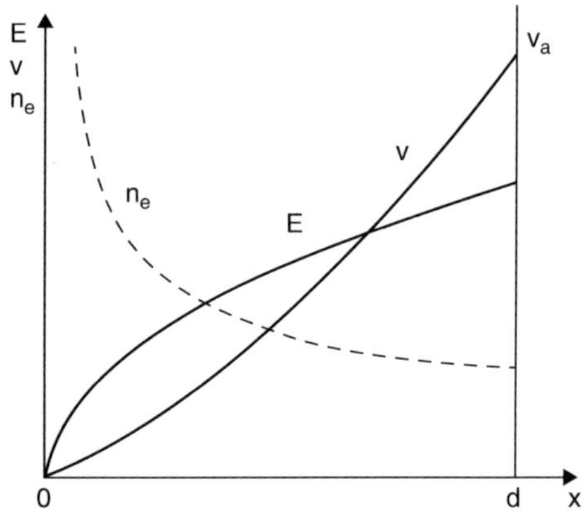

Fig. 1.3 Spatial distributions of the electric field E, potential V, and electron density n_e in a gas, when the electrons are emitted from the cathode at $x = 0$ with zero velocity, assuming a collisional regime with constant mobility μ_e (von Engel 1965)

These very simple expressions are valid in a collisional regime, in which we may use the concept of electron mobility $\mu_e = e/(m\nu_m)$, in which ν_m denotes the electron collision frequency for momentum transfer and m the electron mass. However, at sufficiently low values of pressure such that the electron mean free path, λ, is much larger than the interspacing distance between the electrodes, d, we should replace $\mathbf{J_e} = en_e\mu_e\,\mathbf{E}$ with the equation $\mathbf{J_e} = -\,en_e\mathbf{v_{ed}} = \text{const}$, being $\mathbf{v_{ed}} = \langle\mathbf{v_e}\rangle$ the average electron vector velocity, i.e. the electron drift velocity, obtained from energy conservation $eV = \frac{1}{2}\,mv_{ed}^2$, with $v_{ed} = |\mathbf{v_{ed}}|$ and $v_{ed}(0) = 0$ at the cathode. From these two equations we obtain $en_e = J_e\sqrt{m/2eV}$, and this relation may be inserted into Poisson's equation for the electrical potential $\nabla^2 V = en_e/\epsilon_0$ obtaining

$$\frac{d^2V}{dx^2} = C\,V^{-1/2}, \tag{1.15}$$

with

$$C = \frac{J_e}{\epsilon_0}\sqrt{\frac{m}{2e}} = \text{const.} \tag{1.16}$$

The equation (1.15) has the solution (see Appendix A.1.2)

$$V^{3/2} = \frac{9}{4}\,C\,x^2, \tag{1.17}$$

from which we obtain, assuming $V_0 = 0$ and $E_0 = 0$ at $x = 0$,

$$V(x) = \frac{3}{2}\left(\frac{3m}{4e}\right)^{1/3}\left(\frac{J_e}{\epsilon_0}\right)^{2/3} x^{4/3} \tag{1.18}$$

$$E(x) = \frac{dV}{dx} = 2\left(\frac{3m}{4e}\right)^{1/3}\left(\frac{J_e}{\epsilon_0}\right)^{2/3} x^{1/3} \tag{1.19}$$

$$n_e(x) = \frac{\epsilon_0}{e}\frac{dE}{dx} = \frac{2\epsilon_0}{3e}\left(\frac{3m}{4e}\right)^{1/3}\left(\frac{J_e}{\epsilon_0}\right)^{2/3} x^{-2/3}. \tag{1.20}$$

The dependence on x of the different quantities changes from the collision to the collisionless case. The potential, for instance, changes from $V \propto x^{1.5}$ to $\propto x^{1.33}$. From equations (1.13) and (1.18) we obtain the following relations between the electron current density and the potential V_a at the anode (x = d), respectively, for the collision and the collisionless case

$$J_e = \frac{9}{8} \frac{\epsilon_0 \, \mu_e}{d^3} V_a^2 \tag{1.21}$$

$$J_e = \frac{4}{9} \frac{\epsilon_0}{d^2} \sqrt{\frac{2e}{m}} V_a^{3/2}. \tag{1.22}$$

The equation (1.22) is known in the literature as the Child-Langmuir equation. Comparison of equations (1.21) and (1.22) shows that at sufficiently low pressure J_e is less sensitive to variations of either V_a or d. When the space-charge field effects are not neglected both current-voltage characteristics exhibit a more rapid growth than the linear dependence. In both derivations it is assumed that $E_0 = 0$ and $v_{ed}(0) = 0$ at $x = 0$, which leads to the electron density becomes infinite at the cathode. However, this difficulty can be overcome by allowing for non-null initial velocity of the emitted electrons.

Let us go back to the situation shown in Fig. 1.1b, in which a sufficiently powerful space-charge is now assumed to exist created by irradiation of the gas between two plane parallel electrodes submitted to a constant applied potential difference. Without the applied potential equal numbers of electrons and positive ions are uniformly distributed throughout the volume. However, when the potential is applied the electric field will move the electrons and the positive ions to opposite electrodes. Since their mobilities are different, the number of electrons and ions arriving per time unit at the electrodes and hence the electron and ion current densities will not be equal. This leads that at high enough current densities the electrons start to be repelled and the ions attracted by the opposite electrodes, so that in the front of the cathode an excess of positive charge is observed and correspondingly an excess of negative charge is observed in front of the anode. However, because of their larger mobilities the electrons cause a smaller field and are less accelerated to the anode than the positive ions to the cathode. This leads to the rates of removal of electrons and ions at both electrodes become equal.

Figure 1.4 shows the space-charge distribution $\rho(x)$ and the spatial distributions of the electric field and potential between the two electrodes, assuming $V(0) = 0$ at the cathode $x = 0$. For comparison it is also plotted with broken lines the electric field and potential in the absence of space-charge. Since the potential in the anode is kept constant, the areas enclosed by the $E(x)$ plot are the same. The field in the case of negligible space-charge is constant and the potential a straight line. With increasing space-charge the field, oriented towards the cathode, $\mathbf{E} = -\, E\, \mathbf{e_x}$, is obtained from Poisson's equation $dE/dx = -\, \rho/\epsilon_0$, with $\rho > 0$ near the cathode and $\rho < 0$ near the anode, which gives place to the existence of straight lines with negative and positive slopes, respectively. With respect to the potential it exists a curve with three parts. Near the electrodes the potential changes relatively rapid,

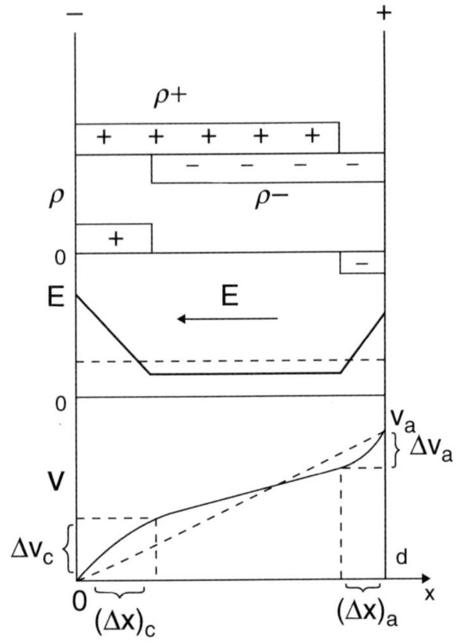

Fig. 1.4 Spatial distribution of space-charge $\rho(x)$, electric field $E(x)$, and potential $V(x)$, in a gas-filled condenser with plane electrodes and uniform ionization by an external irradiating volume source. The electrodes at $x = 0$ and $x = d$ are the cathode and anode, respectively. ΔV_c and ΔV_a are the cathode and anode potential falls, whereas $(\Delta x)_c$ and $(\Delta x)_a$ represent the widths of the two corresponding sheaths. V_a is the applied potential at the anode. The broken lines are for $E(x)$ and $V(x)$ when $\rho = 0$ (von Engel 1965)

which are usually termed as cathode and anode potential falls. The intermediate zone with a much smaller potential variation and a nearly constant field strength goes over into the positive column of a glow discharge.

1.1.3 *Ionization by Multiplication Due to Collisions*

We have seen in Sect. 1.1.1 that when the electric field increases, after a first stage where the electron current density increases, the current reaches the saturation with all electrons produced being captured at the anode. However, if the field strength is increased further the current will rise again. This is because at much larger fields some of the electrons originally formed by irradiation of the cathode or the gas will be accelerated in the field and can reach velocities which enable them to ionize the gas. Thus the discharge current density is no longer constant but rather than depends on the field strength (see Fig. 1.5).

Let us define by α the number of pairs of electrons and ions produced per unity length by the impact of an electron accelerated by the field at any point between the two electrodes, expressed from now on in m^{-1}. A primary electron produces α electrons and ions per unity length of its path in the field direction and these new electrons and ions after to be separated by the field move in opposite directions. The electrons originate then a multiplication process which ends at the anode.

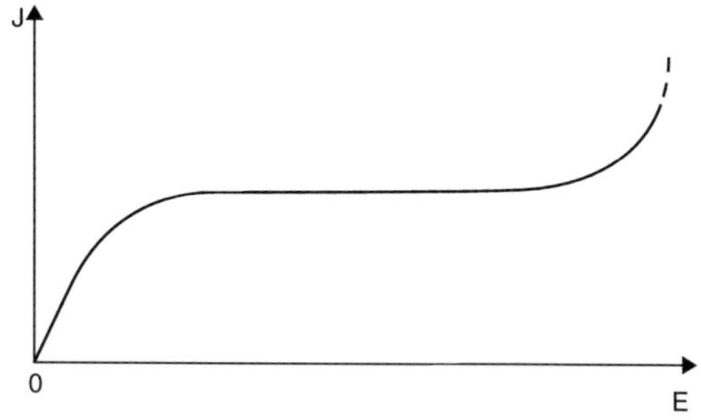

Fig. 1.5 Characteristic of the current density against the field in a gap between two plane electrodes showing the rise of J at large field strengths due to ionization by collisions superimposed on the initial ionization by irradiation of the cathode or the gas (von Engel 1965)

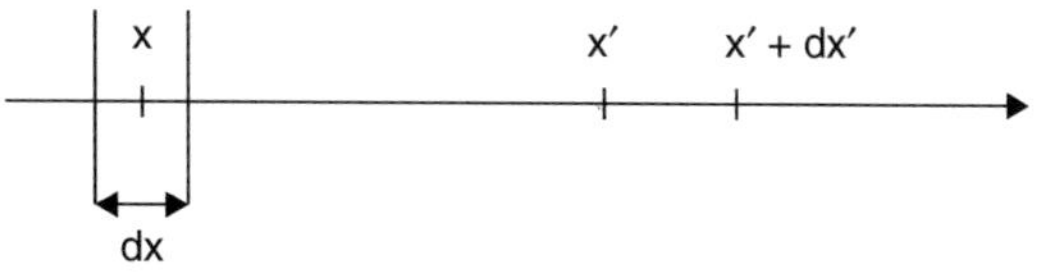

Fig. 1.6 Multiplication grid for the electrons in their path towards the anode

If the number of free paths in the gap is sufficiently large and distributed, we can treat ionization as occurring continuously throughout the space. In the case of the primary electrons leave the cathode and since the number of electron-ion pairs produced along the distance dx is $\alpha\ dx$, for N electrons at any position x with their origin in a sole primary electron leaving the cathode, we have a further multiplication in the length dx with more new $dN = N\alpha\ dx$ electron-ion pairs produced. By integrating this relation between $x = 0$ and $x =$ d, we obtain a final multiplication factor equal to $e^{\alpha d}$.

Let us consider now the case where the initial electrons are created at any point of the interspacing between the two electrodes by irradiating the gas, being R_{ion} the ionization rate (number of primary electrons produced per volume and time units). Thus, at any position x within the length dx, they are created $\Gamma(x) = R_{ion}\ dx$ primary electrons per area and time units. When these electrons arrive at position $x' > x$ they are accompanied by all electrons created by multiplication between x and x' (see Fig. 1.6).

Let it be $\Gamma(x', x)$ the total number of electrons at x', with origin in the electrons created at dx, and $d\Gamma(x') = \Gamma(x', x)\ \alpha\ dx'$ the number of further electrons produced between x' and $x' + dx'$. We have therefore $d\Gamma(x')/\Gamma(x', x) = \alpha\ dx'$ and this equation may be integrated between x and x' allowing to write

$$\Gamma(x', x) = R_{ion}\ e^{\alpha(x'-x)}\ dx. \tag{1.23}$$

At the anode $x' = d$ the number of electrons arrived per area and time units, with origin in the electrons created at dx, is

$$\Gamma(d, x) = R_{ion}\ e^{\alpha(d-x)}\ dx. \tag{1.24}$$

We need now to integrate over all positions x where the initial electrons are primarily created to obtain at the end

$$\Gamma = \int_0^d \Gamma(d,x) = R_{ion} \int_0^d e^{\alpha(d-x)}\, dx$$
$$= \frac{R_{ion}}{\alpha} \left(e^{\alpha d} - 1\right). \tag{1.25}$$

The electron current density is $J_e = e\Gamma$ and since, as we have seen in Sect. 1.1.1, the electron current density of saturation is $J_{es} = edR_{ion}$, we obtain

$$J_e = J_{es} \frac{\left(e^{\alpha d} - 1\right)}{\alpha d}. \tag{1.26}$$

We have seen before that as all initial electrons are created at the cathode the multiplication factor is $e^{\alpha d}$, so that in this case the electron current density is $J_e = J_{es}\, e^{\alpha d}$.

The equation (1.26) shows to exist an exponential growth for the electron current density completely different from the initial linear growth at low electric fields. This multiplication is often called multiplication α and the coefficient α (in m^{-1}) is known as first Townsend's ionization coefficient, with its value depending on the field strength E, the pressure p, and the nature of the gas. Although the coefficient α can only be properly derived by considering the electron kinetics, the ionization may be seen as a process activated by the energy gained from the field $eE\lambda$, with λ denoting the mean free path, in which the activation energy is the ionization threshold energy u_{ion}. This leads to a formula analogous to that of Arrhenius for thermally activated processes (von Engel 1965)

$$\alpha = \frac{1}{\lambda_{ion}} = \frac{k}{\lambda} \exp\left(-\frac{u_{ion}}{eE\lambda}\right), \tag{1.27}$$

being k a constant. Since the mean free path is inversely proportional to pressure, the coefficient α can also be written as

$$\alpha = Ap \exp\left(-\frac{Bp}{E}\right), \tag{1.28}$$

where A and B are two constants dependent on the nature of the gas. Being $<\nu_{ion}>$ the velocity-averaged ionization frequency, and $\mathbf{v_{ed}}$ the electron drift velocity, the coefficient α may also be expressed as $\alpha = <\nu_{ion}>/v_{ed}$.

1.1.4 Multiplication Including Secondary Effects

Up to now the positive ions have been neglected since due to their lower mobility the current of ions directed to the cathode is much smaller than the current of electrons to the anode. However, as we shall see below this is no longer true in a stationary discharge where the currents collected at each electrode have the same value. Nevertheless for sufficiently low space-charge fields this equilibration does not yet occur but it remains the need of analysing the effects produced by the impingement of ions on the cathode.

The effect associated with the impingement of ions on the cathode is the release of secondary electrons through a mechanism that depends on the cathode nature known in the literature as multiplication γ. The release of secondary electrons occurs also due to other processes, such as the arrival of photons at the cathode (δ effect), and due to the arrival of metastable atoms and molecules as well, but here we will consider only the impingement of ions. The number of secondary electrons emitted is equal to the product of the number of positive ions returning to the cathode by a coefficient γ known as second Townsend's ionization coefficient which depends of the cathode type, expressing the number of secondary electrons released by each positive ion arriving at the cathode. However, these secondary electrons ionize the gas in the same way as the primary electrons and the new ions so produced return to the cathode and release more electrons. This multiplication takes place an infinite number of times.

The total current can be calculated in the following manner considering the simplest case where the primary electrons leave the cathode. In this case as seen above each primary electron originates a multiplication α and at the anode arrive $e^{\alpha d}$ electrons. Each primary electron produces thus $(e^{\alpha d} - 1)$ electron-ion pairs along its travel to the anode. Then the $(e^{\alpha d} - 1)$ ions produced go to the cathode and originate $\gamma(e^{\alpha d} - 1)$ secondary electrons. These are multiplied again giving $\gamma(e^{\alpha d} - 1)e^{\alpha d}$ electrons at the anode and $\gamma(e^{\alpha d} - 1)^2$ new ions produced in the interspacing distance. These latter go to the cathode and produce new $\gamma^2(e^{\alpha d} - 1)^2$ electrons. The multiplication factor is the sum of all electrons entering the anode for one primary electron emitted at the cathode

$$\begin{aligned} m &= 1 + (e^{\alpha d} - 1) + \gamma\ (e^{\alpha d} - 1) + \gamma\ (e^{\alpha d} - 1)^2 + \gamma^2\ (e^{\alpha d} - 1)^2 + \ldots.. \\ &= e^{\alpha d}\ \left[1 + \gamma\ (e^{\alpha d} - 1) + \gamma^2\ (e^{\alpha d} - 1)^2 + \ldots..\right] \\ &= \frac{e^{\alpha d}}{1 - \gamma\ (e^{\alpha d} - 1)}. \end{aligned} \tag{1.29}$$

For an initial current released at the cathode Γ_{e0}, the ionization by electron collisions in the gas and secondary electron emission by the impingement of ions on the cathode allows to obtain the final current $\Gamma = m\ \Gamma_{e0}$ at the anode. A similar calculation could also be carried out for the case where the primary electrons are created at any point of the gas and not only at the cathode. The multiplication factor is in this case (see Appendix A.1.3)

$$m = \frac{(1 + \gamma)(e^{\alpha d} - 1)/\alpha d - \gamma\ e^{\alpha d}}{1 - \gamma\ (e^{\alpha d} - 1)}. \tag{1.30}$$

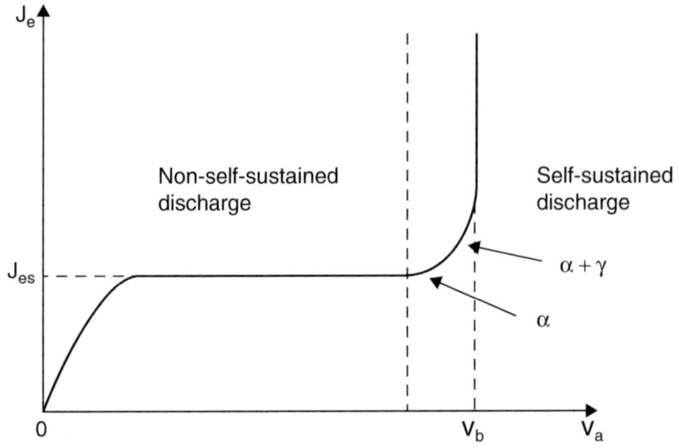

Fig. 1.7 Characteristic current-voltage of a non-self-sustained discharge. The regions α and $\alpha + \gamma$ indicate the multiplication α and the multiplication $\alpha + \gamma$. The discharge becomes self-sustained beyond the breakdown voltage V_b. J_{es} denotes the saturation current when all primary electrons are collected at the anode

Inspection of equations (1.29) and (1.30) shows that both expressions have the same denominator and when it approaches zero, provided this can take place without invalidating the original assumptions, the multiplication factor and hence the current density tends to infinity. Thus the condition for the breakdown of the gas contained in the gap between two plane parallel electrodes is (von Engel 1965)

$$\gamma\ (e^{\alpha d} - 1) = 1. \tag{1.31}$$

It means that the current in the discharge becomes unstable and thus a large current may develop without the presence of an external ionizing source. The discharge thus goes over into a self-sustained discharge. As referred above the first Townsend's ionization coefficient is a function of the electric field so that the condition (1.31) is attained when the applied voltage increases. Such potential is known as breakdown voltage. Figure 1.7 shows the current-voltage characteristic $J_e(V_a)$, with V_a denoting the potential at the anode when $V_c = 0$ is assumed at the cathode, from the very small current at the beginning up to the point where the breakdown occurs.

The condition (1.31) must be seen as a limit of stability only. In deriving equations (1.29) and (1.30) a uniform electric field and hence a constant α coefficient were assumed and it is difficult to see how this could be true if the current was allowed to attain a very large value which produces field distortion by intense space-charges. However, in many cases the field distortion becomes important only when the point of instability is reached. As a matter of fact, the difference between the value of the applied field at which space-charge distortion becomes important and the breakdown field is so small that the condition (1.31) represents a very good approximation for breakdown of gases even at high pressure (von Engel 1965).

Since γ essentially depends on the nature of the cathode, the equation (1.31) can be written under the form of a constant product for αd

$$\alpha d = \ln\left(\frac{1}{\gamma} + 1\right) = \text{const.} \tag{1.32}$$

Because the breakdown field and the voltage are linked each other in planar geometry as $E_b = V_b/d$, we can still write equation (1.32) as

$$V_b = \frac{E_b/p}{\alpha/p} \ln\left(\frac{1}{\gamma} + 1\right). \tag{1.33}$$

As we will see below using the kinetic theory, or simply from equation (1.28), the parameter α/p is a function of E/p only, so that the breakdown voltage becomes a function of E_b/p, that is a function $V_b = V_b(E_b/p)$ or $V_b = V_b(V_b/pd)$. Therefore, we can also express the relation for breakdown under the form $V_b = V_b(pd)$, which is known in the literature as Paschen's law (Druyvesteyn and Penning 1940). We note that equations (1.28) and (1.32) yield to write (Braithwaite 2000)

$$V_b = \frac{Bpd}{\ln(Apd) - \ln[\ln(1/\gamma + 1)]}. \tag{1.34}$$

Equation (1.34) gives the voltage necessary to start a discharge between two electrodes in a gas, as a function of the product of pressure and distance between the electrodes. The voltage necessary for breakdown first decreases up to a minimum as the pressure is reduced and then it gradually increases again. Figure 1.8 shows the breakdown potential as a function of the reduced interelectrode distance *pd* in Ne, Ar, H_2, Hg and air (von Engel 1965).

The minimum in the curve of breakdown potential against *pd* appears due to the following reasons. At low pressures the mean free path is large and the number of collisions with the gas of molecules is low, being hence also small the number of ionizing collisions. On the contrary, at larger pressures the mean free path is small and the number of collisions in the gas is high. Thus the number of collisions of other type rather than ionization collisions is very high and only few electrons acquire enough energy over a mean free path to ionize. Consequently in order to produce sufficient ionization between two electrodes the potential needs to increase either as the pressure decreases or the pressure increases from a minimum breakdown voltage that depends on the nature of the gas. A similar argument would apply for the variation on *d*.

Finally, it is worth noting here that the electric breakdown does not simply occur when the potential between the electrodes exceeds the ionization potential. The electric field has not only to produce ionization in the gas, but also to produce the multiplication of charges at such a rate that the current passing through the gas no longer requires external ionization. When the potential applied over the interelectrode distance is equal to the ionization potential each electron produces an electron-ion pair but no further multiplication of charges occurs.

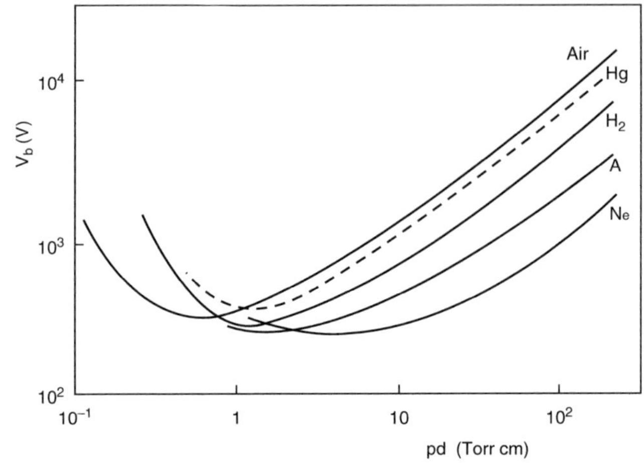

Fig. 1.8 Breakdown potential in Volt as a function of the reduced interelectrode distance *pd* in Torr cm (1 Torr = 133.3 Pa) for Ne, Ar, H_2, Hg and air (von Engel 1965)

1.2 Self-Sustained Discharges

1.2.1 Glow Discharge

As long as the discharge current is sufficiently low in order the space-charge field effects do not exist, typically for discharge current densities $J < 1\,\mu\mathrm{A\,cm}^{-2}$, the breakdown condition is expressed by equation (1.31) and this discharge is usually termed Townsend (or dark) discharge. This discharge is not visible due to the very low concentrations of atomic or molecular radiative species present in the gas being hence only detectable by an amperemeter inserted in the external circuit. However, when the current increases an accumulation of electrons and ions exist in front of the electrodes (see Fig. 1.4) deforming the structure of the field and producing a variation of the first Townsend's ionization coefficient α with the position. Then, the breakdown condition to be considered is

$$\gamma\left[\exp\left(\int_0^d \alpha(x)\,dx\right)-1\right]=1. \qquad (1.35)$$

This equation leads to the existence of smaller values for V_b than those in the absence of space-charge field effects. The voltage-current characteristics with the various types of discharges obtained when the discharge current increases is represented in Fig. 1.9.

Let us analyse now qualitatively the evolution towards a glow discharge (von Engel 1965; Raizer 1991; Lieberman and Lichtenberg 1994). When a long cylindrical glass tube with two plane electrodes at its ends is filled with a gas of

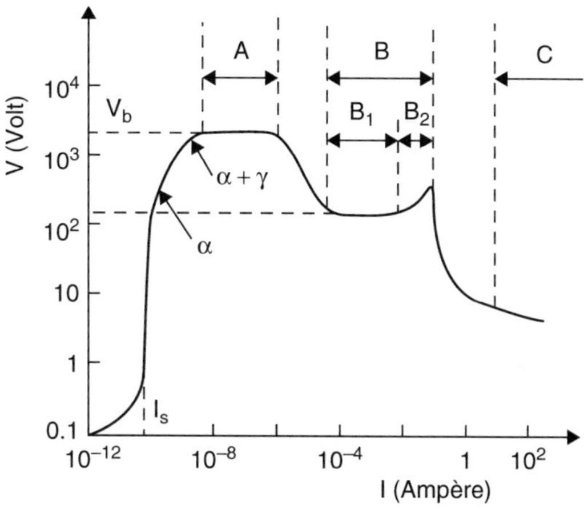

Fig. 1.9 Voltage-current characteristics showing the different types of discharges in a gas between two electrodes: *A* — Townsend discharge; *B* — Glow discharge (B_1 normal; B_2 abnormal); *C* — Arc discharge. V_b and I_s denote the breakdown voltage and the saturation current

order 1 Torr of pressure (i.e. 133.3 Pa) and the electric potential V between the electrodes is slowly increased, a small current of order 1 pA may be detected by an amperemeter inserted in the external circuit. This initial current is due to ionization at the electrodes and of the gas by background radiation of external agents. Then as V increases the ionization by multiplication α, in a first stage, and by $\alpha + \gamma$ mechanisms, in a subsequent stage, starts to take place and the current increases up to a point, corresponding to the breakdown voltage V_b, beyond which the discharge becomes self-sustained (Townsend discharge). At the same time some of the electrons will become attached to the inner glass wall and are partly neutralized by incoming positive ions. The glass wall becomes negative charged while an equal amount of positive charge will stay in the gas volume. However, the majority of positive ions and electrons will flow to the opposite electrodes.

As a result of the charged wall a radial electric field develops which restricts the flow of electrons to the wall. By increasing V still more, space-charge in volume appears leading to a distortion of the axial component of the field. At still larger voltages positive and negative charges are collected in front of the cathode and anode, respectively, and the current becomes large enough to produce a transition from the Townsend discharge to a more complex form known as glow discharge. In the glow discharge the charged regions in front of the electrodes are completely formed. At the same time the radial space-charge field (formed due to the presence of the negative charged wall) produces a reduction of the rate of electron loss by diffusion to the wall and therefore a reduction of the axial electric field necessary for sustaining the discharge. That is the system becomes more operative. In consequence the discharge voltage decreases and the current increases, being this latter mainly determined by the external circuit.

The glow discharge owes its name to the typical luminous glow. The discharge emits light because the electron energy and the number density are high enough to produce excited gas atoms by collisions, which relax to a lower state with emission of photons. The glow discharges have two different regimes: normal and abnormal.

In a normal glow discharge (region B_1 in Fig. 1.9) the voltage is almost constant and independent of the discharge current over several orders of magnitude. This results from the fact that at low currents only a small part of the cathode surface participates in the discharge and the further increase of the current is due to a progressive enlargement of the cathode surface covered by the discharge keeping the discharge current density approximately a constant value. On the contrary, in an abnormal glow discharge the whole surface of the cathode is already covered by the discharge and the only manner of the total discharge current to increase further is by an enhancement of the current density. This requires more energy for the electrons and consequently an increase of the applied voltage (region B_2 in Fig. 1.9). Except for being brighter, the abnormal glow discharge resembles the normal discharge. When the voltage and the discharge current density increase the average ion energy bombarding the cathode surface also increases. The bombardment with ions ultimately heats the cathode causing thermionic emission. Once the cathode is hot enough to emit electrons thermionically, the discharge will change to an arc regime (region C in Fig. 1.9).

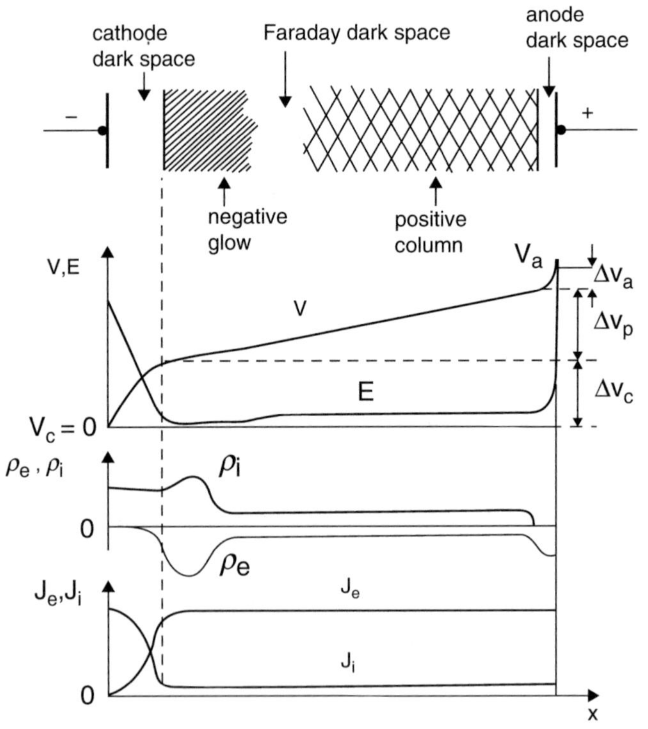

Fig. 1.10 Scheme of the spatial distributions of luminous and dark zones in a glow discharge, electric field E, voltage V, space-charge densities of electrons $\rho_e < 0$ and ions ρ_i, and absolute values of current densities of electrons J_e and ions J_i (von Engel 1965)

Due to both the large number of applications of glow discharges and the interest of studying such system for a better understanding of the discharges in general, let us consider Fig. 1.10. When a direct-current (DC) glow discharge is established, the visible light emitted from the discharge is distributed along the length of the tube as shown in Fig. 1.10. Starting from the cathode (at $x = 0$) it exists sometimes a very narrow dark space (Aston's space) closed to it (not represented in the figure) followed by a thin scarcely luminous layer (not represented in the figure as well), which in turn is followed by a much larger zone called the cathode dark space. Both the Aston's dark space and the cathode glow are not always clearly visible in a glow discharge depending of the pressure and the nature of the gas. A sharp boundary separates the cathode dark space from the next region – the negative glow – which becomes progressively less intense towards the Faraday dark space. At the end of this region appears the positive column. After the positive column in the direction of the anode there is sometimes an anode dark space followed by an anode glow (this latter not represented in the figure) close to the anode itself.

There are a large variety of glow discharges with their appearance depending of the pressure, nature of the gas, dimensions of the vessel, and material of the electrodes. When the distance between the cathode and the anode varies the axial lengths of the other zones rather than the positive column remain unchanged while the length of the positive column varies accordingly. In fact the positive column can be extended to any length provided the voltage for maintaining the discharge is large enough. On the other hand, when the gas pressure is increased above 0.1 Torr the cathode dark space, the negative glow, and the Faraday dark space contract towards the cathode, whereas the anode dark space contracts to the anode. The positive column fills then the remainder of the distance, but it contracts radially as the pressure still increases. From this behaviour we can conclude that the motion of the charged particles in the zones near the electrodes is of a beam-like nature, whereas the motion in the positive column is of the random type.

The transport of current through a glow discharge occurs by the axial motion of electrons and positive ions. The current through the cathode zones can be understood by the inspection of the distributions of the axial electric field E. This field is larger near the cathode and decreases of magnitude towards the negative glow and Faraday dark space. After it remains constant throughout the positive column raising only again near the anode.

Let us consider now an electron emitted from the cathode, for example due to the impingement of a positive ion. This electron is first accelerated in the cathode dark space by a strong field, but initially it executes few ionizing collisions only because its energy is not sufficiently far above the ionization threshold energy. At larger distances from the cathode, though the field has become smaller, the electron has acquired enough energy to produce ionization by electron multiplication. In the negative glow the field is very weak so that only the electrons which have not lost appreciable energy by inelastic collisions will be able to continue the ionization. Besides these fast electrons, a larger number of slow electrons also enter into the negative glow.

Due to the ionization by multiplication α a large number of positive ions are produced in the cathode dark space and negative glow. These ions will move through the cathode dark space gaining energy and impinging the cathode leading to the emission of secondary electrons by multiplication γ. The electrons released from the cathode are then multiplied again by α mechanism and accelerated towards the regions more distant from the cathode. Due to the larger number of elastic and inelastic collisions suffered the random velocity of the electrons is now several orders of magnitude larger than the drift velocity.

In a steady-state glow discharge the current densities at the cathode and the anode need to be equal. Since the mobility of electrons is much larger than that of ions, the ion density near the cathode needs to be much larger than the electron density near the anode. Furthermore in order the two current densities become equal at the electrodes, the ions need to be accelerated in the cathode dark space, along an extension and with an electric field magnitude, that are both larger than the extension and the field magnitude with which the electrons are accelerated in the anode dark space. At the end the conjoint action of all these effects makes the glow discharge to exhibit axial profiles for the electric field, potential, electron and ion charge densities, and electron and ion current densities of the type of those schematically represented in Fig. 1.10.

Near the cathode the electron number density and the electron current density are both vanishingly small, the total current density is produced by the ions only, and these are accelerated by the potential difference ΔV_c along the cathode sheath of width Δx_c. Near the anode a reverse situation occurs but now the potential difference ΔV_a and the sheath width Δx_a are smaller. Typical values for the two sheath potentials are $\Delta V_c = 500\,\text{V}$ and $\Delta V_a = 1\,\text{V}$. The space-charge is approximately zero in the negative glow and in the positive column, $\rho_e + \rho_i \simeq 0$, so that we are in the presence of a plasma, but the individual charge densities of electrons and ions are larger in the negative glow.

In the positive column the axial component of the electric field is found to be almost constant and at sufficiently high pressure two local equilibrations are established at any point: *(i)* the balance between the electron ionization rate by the impact of electrons upon neutral atoms and molecules and the loss rate of electrons by radial diffusion to the wall and electron-ion recombination; *(ii)* the balance between the energy gained from the field and the energy lost in any sort of collisions. As the pressure decreases below a certain value such that the mean free path becomes comparable to the radial dimension of the vessel (i.e. the tube radius), both equilibria become non-local and the electron transport from one point to another needs to be considered in conjoint with the local terms.

Furthermore, due to the large difference between the mobilities of electrons and ions the tube wall becomes negatively charged and a radial space-charge electric field is formed. This field reduces the rate of electron losses to the wall so that the ionization rate necessary to maintain the discharge is significantly reduced. This leads to a reduction of the axial electric field necessary to sustain the discharge. This explains why the magnitude of the axial electric field in the positive column of a glow discharge is small when compared with the electric field in a Townsend discharge. In this latter case the radial space-charge electric field does not exist and consequently the electrons are lost by free diffusion to the wall at a much larger rate. Then, the ionization rate and consequently the electric field need to be larger to maintain the discharge. The positive column is axially uniform but since the charged species are lost in the wall the discharge is not radially uniform. Besides the electrons and ions the positive column becomes an active medium with many excited radiative and metastable species showing a characteristic glow that depends of the pressure and nature of the gas.

The analysis presented here gives us a qualitative picture of how a glow discharge works. However, the complete quantitative description can only be realized in the framework of the microscopic kinetic theory, by using e.g. the electron Boltzmann transport equation, in the case of the positive column, or by using the moments of the Boltzmann equation, i.e. the equations of continuity, momentum conservation, and energy conservation for both electrons and ions, in the case of the electrode regions. The analysis based on the Boltzmann equation will be detailedly developed along the various chapters of Part I of this book, whereas the situation addressed with the cathode region will be only briefly referred in this chapter in Sect. 1.3, devoted to physical boundaries to the plasma, since it does not constitute the main purpose of this textbook.

1.2.2 Corona Discharge

A different type of discharges with interest to be referred here occurs in regions of high electric fields near sharp points or along wires in air raised to high electrical potentials with respect to their surroundings, in which the breakdown occurs (von Engel 1965; Roth 1995). One can regard the corona discharge as a Townsend or a glow discharge without positive or negative regions.

Thus, let us consider, as an example, a very long cylinder of radius r_a, with a thin wire of radius $r_c \ll r_a$ at the centre. The inner wire is assumed grounded at zero potential, while the space between r_c and r_a is filled of gas. Since $V(r) > 0$ for $r > r_c$, the electric field is directed towards the thin wire (see Fig. 1.11) and the wire acts as an electron-emitting cathode filament, whereas the cylinder of radius r_a acts as an anode.

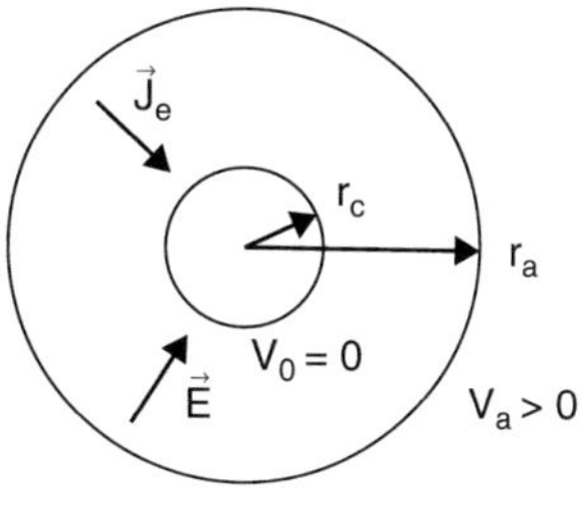

Fig. 1.11 Corona discharge in cylindrical geometry, consisting of an axial fine wire of radius r_c and a cylindrical outer electrode of radius r_a. The wire is maintained at grounded potential, while the external cylinder is at potential $V_a > 0$

In the limiting case where the space-charge density in the interspace between the electrodes is vanishingly small, Poisson's equation writes as

$$\nabla^2 V = \frac{1}{r}\frac{d}{dr}\left(r\,\frac{dV}{dr}\right) = 0, \tag{1.36}$$

allowing to obtain at $r_c \le r \le r_a$

$$V(r) = V_a\,\frac{\ln(r/r_c)}{\ln(r_a/r_c)}, \tag{1.37}$$

whereas for the electric field $\mathbf{E} = -\,E(r)\,\mathbf{e_r}$, we obtain

$$E(r) = \frac{dV}{dr} = \frac{V_a}{r\,\ln(r_a/r_c)} \tag{1.38}$$

and therefore

$$E(r) = \frac{r_c}{r}\,E_0, \tag{1.39}$$

from which we may conclude that $E(r_c) = E_0 \to \infty$ when $r_c \to 0$.

When ionization takes place in an extremely narrow region around the inner wire, the positive ions are attracted by it and neutralized, while the electrons are repelled and travel in the radial cylindrical symmetrical field to the outer cylinder. Assuming that there is no further ionization due to collisions in the gas, the current per length unit i/l remains constant in the interspace electrodes being given by

$$i/l = 2\pi r\,J_e(r) = 2\pi r\,en_e(r)\,\mu_e E(r), \tag{1.40}$$

with $\mathbf{J_e} = -\,J_e(r)\,\mathbf{e_r}$ denoting the electron current density, $n_e(r)$ the electron density, $\mu_e > 0$ the electron mobility, and e the absolute value of the electron charge. Then, the electric field is obtained from Poisson's equation

$$\frac{1}{r}\frac{d}{dr}(r\,E) = \frac{en_e}{\epsilon_0} = \frac{i/l}{2\pi\epsilon_0\mu_e\,rE}, \tag{1.41}$$

which can be written as

$$(rE)\,\frac{d}{dr}(rE) = \frac{i/l}{2\pi\epsilon_0\mu_e}\,r \tag{1.42}$$

and integrated from r_c, where $E(r_c) = E_0$, up to r position, yielding to obtain

$$E(r) = \sqrt{\left(\frac{r_c}{r}\,E_0\right)^2 + \frac{i/l}{2\pi\epsilon_0\mu_e}\left(1 - \left(\frac{r_c}{r}\right)^2\right)}. \tag{1.43}$$

The electric field is modified relatively to the previous result (1.39) where the space-charge density is vanishingly small.

At radii far way from the inner wire $r \gg r_c$ and large discharge currents, we have approximately

$$E \simeq \sqrt{\frac{i/l}{2\pi\epsilon_0\mu_e}} = \text{const} \tag{1.44}$$

and the field strength is constant and independent of position. It rises with increasing current and decreasing mobility, i.e. with increasing pressure. This shows that a uniform field can be produced over a distant region of a cylindrical gap in the presence of a strong space-charge density which distorts the originally hyperbolic field distribution (1.39). In turn, equation (1.40) shows that the electron density varies as $n_e(r) \propto 1/r$.

Equation (1.43) can be integrated between r_c and r to obtain the electric potential with respect to the potential $V_0 = 0$ at the wire

$$V(r) = \sqrt{\frac{i/l}{2\pi\epsilon_0\mu_e}} \int_{r_c}^{r} \sqrt{1 + \left(\frac{2\pi\epsilon_0\mu_e}{i/l} E_0^2 - 1\right)\frac{r_c^2}{r^2}}\, dr. \tag{1.45}$$

Introducing the dimensionless variable

$$\alpha = \frac{r}{\chi\, r_c}, \tag{1.46}$$

in which

$$\chi = \sqrt{\frac{2\pi\epsilon_0\mu_e}{i/l} E_0^2 - 1}, \tag{1.47}$$

we may write

$$V(r) = \sqrt{\frac{i/l}{2\pi\epsilon_0\mu_e}}\, \chi\, r_c \int_{\alpha_c}^{\alpha} \sqrt{1 + \frac{1}{\alpha^2}}\, d\alpha. \tag{1.48}$$

The solution of this primitive is known and it yields to write

$$V(r) = \sqrt{\frac{i/l}{2\pi\epsilon_0\mu_e}}\, \chi\, r_c \left\{ \sqrt{1+\alpha^2} - \sqrt{1+\alpha_c^2} + \ln\left(\frac{\alpha_c}{\alpha}\, \frac{\sqrt{1+\alpha^2}-1}{\sqrt{1+\alpha_c^2}-1}\right)\right\} \tag{1.49}$$

and since α and α_c are $\ll 1$ (i.e. $E_0 \gg \sqrt{(i/l)/2\pi\epsilon_0\mu_e}$), we may use the expansion $\sqrt{1+\alpha} \simeq 1 + \alpha/2$ and write

$$V(r) \simeq \frac{1}{2}\left(\frac{i/l}{2\pi\epsilon_0\mu_e}\right)\frac{r_c}{E_0}\left(\frac{r^2}{r_c^2}-1\right)+E_0\, r_c\, \ln\left(\frac{r}{r_c}\right). \tag{1.50}$$

The second term on the right-hand side of equation (1.50) is the dominant contribution to the potential (1.37), and the first term is the first-order modification due to the presence of the space-charge. This equation makes it possible to obtain the surface electric field on the wire E_0, from a knowledge of the coronal current i/l, the applied potential V_a on the outer cylinder, and the geometric parameters of the coaxial configuration.

Let us consider now the case of sufficiently low pressure in which a collisionless model needs to be used. In this case instead of equation (1.40) we have

$$i/l = 2\pi r\, J_e(r) = 2\pi r\, en_e(r)\, v_{ed}(r), \tag{1.51}$$

being $v_{ed}(r)$ the absolute value of the drift velocity of electrons emitted from the wire. If they are emitted with negligible initial velocity, their velocity at radius r can be obtained from energy conservation

$$\frac{1}{2}\, mv_{ed}^2(r) = eV(r) \tag{1.52}$$

and using equations (1.51) and (1.52), we obtain

$$en_e(r) = \frac{i/l}{2\pi r}\sqrt{\frac{m}{2eV(r)}}. \tag{1.53}$$

Substituting this equation into Poisson's equation

$$\frac{1}{r}\frac{d}{dr}\left(r\frac{dV}{dr}\right) = \frac{en_e}{\epsilon_0} = \frac{i/l}{2\pi\epsilon_0 r}\sqrt{\frac{m}{2eV(r)}}, \tag{1.54}$$

we find

$$\frac{d}{dr}\left(r\frac{dV}{dr}\right) = \frac{C}{\sqrt{V(r)}}, \tag{1.55}$$

with

$$C = \frac{i/l}{2\pi\epsilon_0}\sqrt{\frac{m}{2e}} = \text{const.} \tag{1.56}$$

Equation (1.55) is a second-order, nonlinear differential equation with no known exact analytical solution.

Defining the dimensionless radial variable

$$x = \ln\left(\frac{r}{r_c}\right) \tag{1.57}$$

and the parameter β such as

$$\beta^2 = \frac{8\pi\epsilon_0}{9\,i/l}\sqrt{\frac{2e}{m}}\,\frac{V^{3/2}}{r}, \tag{1.58}$$

equation (1.55) can be rewritten under the form (see Appendix A.1.4)

$$3\beta\,\frac{d^2\beta}{dx^2} + \left(\frac{d\beta}{dx}\right)^2 + 4\beta\,\frac{d\beta}{dx} + \beta^2 = 1. \tag{1.59}$$

At the first sight equation (1.59) does not seem to be a large improvement relatively to equation (1.55), but it has the important characteristic of containing no parameter other than x and β, so that a solution for β against x is universal, whereas a solution to equation (1.55) would require a family of curves, with i/l as a parameter. For large values of r/r_c, the following approximation is obtained neglecting the $(d\beta/dx)^2$ term (Roth 1995)

$$\beta \simeq 1 + 0.9769\left(\frac{r_c}{r}\right)^{2/3}\sin\left[1.0854\,\log_{10}\!\left(\frac{r}{11.93\,r_c}\right)\right]. \tag{1.60}$$

1.3 Physical Boundary to a Plasma

1.3.1 Debye Length

Before to analyse the effects produced by a boundary to a plasma let us introduce firstly an important characteristic of a plasma that is the shielding of the electric potential applied to it. In order to characterize this shielding it is used a parameter of the plasma called Debye length which provides a measure of the distance over which the influence of the electric field of an individual charged particle is felt by the other charged particles present in the plasma. This concept is usually given at the beginning of any textbook on Plasma Physics (see e.g. Chen 1984).

The Debye length can be estimated as a function of the characteristic quantities of the plasma such as the electron density and temperature. Due to the larger mass of ions we may assume for simplicity that they do not move significantly on the time scale of electron movement. Assuming the electric potential near a reference ion as $V(\mathbf{r}) > 0$ and that it monotonously decreases to zero in direction to the background plasma, the electron density around an ion test is given by (as it will be shown in Sect. 3.1.5)

$$n_e(\mathbf{r}) = n_{e\infty}\,\exp\left(\frac{e\,V(\mathbf{r})}{k_B T_e}\right), \tag{1.61}$$

where e denotes the absolute value of the electron charge, k_B the Boltzmann constant, and T_e the electron temperature. The electrons are attracted to regions of higher potential so that the electron density in the vicinity of an ion is higher than in the neutral background plasma where $n_{e\infty} = n_{i\infty}$. To determine the potential and density variations we use Poisson's equation

$$\nabla^2 V(\mathbf{r}) = -\frac{e}{\epsilon_0}\,(n_{i\infty} - n_e(\mathbf{r}))\,, \tag{1.62}$$

which may be written under the following form in symmetrical spherical geometry, assuming $eV(r)/(k_B T_e) \ll 1$

$$\frac{1}{r}\frac{d^2}{dr^2}\,(r\ V(r)) = \frac{e n_{e\infty}}{\epsilon_0}\left[\frac{e\ V(r)}{k_B T_e} + \frac{1}{2}\left(\frac{e\ V(r)}{k_B T_e}\right)^2 + \ldots\right]. \tag{1.63}$$

No simplification is possible for the region near the ion, where $eV(r)/(k_B T_e)$ is large. However, this region does not significantly contribute to the thickness of the region where the non-null charge density exists. We may therefore write equation (1.63) keeping only the linear term as

$$\frac{d^2}{dr^2}\,(r\ V(r)) = \frac{1}{\lambda_D^2}\,(r\ V(r)) \tag{1.64}$$

having defined the quantity λ_D called Debye length as

$$\lambda_D = \sqrt{\frac{\epsilon_0 k_B T_e}{e^2 n_{e\infty}}}. \tag{1.65}$$

The solution of equation (1.64) with physical significance is called Yukawa potential

$$V(r) = \frac{A}{r}\ \exp\left(-\frac{r}{\lambda_D}\right), \tag{1.66}$$

being $A = Ze/(4\pi\epsilon_0)$ in the case of an ion of charge Ze. The potential rapidly vanishes when $r > \lambda_D$, so that the Debye length defines the radius of a cloud around each ion outside which the plasma will tend to remain neutral. We note that λ_D decreases as the electron density increases which means a greater efficiency of the shielding. On the contrary, with increasing T_e the radius of the non-neutral region increases since as high the electrons are energetic they may escape from the influence of the potential.

1.3.2 *Bohm Criterion*

As referred in Sect. 1.2.1, when a surface, or a wall, is placed in front of a plasma the surface becomes negatively charged due to the more rapid velocity of electrons, and then a space-charge electric field is formed in the direction to the surface, which accelerates the ions and decelerates the electrons. The movement of the electrons to the surface can be expressed by the equation for the electron particle current density, which is, in fact, the equation for the electron momentum transfer (see Sect. 3.1.5)

$$\mathbf{\Gamma_e} = -\, D_e\, \nabla n_e \;-\; n_e\, \mu_e\, \mathbf{E_s}, \tag{1.67}$$

in which the dominant terms are the electron free diffusion current towards the surface and the electron current in the direction of the homogeneous plasma due to the space-charge electric field $\mathbf{E_s}$. D_e and μ_e are the electron free diffusion coefficient and the electron mobility, respectively. As we will see later on in Chap. 5, the two parameters are linked each other by $D_e/\mu_e = k_B T_e/e$, in the case of a Maxwellian electron velocity distribution at temperature T_e, with k_B and e denoting the Boltzmann constant and the absolute value of electron charge. In present conditions both terms in equation (1.67) nearly exactly cancel each other being vanishing small the net current density to the surface, $\mathbf{\Gamma_e} \simeq 0$, so that replacing $\mathbf{E_s} = -\, \nabla V$, we obtain using slab geometry

$$\frac{1}{n_e}\,\frac{dn_e}{dx} \simeq \frac{e}{k_B T_e}\,\frac{dV}{dx}. \tag{1.68}$$

Integrating this equation from the homogeneous plasma at $x = 0$, in which $V(0) = 0$ and $n_e^0 = n_i^0$, up to the surface at x position in which $V(x) < 0$, and assuming the electron temperature homogeneous, we obtain the Maxwell-Boltzmann law for the electron density such as it has been already used in (1.61)

$$n_e(x) = n_e^0\, \exp\left(\frac{eV}{k_B T_e}\right). \tag{1.69}$$

On the contrary, in the case of the ions the net current density at the surface is not at all vanishing because the ions are accelerated by the conjoint action of the diffusion term and the space-charge electric field. The movement of the ions is then obtained from the equation for momentum transfer of ions in which the dominant terms are now (see Sect. 3.1.5) the inertia and the space-charge electric field terms

$$n_i\, M\, (\mathbf{v_{id}}\,.\nabla)\mathbf{v_{id}} \simeq e\, n_i\, \mathbf{E_s}, \tag{1.70}$$

in which $\mathbf{v_{id}}$ is the ion drift velocity and M the ion mass. Thus, we may write in slab geometry

$$\frac{1}{2}\, M\, \frac{dv_{id}^2}{dx} \simeq -\, e\, \frac{dV}{dx} \tag{1.71}$$

and this equation can be integrated from v_{id}^0, at $x = 0$, up to v_{id}, at x position, yielding

$$\frac{1}{2}\, M\, (v_{id}^2 - v_{id}^{0\,2}) \simeq -\, e\, V. \tag{1.72}$$

In writing equation (1.70) we have assumed that the sheath in front of the surface is collisionless so that the continuity equation $(\nabla . \mathbf{J_i}) = 0$ imposes the constancy of the ion current density: $J_i = en_i v_{id} =$ const. Making the assumption that the ions coming from the plasma reach the plasma-sheath boundary with the drift velocity v_{id}^b and that the origin of the potential is defined at this boundary, we may write equations (1.69) and (1.72) as

$$n_e(x) = n_e^b \exp\left(\frac{eV}{k_B T_e}\right) \tag{1.73}$$

$$\frac{1}{2} M \left(v_{id}^2 - v_{id}^{b\,2}\right) \simeq - e\, V. \tag{1.74}$$

To these equations we must still add the constancy of the ion current density and Poisson's equation

$$n_i\, v_{id} = n_i^b\, v_{id}^b \tag{1.75}$$

$$\frac{d^2 V}{dx^2} = - \frac{e}{\epsilon_0} (n_i - n_e). \tag{1.76}$$

Replacing equations (1.73) and (1.75) into equation (1.76). and assuming $n_e^b = n_i^b$, we obtain

$$\frac{d^2\eta}{dx^2} = \frac{1}{\lambda_D^2} \left(\frac{v_{id}^b}{v_{id}} - e^{-\eta}\right), \tag{1.77}$$

with

$$\eta = - \frac{e\, V}{k_B T_e} \tag{1.78}$$

and where

$$\lambda_D = \sqrt{\frac{\epsilon_0 k_B T_e}{e^2 n_e^b}} \tag{1.79}$$

denotes the Debye length at the plasma-sheath boundary.

Using now equation (1.74) into (1.77), we still obtain

$$\frac{d^2\eta}{dx^2} = \frac{1}{\lambda_D^2} \left[\left(1 + \frac{2\eta}{y^2}\right)^{-1/2} - e^{-\eta}\right], \tag{1.80}$$

with

$$y = v_{id}^b \sqrt{\frac{M}{k_B T_e}}. \tag{1.81}$$

Near the boundary the absolute value of the potential is small and therefore $2\eta/y^2 \ll 1$ and $\eta \ll 1$, so that the right-hand side member of equation (1.80) can be expanded as a power series in η (Franklin 1976)

$$\frac{d^2\eta}{dx^2} \simeq \frac{1}{\lambda_D^2}\,\eta\left(1-\frac{1}{y^2}\right). \tag{1.82}$$

Equation (1.82) shows that in order η will be a non-oscillatory function of distance, we need to have $y \geq 1$, and hence

$$v_{id}^b \geq \sqrt{\frac{k_B T_e}{M}}. \tag{1.83}$$

The condition (1.83) is usually referred to as the Bohm criterion. With a solution satisfying this condition η is a monotonically increasing function within a scale length equal to the Debye length.

1.3.3 Floating Sheath

Because of the large difference between the mobilities of electrons and ions a surface boundary to the plasma charges very rapidly with a negative floating potential with respect to the plasma. The magnitude of this steady-state floating potential $V_F < 0$ can be easily derived assuming the equality between the electron and ion current densities at the wall. Let us consider Fig. 1.12 showing the sheath between the plasma and the wall. The origin $x = 0$ in which $V(0) = 0$ and $n_e^0 = n_i^0$ is localized in the interior of the plasma (contrary to the previous section where it was in the plasma-sheath boundary). In the direction towards the wall the electric potential $V(x)$ and the ion density $n_i(x)$ monotonously decrease, whereas the electron density $n_e(x)$ is assumed to decrease first with the same rate as $n_i(x)$ up to the plasma-sheath boundary and beyond this point the plasma practically ceases to exist and a positive space-charge region (the ion sheath) is built up. The ion sheath can be observed as a dark space since most gas phase excitation processes involve electron collisions and the optical emission of the discharge is caused by radiative decay of short-living excited species.

The current density of electrons to the wall is due to their thermal movement (see Appendix A.1.1)

$$J_e^W = \frac{e\, n_e^W <v_e>}{4}, \tag{1.84}$$

with $< v_e > = \sqrt{8k_BT_e/(\pi m)}$ denoting the average velocity of electrons at the temperature T_e, and n_e^W is the small electron density at the wall (1.73)

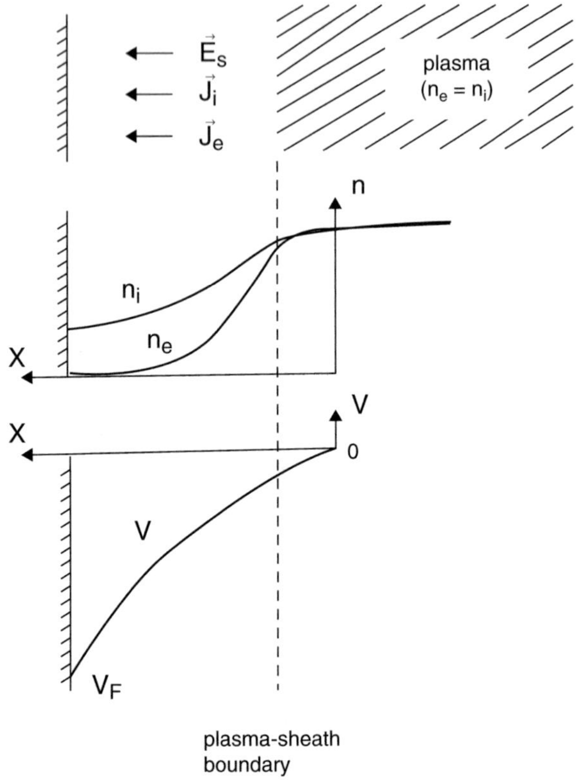

Fig. 1.12 Scheme of the ion sheath built up between the plasma and the floating wall. The broken line is the plasma-sheath boundary

$$n_e^W = n_e^0 \exp\left(\frac{e\,V_F}{k_B T_e}\right). \tag{1.85}$$

It is worth noting that in (1.73) the origin of the potential was in the plasma-sheath boundary whereas here it is localized in the plasma. In both cases $V(x) < 0$.

In the case of the ions and because the drift velocity at the boundary v_{id}^b is much larger than the average ion velocity $<v_i>= \sqrt{8k_BT_i/(\pi M)}$, the ions enter the sheath with a velocity determined by the Bohm criterion (1.83). Since in the plasma $v_{id}^0 = 0$ the potential at the boundary is determined from energy conservation

$$\frac{1}{2}\,M\,v_{id}^2 = -\,e\,V, \tag{1.86}$$

and we obtain $V^b = -\,k_BT_e/(2e)$ at $v_{id}^b = \sqrt{k_BT_e/M}$. The electron density (1.85) at the boundary is hence

$$n_e^b = n_e^0 \exp\left(-\frac{1}{2}\right) \tag{1.87}$$

and because $n_i^b = n_e^b$, we obtain the ion current density at the boundary given by

$$J_i^b = e\, n_i^b\, v_{id}^b = e\, n_e^0\, \exp\left(-\frac{1}{2}\right)\sqrt{\frac{k_B T_e}{M}}. \tag{1.88}$$

Using a derivation close to that indicated in van Roosmalen and de Vries (1985) we assume that in a collisionless sheath the continuity of the ion current density $dJ_i/dx = 0$ imposes the constancy of the ion current density in the sheath, so that the floating potential at the wall can be determined by equalling the electron and ion current densities (1.84) and (1.88)

$$\frac{e\, n_e^0}{4}\sqrt{\frac{8k_B T_e}{\pi m}}\, \exp\left(\frac{e\, V_F}{k_B T_e}\right) = e\, n_e^0\, \exp\left(-\frac{1}{2}\right)\sqrt{\frac{k_B T_e}{M}},$$

obtaining

$$V_F = -\frac{k_B T_e}{2e}\left[1 + \ln\left(\frac{M}{2\pi m}\right)\right], \tag{1.89}$$

while for the electron density at the wall we get

$$n_e^W = n_e^0\, \exp\left(-\frac{1}{2}\right)\sqrt{\frac{2\pi m}{M}}. \tag{1.90}$$

In the case of argon plasma with the electrons at temperature $k_B T_e = 2\,\mathrm{eV}$ and density $n_e^0 = 10^{16}\,\mathrm{m}^{-3}$, we obtain $V_F = -10.4\,\mathrm{V}$, $v_{id}^b = 2.20 \times 10^3\,\mathrm{m\,s}^{-1}$, and $J_i = 2.14\,\mathrm{A\,m}^{-2}$.

Let us estimate now the ion sheath thickness. Since the sheath is collisionless the ion current density to the wall is related with the floating potential by the Child-Langmuir equation (1.22) valid for a collisionless space-charge limited current which under present conditions is written as

$$J_i = \frac{4}{9}\frac{\epsilon_0}{d^2}\sqrt{\frac{2e}{M}}\,|V_F|^{3/2}, \tag{1.91}$$

with d denoting the ion sheath thickness. Using equations (1.88) and (1.89), we obtain

$$d = \frac{\sqrt{2}}{3}\exp\left(\frac{1}{4}\right)\lambda_D\left[1 + \ln\left(\frac{M}{2\pi m}\right)\right]^{3/4}, \tag{1.92}$$

where $\lambda_D = \sqrt{\epsilon_0 k_B T_e/(e^2 n_e^0)}$ is the Debye length. For the above standard conditions we get $\lambda_D = 1.1 \times 10^{-4}\,\mathrm{m}$ and $d = 3.7 \times 10^{-4}\,\mathrm{m}$.

1.3.4 Cathode Region

In the previous section we analyzed the interaction with a perfect floating wall in which there is not current extracted from the surface and a negative potential exists with respect to the plasma. Here, we will describe the cathode sheath in an abnormal glow discharge. The term abnormal indicates the situation where the negative glow covers the cathode completely, whereas in a normal glow discharge only part of the cathode participates in the discharge. Raising the current the electrode coverage increases and the glow discharge changes from normal to abnormal.

A typical cathode potential is $V_c = -500\,\mathrm{V}$ and with voltages of this order the electrons cannot reach the cathode surface, whereas the positive ions are strongly accelerated to the cathode and a net current is drawn. In the other side of the discharge the reverse situation occurs being the potential only slightly less negative than the floating potential V_F, in order to compensate the cathode current with a net electron current to the anode. As in previous section, in the case of a collisionless cathode sheath the current density of positive ions is given by the Child-Langmuir law (1.22) and (1.91)

$$J_i = \frac{4}{9}\frac{\epsilon_0}{d^2}\sqrt{\frac{2e}{M}}\,|V_c|^{3/2} \tag{1.93}$$

since this derivation holds for all positive d values of a collisionless sheath. For the previous conditions of a plasma with $k_B T_e = 2\,\mathrm{eV}$ and $n_e^0 = 10^{16}\,\mathrm{m^{-3}}$, in which $J_i = 2\,\mathrm{A\,m^{-2}}$, and a cathode potential $V_c = -500\,\mathrm{V}$, we obtain $d = 7.0 \times 10^{-3}\,\mathrm{m}$ for the sheath thickness.

Using this simple model the potential, the electric field, and the density of ions can be derived using a similar procedure as in Sect. 1.1.2 valid for whatever the collisionless sheath. Assuming the reference $x = 0$ at the neutral plasma, we obtain as x increases in direction to the cathode

$$V(x) = V_c\left(\frac{x}{d}\right)^{4/3} \tag{1.94}$$

$$E(x) = \frac{4}{3}\frac{|V_c|}{d^{4/3}}\,x^{1/3} \tag{1.95}$$

$$n_i(x) = \frac{4\epsilon_0}{9e}\frac{|V_c|}{d^{4/3}}\,x^{-2/3}. \tag{1.96}$$

A plot of these quantities is shown in Fig. 1.13. At the cathode we have $E_c = 9.6 \times 10^4\,\mathrm{V\,m^{-1}}$ and $n_i^c = 2.5 \times 10^{14}\,\mathrm{m^{-3}}$. Obviously these expressions must be considered for distances larger than the plasma-sheath boundary, in which the potential is $V^b = -\,k_B T_e/(2e)$ obtained from the Bohm criterion. At the conditions referred here $V^b = -1\,\mathrm{V}$ and $n_e^b = n_i^b = 6.1 \times 10^{15}\,\mathrm{m^{-3}}$.

For an ion sheath thickness $d = 7.0 \times 10^{-3}\,\mathrm{m}$ the concept of collisionless sheath may be doubted. The collision cross section of Ar^+ ions in Ar is of the order of $\sigma_{io} = 5 \times 10^{-19}\,\mathrm{m^2}$, which for $p = 20\,\mathrm{Pa}$ and $T_o = 300\,\mathrm{K}$ implies an ion mean

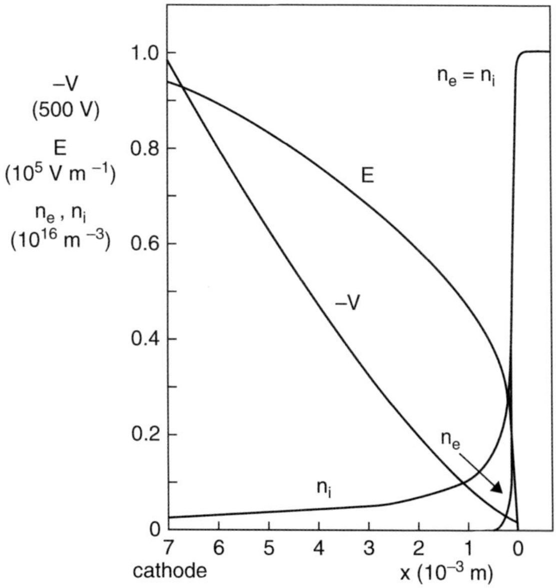

Fig. 1.13 Potential ($-V$) in 500 V, electric field (E) in $10^5\,\mathrm{V\,m^{-1}}$, and ion and electron densities (n_i and n_e) in $10^{16}\,\mathrm{m^{-3}}$, in the collisionless cathode sheath of a glow discharge, with $k_B T_e = 2\,\mathrm{eV}$ and $n_e^0 = n_i^0 = 10^{16}\,\mathrm{m^{-3}}$ (van Roosmalen and de Vries 1985)

free path of only $\lambda_i = 1/(n_o \sigma_{io}) \simeq 4 \times 10^{-4}\,\mathrm{m}$. In this case instead of a gain of energy equal to the cathode potential for all ions a wide range of energies for the ion flux impinging the cathode should be considered. The expressions to be considered in this case are those derived before for a mobility dominated electron movement (1.12), (1.13) and (1.14) modified now for ions. According to this model the negative electrical potential is given by

$$V(x) = -\frac{2\sqrt{2}}{3}\sqrt{\frac{J_i}{\epsilon_0 \mu_i}}\,x^{3/2}. \tag{1.97}$$

The equation (1.97) is qualitatively quite similar to the expression derived for the collisionless case. The voltage varies now as $x^{1.5}$ instead of $x^{1.33}$ which does not introduce significant differences. For the typical case we are considering here even the sheath thickness has about the same value. Estimating the mobility of energetic Ar^+ ions at $p = 20$ Pa as $\mu_i = 0.5\,\mathrm{m^2\,V^{-1}\,s^{-1}}$ (van Roosmalen and de Vries 1985), we obtain $d = 8.5 \times 10^{-3}\,\mathrm{m}$ for $V_c = -500\,\mathrm{V}$ and $J_i = 2\,\mathrm{A\,m^{-2}}$. We may conclude then that this pressure value is in a transition zone for which both models give reasonable results. One has to go to either much higher or much lower pressures to discriminate between the two models.

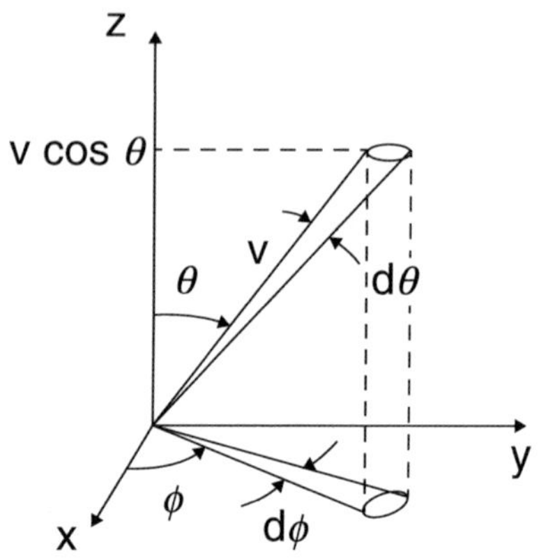

Fig. 1.14 Particles within the angle solid $d\Omega = \sin\theta\ d\theta\ d\phi$ and at a distance $dz = v\ \cos\theta\ dt$ of the plane xOy, impinging on it during the time interval dt

Appendices

A.1.1 Particle Current Density Upon a Surface

The expression of the particle current density upon a surface due to the random movement used in equations (1.1) and (1.84) can be derived as follows. Let us consider a certain type of particles (electrons, ions, atoms) in which the number of particles, per volume unit, with the absolute value of velocity between v and $v + dv$ is

$$dn = f(v)\ 4\pi v^2\ dv, \tag{1.98}$$

being $f(v)$ the particle velocity distribution normalized such that

$$n = \int_0^\infty f(v)\ 4\pi v^2\ dv, \tag{1.99}$$

with n denoting the particle number density. The particles within the angle solid $d\Omega = \sin\theta\ d\theta\ d\phi$ will hit the plane xOy in the time interval dt if they lie at a distance $dz = v\ \cos\theta\ dt$ of the plane (see Fig. 1.14). Since the fractional number of particles within the angle solid is $d\Omega/4\pi$, the number of particles, per surface unit, with velocities between v and $v + dv$ and within the angle solid $d\Omega$, that impinge the plane xOy in the time interval dt is

$$f(v)\ 4\pi v^2\ dv\ \frac{d\Omega}{4\pi}\ v\ \cos\theta\ dt. \tag{1.100}$$

Integrating first over all velocities

$$\cos\theta\ \frac{d\Omega}{4\pi}\ dt \int_0^\infty v\, f(v)\ 4\pi v^2\ dv, \tag{1.101}$$

we obtain

$$\cos\theta \, \frac{d\Omega}{4\pi} \, dt \, n <v>, \tag{1.102}$$

being

$$<v>= \frac{1}{n} \int_0^\infty v \, f(v) \, 4\pi v^2 \, dv \tag{1.103}$$

the average particle velocity. Integrating now over all directions in the semispace $z \geq 0$, we obtain the number of collisions per surface unit in the time interval dt

$$\frac{n <v>}{4\pi} \, dt \int_0^{\pi/2} \cos\theta \, \sin\theta \, d\theta \int_0^{2\pi} d\phi = \frac{n <v>}{4} \, dt. \tag{1.104}$$

Thus, the number of collisions per surface and time units, i.e. the particle current density is

$$\Gamma = \frac{n <v>}{4}. \tag{1.105}$$

A.1.2 Solution of Equation (1.15)

Let us start by multiplying both members of equation (1.15) by $2 \, dV/dx$ and write it under the form

$$\frac{d}{dx}\left[\left(\frac{dV}{dx}\right)^2\right] = 2 \, C \, V^{-1/2} \, \frac{dV}{dx}. \tag{1.106}$$

This equation is easily integrated to get

$$\left(\frac{dV}{dx}\right)^2 = 4 \, C \, V^{1/2} + K, \tag{1.107}$$

in which $K = 0$ due to the condition at the cathode $E_0 = (dV/dx)_0 = 0$. We have therefore

$$\frac{dV}{dx} = 2 \, \sqrt{C} \, V^{1/4} \tag{1.108}$$

and now this equation may be integrated again

$$\int_0^V V^{-1/4} dV = 2 \, \sqrt{C} \int_0^x dx \tag{1.109}$$

allowing to obtain

$$\frac{V^{3/4}}{3/4} = 2\,\sqrt{C}\,x, \tag{1.110}$$

which at the end takes the form of equation (1.17)

$$V^{3/2} = \frac{9}{4}\,C\,x^2. \tag{1.111}$$

A.1.3 Multiplication Factor (1.30)

Each primary electron originates a multiplication α and in this case they arrive $(e^{\alpha d} - 1)/\alpha d$ electrons at the anode as given by equation (1.26). Thus they are created $[(e^{\alpha d} - 1)/\alpha d] - 1$ electron-ion pairs and these ions extract from the cathode $\gamma\ \{[(e^{\alpha d} - 1)/\alpha d] - 1\}$ secondary electrons. After the secondary electrons create more $\gamma\ \{[(e^{\alpha d} - 1)/\alpha d] - 1\}\ (e^{\alpha d} - 1)$ ions during their travel to the anode and these extract more $\gamma^2\ \{[(e^{\alpha d} - 1)/\alpha d] - 1\}\ (e^{\alpha d} - 1)$ electrons from the cathode. The sum of all electrons entering the anode per each primary electron is

$$\begin{aligned} m = 1 &+ \left(\frac{e^{\alpha d} - 1}{\alpha d} - 1\right) + \gamma\ \left(\frac{e^{\alpha d} - 1}{\alpha d} - 1\right) \\ &+ \gamma\ \left(\frac{e^{\alpha d} - 1}{\alpha d} - 1\right)\left(e^{\alpha d} - 1\right) + \gamma^2\ \left(\frac{e^{\alpha d} - 1}{\alpha d} - 1\right)\left(e^{\alpha d} - 1\right) \\ &+ \gamma^2\ \left(\frac{e^{\alpha d} - 1}{\alpha d} - 1\right)\left(e^{\alpha d} - 1\right)^2 + \ \ldots\ldots\ldots\ldots \end{aligned} \tag{1.112}$$

This series may hence be written under the form

$$\begin{aligned} m &= \frac{e^{\alpha d} - 1}{\alpha d} + \left(\frac{e^{\alpha d} - 1}{\alpha d} - 1\right)\ \gamma\ e^{\alpha d}\left(1 + \ \gamma\ (e^{\alpha d} - 1) + \ldots\ldots.\right) \\ &= \frac{e^{\alpha d} - 1}{\alpha d} + \left(\frac{e^{\alpha d} - 1}{\alpha d} - 1\right)\ \gamma\ e^{\alpha d}\ \frac{1}{1 - \gamma\ (e^{\alpha d} - 1)}, \end{aligned} \tag{1.113}$$

alloing to obtain at the end the multiplication factor

$$m = \frac{(1 + \gamma)(e^{\alpha d} - 1)/\alpha d - \gamma\ e^{\alpha d}}{1 - \gamma\ (e^{\alpha d} - 1)}. \tag{1.114}$$

A.1.4 Demonstration of Equation (1.59)

Defining the variable x as

$$x = \ln\left(\frac{r}{r_c}\right), \tag{1.115}$$

we have

$$\frac{dx}{dr} = \frac{1}{r} \tag{1.116}$$

and the first member of equation (1.55) writes as

$$\frac{d}{dr}\left(r\,\frac{dV}{dr}\right) = \frac{1}{r}\,\frac{d^2V}{dx^2}. \tag{1.117}$$

Using now the variable β^2 defined by equation (1.58), equation (1.55) takes the form

$$\frac{d^2V}{dx^2} = \frac{4}{9}\,\frac{V}{\beta^2}. \tag{1.118}$$

Differentiating $\ln \beta^2$, we obtain

$$\frac{d}{dx}\left(\ln \beta^2\right) = \frac{3}{2}\,\frac{1}{V}\,\frac{dV}{dx} - 1 \tag{1.119}$$

and therefore

$$\frac{dV}{dx} = \frac{2}{3}\,V\left(1 + \frac{2}{\beta}\,\frac{d\beta}{dx}\right). \tag{1.120}$$

Differentiating again with respect to x, and using equation (1.120) to eliminate dV/dx, we obtain the following equation

$$\frac{d^2V}{dx^2} = \frac{4}{9}\,V\left(1 + \frac{4}{\beta}\,\frac{d\beta}{dx} + \frac{1}{\beta^2}\left(\frac{d\beta}{dx}\right)^2 + \frac{3}{\beta}\,\frac{d^2\beta}{dx^2}\right) \tag{1.121}$$

and substituting it in equation (1.118), we obtain at the end equation (1.59)

$$1 = \beta^2 + 4\beta\,\frac{d\beta}{dx} + \left(\frac{d\beta}{dx}\right)^2 + 3\beta\,\frac{d^2\beta}{dx^2}. \tag{1.122}$$

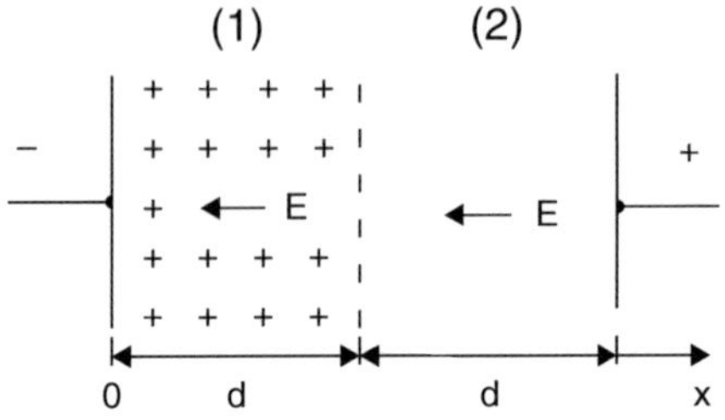

Fig. 1.15 Plane parallel electrodes limiting a gap with positive ions of density ρ uniformly distributed in one-half of the interspace (von Engel 1965)

Exercises

Exercise 1.1. An electrical potential difference V is applied to a plane parallel gap of width $2d$ which contains a constant uniformly distributed space-charge of density ρ between the plane $x = 0$ and a plane at distance d from it, while the remaining space is entirely free of space-charge (see Fig. 1.15). Determine the expressions for the potential and the electric field in the gap of width $2d$ (von Engel 1965).

Resolution: Assuming $\mathbf{E} = -E\,\mathbf{e_x}$, the Poisson's equation writes $dE/dx = -\rho/\epsilon_0$, so that we obtain at the regions $0 < x < d$ and $d < x < 2d$, respectively,

$$E_1(x) = -\frac{\rho}{\epsilon_0}\,x + C_1$$

$$E_2(x) = C_2,$$

while for the electric potential, we obtain from $E = dV/dx$

$$V_1(x) = -\frac{\rho}{2\epsilon_0}\,x^2 + C_1\,x$$

$$V_2(x) = C_2\,x + C_3,$$

having assumed $V_1(0) = 0$.

Because the separation surface $x = d$ is not electrically charged, we have $E_1 = E_2$ at the boundary, which together with the continuity of the potential, allows to write

$$-\frac{\rho}{\epsilon_0}\,d + C_1 = C_2$$

$$-\frac{\rho}{2\epsilon_0}\,d^2 + C_1\,d = C_2\,d + C_3.$$

Because at the plane $x = 2d$, we have $V_2(2d) = C_2\,2d + C_3 = V$, the electrical potential is

$$V_1(x) = -\frac{\rho}{2\epsilon_0}\,x^2 + \left(\frac{V}{2d} + \frac{3}{4}\,\frac{\rho d}{\epsilon_0}\right)x$$

$$V_2(x) = \left(\frac{V}{2d} - \frac{\rho d}{4\epsilon_0}\right)x + \frac{\rho d^2}{2\epsilon_0},$$

while for the electrical field, we find

$$E_1(x) = -\frac{\rho}{\epsilon_0}x + \left(\frac{V}{2d} + \frac{3}{4}\frac{\rho d}{\epsilon_0}\right)$$

$$E_2 = \left(\frac{V}{2d} - \frac{\rho d}{4\epsilon_0}\right) = \text{const.}$$

We note that $E_2 < 0$ for $\rho > 2\epsilon_0 V/d^2$.

Exercise 1.2. Consider two infinite plane parallel in a gas at pressure such that the mean free path of ions is much smaller than the electrode separation, $\lambda_i \ll d$. One electrode at $x = d$ emits an unlimited number of positive ions, while the other at $x = 0$ emits electrons at the same rate. If the back scattering and the charge multiplication in the interelectrode spacing are neglected, determine the spatial distributions of the electric field, potential, and ion and electron densities.

Resolution: Since $\lambda_i \ll d$ we must use a collision model and find the same situation as in equations (1.10) to (1.14). Assuming $\mathbf{E} = -E\,\mathbf{e_x}$ and $\mathbf{J} = -J\,\mathbf{e_x}$, with $J = J_e + J_i$, in order E and J are positive, Poisson's equation writes as

$$\frac{dE}{dx} = \frac{1}{\epsilon_0 E}\left(\frac{J_e}{\mu_e} - \frac{J_i}{\mu_i}\right).$$

Because of $\mu_e \gg \mu_i$ and $J_i = J_e = J/2$, we have approximately

$$E\,\frac{dE}{dx} \simeq -\frac{J}{2\epsilon_0\mu_i}.$$

This equation can be integrated now from $x = d$ (where $n_i = \infty$ and $E = 0$) allowing to obtain the following expression for the electric field

$$E(x) = \sqrt{\frac{J}{\epsilon_0\mu_i}}\,(d-x)^{1/2},$$

while for the potential, with $V(0) = 0$, we find

$$V(x) = \frac{2}{3}\sqrt{\frac{J}{\epsilon_0\mu_i}}\left[d^{3/2} - (d-x)^{3/2}\right].$$

Then, the densities of charged species are

$$n_i(x) = \frac{J}{2e\mu_i E(x)} = \frac{1}{2e}\sqrt{\frac{\epsilon_0 J}{\mu_i}}\,(d-x)^{-1/2}$$

$$n_e(x) = \frac{J}{2e\mu_e E(x)} = \frac{\mu_i}{\mu_e}\,n_i(x).$$

Exercise 1.3. Consider a gas uniformly irradiated with X-rays contained in a long concentric cylindre of radius r_0 and r_1, with an applied potential $(V_1 - V_0)$ of such a value that multiplication of charges does not occur. Find the relation between the current per unity length i/l and the potential difference $(V_1 - V_0)$, assuming the density of electrons and ions uniformly distributed, with $n_e = n_i$, the pressure high enough to use a collisional model, and neglecting space-charge field effects.

Resolution: Assuming that there is no multiplication of charges and because of the large difference between the mobilities of electrons of ions, we have

$$i/l = 2\pi r\, e\Gamma_e(r) = \text{const},$$

with $\Gamma_e(r) = n_e\mu_e\, E(r)$ denoting the electron particle current density. Since in the absence of space-charge field effects equation (1.38) yields to write

$$r\,E(r) = \frac{(V_1 - V_0)}{\ln(r_1/r_0)},$$

we obtain

$$i/l = 2\pi\, en_e\mu_e\,\frac{(V_1 - V_0)}{\ln(r_1/r_0)}.$$

References

E. Badareu, I. Popescu, Gas Ionisés – Décharges Électriques dans les Gaz (Éditions Dunod, Paris, 1968)

N.St.J. Braithwaite, Introduction to gas discharges. Plasma Sources Sci. Technol. **9**, 517–527 (2000)

F.C. Chen, *Introduction to Plasma Physics and Controlled Fusion – Volume 1: Plasma Physics*, 2nd edn. (Plenum Press, New York/London, 1984)

M.J. Druyvesteyn, F.M. Penning, The mechanism of electrical discharges in gases of low pressure. Rev. Mod. Phys. **12**, 87–174 (1940); Erratum: **13**, 72–73 (1941)

R.N. Franklin, *Plasma Phenomena in Gas Discharges* (Clarendon Press, Oxford, 1976)

M.A. Lieberman, A.J. Lichtenberg, *Principles of Plasma Discharges and Materials Processing* (Wiley, New York, 1994)

Y.P. Raizer, *Gas Discharge Physics* (Springer, Berlin, 1991)

J.R. Roth, *Industrial Plasma Engineering* (Institute of Physics Publishing (IOP), Bristol/ Philadelphia, 1995)

A.J. van Roosmalen, C.A.M. de Vries, Basics of gas discharges, in *Advanced Course on Low-Pressure Plasmas Technology and Applications* (Eindhoven, 1985), Edited by P. Massee and W.F.H. Merck

A. von Engel, *Ionized Gases*, 2nd edn. (Clarendon Press, Oxford, 1965)

Chapter 2
Kinetic Description of a Plasma

This chapter is devoted to the Boltzmann transport equation. The Boltzmann equation has been initially established for a dilute gas of neutral molecules and it provides the standard kinetic approach to describe the microscopic evolution of a gas to equilibrium. The Boltzmann equation can be applied to the description of a medium in which the dominant interactions are of short-range type so that it is suitable to describe the evolution of a plasma determined by electron-molecule and ion-molecule collisions. The Boltzmann equation fails however when applied to long-range Coulomb interactions. The collisional terms for both elastic and inelastic electron-molecule collisions are then consistently derived. The chapter ends with the establishment of the equations for the moments of the Boltzmann equation, i.e. the fluid equations, for electron-neutral interactions.

2.1 Boltzmann Transport Equation for Molecules

2.1.1 Collisionless Boltzmann Equation

The system under consideration here is a dilute gas of molecules, such as usually considered in the classical kinetic theory of gases. The temperature is assumed sufficiently high and the density sufficiently low for each molecule may be considered a classical particle with a rather well defined position and momentum. Furthermore, the molecules interact with each other through collisions whose nature may be specified through the scattering cross section (Chapman and Cowling 1939; Huang 1963).

J.M.A.H. Loureiro, J. de Amorim Filho, *Kinetics and Spectroscopy of Low Temperature Plasmas*, Graduate Texts in Physics,
DOI 10.1007/978-3-319-09253-9_2

The analysis is based on the statistical role played by the particle distribution function of a system with one kind of molecules, $f(\mathbf{r}, \mathbf{v}, t)$, defined so that

$$f(\mathbf{r}, \mathbf{v}, t)\ \mathbf{dr}\ \mathbf{dv} \tag{2.1}$$

represents the number of particles which, at the instant t, have positions lying within a space-volume element $\mathbf{dr}$ at the extremity of vector $\mathbf{r}$, and velocities lying within a velocity-volume element $\mathbf{dv}$ at the extremity of vector $\mathbf{v}$. The volume elements $\mathbf{dr}$ and $\mathbf{dv}$ are large enough to contain a very large number of molecules but small as compared to the macroscopic dimensions of the whole gas. The distribution function changes with time, because the molecules continuously enter and leave a given double volume element $\mathbf{dr}\ \mathbf{dv}$.

Having a system with N molecules in a volume V, the normalization of the distribution function is as follows

$$\int_{\mathbf{r}}\int_{\mathbf{v}} f(\mathbf{r}, \mathbf{v}, t)\ \mathbf{dr}\ \mathbf{dv} \ = \ N. \tag{2.2}$$

If the molecules are uniformly distributed in space, so that f is independent of $\mathbf{r}$, we obtain

$$\int_{\mathbf{v}} f(\mathbf{v}, t)\ \mathbf{dv} \ = \ \frac{N}{V}. \tag{2.3}$$

On the contrary, if the distribution is not uniform in space, we may define the gas number density $n(\mathbf{r}, t)$ by

$$\int_{\mathbf{v}} f(\mathbf{r}, \mathbf{v}, t)\ \mathbf{dv} \ = \ n(\mathbf{r}, t). \tag{2.4}$$

Let us start by considering first the gas in the absence of collisions. A given molecule with the coordinates $(\mathbf{r}, \mathbf{v})$ at the instant t will have the coordinates $(\mathbf{r} + \mathbf{v}\ dt, \mathbf{v} + \mathbf{a}\ dt)$ at the instant $t + dt$, with $\mathbf{a} = \mathbf{F}/m$ denoting the particle acceleration, $\mathbf{F}$ is the external force acting on the molecule, and m denotes its mass. Thus all the molecules that are in the double volume element $\mathbf{dr}\ \mathbf{dv}$, at the extremity of $(\mathbf{r}, \mathbf{v})$ and at the instant t, will be in a volume element $\mathbf{dr}'\ \mathbf{dv}'$, at the extremity of $(\mathbf{r} + \mathbf{v}\ dt, \mathbf{v} + \mathbf{F}/m\ dt)$ at the instant $t + dt$. Therefore, when the collisions do not exist we may write the following equality

$$f\left(\mathbf{r} + \mathbf{v}\ dt, \mathbf{v} + \frac{\mathbf{F}}{m}\ dt, t + dt\right)\ \mathbf{dr}'\ \mathbf{dv}' \ = \ f(\mathbf{r}, \mathbf{v}, t)\ \mathbf{dr}\ \mathbf{dv}. \tag{2.5}$$

Since $\mathbf{dr}\ \mathbf{dv} = \mathbf{dr}'\ \mathbf{dv}'$ according to Liouville's theorem (see Appendix A.2.1), we still have

$$f\left(\mathbf{r} + \mathbf{v}\ dt, \mathbf{v} + \frac{\mathbf{F}}{m}\ dt, t + dt\right) \ = \ f(\mathbf{r}, \mathbf{v}, t). \tag{2.6}$$

Let us consider now the case in which the collisions exist in the gas. The equation (2.6) must be modified in order to include the introduction and removal of molecules from a given double volume element **dr dv** due to the collisions

$$f\left(\mathbf{r}+\mathbf{v}\,dt, \mathbf{v}+\frac{\mathbf{F}}{m}\,dt, t+dt\right) = f(\mathbf{r},\mathbf{v},t) + \left(\frac{\partial f}{\partial t}\right)_{coll} dt, \tag{2.7}$$

in which $(\partial f/\partial t)_{coll}$ represents the time rate of change of the particle distribution function $f(\mathbf{r}, \mathbf{v}, t)$ due to collisions. Expanding the left-hand side member to the first order in dt, when this time interval is vanishingly small, and subtracting it from the first right-hand side term, we obtain the following equation of motion for the particle distribution function

$$\frac{\partial f}{\partial t} + \left(\mathbf{v}.\,\frac{\partial f}{\partial \mathbf{r}}\right) + \left(\frac{\mathbf{F}}{m}.\,\frac{\partial f}{\partial \mathbf{v}}\right) = \left(\frac{\partial f}{\partial t}\right)_{coll}. \tag{2.8}$$

Alternatively equation (2.8) may be written under the following form using the gradient operators with respect to **r** and **v**

$$\frac{\partial f}{\partial t} + (\mathbf{v}.\nabla_{\mathbf{r}} f) + \left(\frac{\mathbf{F}}{m}.\nabla_{\mathbf{v}} f\right) = \left(\frac{\partial f}{\partial t}\right)_{coll}. \tag{2.9}$$

When **F** does not depend on **v**, the equation (2.9) may still be written under the form

$$\frac{\partial f}{\partial t} + \nabla_{\mathbf{r}}.(f\,\mathbf{v}) + \nabla_{\mathbf{v}}.\left(f\,\frac{\mathbf{F}}{m}\right) = \left(\frac{\partial f}{\partial t}\right)_{coll}, \tag{2.10}$$

showing that the second and the third terms of the left-hand side member represent the divergence of the components of the current of molecules, respectively, in the space of positions and in the space of velocities.

The meaning of the equation (2.8) becomes clear if one notes that the left-hand side member, df/dt, represents the total (or the convective) derivate of f in the double, or phase space. The total derivative df/dt can be interpreted as the rate of change as seen in a frame moving with the molecule in $(\mathbf{r}, \mathbf{v})$ space. The equation of motion simply says that df/dt is zero in the absence of collisions. The collisions have the effect of removing a particle from one element of phase space and introducing it in another. Obviously, this equation is not meaningful until the term $(\partial f/\partial t)_{coll}$ may be explicitly specified. It is in specifying this term that the assumption that the system is a dilute gas of molecules becomes relevant. Here, we still note that the collision term may be expressed as follows

$$\left(\frac{\partial f}{\partial t}\right)_{coll} = R_{in} - R_{out}, \tag{2.11}$$

in which R_{in} **dr dv** dt represents the number of collisions occurring during the time interval dt in which one molecule enters into the double volume element **dr dv**, while R_{out} **dr dv** dt represents the number of collisions occurring during the same time interval in which one molecule leaves **dr dv**. Here, we are implicitly assuming that if a molecule enters or leaves the double volume element **dr dv** as a result of a collision, none of its partners enters or leaves this volume element in the same time interval dt. This error is negligible because of the smallness of the double volume element **dr dv**.

2.1.2 Binary Collisions

Let us consider here an elastic collision between two molecules of equal mass. The molecules have well defined positions and velocities so that the initial and final states of the collision may be described classically. If the velocities of the incoming molecules are $\mathbf{v_1}$ and $\mathbf{v_2}$, and the velocities of the outgoing molecules are $\mathbf{v'_1}$ and $\mathbf{v'_2}$, from the conservation of momentum and energy we can write

$$\mathbf{v_1} + \mathbf{v_2} = \mathbf{v'_1} + \mathbf{v'_2} \tag{2.12}$$

$$v_1^2 + v_2^2 = v_1'^2 + v_2'^2, \tag{2.13}$$

where $v_k = |\mathbf{v_k}|$, with $k \equiv (1, 2)$, denotes the absolute values of the velocities. As it is well known from any textbook on mechanics, when the new variables for the centre-of-mass velocity and relative velocity are introduced

$$\mathbf{V} = \frac{1}{2}(\mathbf{v_1} + \mathbf{v_2}) \tag{2.14}$$

$$\mathbf{v} = \mathbf{v_2} - \mathbf{v_1} \tag{2.15}$$

and similar variables $\mathbf{V}'$ and $\mathbf{v}'$ are defined, the system of equations (2.12) and (2.13) can be rewritten as

$$\mathbf{V} = \mathbf{V}' \tag{2.16}$$

$$v = v', \tag{2.17}$$

where $v = |\mathbf{v}|$ is the absolute value of the relative velocity. The collision is represented geometrically in Fig. 2.1. The relative velocity $\mathbf{v}$ merely rotates to $\mathbf{v}'$, keeping constant its magnitude. The collision is completely determined by specifying $\mathbf{V}$, $\mathbf{v}$, and the scattering angles χ and ϕ of $\mathbf{v}'$ with respect to $\mathbf{v}$.

When $\mathbf{V}$ and $\mathbf{v}$ are slightly changed to $\mathbf{V} + \mathbf{dV}$ and $\mathbf{v} + \mathbf{dv}$, respectively, with χ and ϕ kept constant, and $\mathbf{V}'$ and $\mathbf{v}'$ changed to $\mathbf{V}' + \mathbf{dV}'$ and $\mathbf{v}' + \mathbf{dv}'$ it is easy to verify that the following equality holds

$$d^3V\, d^3v = d^3V'\, d^3v'. \tag{2.18}$$

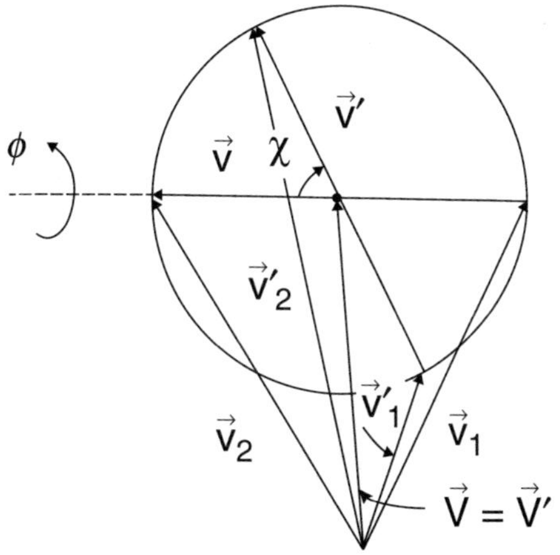

Fig. 2.1 Geometry of an elastic collision in velocity space. $\mathbf{v_1}$ and $\mathbf{v_2}$ are the velocities of the two molecules in the laboratory system, whereas $\mathbf{V}$ and $\mathbf{v}$ are the centre-of-mass and the relative velocities (Huang 1963)

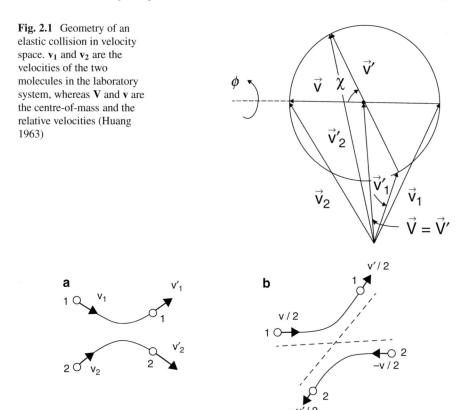

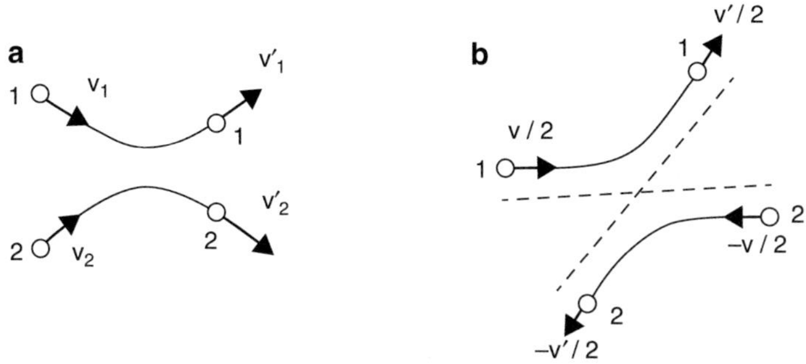

Fig. 2.2 (**a**) Scheme of a collision in the laboratory coordinate system between two molecules with initial $\mathbf{v_1}$ and $\mathbf{v_2}$ and final $\mathbf{v}_1'$ and $\mathbf{v}_2'$ velocities. (**b**) Collision view from the centre-of-mass system in which two molecules with the relative velocity $\mathbf{v}$ interact each other outcoming with the relative velocity $\mathbf{v}'$ (Huang 1963)

Here $d^3V = dV_x dV_y dV_z$ and $d^3v = dv_x dv_y dv_z$ denote equivalent representations for $\mathbf{dV}$ and $\mathbf{dv}$. Using the velocities of molecules in the laboratory coordinate system, we may also write

$$d^3v_1\, d^3v_2 \;=\; d^3v_1'\, d^3v_2'. \tag{2.19}$$

The velocity of the centre-of-mass system $\mathbf{V}$ is not important here. In fact, if we translate the coordinate system with a uniform velocity $\mathbf{V}$, in the new coordinate system only the relative velocities $\mathbf{v}$ and $\mathbf{v}'$ need to be considered. Such a coordinate system is called the centre-of-mass system. The collision processes as viewed in the laboratory coordinate system and as viewed in the centre-of-mass system are shown in Fig. 2.2a, b.

In the centre-of-mass system it suffices to consider only one of the molecules, because its partner always moves oppositely. Thus the problem reduces to the

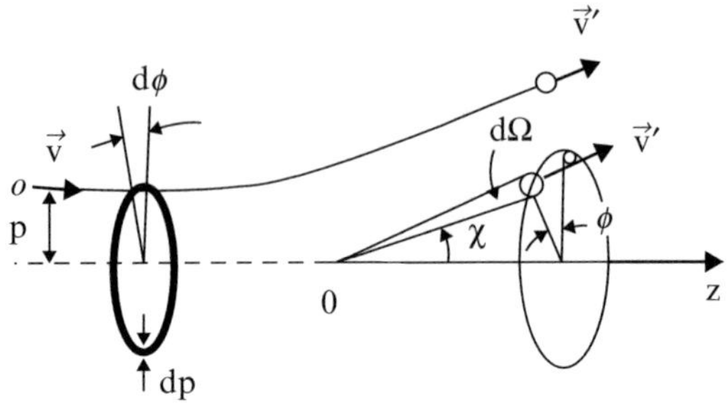

Fig. 2.3 Scattering of a molecule by a fixed centre of forces specified by the impact parameter p (Huang 1963)

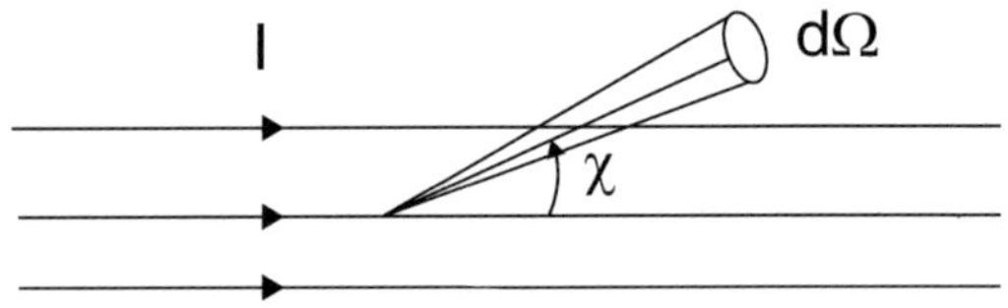

Fig. 2.4 Scattering of a incident beam of molecules by a centre of forces within the solid-angle element $d\Omega$

scattering of a molecule by a fictitious fixed centre of forces, represented by the point O in Fig. 2.3, in which p is the impact parameter.

The molecule approaches O with velocity $\mathbf{v}$, whose perpendicular distance to O is the impact parameter p. If the system of reference is chosen with O located at the origin of the coordinate system, with the z axis parallel to the velocity $\mathbf{v}$, and because $|\mathbf{v}'| = |\mathbf{v}|$, the final state of the molecule is specified by two scattering angles χ and ϕ, with χ denoting the angle between $\mathbf{v}'$ and the z axis, and ϕ the azimuthal angle of $\mathbf{v}'$ around the z axis. The two angles are collectively denoted by Ω, with $d\Omega = \sin\chi \; d\chi \; d\phi$ representing the solid-angle element. This totally completes the kinetic description of a binary collision.

The dynamical aspects of a binary collision are contained in the differential cross section $\sigma(v, \Omega)$. The initial velocities $\mathbf{v_1}$ and $\mathbf{v_2}$ of a collision do not uniquely determine the collision because they do not determine the impact parameter. Thus specifying $\mathbf{v_1}$ and $\mathbf{v_2}$ we specify a class of collisions with the same centre-of-mass system. They can be represented in Fig. 2.3 by trajectories corresponding to all possible impact parameters, and thus to all possible scattering angles. We may represent this class of collisions by imagining that a steady beam of particles of initial velocity $\mathbf{v}$, uniformly spread out in space, impinges on the centre of forces O.

Figure 2.4 represents the scattering of an incident beam with I representing the number of molecules crossing a unit area normal to the beam, per time unit. I is called the incident particle flux (expressed in molecules m^{-2} s^{-1}). In the case of molecules with the absolute value of the relative velocity v, the differential cross section $\sigma(v, \Omega)$ is defined so that

$$I \, \sigma(v, \Omega) \; d\Omega \;=\; I \, \sigma(v, \chi, \phi) \; \sin\chi \; d\chi \; d\phi \tag{2.20}$$

represents the number of molecules scattered, per time unit, in a direction lying within the solid-angle element $d\Omega$.

The differential cross section has hence the dimensions of an area. Using the concept of impact parameter Fig. 2.3 shows that the expression (2.20) can also be written under the form

$$I\,\sigma(v,\Omega)\;d\Omega \;=\; I\;p\;dp\;d\phi. \tag{2.21}$$

If the scattering is independent of the azimuthal angle ϕ, we simply obtain

$$I\,\sigma(v,\chi)\;\sin\chi\;d\chi \;=\; I\;p\;dp. \tag{2.22}$$

Then the total cross section is obtained by integrating $\sigma(v,\Omega)$ over all solid-angle elements

$$\sigma(v) \;=\; \int_{\Omega} \sigma(v,\Omega)\;d\Omega\,. \tag{2.23}$$

The differential cross section is a directly measured quantity or if the intermolecular potential is known it can also be calculated quantum mechanically. The detailed form of $\sigma(v,\Omega)$ depends of the intermolecular potential describing the interactions in a given particular gas. For our purposes we consider the differential cross section as an input parameter specified when a given gas is chosen.

2.1.3 *Collision Term of the Boltzmann Equation*

The collision term $(\partial f/\partial t)_{coll}$ is derived assuming the following approximations:

(i) The collisions are strictly binary, so that the present derivation is valid only for a sufficiently dilute gas.

(ii) The interactions between molecules are described by central forces, isotropic and of short-range, so that their effects are felt only in the interior of a small volume in the space of positions, i.e. in the interior of a small sphere of radius r_c called collision sphere. The interactions may be considered hence as true collisions.

(iii) The external forces applied on the molecules are sufficiently weak in order they may be vanished in the interior of the collision sphere. The effect of the external forces on the collision cross section is thus neglected.

(iv) For distances larger than the radius of the collision sphere there are no correlations. The velocity of a molecule is also uncorrelated with its position. This assumption is known as the hypothesis of *molecular chaos*.

The hypothesis *(iv)* is a crucial point in the derivation of the collision term of the Boltzmann equation, as we will discuss below, and it simply states that, in a spatial volume element **dr** at the extremity of vector **r**, the number of pairs of molecules with velocities lying in the velocity volume elements $\mathbf{dv_1}$ at the extremity of $\mathbf{v_1}$ and $\mathbf{dv_2}$ at the extremity of $\mathbf{v_2}$ is

$$f(\mathbf{r}, \mathbf{v_1}, t)\ \mathbf{dr}\ \mathbf{dv_1}\ f(\mathbf{r}, \mathbf{v_2}, t)\ \mathbf{dr}\ \mathbf{dv_2}. \tag{2.24}$$

This assumption is introduced here for simplification but it is not clear if this is a general condition for the description of the state of the gas under consideration. The collision term is hence derived from a model based on the classical kinetic theory of gases, in which the binary collisions occur in limited regions of space and time and where, between two collisions, the trajectory of a molecule is due to the action of external forces only.

Let us derive an explicit expression for the collision term $(\partial f/\partial t)_{coll}$. The rate of decrease of the velocity distribution of molecules whose velocity lies in $\mathbf{dv_1}$ at the extremity of $\mathbf{v_1}$ due to collisions, denoted by R_{out} in equation (2.11), can be obtained by considering that in the same spatial volume element $\mathbf{dr}$ at the extremity of $\mathbf{r}$, there are molecules of any velocity $\mathbf{v_2}$ impinging as a beam on the molecules of velocity $\mathbf{v_1}$. The flux of this incident beam (in molecules m^{-2} s^{-1}) is hence

$$I_{21} = f(\mathbf{r}, \mathbf{v_2}, t)\ \mathbf{dv_2}\ |\mathbf{v_2} - \mathbf{v_1}|, \tag{2.25}$$

so that number of collisions of the type $\{\mathbf{v_1}, \mathbf{v_2}\} \rightarrow \{\mathbf{v}'_1, \mathbf{v}'_2\}$ occurring in $\mathbf{dr}$ during the time interval dt, leading to the exit of a molecule with velocity $\mathbf{v_1}$ from $\mathbf{dv_1}$, is

$$I_{21}\ \sigma(|\mathbf{v_2}-\mathbf{v_1}|, \Omega)\ d\Omega\ dt = f(\mathbf{r}, \mathbf{v_2}, t)\ \mathbf{dv_2}\ |\mathbf{v_2}-\mathbf{v_1}|\ \sigma(|\mathbf{v_2}-\mathbf{v_1}|, \Omega)\ d\Omega\ dt, \tag{2.26}$$

where $\sigma(|\mathbf{v_2} - \mathbf{v_1}|, \Omega)$ is the differential cross section in the centre-of-mass system and $\Omega \equiv (\chi, \phi)$ represents the deviation angle between $\mathbf{v}'_2 - \mathbf{v}'_1$ and $\mathbf{v_2} - \mathbf{v_1}$. The rate R_{out} is obtained by integrating (2.26) over all velocities $\mathbf{v_2}$ and all angles Ω and by multiplying the spatial density of molecules in $\mathbf{dv_1}$, so that we obtain

$$f(\mathbf{r}, \mathbf{v_1}, t)\ \mathbf{dv_1} \int_{\Omega} \int_{\mathbf{v_2}} f(\mathbf{r}, \mathbf{v_2}, t)\ |\mathbf{v_2} - \mathbf{v_1}|\ \sigma(|\mathbf{v_2} - \mathbf{v_1}|, \Omega)\ d\Omega\ \mathbf{dv_2}\ dt. \tag{2.27}$$

The expression (2.27) gives the number of collisions, per volume unit of the space of positions, that result in the removal of a molecule from the velocity element $\mathbf{dv_1}$ in the time interval dt, that is it represents $R_{out}\ \mathbf{dv_1}\ dt$, so that the time rate for the change of f is

$$R_{out} = f(\mathbf{r}, \mathbf{v_1}, t) \int_{\Omega} \int_{\mathbf{v_2}} f(\mathbf{r}, \mathbf{v_2}, t)\ |\mathbf{v_2} - \mathbf{v_1}|\ \sigma(|\mathbf{v_2} - \mathbf{v_1}|, \Omega)\ d\Omega\ \mathbf{dv_2}. \tag{2.28}$$

In a similar manner we can calculate now R_{in} defined in (2.11), in which a collision of the type $\{\mathbf{v}'_1, \mathbf{v}'_2\} \rightarrow \{\mathbf{v_1}, \mathbf{v_2}\}$ produces an entrance of a molecule into the volume element $\mathbf{dv_1}$. In this case we start by considering the flux of incident molecules $\mathbf{v}'_2$ impinging on a molecule $\mathbf{v}'_1$

$$I'_{21} = f(\mathbf{r}, \mathbf{v}'_2, t)\, \mathbf{dv}'_2\, |\mathbf{v}'_2 - \mathbf{v}'_1|. \tag{2.29}$$

The number of collisions occurring in volume element **dr** in the time interval dt is

$$f(\mathbf{r}, \mathbf{v}'_2, t)\, \mathbf{dv}'_2\, |\mathbf{v}'_2 - \mathbf{v}'_1|\, \sigma(|\mathbf{v}'_2 - \mathbf{v}'_1|, \Omega)\, d\Omega\, dt. \tag{2.30}$$

By integrating now over all velocities $\mathbf{v}'_2$ and all angles and by multiplying the spatial density of molecules $\mathbf{v}'_1$, we obtain

$$f(\mathbf{r}, \mathbf{v}'_1, t)\, \mathbf{dv}'_1 \int_\Omega \int_{\mathbf{v}'_2} f(\mathbf{r}, \mathbf{v}'_2, t)\, |\mathbf{v}'_2 - \mathbf{v}'_1|\, \sigma(|\mathbf{v}'_2 - \mathbf{v}'_1|, \Omega)\, d\Omega\, \mathbf{dv}'_2\, dt. \tag{2.31}$$

However, in accordance with equation (2.11) the expression (2.31) represents $R_{in}\, \mathbf{dv}_1\, dt$.

Since these collisions are reverse from $\{\mathbf{v}_1, \mathbf{v}_2\} \to \{\mathbf{v}'_1, \mathbf{v}'_2\}$, we still have

$$|\mathbf{v}'_2 - \mathbf{v}'_1| = |\mathbf{v}_2 - \mathbf{v}_1| \tag{2.32}$$

$$\mathbf{dv}'_2\, \mathbf{dv}'_1 = \mathbf{dv}_2\, \mathbf{dv}_1, \tag{2.33}$$

so that the rate for the entrance of a molecule in $\mathbf{dv}_1$ is

$$R_{in} = f(\mathbf{r}, \mathbf{v}'_1, t) \int_\Omega \int_{\mathbf{v}_2} f(\mathbf{r}, \mathbf{v}'_2, t)\, |\mathbf{v}_2 - \mathbf{v}_1|\, \sigma(|\mathbf{v}_2 - \mathbf{v}_1|, \Omega)\, d\Omega\, \mathbf{dv}_2. \tag{2.34}$$

Combining the results for R_{out} and R_{in} given by equations (2.28) and (2.34), we obtain the following collision term

$$\left(\frac{\partial f_1}{\partial t}\right)_{coll} = \int_\Omega \int_{\mathbf{v}_2} (f'_2 f'_1 - f_2 f_1)\, |\mathbf{v}_2 - \mathbf{v}_1|\, \sigma(|\mathbf{v}_2 - \mathbf{v}_1|, \Omega)\, d\Omega\, \mathbf{dv}_2, \tag{2.35}$$

where $\sigma(|\mathbf{v}_2 - \mathbf{v}_1|, \Omega)$ is the differential cross section for the collision $\{\mathbf{v}_1, \mathbf{v}_2\} \to \{\mathbf{v}'_1, \mathbf{v}'_2\}$ and the following abbreviations have been used: $f_1 \equiv f(\mathbf{r}, \mathbf{v}_1, t)$, etc. Substituting (2.35) in (2.8), we obtain then the Boltzmann transport equation able to describe the evolution of the molecules with velocity $\mathbf{v}_1$

$$\frac{\partial f_1}{\partial t} + \left(\mathbf{v}_1 . \frac{\partial f_1}{\partial \mathbf{r}}\right) + \left(\frac{\mathbf{F}}{m} . \frac{\partial f_1}{\partial \mathbf{v}_1}\right) =$$

$$\int_\Omega \int_{\mathbf{v}_2} (f'_2 f'_1 - f_2 f_1)\, |\mathbf{v}_2 - \mathbf{v}_1|\, \sigma(|\mathbf{v}_2 - \mathbf{v}_1|, \Omega)\, d\Omega\, \mathbf{dv}_2, \tag{2.36}$$

which is a nonlinear integro-differential equation for the unknown function f_1. This equation is irreversible with respect to the time, which is a paradoxical result, since it violates the reversibility of molecular dynamics and the Liouville equation from which, in principle, the Boltzmann equation seems to be derived (Huang 1963; Liboff 1969; Sone 2007).

2.1.4 Validity of the Boltzmann Equation

At the end of nineteenth century a large controversy raised about the validity of Boltzmann equation through a famous paradox presented by Loschmidt. He objected that it should not be possible to deduce an irreversible equation from a time-symmetric dynamics using a time-symmetric formalism. The explanation is that the Boltzmann equation is based on the assumption of molecular chaos and this assumption breaks the time reversal symmetry. The apparent paradox is conveniently explained when we derive the collision integral of the Boltzmann equation from a more structured analysis. This derivation should be conducted with basis on the Liouville equation and it allows to obtain a system of N coupled equations for a system of N particles known as BBGKY (Born-Bogolioubov-Green-Kirkwood-Yvon) hierarchy (Delcroix 1963, 1966; Liboff 1969; Nicholson 1983; Delcroix and Bers 1994).

Let us define as F a function that contains the maximum of information about a given system. The probability of the states of a system in which the particle 1 has its position lying within a space-volume element $\mathbf{dr_1}$ at the extremity of vector $\mathbf{r_1}$ and its velocity lying within the velocity-volume $\mathbf{dv_1}$ at the extremity of vector $\mathbf{v_1}$, being the states of particles $2, \ldots, N$ whatever, may be obtained by integrating over all space of positions $\mathbf{r_2}, \ldots, \mathbf{r_N}$ and over all space of velocities $\mathbf{v_2}, \ldots, \mathbf{v_N}$

$$\mathbf{dr_1}\,\mathbf{dv_1} \int_{\mathbf{r_2}} \int_{\mathbf{v_2}} \ldots \int_{\mathbf{r_N}} \int_{\mathbf{v_N}} F\,\mathbf{dr_2}\,\mathbf{dv_2} \ldots \mathbf{dr_N}\,\mathbf{dv_N}. \tag{2.37}$$

The numbering of particles introduced to define F is artificial and arbitrary, because the particles are undistinguished.

The probable number $<dN>_1$ of particles that are in the interior of the double volume element $\mathbf{dr_1}\,\mathbf{dv_1}$ is equal to the above probability, valid for any particle, multiplied by the number of particles N, whose result may be written under the form

$$<dN_1> = f_1\,\mathbf{dr_1}\,\mathbf{dv_1}\,, \tag{2.38}$$

being $f_1 \equiv f(\mathbf{r_1}, \mathbf{v_1}, t)$ the distribution function of the positions and velocities of one particle given by

$$f_1 = N \int_{\mathbf{r_2}} \int_{\mathbf{v_2}} \ldots \int_{\mathbf{r_N}} \int_{\mathbf{v_N}} F\,\mathbf{dr_2}\,\mathbf{dv_2} \ldots \mathbf{dr_N}\,\mathbf{dv_N}. \tag{2.39}$$

On the other hand, the density of one particle $n_1 \equiv n(\mathbf{r_1}, t)$ is obtained by integrating f_1 over the space of velocities $\mathbf{v_1}$

$$n_1 = \int_{\mathbf{v_1}} f_1 \, \mathbf{dv_1} \tag{2.40}$$

and it allows to determine the probable number of particles placed in the volume element $\mathbf{dr_1}$ whatever their velocities through the expression

$$<dN_1'> = n_1 \, \mathbf{dr_1}. \tag{2.41}$$

Let us consider now two volume elements $\mathbf{dr_1}$ and $\mathbf{dr_2}$ in the space of positions and two volume elements $\mathbf{dv_1}$ and $\mathbf{dv_2}$ in the space of velocities, and let us consider further the probable number of pairs of particles, in which the first particle is placed in the double volume element $\mathbf{dr_1}\,\mathbf{dv_1}$ and the second particle in the double volume element $\mathbf{dr_2}\,\mathbf{dv_2}$. This probably number $<dN_{12}>$ is equal to the probability of the system to be in a state where two given particles, for instance the particles 1 and 2 satisfying imposed conditions, are placed in $\mathbf{dr_1}\,\mathbf{dv_1}\,\mathbf{dr_2}\,\mathbf{dv_2}$, multiplied by the number of possible pairs, that is by $N\,(N-1)$. We may therefore write

$$<dN_{12}> = f_{12} \, \mathbf{dr_1}\,\mathbf{dv_1}\,\mathbf{dr_2}\,\mathbf{dv_2}, \tag{2.42}$$

being $f_{12} \equiv f(\mathbf{r_1}, \mathbf{v_1}, \mathbf{r_2}, \mathbf{v_2}, t)$ the double distribution function defined by

$$f_{12} = N\,(N-1) \int_{\mathbf{r_3}} \int_{\mathbf{v_3}} \cdots \int_{\mathbf{r_N}} \int_{\mathbf{v_N}} F \, \mathbf{dr_3}\,\mathbf{dv_3} \ldots \mathbf{dr_N}\,\mathbf{dv_N}, \tag{2.43}$$

and where F is the function that contains all the information about the system. By integrating now f_{12} over all velocities we obtain the double density $n_{12} \equiv n(\mathbf{r_1}, \mathbf{r_2}, t)$

$$n_{12} = \int_{\mathbf{v_1}} \int_{\mathbf{v_2}} f_{12} \, \mathbf{dv_1}\,\mathbf{dv_2}. \tag{2.44}$$

From expressions (2.39) and (2.43) we obtain

$$\int_{\mathbf{r_2}} \int_{\mathbf{v_2}} f_{12} \, \mathbf{dr_2}\,\mathbf{dv_2} = (N-1)\, f_1 \tag{2.45}$$

and from (2.40) and (2.44)

$$\int_{\mathbf{r_2}} n_{12} \, \mathbf{dr_2} = (N-1)\, n_1. \tag{2.46}$$

Since

$$N = \int_{\mathbf{r_1}} \int_{\mathbf{v_1}} f_1 \, \mathbf{dr_1} \, \mathbf{dv_1} = \int_{\mathbf{r_2}} \int_{\mathbf{v_2}} f_2 \, \mathbf{dr_2} \, \mathbf{dv_2}, \tag{2.47}$$

we still have from (2.45)

$$N \int_{\mathbf{r_2}} \int_{\mathbf{v_2}} f_{12} \, \mathbf{dr_2} \, \mathbf{dv_2} = (N-1) \, f_1 \int_{\mathbf{r_2}} \int_{\mathbf{v_2}} f_2 \, \mathbf{dr_2} \, \mathbf{dv_2}. \tag{2.48}$$

If there are no correlations, we obtain

$$N \, f_{12} = (N-1) \, f_1 \, f_2 \tag{2.49}$$

and for high values of N, we have approximately

$$f_{12} \simeq f_1 \, f_2 \tag{2.50}$$

and

$$n_{12} \simeq n_1 \, n_2. \tag{2.51}$$

If the equalities (2.50) and (2.51) do not satisfy it is because there are binary correlations in result of interacting forces.

The above definitions may be still generalized to triple velocity distributions $f_{123} \equiv f(\mathbf{r_1}, \mathbf{v_1}, \mathbf{r_2}, \mathbf{v_2}, \mathbf{r_3}, \mathbf{v_3}, t)$ and triple densities $n_{123} \equiv n(\mathbf{r_1}, \mathbf{r_2}, \mathbf{r_3}, t)$, as follows

$$f_{123} = N \, (N-1) \, (N-2) \int_{\mathbf{r_4}} \int_{\mathbf{v_4}} \ldots \int_{\mathbf{r_N}} \int_{\mathbf{v_N}} F \, \mathbf{dr_4} \, \mathbf{dv_4} \ldots \mathbf{dr_N} \, \mathbf{dv_N} \tag{2.52}$$

$$n_{123} = \int_{\mathbf{v_1}} \int_{\mathbf{v_2}} \int_{\mathbf{v_3}} f_{123} \, \mathbf{dv_1} \, \mathbf{dv_2} \, \mathbf{dv_3}. \tag{2.53}$$

These triple functions are related with the double functions through the expressions

$$\int_{\mathbf{r_3}} \int_{\mathbf{v_3}} f_{123} \, \mathbf{dr_3} \, \mathbf{dv_3} = (N-2) \, f_{12} \tag{2.54}$$

$$\int_{\mathbf{r_3}} n_{123} \, \mathbf{dr_3} = (N-2) \, n_{12}. \tag{2.55}$$

Let us establish now the kinetic equations governing the distribution functions f's of different order. Starting from the Liouville equation (Delcroix 1963, 1966; Liboff 1969; Delcroix and Bers 1994)

$$\frac{\partial F}{\partial t} + \sum_{i=1}^{N} \left(\mathbf{v_i} . \frac{\partial F}{\partial \mathbf{r_i}} \right) + \sum_{i=1}^{N} \left(\frac{\mathbf{F_i}}{m} . \frac{\partial F}{\partial \mathbf{v_i}} \right) = - \sum_{i,j=1 (i \neq j)}^{N} \left(\frac{\mathbf{F_{ij}}}{m} . \frac{\partial F}{\partial \mathbf{v_i}} \right), \tag{2.56}$$

where $\mathbf{F_i}$ and $\mathbf{F_{ij}}$ denote the external and the interacting forces acting upon the particle i belonging to a set of N particles. Multiplying this equation by N and $\mathbf{dr_2}, \mathbf{dv_2}, \ldots, \mathbf{dr_N}, \mathbf{dv_N}$, and integrating over these variables, we obtain

$$\frac{\partial f_1}{\partial t} + \left(\mathbf{v_1} . \frac{\partial f_1}{\partial \mathbf{r_1}}\right) + \left(\frac{\mathbf{F_1}}{m} . \frac{\partial f_1}{\partial \mathbf{v_1}}\right) = - N \sum_{j=2}^{N} \int_{\mathbf{r_2}} \int_{\mathbf{v_2}} \ldots \int_{\mathbf{r_N}} \int_{\mathbf{v_N}} \left(\frac{\mathbf{F_{1j}}}{m} . \frac{\partial F}{\partial \mathbf{v_1}}\right) \mathbf{dr_2}\, \mathbf{dv_2} \ldots \mathbf{dr_N}\, \mathbf{dv_N}, \quad (2.57)$$

being f_1 the distribution function of one particle (2.39). The right-hand side member of equation (2.57) can still be written as follows

$$- N\,(N-1) \int_{\mathbf{r_2}} \int_{\mathbf{v_2}} \ldots \int_{\mathbf{r_N}} \int_{\mathbf{v_N}} \left(\frac{\mathbf{F_{12}}}{m} . \frac{\partial F}{\partial \mathbf{v_1}}\right) \mathbf{dr_2}\, \mathbf{dv_2} \ldots \mathbf{dr_N}\, \mathbf{dv_N} \quad (2.58)$$

and making use of the definition of the double distribution function f_{12} (2.43), we still have

$$- \int_{\mathbf{r_2}} \int_{\mathbf{v_2}} \left(\frac{\mathbf{F_{12}}}{m} . \frac{\partial f_{12}}{\partial \mathbf{v_1}}\right) \mathbf{dr_2}\, \mathbf{dv_2}. \quad (2.59)$$

The equation (2.57) for the evolution of the distribution function f_1 writes then as follows

$$\frac{\partial f_1}{\partial t} + \left(\mathbf{v_1} . \frac{\partial f_1}{\partial \mathbf{r_1}}\right) + \left(\frac{\mathbf{F_1}}{m} . \frac{\partial f_1}{\partial \mathbf{v_1}}\right) = - \int_{\mathbf{r_2}} \int_{\mathbf{v_2}} \left(\frac{\mathbf{F_{12}}}{m} . \frac{\partial f_{12}}{\partial \mathbf{v_1}}\right) \mathbf{dr_2}\, \mathbf{dv_2}. \quad (2.60)$$

The second term on the left-hand side member expresses the influence of diffusion phenomena, with $\partial f_1/\partial \mathbf{r_1}$ representing the gradient of f_1 in the space of positions, while the third term expresses the action of external forces, with $\mathbf{F_1}/m$ representing the particle acceleration, and $\partial f_1/\partial \mathbf{v_1}$ the gradient of f_1 in the velocity space. In the right-hand side member, $\mathbf{F_{12}}$ represents the interaction forces of particle 1 with any other particle.

By making the same calculation but with less one integration, we obtain the equation for the evolution of the double distribution function f_{12} (Delcroix 1963, 1966)

$$\frac{\partial f_{12}}{\partial t} + \left(\mathbf{v_1} . \frac{\partial f_{12}}{\partial \mathbf{r_1}}\right) + \left(\mathbf{v_2} . \frac{\partial f_{12}}{\partial \mathbf{r_2}}\right) + \left(\frac{\mathbf{F_1} + \mathbf{F_{12}}}{m} . \frac{\partial f_{12}}{\partial \mathbf{v_1}}\right) + \left(\frac{\mathbf{F_2} + \mathbf{F_{21}}}{m} . \frac{\partial f_{12}}{\partial \mathbf{v_2}}\right)$$
$$= - \int_{\mathbf{r_3}} \int_{\mathbf{v_3}} \left[\left(\frac{\mathbf{F_{13}}}{m} . \frac{\partial f_{123}}{\partial \mathbf{v_1}}\right) + \left(\frac{\mathbf{F_{23}}}{m} . \frac{\partial f_{123}}{\partial \mathbf{v_2}}\right)\right] \mathbf{dr_3}\, \mathbf{dv_3}, \quad (2.61)$$

where the right-hand side term represents the effect of triple interactions.

The equations (2.60) and (2.61) are the first two equations of a system of infinity equations, as a matter of fact of $(N - 1)$ equations, in which to determine f_1 we need to known previously f_{12}, and to determine f_{12} we need to know f_{123}. We might write the equation for f_{123} but it still would include f_{1234}. Starting from the Liouville equation we obtain then a coupled system of equations known as BBGKY hierarchy. For practical purposes, we need to close the system making an hypothesis on the distribution functions. The simplest solution of BBGKY system consists in considering only double interactions, by making $f_{123} = 0$ in the equation for f_{12}, which is surely valid only for a dilute gas.

By comparing the right-hand side member of equation (2.60) with the collision integral of the Boltzmann equation (2.36), we may verify at which point the time-symmetric formalism of mechanics, described by the Liouville equation, was broken and transformed to the irreversible description used in thermodynamics. The Boltzmann equation has been deducted assuming the hypothesis of molecular chaos, i.e. that the correlations between two or more particles are neglected everywhere. We can recall here that the assumption of molecular chaos states the following: If $f(\mathbf{r}, \mathbf{v}, t)$ is the probability of finding a particle with velocity $\mathbf{v}$, at position $\mathbf{r}$ and time t, the probability of simultaneously finding a molecule with velocity $\mathbf{v}$ and another with velocity $\mathbf{v}'$, at position $\mathbf{r}$ and time t, is $f(\mathbf{r}, \mathbf{v}, t)\ f(\mathbf{r}, \mathbf{v}', t)$. This assumption concerns with the absence of correlations and has noting to say about the form of the actual distribution function of the real gas. Thus a true state of the gas possessing a given distribution function may or may not satisfy the assumption of molecular chaos. Furthermore, the molecular collisions which are responsible for the change of the velocity distribution can create the molecular chaos, when it does not exist, and destroy the molecular chaos, when it exists (Huang 1963).

Thus, the Boltzmann transport equation is not a rigorous consequence of molecular dynamics. The latter is invariant under thermal reversal, but the former is not. The Boltzmann equation is only valid for a gas at the instants when the gas is in a state of molecular chaos. But the collisions can destroy the molecular chaos once established. The Boltzmann equation is not rigorously valid hence for all times. The actual distribution function does not satisfy the Boltzmann equation at all the instants. It satisfies the Boltzmann equation only at the instants when the assumption of molecular chaos is valid. If the molecular chaos is a condition valid most of the time, the Boltzmann equation may be regarded as valid in a statistical sense.

In the derivation of the Boltzmann equation we have defined a collision sphere, of radius r_c, in the interior of which the interactions may only take place and the effects of external forces are not felt. On the contrary, in the exterior of this sphere it is assumed that there are no correlations and that the particles are governed by mechanical laws under the action of the external forces. Thus, the Boltzmann equation provides an average description of the approach to equilibrium. It gives the evolution within a time-scale Δt much larger than the time of collision $t_c = r_c/v_1$, with $v_1 = |\mathbf{v_1}|$ denoting the absolute value of velocity, but it does not give the description in the interior of the collision sphere.

It can be shown that the equilibrium distribution function is the solution of the Boltzmann equation and that this function is independent of time or, which is the same, it is the limiting form of the distribution function as the time tends to infinity (see e.g. Huang 1963). Let us assume then that there are no external forces in (2.36). In this case, it is consistent to assume further that the distribution function is independent of $\mathbf{r}$. The equilibrium distribution function, denoted by $f_0(\mathbf{v_1}) \equiv f_{10}$, is the solution of the equation when $\partial f(\mathbf{v_1}, t)/\partial t = 0$. According to the Boltzmann equation (2.36), f_{10} satisfies the integral relation

$$\int_{\Omega} \int_{\mathbf{v_2}} (f'_{20} f'_{10} - f_{20} f_{10}) \, |\mathbf{v_2} - \mathbf{v_1}| \, \sigma(|\mathbf{v_2} - \mathbf{v_1}|, \Omega) \, d\Omega \, \mathbf{dv_2} = 0. \tag{2.62}$$

As long as the differential cross section $\sigma(|\mathbf{v_2} - \mathbf{v_1}|, \Omega)$ is non-vanishing, a sufficient condition for (2.62) is

$$f'_{20} f'_{10} - f_{20} f_{10} = 0, \tag{2.63}$$

for any possible collision $\{\mathbf{v_1}, \mathbf{v_2}\} \rightarrow \{\mathbf{v'_1}, \mathbf{v'_2}\}$. We thus arrive at the interesting conclusion that the equilibrium distribution $f_0(\mathbf{v_1})$ is independent of the cross section.

The equilibrium distribution function $f_0(\mathbf{v_1})$ for a gas of one type of molecules of mass M, labelled here $f(\mathbf{v})$ for simplification, is a solution of (2.63) and this distribution is the so-called Maxwell-Boltzmann distribution (Huang 1963; Bittencourt 2004)

$$f(\mathbf{v}) = n \left(\frac{M}{2\pi k_B T} \right)^{3/2} \exp\left(- \frac{M \, |\mathbf{v} - \mathbf{v_d}|^2}{2 k_B T} \right), \tag{2.64}$$

being n the particle number density

$$n = \frac{N}{V} = \int_{\mathbf{v}} f(\mathbf{v}) \, \mathbf{dv}, \tag{2.65}$$

$\mathbf{v_d} = <\mathbf{v}>$ the drift or average vector velocity

$$<\mathbf{v}> = \frac{1}{n} \int_{\mathbf{v}} \mathbf{v} \, f(\mathbf{v}) \, \mathbf{dv}, \tag{2.66}$$

and T the temperature of molecules

$$\frac{3}{2} k_B T = <E> = \frac{1}{n} \int_{\mathbf{v}} E \, f(\mathbf{v}) \, \mathbf{dv}, \tag{2.67}$$

being $E = (1/2) M v^2$ the particle kinetic energy, and k_B the Boltzmann constant. The velocity distribution (2.64) gives the probability of finding a molecule with velocity $v = |\mathbf{v}|$ in the gas, under equilibrium conditions. We have noted the interesting fact of the Maxwell-Boltzmann distribution to be independent of the detailed form of the molecular interactions, as long as they exist.

2.2 Boltzmann Transport Equation for Plasmas

2.2.1 Application to Charged Species

The Boltzmann transport equation has been initially established for a dilute gas of neutral molecules, but it may also be used in weakly ionized classical kinetic plasmas to describe the electron-molecule and ion-molecule collisions. A plasma is called classical kinetics when the electron number density n_e is low enough to exist a high number of electrons around each ion in the interior of the Debye sphere, i.e. when $d_e \ll \lambda_D$, with $d_e \sim n_e^{-1/3}$ denoting the average distance between two electrons and λ_D the Debye length given by

$$\lambda_D = \sqrt{\frac{\epsilon_0 k_B Te}{e^2 n_e}}. \tag{2.68}$$

As seen in Sect. 1.3.1, the Debye length provides a measure of the distance over which the influence of the field of a charged particle is felt by the other particles. The Debye length defines the radius of a cloud around each charge particle outside which the plasma will tend to remain neutral. The condition for a classical kinetic plasma is hence

$$n_e << \left(\frac{\epsilon_0\, k_B Te}{e^2}\right)^3. \tag{2.69}$$

In the case of $k_B T_e = 1\,\text{eV}$ (i.e. $T_e = 11{,}604\,\text{K}$), we have $n_e \ll 1.68 \times 10^{23}\,\text{m}^{-3}$.

Under such conditions, the electrons produce a screen effect around each ion and the coulomb interactions are vanished for distances larger than the Debye length having the electric potential the form of the Yukawa potential

$$V(r) = \frac{Ze}{4\pi\epsilon_0 r} \exp\left(-\frac{r}{\lambda_D}\right). \tag{2.70}$$

The electron-ion interactions are of short distance due to this screening and they take place only as the electron trajectory with respect to an ion has an impact parameter with $p < \lambda_D$. Such plasmas are termed classical kinetic in parallel with the classical kinetic theory of gases, in which collective effects do not exist. In this category are included the plasmas created by conventional discharges. For instance, at $k_B T_e = 1\,\text{eV}$ and $n_e = 10^{16}\,\text{m}^{-3}$, we obtain $\lambda_D = 7.43 \times 10^{-5}\,\text{m}$ and $d_e = 2.15 \times 10^{-6}\,\text{m}$. The number of electrons in the interior of the Debye sphere is $N_e = \frac{4}{3}\pi\lambda_D^3 n_e = 1.72 \times 10^4$. However, even in the case of classical kinetic plasmas, the Boltzmann equation cannot be used to describe the interactions between charged species, because these interactions do not occur through true collisions, but via a continuous addition of small deviations in the interior of the Debye sphere. The kinetic equation to be used for interactions of this type is the Fokker-Planck equation (see e.g. Allis 1956). The Boltzmann equation may be used hence to describe non-Coulomb interactions, such as electron-molecule and ion-molecule collisions (Allis 1956; Nicholson 1983).

Furthermore, the ionization degree n_e/n_o, with $n_e = n_i$, and being n_e, n_i and n_0 the density of electrons, ions and molecules, must be low enough, in order the effects produced by the electron-electron, electron-ion and ion-ion interactions upon the trajectories of charged species are vanishingly small. The trajectories are then determined by collisions with neutral species only.

In a plasma there are different species present and we need to establish different Boltzmann equations. In the case of electrons, for instance, the kinetic equation yielding to determine the electron distribution function takes the form

$$\frac{\partial f_e}{\partial t} + \left(\mathbf{v_e}.\frac{\partial f_e}{\partial \mathbf{r}}\right) + \left(\frac{\mathbf{F_e}}{m}.\frac{\partial f_e}{\partial \mathbf{v_e}}\right) = \left(\frac{\partial f_e}{\partial t}\right)_{coll}, \tag{2.71}$$

in which the collision term includes, in principle, the contributions of electron-electron (e-e), electron-ion (e-i), and electron-molecule (e-o) interactions, but where the third term is dominant

$$\left(\frac{\partial f_e}{\partial t}\right)_{coll} = \left(\frac{\partial f_e}{\partial t}\right)_{e-e} + \left(\frac{\partial f_e}{\partial t}\right)_{e-i} + \left(\frac{\partial f_e}{\partial t}\right)_{e-o} \simeq \left(\frac{\partial f_e}{\partial t}\right)_{e-o}. \tag{2.72}$$

Thus, in a weakly or medium ionized plasma, the collision terms for interactions between charged species are discarded in the kinetic equation for electrons, as compared with the term for electron-neutral collisions.

On the contrary, in the case of molecules the collisions between two molecules are the dominant interactions, which associated with the absence of spatial inhomogeneities, i.e. null gradients in the space of positions, and null external forces acting upon the molecules, allows to obtain an equation in which the interactions occurs among one type of particles only

$$\frac{\partial F_o}{\partial t} = \left(\frac{\partial F_o}{\partial t}\right)_{o-o}. \tag{2.73}$$

Assuming further the gas in equilibrium with the walls of the container, we obtain the Maxwell-Boltzmann distribution at temperature T_o as solution for this equation

$$F_o(\mathbf{v_o}) = n_o \left(\frac{M}{2\pi k_B T_o}\right)^{3/2} \exp\left(-\frac{M|\mathbf{v_o}|^2}{2k_B T_o}\right), \tag{2.74}$$

in which n_o is the gas number density and M the mass of molecules.

2.2.2 Vlasov Equation

In plasmas moderately ionized the electron density may not be so low that the long-distance interactions may be totally discarded, due to an incomplete screening produced by the electrons. In this case the interactions do not act as true collisions, since there are appreciable distances involved. Further, if the correlations at such distances are neglected, we may simply assume that the double distribution function is $f_{12} = f_1 f_2$. This condition introduced in (2.59) allows to write for the interaction term

$$\left(\frac{\partial f_1}{\partial t}\right)_{int} = - \int_{\mathbf{r}_2} \int_{\mathbf{v}_2} \left(\frac{\mathbf{F}_{12}}{m} \cdot \frac{\partial f_{12}}{\partial \mathbf{v}_1}\right) \mathbf{dr}_2\, \mathbf{dv}_2$$
$$= - \left(\frac{\partial f_1}{\partial \mathbf{v}_1} \cdot \frac{1}{m} \int_{\mathbf{r}_2} \int_{\mathbf{v}_2} \mathbf{F}_{12} f_2\, \mathbf{dr}_2\, \mathbf{dv}_2\right). \quad (2.75)$$

Introducing now the space-charge force defined by

$$\mathbf{F}_1' = \int_{\mathbf{r}_2} \int_{\mathbf{v}_2} \mathbf{F}_{12} f_2\, \mathbf{dr}_2\, \mathbf{dv}_2, \quad (2.76)$$

or, in the case of a velocity independent force, by

$$\mathbf{F}_1' = \int_{\mathbf{r}_2} \mathbf{F}_{12}\, n_2\, \mathbf{dr}_2, \quad (2.77)$$

we may write then the interaction term (2.75) as

$$\left(\frac{\partial f_1}{\partial t}\right)_{int} = - \left(\frac{\partial f_1}{\partial \mathbf{v}_1} \cdot \frac{\mathbf{F}_1'}{m}\right). \quad (2.78)$$

This term has the same form as that due to the external forces in the left-hand side member of equation (2.60), so that as long-range interactions only exist the kinetic equation takes the form

$$\frac{\partial f_1}{\partial t} + \left(\mathbf{v}_1 \cdot \frac{\partial f_1}{\partial \mathbf{r}_1}\right) + \left(\frac{\mathbf{F}_1 + \mathbf{F}_1'}{m} \cdot \frac{\partial f_1}{\partial \mathbf{v}_1}\right) = 0. \quad (2.79)$$

The equation (2.79) known as Vlasov equation is a collisionless kinetic equation in which both the external forces applied to the charged particles and the space-charge forces resulting from Coulomb interactions are included. It allows to study, in a self-consistent manner, the collective movement of charged particles in moderately ionized plasmas, where the distributions are strongly non-Maxwellian, and therefore a fluid description is not accurate.

2.3 Collision Phenomena in Plasmas

2.3.1 *Electron-Neutral Elastic Collisions*

Let us consider a binary collision $\{\mathbf{v_e}, \mathbf{v_o}\} \rightarrow \{\mathbf{v'_e}, \mathbf{v'_o}\}$ between an electron and a neutral molecule. In a collision, the positions of the particles do not change appreciably but their velocities may change a lot. Here, we do not consider the modification of velocities during the collision but only the velocities before and after the particles are within the range of their interaction forces. Introducing the position and the velocity of the centre-of-mass system, we get

$$\mathbf{R} = \frac{m\,\mathbf{r_e} + M\,\mathbf{r_o}}{m + M} \tag{2.80}$$

$$\mathbf{V} = \frac{\mathbf{dR}}{dt} = \frac{m}{m + M}\,\mathbf{v_e} + \frac{M}{m + M}\,\mathbf{v_o}. \tag{2.81}$$

where $\mathbf{v_e} = \mathbf{dr_e}/dt$ and $\mathbf{v_o} = \mathbf{dr_o}/dt$ are the electron and molecule velocities. Introducing now the relative velocity $\mathbf{v} = \mathbf{v_e} - \mathbf{v_o}$, the equation (2.81) can take one of the forms

$$\mathbf{v_e} - \mathbf{V} = \frac{M}{m + M}\,\mathbf{v} \tag{2.82}$$

$$\mathbf{v_o} - \mathbf{V} = -\frac{m}{m + M}\,\mathbf{v}. \tag{2.83}$$

In the same manner, we may write for the velocities after the collision

$$\mathbf{v'_e} - \mathbf{V'} = \frac{M}{m + M}\,\mathbf{v'} \tag{2.84}$$

$$\mathbf{v'_o} - \mathbf{V'} = -\frac{m}{m + M}\,\mathbf{v'}, \tag{2.85}$$

where $\mathbf{v'} = \mathbf{v'_e} - \mathbf{v'_o}$ and $\mathbf{V'} = \mathbf{V}$. During the collision the velocity of the centre-of-mass, $\mathbf{V}$, remains fixed and if the collision is elastic, the relative velocity has the same magnitude before and after the collision, $|\mathbf{v'}| = |\mathbf{v}|$, but it changes in direction by an angle χ called scattering angle in centre-of-mass system (see Fig. 2.5 and Appendix A.2.2). The centre-of-mass system divides the relative velocity inversely proportional to the masses, so that the velocities of electrons and molecules remain over two spheres in velocity space, centred in the centre-of-mass velocity, with the radii

$$\frac{M}{m + M}\,|\mathbf{v}| \quad \text{and} \quad \frac{m}{m + M}\,|\mathbf{v}|. \tag{2.86}$$

The radius of molecules' sphere is very small compared with that of electrons.

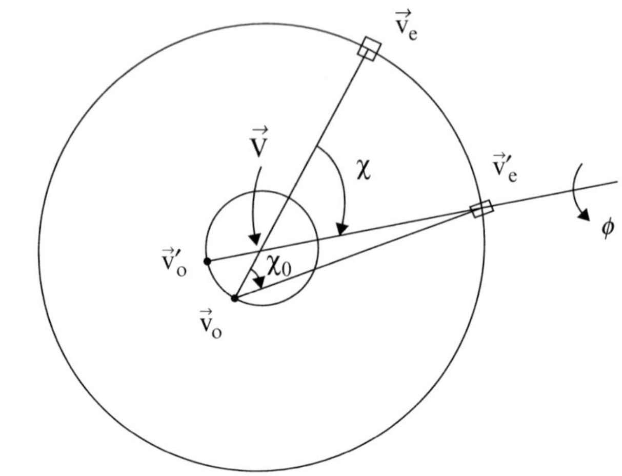

Fig. 2.5 Geometry of an elastic collision between an electron and a molecule. If the molecule is at rest before the collision the origin of the system is at $\mathbf{v_o}$. The circles are spheres centred in $\mathbf{V} = \mathbf{V}'$ (Allis 1956)

From equations (2.83) and (2.84), we still have

$$\mathbf{v}'_\mathbf{e} - \mathbf{v_o} = \frac{M}{m+M}\,\mathbf{v}' + \frac{m}{m+M}\,\mathbf{v} \tag{2.87}$$

and using the equality $|\mathbf{v}| = |\mathbf{v}'|$, we obtain

$$\begin{aligned} |\mathbf{v}'_\mathbf{e} - \mathbf{v_o}|^2 &= (\mathbf{v}'_\mathbf{e} - \mathbf{v_o}).\,(\mathbf{v}'_\mathbf{e} - \mathbf{v_o}) \\ &= \frac{M^2}{(m+M)^2}\,|\mathbf{v}|^2 + \frac{m^2}{(m+M)^2}\,|\mathbf{v}|^2 + \frac{2mM}{(m+M)^2}\,(\mathbf{v}.\mathbf{v}') \\ &= |\mathbf{v}|^2 - \frac{2mM}{(m+M)^2}\,|\mathbf{v}|^2\,(1-\cos\chi), \end{aligned} \tag{2.88}$$

with χ denoting the scattering angle between the relative velocities $\mathbf{v}$ and $\mathbf{v}'$.

If the molecule is at rest before the collision, the origin of velocity coordinates should be taken at $\mathbf{v_0}$ and we have $\mathbf{v} = \mathbf{v_e}$. The electron velocities before and after the collision are then given by $\overline{\mathbf{v_o v_e}}$ and $\overline{\mathbf{v_o v'_e}}$ in Fig. 2.5. From equation (2.88) we obtain in this case

$$v_e'^{\,2} = v_e^2 - \frac{2mM}{(m+M)^2}\,v_e^2\,(1-\cos\chi) \tag{2.89}$$

and being $u = \frac{1}{2}mv_e^2$ the initial kinetic energy of the electron, the small fractional energy lost in striking a molecule at rest, $\Delta u = u - u'$, is

$$\frac{\Delta u}{u} = \frac{2mM}{(m+M)^2}\,(1-\cos\chi). \tag{2.90}$$

This energy goes into the recoil energy of the molecule.

Another useful relation from the triangle $\mathbf{v_o}$, $\mathbf{V}$ and $\mathbf{v'_e}$ gives the scattering angle in the laboratory system, χ_0, in terms of the scattering angle in the centre-of-mass system, χ, (see Appendix A.2.3)

$$\sin^2 \chi_0 = \frac{\sin^2 \chi}{1 + (m/M)^2 + (2m/M)\cos\chi}. \tag{2.91}$$

For practical purposes $\chi_0 \simeq \chi$ because of $m/M << 1$.

The scattering cross section $\sigma(v, \Omega) \equiv \sigma(v, \chi, \phi)$ is deduced from the force law as a function of $v = |\mathbf{v}|$ and χ, and it is measured experimentally in terms of $|\mathbf{v_e}|$ and χ_0. The dependence on the azimuthal angle ϕ is never known so that it is assumed that σ does not depend on ϕ.

2.3.2 Electron-Neutral Inelastic Collisions

Let us consider now an inelastic binary collision expressed through the reaction

$$X_1 + X_2 \rightarrow X_3 + X_4, \tag{2.92}$$

in which a mono-energetic beam of particles (1) impinges on a sample of particles (2) at rest in a given system (see Fig. 2.6). From energy conservation, we have

$$E_1 + E_2 \rightarrow E_3 + E_4 + \Delta E, \tag{2.93}$$

being ΔE the energy threshold of the reaction.

The number of reactions of this type produced by one particle (1) striking on a sample of particles (2) of width dx is

$$n_2\, \sigma(v)\, dx, \tag{2.94}$$

being n_2 the density of particles (2) and $\sigma(v)$ the collision cross section. Here, $\sigma(v)$ is the total cross section, function of the absolute value of the relative velocity $|\mathbf{v}| = |\mathbf{v_1} - \mathbf{v_2}|$, i.e. it does not specify the final scattering angles of the particles and, therefore, it does not specify as the final energy is distributed between the two product particles. If the particles (2) are not in rest but instead they have the velocity $\mathbf{v_2}$, the number of reactions produced by one particle (1) impinging the particles (2)

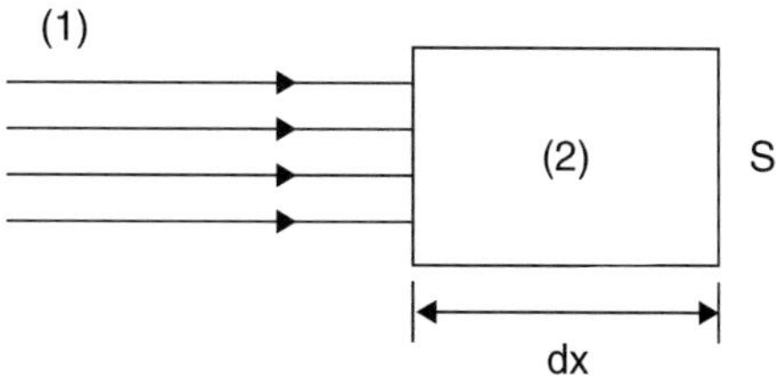

Fig. 2.6 Mono-energetic beam of particles *(1)* impinging on a sample of particles *(2)* of width dx and area S

with velocities $\mathbf{dv_2}$ defined in the extremity of the velocity $\mathbf{v_2}$ is given by

$$f(\mathbf{v_2})\ \mathbf{dv_2}\ \sigma(v)\ dx, \tag{2.95}$$

being $f(\mathbf{v_2})$ the velocity distribution of particles (2).

Let us consider now a beam of particles with velocity distribution $f(\mathbf{v_1})$. The number of particles of velocities $\mathbf{dv_1}$ defined in the extremity of velocity $\mathbf{v_1}$ crossing a sample of particles (2) of area S in the time dt is

$$f(\mathbf{v_1})\ \mathbf{dv_1}\ v\ S\ dt, \tag{2.96}$$

where v is the absolute relative velocity, and hence the number of probable reactions produced in the volume of the sample $\mathbf{dr} = S\ dx$, in the time interval dt, by the crossing of the two beams is

$$dN = f(\mathbf{v_1})\ \mathbf{dv_1}\ v\ S\ dt\, f(\mathbf{v_2})\ \mathbf{dv_2}\ \sigma(v)\ dx. \tag{2.97}$$

The number of reactions produced in the sample, per volume and time units, is

$$\frac{dN}{\mathbf{dr}\ dt} = v\ \sigma(v)\, f(\mathbf{v_1})\, f(\mathbf{v_2})\ \mathbf{dv_1}\ \mathbf{dv_2}. \tag{2.98}$$

If we assume now that the particles (1) and (2) may have different velocities, we obtain for the total rate of reaction (2.92), in $\mathrm{m^{-3}\,s^{-1}}$

$$\begin{aligned} R &= \int_{\mathbf{v_1}}\int_{\mathbf{v_2}} v\ \sigma(v)\, f(\mathbf{v_1})\, f(\mathbf{v_2})\ \mathbf{dv_1}\ \mathbf{dv_2} \\ &= n_1\ n_2\ C \end{aligned} \tag{2.99}$$

being C the collisional rate coefficient (in $\mathrm{m^3\,s^{-1}}$) of the reaction defined as

$$C = <v\ \sigma(v)> = \frac{1}{n_1\ n_2}\int_{\mathbf{v_1}}\int_{\mathbf{v_2}} v\ \sigma(v)\, f(\mathbf{v_1})\, f(\mathbf{v_2})\ \mathbf{dv_1}\ \mathbf{dv_2}. \tag{2.100}$$

In equation (2.98) the produced particles (3,4) may share the final energy $E_3 + E_4$ through different forms. They depend of the scattering angles of the two particles. Let us define as $\sigma(v, \chi, \phi)$ the differential cross section associated with the inelastic reaction (2.92), in which two particles (1,2) with the relative velocity $\mathbf{v} = \mathbf{v_1} - \mathbf{v_2}$ are transformed into two particles (3,4) with relative velocity $\mathbf{v}' = \mathbf{v}_3' - \mathbf{v}_4'$, being $|\mathbf{v}'| < |\mathbf{v}|$, and where χ and ϕ are the scattering angles of velocity $\mathbf{v}'$ with respect to $\mathbf{v}$. The probable number of reactions, per volume and time units, in which after the collision the two product particles are in the scattered solid-angle element $d\Omega = \sin\chi\ d\chi\ d\phi$, is

$$\frac{dN(\Omega)}{\mathbf{dr}\,dt} = v\,\sigma(v,\Omega)\,f(\mathbf{v_1})\,f(\mathbf{v_2})\,\mathbf{dv_1}\,\mathbf{dv_2}\,d\Omega. \tag{2.101}$$

In the case of an inelastic collision with velocities $\{\mathbf{v_1}, \mathbf{v_2}\} \rightarrow \{\mathbf{v}'_3, \mathbf{v}'_4\}$, we have from equations (2.82) and (2.83)

$$\mathbf{v_1} - \mathbf{V} = \frac{m_2}{m_1 + m_2}\,\mathbf{v} \tag{2.102}$$

$$\mathbf{v_2} - \mathbf{V} = -\,\frac{m_1}{m_1 + m_2}\,\mathbf{v}, \tag{2.103}$$

being $\mathbf{V}$ the velocity of the centre-of-mass system. Before the collision the two particles have the momenta $m_1(\mathbf{v_1} - \mathbf{V})$ and $m_2(\mathbf{v_2} - \mathbf{V})$, with equal absolute values and reverse directions, which may be expressed as

$$m_1\,(\mathbf{v_1} - \mathbf{V}) = -\,m_2\,(\mathbf{v_2} - \mathbf{V}) = \mu\,\mathbf{v}, \tag{2.104}$$

being $\mu = m_1 m_2/(m_1 + m_2)$ the reduced mass. After the collision the situation is likewise being now

$$m_3\,(\mathbf{v}'_3 - \mathbf{V}') = -\,m_4\,(\mathbf{v}'_4 - \mathbf{V}') = \mu'\,\mathbf{v}', \tag{2.105}$$

with $\mu' = m_3 m_4/(m_3 + m_4)$ (see Fig. 2.7). The relative velocity after the collision is determined by energy conservation

$$\frac{1}{2}\,\mu'\,|\mathbf{v}'|^2 = \frac{1}{2}\,\mu\,|\mathbf{v}|^2 - \Delta E. \tag{2.106}$$

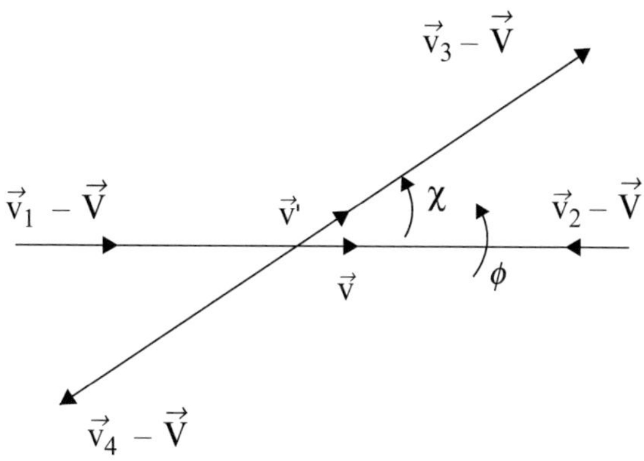

Fig. 2.7 Inelastic collision $\{X_1, X_2\} \rightarrow \{X_3, X_4\}$ shown in the centre-of-mass system (Delcroix 1963, 1966)

2.3.3 Collision Term for Inelastic Collisions

Using a similar procedure used to derive the collision term of the Boltzmann equation for elastic collisions (2.35), we may derive now a similar term for the inelastic collisions as defined through reaction (2.92). The number of collisions, per volume unit in the space of positions and time unit, that result in the removal of a molecule from the velocities $\mathbf{dv_1}$ at the extremity of $\mathbf{v_1}$ is given by

$$\left(\frac{\partial f_1}{\partial t}\right)_{coll} \mathbf{dv_1} = -f(\mathbf{r}, \mathbf{v_1}, t)\, \mathbf{dv_1} \int_\Omega \int_{\mathbf{v_2}} f(\mathbf{r}, \mathbf{v_2}, t)\, v\, \sigma_{12}^{34}(v, \Omega)\, d\Omega\, \mathbf{dv_2}, \quad (2.107)$$

where $\sigma_{12}^{34}(v, \Omega)$ is the differential cross section of reaction (2.92) and $v = |\mathbf{v}| = |\mathbf{v_1} - \mathbf{v_2}|$ is the absolute value of the relative velocity.

Let us consider now the reverse reaction

$$X_3 + X_4 \rightarrow X_1 + X_2. \quad (2.108)$$

The number of collisions, per volume and time units, that result in the entrance of a molecule into the velocities $\mathbf{dv_1}$ is

$$\left(\frac{\partial f_1}{\partial t}\right)_{coll} \mathbf{dv_1} = f(\mathbf{r}, \mathbf{v}_3', t)\, \mathbf{dv}_3' \int_\Omega \int_{\mathbf{v}_4'} f(\mathbf{r}, \mathbf{v}_4', t)\, v'\, \sigma_{34}^{12}(v', -\Omega)\, (-d\Omega)\, \mathbf{dv}_4', \quad (2.109)$$

being now $\sigma_{34}^{12}(v', -\Omega)$ the differential cross section of reaction (2.108) and $v' = |\mathbf{v}'| = |\mathbf{v}_3' - \mathbf{v}_4'|$ the absolute value of the relative velocity of particles (3) and (4). However, from micro-reversibility, or detailed balance, between the direct and reverse processes, we may write the following relation

$$v'\, \sigma_{34}^{12}(v', -\Omega)\, (-d\Omega)\, \mathbf{dv}_3'\, \mathbf{dv}_4' = v\, \sigma_{12}^{34}(v, \Omega)\, d\Omega\, \mathbf{dv_1}\, \mathbf{dv_2}, \quad (2.110)$$

so that including this relation in equation (2.109), we obtain the following collision term gathering equations (2.107) and (2.109)

$$\left(\frac{\partial f_1}{\partial t}\right)_{coll} = \int_\Omega \int_{\mathbf{v_2}} (f_3' f_4' - f_1 f_2)\, v\, \sigma_{12}^{34}(v, \Omega)\, d\Omega\, \mathbf{dv_2}, \quad (2.111)$$

in which the following abbreviations have been used: $f_1 \equiv f(\mathbf{r}, \mathbf{v_1}, t)$, etc.

In the case of a collision where the partners (1) and (3) are the same

$$X_1 + X_2 \rightleftharpoons X_1 + X_4, \quad (2.112)$$

the collision term (2.111) needs to be modified to take into account the indiscernibility of particles (1) and (3). In this case, besides the transitions $\{\mathbf{v_1}, \mathbf{v_2}\} \rightleftharpoons \{\mathbf{v}_1', \mathbf{v}_4'\}$, we must also include the transitions $\{\mathbf{v_1}, \mathbf{v_4}\} \rightleftharpoons \{\mathbf{v}_1'', \mathbf{v}_2''\}$. The collision term takes hence the form

$$\left(\frac{\partial f_1}{\partial t}\right)_{coll} = \int_\Omega \int_{\mathbf{v_2}} (f_1' f_4' - f_1 f_2)\, v_{12}\, \sigma_{12}^{14}(v_{12}, \Omega)\, d\Omega\, \mathbf{dv_2} + \int_\Omega \int_{\mathbf{v_4}} (f_1'' f_2'' - f_1 f_4)\, v_{14}\, \sigma_{14}^{12}(v_{14}, \Omega)\, d\Omega\, \mathbf{dv_4}, \quad (2.113)$$

being the relative velocities to be considered in the two terms $\mathbf{v_{12}} = \mathbf{v_1} - \mathbf{v_2}$ and $\mathbf{v_{14}} = \mathbf{v_1} - \mathbf{v_4}$, respectively.

Finally, let us consider now the direct and reverse electron-molecule inelastic collisions

$$\mathrm{e} + \mathrm{X} \rightleftharpoons \mathrm{e} + \mathrm{X}^*, \quad (2.114)$$

with X and X* denoting the ground-state and an excited-state of a given molecule. We may assume $m_2 = m_4 \simeq \infty$ in reaction (2.113) and therefore $\mathbf{v_2} = \mathbf{v_4} = 0$. The molecule acts as a fixed centre of interactions and the centre-of-mass coincides with this fixed centre. We may replace then the velocity distributions f_2 and f_4 with delta functions $n_2\, \delta(\mathbf{v_2})$ and $n_4\, \delta(\mathbf{v_4})$ having, for instance, for particle (2)

$$\int_{\mathbf{v_2}} f_2\, \mathbf{dv_2} = \int_{\mathbf{v_2}} n_2\, \delta(\mathbf{v_2})\, \mathbf{dv_2} = n_2, \quad (2.115)$$

being the relative velocities equal to $\mathbf{v_1}$ in both terms of equation (2.113). The collision term (2.113) takes hence the form

$$\left(\frac{\partial f_1}{\partial t}\right)_{coll} = \int_\Omega \int_{\mathbf{v_2}} f_1'\, n_4\, \delta(\mathbf{v_4'})\, v_1\, \sigma_{12}^{14}(v_1, \Omega)\, d\Omega\, \mathbf{dv_2} - n_2 \int_\Omega f_1\, v_1\, \sigma_{12}^{14}(v_1, \Omega)\, d\Omega + \int_\Omega \int_{\mathbf{v_4}} f_1''\, n_2\, \delta(\mathbf{v_2''})\, v_1\, \sigma_{14}^{12}(v_1, \Omega)\, d\Omega\, \mathbf{dv_4} - n_4 \int_\Omega f_1\, v_1\, \sigma_{14}^{12}(v_1, \Omega)\, d\Omega. \quad (2.116)$$

Introducing now the total cross sections in the second and fourth terms

$$\int_\Omega \sigma_{12}^{14}(v_1, \Omega)\, d\Omega = \sigma_{12}^{14}(v_1) \quad (2.117)$$

$$\int_\Omega \sigma_{14}^{12}(v_1, \Omega)\, d\Omega = \sigma_{14}^{12}(v_1) \quad (2.118)$$

and the following micro-reversibility relations in the first and third terms

$$v_1\, \sigma_{12}^{14}(v_1, \Omega)\, d\Omega\, \mathbf{dv_1}\, \mathbf{dv_2} = v_1'\, \sigma_{14}^{12}(v_1', -\Omega)\, (-d\Omega)\, \mathbf{dv_1'}\, \mathbf{dv_4'} \quad (2.119)$$

$$v_1\, \sigma_{14}^{12}(v_1, \Omega)\, d\Omega\, \mathbf{dv_1}\, \mathbf{dv_4} = v_1''\, \sigma_{12}^{14}(v_1'', -\Omega)\, (-d\Omega)\, \mathbf{dv_1''}\, \mathbf{dv_2''}, \quad (2.120)$$

we may rewrite equation (2.116) as follows

$$\left(\frac{\partial f_1}{\partial t}\right)_{coll} = \int_\Omega \int_{\mathbf{v}_4'} f_1' \, n_4 \, \delta(\mathbf{v}_4') \, v_1' \, \sigma_{14}^{12}(v_1', \Omega) \, d\Omega \, \frac{\mathbf{dv}_1'}{\mathbf{dv}_1} \, \mathbf{dv}_4' \; - \; n_2 f_1 \, v_1 \, \sigma_{12}^{14}(v_1)$$
$$+ \int_\Omega \int_{\mathbf{v}_2''} f_1'' \, n_2 \, \delta(\mathbf{v}_2'') \, v_1'' \, \sigma_{12}^{14}(v_1'', \Omega) \, d\Omega \, \frac{\mathbf{dv}_1''}{\mathbf{dv}_1} \, \mathbf{dv}_2'' \; - \; n_4 f_1 \, v_1 \, \sigma_{14}^{12}(v_1). \tag{2.121}$$

Taking into account that in the first and third terms of equation (2.121) we still have

$$\int_{\mathbf{v}_4'} n_4 \, \delta(\mathbf{v}_4') \, \mathbf{dv}_4' = n_4 \quad \text{and} \quad \int_{\mathbf{v}_2''} n_2 \, \delta(\mathbf{v}_2'') \, \mathbf{dv}_2'' = n_2 \tag{2.122}$$

and the following relationships between the velocity volume elements associated through an inelastic collision

$$\mathbf{dv}_1' = \frac{v_1'}{v_1} \, \mathbf{dv}_1 \quad \text{and} \quad \mathbf{dv}_1'' = \frac{v_1''}{v_1} \, \mathbf{dv}_1, \tag{2.123}$$

the equation (2.121) may still be rewritten with the form

$$\left(\frac{\partial f_1}{\partial t}\right)_{coll} = n_4 \, \frac{v_1'^{\,2}}{v_1} \int_\Omega f_1' \, \sigma_{14}^{12}(v_1', \Omega) \; d\Omega \; - \; n_2 \, v_1 f_1 \, \sigma_{12}^{14}(v_1)$$
$$+ \, n_2 \, \frac{v_1''^{\,2}}{v_1} \int_\Omega f_1'' \, \sigma_{12}^{14}(v_1'', \Omega) \; d\Omega \; - \; n_4 \, v_1 f_1 \, \sigma_{14}^{12}(v_1). \tag{2.124}$$

The four terms in equation (2.124) refer by order to the following transitions

$$\{\mathbf{v}_1', \mathbf{v}_4\} \rightarrow \{\mathbf{v}_1, \mathbf{v}_2\}$$
$$\{\mathbf{v}_1, \mathbf{v}_2\} \rightarrow \{\mathbf{v}_1', \mathbf{v}_4\}$$
$$\{\mathbf{v}_1'', \mathbf{v}_2\} \rightarrow \{\mathbf{v}_1, \mathbf{v}_4\}$$
$$\{\mathbf{v}_1, \mathbf{v}_4\} \rightarrow \{\mathbf{v}_1'', \mathbf{v}_2\}.$$

To attribute significance we may assume that the reaction (1,2) $\rightarrow$ (1,4) is endothermic, with the species (1) being an electron

$$\mathrm{e}(u) + \mathrm{X}_2 \rightleftharpoons \mathrm{e}(u - \Delta u) + \mathrm{X}_4, \tag{2.125}$$

in which a molecule in state X_2 goes to a higher energy state X_4, suffering the electron an inelastic collision, whereas the reaction (1,4) $\rightarrow$ (1,2) is exothermic, suffering the electron a superelastic collision, also called an inelastic collision of

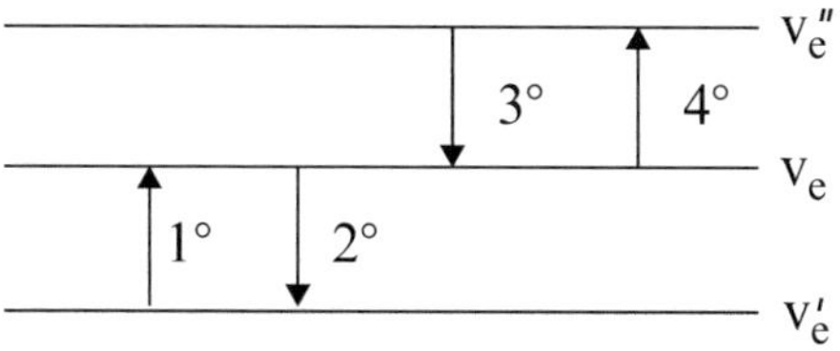

Fig. 2.8 Decrease and increase of the electron velocity as a result of an inelastic ($v_e \rightarrow v'_e$; $v''_e \rightarrow v_e$) or of a superelastic ($v'_e \rightarrow v_e$; $v_e \rightarrow v''_e$) collision, respectively (Delcroix 1963, 1966)

second kind. In the two collisions the electron of energy u loses or gains the energy Δu, respectively.

Let us consider now a given molecule with different quantum states $\ldots, i, \ldots, j, \ldots$. If among them we isolate the states i and j, being $u_{ij} = (E_j - E_i) > 0$, and $v' < v < v''$ denoting three different values of electron velocity obeying to

$$v''^2 - v^2 = v^2 - v'^2 = \frac{2}{m} u_{ij}, \tag{2.126}$$

the collision term for inelastic collisions (2.124) may be written under the form, in the case of the electron distribution function $f_e \equiv f(\mathbf{v_e})$,

$$\left(\frac{\partial f_e}{\partial t}\right)_{coll} = \sum_{i,j} \left\{ n_j \frac{v_e'^2}{v_e} \int_\Omega f'_e \, \sigma_{ji}(v'_e, \Omega) \; d\Omega \; - \; n_i \, v_e \, f_e \, \sigma_{ij}(v_e) \right\}$$
$$+ \sum_{i,j} \left\{ n_i \frac{v_e''^2}{v_e} \int_\Omega f''_e \, \sigma_{ij}(v''_e, \Omega) \; d\Omega \; - \; n_j \, v_e \, f_e \, \sigma_{ji}(v_e) \right\}, \tag{2.127}$$

being $f'_e \equiv f(\mathbf{v'_e})$ and $f''_e \equiv f(\mathbf{v''_e})$ and where the sum indicates the possibility of different reactions to be considered.

The physical meaning of the four terms in equation (2.127) can be well understood through the diagram shown in Fig. 2.8.

The vertical rows from left to right represent by order the four terms of equation (2.127). The down rows represent the inelastic collisions in which an electron diminishes the absolute value of its velocity from $v_e \rightarrow v'_e$ and from $v''_e \rightarrow v_e$, whereas the up rows represent the reverse superelastic collisions, in which an electron increases the absolute value of its velocity from $v'_e \rightarrow v_e$ and from $v_e \rightarrow v''_e$. The second and the fourth terms are loss terms for the electrons of velocity v_e, while the first and the third terms are gain terms. In this latter case, due to collisions with colder electrons and hotter electrons, respectively.

In the case of the terms taking into account the exit of an electron from velocity v_e, either due to an inelastic collision (second term) or a superelastic collision (fourth term), we do not need to specify the form as the electron enters into the new velocity element, so that the total cross section is enough. On the contrary, in the case of the terms describing an entrance at velocity v_e, due to a superelastic collision (first term) or due to an inelastic collision (third term), we need to specify the orientation of the velocity vector before the collision, so that the differential cross section is needed. For instance, $\sigma_{ij}(v_e'', \Omega)$ is the differential cross section of the inelastic collision for an electron with initial velocity $\mathbf{v_e''}$ and final velocity $\mathbf{v_e}$, with $v_e < v_e''$, in which a rotation from $\mathbf{v_e''}$ to $\mathbf{v_e}$ occurs.

2.3.4 Rate Coefficients for Direct and Reverse Processes

In the case of inelastic and superelastic electron-molecule collisions of the type

$$\mathrm{e} + \mathrm{X}_i \rightleftharpoons \mathrm{e} + \mathrm{X}_j, \tag{2.128}$$

with X_i denoting a molecule in a state with the level-energy E_i and X_j the same molecule in a upper state with the level-energy $E_j > E_i$, and since we have $|\mathbf{v_e}| \gg |\mathbf{v_o}|$, and therefore $|\mathbf{v}| \simeq |\mathbf{v_e}|$, we obtain from equation (2.100) the following expression for the rate coefficient of this inelastic process by electron impact

$$C_{ij} = <v_e\, \sigma_{ij}(v_e)> = \frac{1}{n_e\, n_i} \int_{\mathbf{v_e}} \int_{\mathbf{v_i}} v_e\, \sigma_{ij}(v_e)\, f(\mathbf{v_e})\, F(\mathbf{v_i})\, \mathbf{dv_e}\, \mathbf{dv_i}. \tag{2.129}$$

Since for the heavy species we have

$$\int_{\mathbf{v_i}} F(\mathbf{v_i})\, \mathbf{dv_i} = n_i, \tag{2.130}$$

we still obtain

$$C_{ij} = <v_e\, \sigma_{ij}(v_e)> = \frac{1}{n_e} \int_{\mathbf{v_e}} v_e\, \sigma_{ij}(v_e)\, f(\mathbf{v_e})\, \mathbf{dv_e}. \tag{2.131}$$

We will show in Chap. 3 that because of the integrand function $v_e\ \sigma_{ij}(v_e)$ is an isotropic function of the electron velocity, depending hence on the absolute value of the velocity only, the electron rate coefficient C_{ij} is given by

$$C_{ij} = \frac{1}{n_e} \int_{v_{ij}}^{\infty} v_e\, \sigma_{ij}(v_e)\, f_e^0(v_e)\, 4\pi v_e^2\, dv_e, \tag{2.132}$$

with $f_e^0(v_e)$ representing the isotropic component of the electron velocity distribution, i.e. the component that depends on the absolute value of the velocity only, and $v_{ij} = \sqrt{2u_{ij}/m} = \sqrt{2(E_j - E_i)/m}$ is the change of velocity in result of the inelastic collision.

Under thermodynamic equilibrium conditions it should be verified by micro-reversibility

$$n_e\, n_i\, C_{ij} \;=\; n_e\, n_j\, C_{ji}, \tag{2.133}$$

being n_i and n_j the populations of the states X_i and X_j, respectively, and C_{ij} and C_{ji} the electron rate coefficients of the direct and reverse processes (2.128). Since at equilibrium conditions the populations n_i and n_j follow a Maxwell-Boltzmann distribution at the common temperature $T_e = T_o$, we may write

$$\frac{n_j}{n_i} \;=\; \frac{g_j}{g_i}\, \exp\left(-\,\frac{u_{ij}}{k_B T_e}\right), \tag{2.134}$$

with g_i and g_j denoting the statistical weights, or degenerescency degrees (number of distinguished quantum states), of the states of energy E_i and E_j, respectively. Matching equations (2.133) and (2.134) we obtain the following relation between the superelastic and the inelastic electron rate coefficients

$$\frac{C_{ji}}{C_{ij}} \;=\; \frac{g_i}{g_j}\, \exp\left(\frac{u_{ij}}{k_B T_e}\right). \tag{2.135}$$

The rate coefficients are hence larger for the superelastic processes.

At the first sight, we could think that equation (2.135) is valid at thermodynamic equilibrium only. However, the sole condition at which this equation is obliged to fulfill is the velocity distribution of electrons to be a Maxwellian at a given temperature T_e, being hence possible to have $T_e \neq T_o$. As a matter of fact, looking at the rate coefficient of the inelastic process given by equation (2.132), we observe that it only acts the collision cross section, which is a quantity that has noting to do with the macroscopic properties of the medium. Thus, when the velocity distribution of electrons is Maxwellian, the equation (2.135) is verified although the electrons may not be in equilibrium with the molecules.

However, it is still possible to derive from equation (2.135) a relationship between the electron cross sections for the direct and reverse reactions, which obviously does not depend on the electron velocity distribution. Starting by expressing the equation (2.132) in terms of the electron energy $u = \frac{1}{2}\, m v_e^2$, we obtain

$$C_{ij} \;=\; \frac{1}{n_e}\, \frac{8\pi}{m^2} \int_{u_{ij}}^{\infty} u\, \sigma_{ij}(u)\, f_e^0(u)\, du \tag{2.136}$$

and therefore the expression for the ratio C_{ji}/C_{ij} between the reverse and direct rate coefficients can be written as follows

$$\begin{aligned}\frac{C_{ji}}{C_{ij}} &= \frac{\int_0^{\infty} u\, \sigma_{ji}(u)\, f_e^0(u)\, du}{\int_{u_{ij}}^{\infty} u\, \sigma_{ij}(u)\, f_e^0(u)\, du} \\ &= \frac{\int_0^{\infty} u\, \sigma_{ji}(u)\, f_e^0(u)\, du}{\int_0^{\infty} (u + u_{ij})\, \sigma_{ij}(u + u_{ij})\, f_e^0(u + u_{ij})\, du}.\end{aligned} \tag{2.137}$$

In the case of a Maxwellian distribution the following dependence exists

$$f_e^0(u) \propto \exp\left(-\frac{u}{k_B T_e}\right), \tag{2.138}$$

so that the ratio (2.137) takes the form

$$\frac{C_{ji}}{C_{ij}} = \frac{\int_0^\infty u\,\sigma_{ji}(u)\ \exp(-u/(k_B T_e))\,du}{\int_0^\infty (u+u_{ij})\ \sigma_{ij}(u+u_{ij})\ \exp(-u/(k_B T_e))\,du}\ \exp\left(\frac{u_{ij}}{k_B T_e}\right). \tag{2.139}$$

Since the ratio C_{ji}/C_{ij} also fulfills equation (2.135) and the integrals of equation (2.139) are extended over the same domain, we obtain the following relationship from the integrand functions of this latter equation

$$g_j\ u\ \sigma_{ji}(u) = g_i\ (u+u_{ij})\ \sigma_{ij}(u+u_{ij}). \tag{2.140}$$

Equation (2.140) expresses a relationship between the cross sections of direct and reverse processes known as Klein-Rosseland relation. It is a microscopic universal relation, independent hence of the macroscopic properties of the medium.

2.4 Moments of the Boltzmann Equation

2.4.1 Conservation Laws and Continuity Equation

As we have seen in the previous sections, the electron distribution function for a plasma out of equilibrium can be obtained by solving the electron Boltzmann transport equation. Then, once the electron distribution is known, we can obtain the macroscopic quantities of physical interest for the plasma, such as the electron number density n_e, the electron drift velocity $\mathbf{v_{ed}} = <\mathbf{v_e}>$, or the electron mean energy $<u>= \frac{1}{2}\ m\ <v_e^2>$. However, the solution of the Boltzmann equation is generally a matter of great difficulty, so that when a given problem only needs the macroscopic variables it is preferable to use an analysis based on the moments of the Boltzmann equation, i.e. the fluid equations, without having to solve the Boltzmann equation (Huang 1963; Delcroix 1963, 1966; Liboff 1969; Cherrington 1980; Bittencourt 2004). The resulting equations are significantly simpler to use, since they describe the plasma from a macroscopic point of view.

The moments of the Boltzmann equations are conservation laws for any quantity $X(\mathbf{r}, \mathbf{v_e}, t)$ associated with an electron located at position $\mathbf{r}$, velocity $\mathbf{v_e}$, and time t. In order to derive a conservation law for this general quantity X, let us consider equation (2.71), with the collision term with electron-neutral collisions only such as indicated in (2.72). Multiplying both members of (2.71) by X, we can write

$$X \frac{\partial f_e}{\partial t} + \sum_i X\, v_{ei} \frac{\partial f_e}{\partial x_i} + \sum_i X \frac{F_{ei}}{m} \frac{\partial f_e}{\partial v_{ei}} = X \left(\frac{\partial f_e}{\partial t} \right)_{e-o}. \qquad (2.141)$$

Integrating this equation in velocity space and defining the velocity averaged value of X by

$$<X>(\mathbf{r}, t) = \frac{1}{n_e} \int_{\mathbf{v_e}} X(\mathbf{r}, \mathbf{v_e}, t)\, f_e(\mathbf{r}, \mathbf{v_e}, t)\, \mathbf{dv_e}, \qquad (2.142)$$

we obtain for each one the left-hand side terms of equation (2.141)

$$\begin{aligned} \int_{\mathbf{v_e}} X \frac{\partial f_e}{\partial t}\, \mathbf{dv_e} &= \frac{\partial}{\partial t} \left(\int_{\mathbf{v_e}} X f_e\, \mathbf{dv_e} \right) - \int_{\mathbf{v_e}} \frac{\partial X}{\partial t} f_e\, \mathbf{dv_e} \\ &= \frac{\partial}{\partial t} (n_e <X>) - n_e < \frac{\partial X}{\partial t} >; \end{aligned} \qquad (2.143)$$

$$\begin{aligned} \int_{\mathbf{v_e}} X\, v_{ei} \frac{\partial f_e}{\partial x_i}\, \mathbf{dv_e} &= \frac{\partial}{\partial x_i} \left(\int_{\mathbf{v_e}} X\, v_{ei} f_e\, \mathbf{dv_e} \right) - \int_{\mathbf{v_e}} \frac{\partial X}{\partial x_i} v_{ei} f_e\, \mathbf{dv_e} \\ &= \frac{\partial}{\partial x_i} (n_e <X\, v_{ei}>) - n_e < \frac{\partial X}{\partial x_i} v_{ei} >; \end{aligned} \qquad (2.144)$$

$$\begin{aligned} \int_{\mathbf{v_e}} X \frac{F_{ei}}{m} \frac{\partial f_e}{\partial v_{ei}}\, \mathbf{dv_e} &= \int_{\mathbf{v_e}} \frac{\partial}{\partial v_{ei}} \left(X \frac{F_{ei}}{m} f_e\, \mathbf{dv_e} \right) - \int_{\mathbf{v_e}} \frac{\partial}{\partial v_{ei}} \left(X \frac{F_{ei}}{m} \right) f_e\, \mathbf{dv_e} \\ &= -\, n_e < \frac{\partial}{\partial v_{ei}} \left(X \frac{F_{ei}}{m} \right) > . \end{aligned} \qquad (2.145)$$

The first term of equation (2.145) vanishes if $f_e(\mathbf{r}, \mathbf{v_e}, t)$ is assumed to vanish when $v_{ei} \rightarrow \infty$.

The integration in velocity space of equation (2.141) can then be written as follows assuming velocity independent external forces ($\partial F_{ei}/\partial v_{ei} = 0$, although F_{ei} may depend on the components of the drift velocity as in the Lorentz force)

$$\begin{aligned} &\frac{\partial}{\partial t} (n_e <X>) - n_e < \frac{\partial X}{\partial t} > + \sum_i \frac{\partial}{\partial x_i} (n_e <X\, v_{ei}>) - \sum_i n_e < \frac{\partial X}{\partial x_i} v_{ei} > \\ &- \sum_i n_e < \frac{\partial X}{\partial v_{ei}} > \frac{F_{ei}}{m} = \int_{\mathbf{v_e}} X \left(\frac{\partial f_e}{\partial t} \right)_{e-o} \mathbf{dv_e}. \end{aligned} \qquad (2.146)$$

In the case of electrons the independent conserved properties are obtained by making $X = 1$, $X = m v_{ei}$, and $X = \frac{1}{2}\, m v_e^2$. We obtain then the equation for electron number density conservation, i.e. the continuity equation, the equation for momentum conservation, and the equation for energy conservation.

When $X = 1$, we simply obtain

$$\frac{\partial n_e}{\partial t} + \sum_i \frac{\partial}{\partial x_i} (n_e <v_{ei}>) = \int_{\mathbf{v_e}} \left(\frac{\partial f_e}{\partial t} \right)_{e-o} \mathbf{dv_e}, \tag{2.147}$$

which can be written as

$$\frac{\partial n_e}{\partial t} + \nabla.\, (n_e\, \mathbf{v_{ed}}) = \left(\frac{\partial n_e}{\partial t} \right)_{e-o}, \tag{2.148}$$

being $\mathbf{v_{ed}}$ the electron drift velocity. If the collisions do not modify the number of electrons, we have $(\partial n_e/\partial t)_{e-o} = 0$, otherwise as the production of secondary electrons or the electron attachment are included, the right-hand side member of equation (2.148) becomes non-null. In the case of ionizing collisions, the right-hand side member of equation (2.146) takes the following form using equation (2.36)

$$\begin{aligned} \int_{\mathbf{v_e}} X \left(\frac{\partial f_e}{\partial t} \right)_{e-o} \mathbf{dv_e} &= \int_{\mathbf{v_e}} \int_{\Omega} \int_{\mathbf{v_o}} (X' f_e' F_o' - X f_e F_o)\, v_e\, \sigma_{ion}(v_e, \Omega)\, \mathbf{dv_e}\, d\Omega\, \mathbf{dv_o} \\ &= n_o \int_{\mathbf{v_e}} \int_{\Omega} (X' f_e' - X f_e)\, v_e\, \sigma_{ion}(v_e, \Omega)\, \mathbf{dv_e}\, d\Omega . \end{aligned} \tag{2.149}$$

Furthermore, in an ionizing collision $X = 1$ and $X' = 2$, so that we obtain

$$\begin{aligned} \left(\frac{\partial n_e}{\partial t} \right)_{e-o} = \int_{\mathbf{v_e}} \left(\frac{\partial f_e}{\partial t} \right)_{e-o} \mathbf{dv_e} &= n_o \int_{\mathbf{v_e}} \int_{\Omega} v_e\, \sigma_{ion}(v_e, \Omega)\, f_e\, \mathbf{dv_e}\, d\Omega \\ &= n_o \int_{v_e} v_e\, \sigma_{ion}(v_e)\, f_e^0(v_e)\, 4\pi v_e^2\, dv_e, \end{aligned} \tag{2.150}$$

being $\sigma_{ion}(v_e)$ the total electron ionization cross section and, as it was already referred in equation (2.132) and it will be shown later on in Chap. 3, $f_e^0(v_e)$ is the isotropic component of the electron velocity distribution. Considering the ionization frequency $\nu_{ion}(v_e) = n_o v_e \sigma_{ion}(v_e)$, we obtain for equation (2.150)

$$\left(\frac{\partial n_e}{\partial t} \right)_{e-o} = n_e <\nu_{ion}> . \tag{2.151}$$

Finally, introducing the electron particle current density $\mathbf{\Gamma_e} = n_e\, \mathbf{v_{ed}}$, the continuity equation (2.148) takes the form

$$\frac{\partial n_e}{\partial t} + (\nabla.\, \mathbf{\Gamma_e}) = n_e <\nu_{ion}> . \tag{2.152}$$

2.4.2 Equation for Momentum Conservation

The equation of the second moment is the equation for momentum conservation and it can be derived using $X = m\mathbf{v_e}$ in equation (2.146) and integrating in velocity space

$$\frac{\partial}{\partial t}(n_e\, m <\mathbf{v_e}>) + \sum_i \frac{\partial}{\partial x_i}(n_e\, m <\mathbf{v_e}\, v_{ei}>) - \sum_i n_e\, m < \frac{\partial \mathbf{v_e}}{\partial v_{ei}} > \frac{F_{ei}}{m} = \int_{\mathbf{v_e}} m\, \mathbf{v_e} \left(\frac{\partial f_e}{\partial t}\right)_{e-o} \mathbf{dv_e}. \quad (2.153)$$

Since $v_{ei} = (v_{ei} - v_{edi}) + v_{edi}$, being v_{edi} the i-th component of the electron drift velocity $\mathbf{v_{ed}} = <\mathbf{v_e}>$, we may write the second left-hand side term of equation (2.153) under the form

$$\sum_{i,j} \frac{\partial}{\partial x_i} \{n_e\, m <((v_{ei} - v_{edi}) + v_{edi})\ ((v_{ej} - v_{edj}) + v_{edj})>\}$$
$$= \sum_{i,j} \frac{\partial}{\partial x_i} \{n_e\, m\, (<(v_{ei} - v_{edi})\, (v_{ej} - v_{edj})> + v_{edi}\, v_{edj})\}$$
$$= \sum_{i,j} \frac{\partial P_{ij}}{\partial x_i} + \sum_{i,j} m \frac{\partial (n_e\, v_{edi})}{\partial x_i} v_{edj} + \sum_{i,j} n_e\, m\, v_{edi} \frac{\partial v_{edj}}{\partial x_i}, \quad (2.154)$$

where P_{ij} are the components of a dyadic called pressure tensor defined by

$$P_{ij} = n_e\, m\, <(v_{ei} - v_{edi})\, (v_{ej} - v_{edj})>, \quad (2.155)$$

or under vectorial form as

$$\widehat{P} = n_e\, m <(\mathbf{v_e} - \mathbf{v_{ed}})\, (\mathbf{v_e} - \mathbf{v_{ed}})> = n_e\, m\, (<\mathbf{v_e}\, \mathbf{v_e}> - \mathbf{v_{ed}}\, \mathbf{v_{ed}})\,. \quad (2.156)$$

Therefore, using vectorial notation, we may write (2.154) as follows

$$(\nabla.\, \widehat{P}) + m\, (\nabla.\, (n_e\, \mathbf{v_{ed}}))\, \mathbf{v_{ed}} + n_e\, m\, (\mathbf{v_{ed}}.\nabla)\mathbf{v_{ed}}. \quad (2.157)$$

On the other hand, the third left-hand side term of equation (2.153) takes the form

$$-\sum_{i,j} n_e\, m < \frac{\partial v_{ej}}{\partial v_{ei}} > \frac{F_{ei}}{m} = -\sum_{i,j} n_e\, \delta_{ji}\, F_{ei} = -\, n_e\, \mathbf{F_e}. \quad (2.158)$$

Replacing equations (2.157) and (2.158) in equation (2.153), and making use of the continuity equation (2.152) in the first left-hand side term of equation (2.153), we obtain

$$n_e\, m <\nu_{ion}> \mathbf{v_{ed}} \;+\; n_e\, m\, \frac{\partial \mathbf{v_{ed}}}{\partial t} \;+\; (\nabla.\, \widehat{P}) \;+\; n_e\, m\, (\mathbf{v_{ed}}.\nabla)\mathbf{v_{ed}} \;-\; n_e\, \mathbf{F_e} \;=\; \mathbf{I_1}, \tag{2.159}$$

where the right-hand side member $\mathbf{I_1}$ represents a source term in the equation for electron momentum transfer obtained from the collision integral

$$\mathbf{I_1} \;=\; \int_{\mathbf{v_e}} m\, \mathbf{v_e} \left(\frac{\partial f_e}{\partial t}\right)_{e-o} \mathbf{dv_e}, \tag{2.160}$$

such as in the continuity equation we had

$$I_0 \;=\; \int_{\mathbf{v_e}} \left(\frac{\partial f_e}{\partial t}\right)_{e-o} \mathbf{dv_e} \;=\; n_e <\nu_{ion}> . \tag{2.161}$$

The term $\mathbf{I_1}$ may have a contribution from space-charge forces, as in the Vlasov equation (see Sect. 2.2.2), and a term due to the contribution of true collisions:

$$\mathbf{I_1} \;=\; n_e\, \mathbf{F_s} \;+\; \mathbf{I_{1(coll)}}. \tag{2.162}$$

In the case of weak anisotropies in the electron velocity, the off-diagonal elements of the pressure tensor $\widehat{P}$ can be neglected, so that we may write the divergence term in equation (2.159) under the form

$$(\nabla.\, \widehat{P}) \;=\; \sum_i \frac{\partial P_{ii}}{\partial x_i}\, \mathbf{e_i}, \tag{2.163}$$

with the diagonal elements of the pressure tensor (2.155) assumed equal

$$P_{ii} \;=\; n_e\, m < (v_{ei} - v_{edi})^2 > \;=\; \frac{1}{3}\, n_e\, m < |\mathbf{v_e} - \mathbf{v_{ed}}|^2 > . \tag{2.164}$$

If further the electron velocity distribution is Maxwellian at temperature T_e, we still have

$$\frac{1}{2}\, m < |\mathbf{v_e} - \mathbf{v_{ed}}|^2 > \;=\; \frac{3}{2}\, k_B T_e, \tag{2.165}$$

so that we obtain for the divergence of the pressure tensor

$$(\nabla.\, \widehat{P}) \;=\; \nabla p_e, \tag{2.166}$$

being $p_e = n_e k_B T_e$ the electron scalar pressure. With equations (2.162) and (2.166) the equation (2.159) takes the form

$$n_e\, m \left(\frac{\partial}{\partial t} + (\mathbf{v_{ed}}.\nabla) + <\nu_{ion}> \right) \mathbf{v_{ed}} \;+\; \nabla p_e \;-\; n_e\,(\mathbf{F_e} + \mathbf{F_s}) \;=\; \mathbf{I_{1(coll)}}. \quad (2.167)$$

2.4.3 Equation for Energy Conservation

Using the electron kinetic energy $u = \frac{1}{2}\, m v_e^2$ as variable X in equation (2.146), we obtain the equation for energy conservation of electrons

$$\frac{\partial}{\partial t}\left(n_e \frac{1}{2}\, m <v_e^2> \right) + \sum_i \frac{\partial}{\partial x_i}\left(n_e \frac{1}{2}\, m <v_e^2\, v_{ei}> \right)$$
$$- \sum_i n_e \frac{1}{2}\, m < \frac{\partial v_e^2}{\partial v_{ei}} > \frac{F_{ei}}{m} = \int_{\mathbf{v_e}} \frac{1}{2}\, m\, v_e^2 \left(\frac{\partial f_e}{\partial t} \right)_{e-o} \mathbf{dv_e}. \quad (2.168)$$

The first left-hand side term is simply $\partial(n_e <u>)/\partial t$, while for the second left-hand side term we may define a total heat flux vector by

$$\mathbf{q_e} \;=\; \frac{1}{2}\, n_e\, m <v_e^2\, \mathbf{v_e}>, \quad (2.169)$$

and writing this term under the form of a vector divergence, $(\nabla.\, \mathbf{q_e})$. In concerning now the third left-hand side term, we may write in case of velocity independent external forces

$$- \sum_{i,j} n_e \frac{1}{2} <\frac{\partial v_{ej}^2}{\partial v_{ei}}> F_{ei} \;=\; - \sum_{i,j} n_e\, \delta_{ji} <v_{ej}> F_{ei} \;=\; - n_e\, (\mathbf{v_{ed}}.\, \mathbf{F_e}). \quad (2.170)$$

We may write then equation (2.168) under the form

$$\frac{\partial}{\partial t}(n_e <u>) \;+\; (\nabla.\, \mathbf{q_e}) \;-\; n_e\, (\mathbf{v_{ed}}.\, \mathbf{F_e}) \;=\; I_2, \quad (2.171)$$

with

$$I_2 \;=\; \int_{\mathbf{v_e}} u \left(\frac{\partial f_e}{\partial t} \right)_{e-o} \mathbf{dv_e}. \quad (2.172)$$

For the analysis it is sometimes convenient to separate the heat flux into an agitation and a convective component. Thus taking into consideration that

$$v_e^2 \;=\; |\mathbf{v_e} - \mathbf{v_{ed}}|^2 + \;2\,((\mathbf{v_e} - \mathbf{v_{ed}}).\, \mathbf{v_{ed}}) \;+\; v_{ed}^2 \quad (2.173)$$

and $<((\mathbf{v_e} - \mathbf{v_{ed}}).\ \mathbf{v_{ed}})>= 0$, we may write

$$<u> = \frac{1}{2}\, m <v_e^2> = \frac{1}{2}\, m <|\mathbf{v_e} - \mathbf{v_{ed}}|^2> + \frac{1}{2}\, m\, v_{ed}^2$$
$$= \frac{3}{2}\, k_B\, T_e + \frac{1}{2}\, m\, v_{ed}^2. \tag{2.174}$$

Defining the agitation heat flux (flux of the random or thermal energy across a surface element moving with the drift velocity $\mathbf{v_{ed}}$) as

$$\mathbf{q_e^*} = \frac{1}{2}\, n_e\, m <|\mathbf{v_e} - \mathbf{v_{ed}}|^2\, (\mathbf{v_e} - \mathbf{v_{ed}})>$$
$$= \frac{1}{2}\, n_e\, m <\left(v_e^2 - 2\,(\mathbf{v_e}.\ \mathbf{v_{ed}}) + v_{ed}^2\right)\, (\mathbf{v_e} - \mathbf{v_{ed}})>, \tag{2.175}$$

we obtain

$$\mathbf{q_e^*} = \mathbf{q_e} - \frac{1}{2}\, n_e\, m <v_e^2>\, \mathbf{v_{ed}} - n_e\, m\, ((<\mathbf{v_e}\, \mathbf{v_e}> - \mathbf{v_{ed}}\, \mathbf{v_{ed}}) .\ \mathbf{v_{ed}})$$
$$= \mathbf{q_e} - n_e <u>\, \mathbf{v_{ed}} - (\widehat{P}.\ \mathbf{v_{ed}}). \tag{2.176}$$

The heat current may be expressed in terms of the temperature gradient as $\mathbf{q_e^*} = -\, k_T\, \nabla T_e$, being k_T the thermal conductivity assumed independent of the temperature.

The equation (2.171) takes hence the following form using equation (2.176) and neglecting the dependence of the thermal conductivity on the temperature

$$\frac{\partial}{\partial t}\left(n_e\left(\frac{3}{2}\, k_B\, T_e + \frac{1}{2}\, m\, v_{ed}^2\right)\right) + \nabla.\left(n_e\left(\frac{3}{2}\, k_B\, T_e + \frac{1}{2}\, m\, v_{ed}^2\right)\mathbf{v_{ed}}\right)$$
$$-\, k_T\, \nabla^2 T_e + \nabla.\, (\widehat{P}.\ \mathbf{v_{ed}}) = n_e\, (\mathbf{v_{ed}}.\ \mathbf{F_e}) + I_2. \tag{2.177}$$

Finally, in case of weak velocity anisotropies we still have

$$\nabla.\, (\widehat{P}.\ \mathbf{v_{ed}}) = \nabla.\, (n_e\, k_B\, T_e\, \mathbf{v_{ed}}) \tag{2.178}$$

and therefore

$$\frac{\partial}{\partial t}\left(\frac{3}{2}\, p_e + \frac{1}{2}\, n_e\, m\, v_{ed}^2\right) + \nabla.\left(\left(\frac{5}{2}\, p_e + \frac{1}{2}\, n_e\, m\, v_{ed}^2\right)\mathbf{v_{ed}}\right)$$
$$-\, k_T\, \nabla^2 T_e = n_e\, (\mathbf{v_{ed}}.\ \mathbf{F_e}) + I_2. \tag{2.179}$$

At steady-state conditions and sufficiently high neutral gas pressure values, the left-hand side member of equation (2.179) vanishes and the energy conservation is determined by the local balance between the work per time and volume units realized by the external forces (i.e. the electric field), $n_e\ (\mathbf{v_{ed}}.\ \mathbf{F_e}) = (\mathbf{J_e}.\ \mathbf{E})$, being $\mathbf{J_e} = -\ en_e\ \mathbf{v_{ed}}$ the electron current density, and the power lost per volume unit in collisions, $-\ I_2$.

Appendices

A.2.1 Liouville Relation $dq\ dp = dq'\ dp'$

Let us consider a transformation to pass from the variables $q(t)$ and $p(t)$, describing a system in a given instant t, to the variables $q'(t + dt)$ and $p'(t + dt)$, describing the same system at the instant $t + dt$. The evolutions in time of q and p are given by Hamilton's equations

$$\dot{q} = \frac{\partial H}{\partial p} \quad \text{and} \quad \dot{p} = -\frac{\partial H}{\partial q}, \tag{2.180}$$

so that we may write

$$q' = q + \dot{q}\ dt = q + \frac{\partial H}{\partial p}\ dt \tag{2.181}$$

$$p' = p + \dot{p}\ dt = p - \frac{\partial H}{\partial q}\ dt. \tag{2.182}$$

Differentiating q' and p' in terms of q and p we obtain

$$dq' = dq + \frac{\partial}{\partial q}\left(\frac{\partial H}{\partial p}\ dt\right) dq + \frac{\partial}{\partial p}\left(\frac{\partial H}{\partial p}\ dt\right) dp \tag{2.183}$$

$$dp' = dp + \frac{\partial}{\partial q}\left(-\frac{\partial H}{\partial q}\ dt\right) dq + \frac{\partial}{\partial p}\left(-\frac{\partial H}{\partial q}\ dt\right) dp, \tag{2.184}$$

so that we may write in accordance with Jacobi's theorem

$$dq'dp' = J\ dq\ dp, \tag{2.185}$$

where J is the determinant of the transformation

$$J = \begin{vmatrix} 1 + \frac{\partial^2 H}{\partial q \partial p}\ dt & \frac{\partial^2 H}{\partial p^2}\ dt \\ -\frac{\partial^2 H}{\partial q^2}\ dt & 1 - \frac{\partial^2 H}{\partial p \partial q}\ dt \end{vmatrix} \tag{2.186}$$

and therefore

$$J = 1 + \left(\frac{\partial^2 H}{\partial q \partial p} - \frac{\partial^2 H}{\partial p \partial q} \right) dt + \mathcal{O}(dt^2)$$
$$= 1 + \mathcal{O}(dt^2). \tag{2.187}$$

Neglecting the terms of $\mathcal{O}(dt^2)$ order, we have $J = 1$, so that

$$dq'dp' = dq\, dp. \tag{2.188}$$

This proof is equivalent to calculate the Poisson brackets of variables q' and p' with respect to q and p

$$[q', p']_{qp} = \frac{\partial(q', p')}{\partial(q, p)} = \frac{\partial q'}{\partial q} \frac{\partial p'}{\partial p} - \frac{\partial q'}{\partial p} \frac{\partial p'}{\partial q} \tag{2.189}$$

and verifying thus that

$$[q', p']_{qp} = 1. \tag{2.190}$$

This shows that the transformation $(q, p) \rightarrow (q', p')$ is canonical and the condition $dq'dp' = dq\, dp$ is satisfied (see e.g. Goldstein 1980).

A.2.2 Demonstration of $\mathbf{V} = \mathbf{V}'$ *and* $|\mathbf{v}| = |\mathbf{v}'|$ *in Sect. 2.3.1*

From equation (2.81) we may write before and after a given collision, respectively

$$(m + M)\, \mathbf{V} = m\, \mathbf{v_e} + M\, \mathbf{v_o} \tag{2.191}$$
$$(m + M)\, \mathbf{V}' = m\, \mathbf{v}'_\mathbf{e} + M\, \mathbf{v}'_\mathbf{o}. \tag{2.192}$$

The right-hand side members of these two expressions are equal due to momentum conservation, so that the left-hand side members are also equal, and hence $\mathbf{V} = \mathbf{V}'$.

On the other hand, from equations (2.82) and (2.83), we may write

$$v_e^2 = V^2 + \left(\frac{M}{m + M} \right)^2 v^2 + 2\, \frac{M}{m + M}\, (\mathbf{V}.\, \mathbf{v}) \tag{2.193}$$

$$v_o^2 = V^2 + \left(\frac{m}{m + M} \right)^2 v^2 - 2\, \frac{m}{m + M}\, (\mathbf{V}.\, \mathbf{v}) \tag{2.194}$$

and from equations (2.84) and (2.85), with $\mathbf{V} = \mathbf{V}'$, we get similar expressions for $v_e'^2$ and $v_o'^2$. Substituting these expressions into the equation for energy conservation

$$m\, v_e^2 + M\, v_0^2 = m\, v_e'^2 + M\, v_0'^2, \tag{2.195}$$

we rapidly obtain $v = v'$.

A.2.3 Demonstration of Equation (2.91)

Let us consider the triangle $\mathbf{v_o}$, $\mathbf{V}$ and $\mathbf{v'_e}$ shown in Fig. 2.9 obtained from Fig. 2.5. We easily obtain

$$\begin{aligned} |\mathbf{v'_e} - \mathbf{V}|^2 &= |\mathbf{V} - \mathbf{v_o}|^2 \sin^2 \chi_0 + \left(|\mathbf{v'_e} - \mathbf{v_o}| - |\mathbf{V} - \mathbf{v_o}| \cos \chi_0\right)^2 \\ &= |\mathbf{v'_e} - \mathbf{v_o}|^2 + |\mathbf{V} - \mathbf{v_o}|^2 - 2\, |\mathbf{v'_e} - \mathbf{v_o}|\, |\mathbf{V} - \mathbf{v_o}| \cos \chi_0. \end{aligned} \tag{2.196}$$

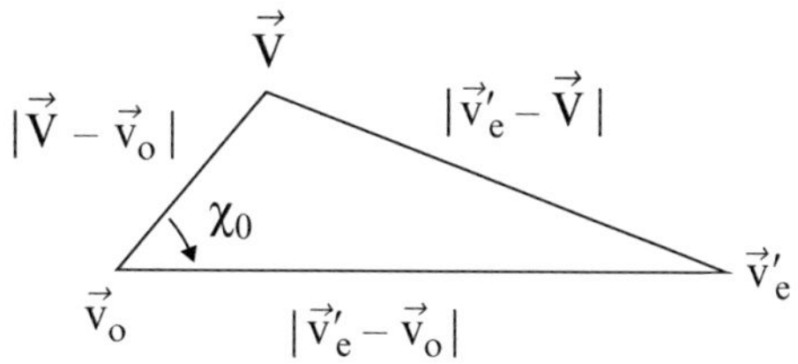

Fig. 2.9 Triangle obtained from Fig. 2.5, in which χ_0 is the scattering angle in the laboratory system

Substituting equations (2.84) and (2.85), with $\mathbf{V}' = \mathbf{V}$ and $|\mathbf{v}'| = |\mathbf{v}|$, we may write

$$\left(\frac{M}{m+M}\right)^2 |\mathbf{v}|^2 = |\mathbf{v'_e} - \mathbf{v_o}|^2 + \left(\frac{m}{m+M}\right)^2 |\mathbf{v}|^2 - 2\, |\mathbf{v'_e} - \mathbf{v_o}| \left(\frac{m}{m+M}\right) |\mathbf{v}| \cos \chi_0. \tag{2.197}$$

Inserting now (2.88), we still have

$$\cos \chi_0 = \frac{m + M \cos \chi}{\sqrt{m^2 + M^2 + 2\, mM \cos \chi}}, \tag{2.198}$$

from which we obtain equation (2.91).

Exercises

Exercise 2.1. Determine the electron rate coefficients of the inelastic electron-molecule processes for excitation of a given state j from state i, with u_{ij} denoting the threshold-energy:

(a) $\sigma_{ij}(u) = 0$, for $u < u_{ij}$, and $\sigma_{ij}(u) = a_{ij} = \text{const}$, for $u > u_{ij}$;
(b) $\sigma_{ij}(u) = 0$, for $u \leq u_{ij}$, and $\sigma_{ij}(u) = a_{ij}\ (u - u_{ij})$, for $u \geq u_{ij}$,

in the case of electrons with a Maxwellian velocity distribution at temperature T_e.

Resolution: For electrons following a Maxwellian velocity distribution at temperature T_e, the isotropic component of the velocity distribution is

$$f_e^0(v_e) \;=\; n_e \left(\frac{m}{2\pi k_B T_e}\right)^{3/2} \exp\left(-\frac{m v_e^2}{2k_B T_e}\right),$$

obeying to the normalization condition

$$\int_0^\infty f_e^0(v_e)\; 4\pi v_e^2\; dv_e \;=\; n_e,$$

whereas the electron rate coefficient (2.132) is given by

$$C_{ij} \;= <v_e\; \sigma_{ij}(v_e)> \;=\; \frac{1}{n_e}\int_{v_{ij}}^\infty v_e\; \sigma_{ij}(v_e)\, f_e^0(v_e)\; 4\pi v_e^2\; dv_e,$$

being $v_{ij} = \sqrt{2u_{ij}/m}$.

Rewriting these expressions in terms of the electron energy $u = \frac{1}{2} m v_e^2$, we have

$$f_e^0(u) \;=\; n_e \left(\frac{m}{2\pi k_B T_e}\right)^{3/2} \exp\left(-\frac{u}{k_B T_e}\right)$$

and

$$C_{ij} \;=\; \frac{1}{n_e}\,\frac{8\pi}{m^2}\int_{u_{ij}}^\infty u\; \sigma_{ij}(u)\, f_e^0(u)\; du,$$

from which we obtain the following expressions for the two cases proposed:

(a)

$$C_{ij} \;=\; \sqrt{\frac{8k_B T_e}{\pi m}}\; a_{ij}\left(1 + \frac{u_{ij}}{k_B T_e}\right) \exp\left(-\frac{u_{ij}}{k_B T_e}\right)\ ;$$

(b)

$$C_{0j} = \sqrt{\frac{8k_BT_e}{\pi m}}\, 2k_BT_e\, a_{ij}\left(1+\frac{u_{ij}}{2k_BT_e}\right)\exp\left(-\frac{u_{ij}}{k_BT_e}\right).$$

Exercise 2.2. Using the Klein-Rosseland relation (2.140) on the cross sections of Exercise 2.1 show that the electron rate coefficients of direct and reverse processes obey the equation (2.135).

Resolution: Using the relation (2.140) we obtain the reverse cross sections of Exercise 2.1: *(a)* $\sigma_{ji}(u) = a_{ij}\ ((u+u_{ij})/u)$, for $u > 0$; *(b)* $\sigma_{ji}(u) = a_{ij}\ (u+u_{ij})$, for $u \geq 0$. Calculating then the electron rate coefficient of the superelastic process

$$C_{ji} = \frac{1}{n_e}\frac{8\pi}{m^2}\int_0^\infty u\,\sigma_{ji}(u)\, f_e^0(u)\, du,$$

we obtain

$$\frac{C_{ji}}{C_{ij}} = \exp\left(\frac{u_{ij}}{k_BT_e}\right)$$

in both situations.

Exercise 2.3. The cross sections for electron impact excitation of a dipole-allowed transition at high energies is of the form (Massey and Burhop 1969)

$$\sigma_{ij}(u) = \frac{a_{ij}}{u}\ln\left(b_{ij}\frac{u}{u_{ij}}\right),$$

being $u = \frac{1}{2}mv_e^2$ the electron energy, u_{ij} the threshold-energy, and a_{ij} and b_{ij} two constants dependent of the gas. Determine the expression for the corresponding electron rate coefficient in the case of a Maxwellian velocity distribution at temperature T_e.

Resolution: In the case of a Maxwellian velocity distribution, the expression for the rate coefficient is as follows

$$C_{ij} = \sqrt{\frac{8}{\pi m}}\frac{1}{(k_BT_e)^{3/2}}\int_{u_{ij}}^\infty u\,\sigma_{ij}(u)\,\exp\left(-\frac{u}{k_BT_e}\right)du,$$

so that assuming, as in Exercise 2.1-*(ii)*, a linear dependence $\sigma_{ij}(u) = c_{ij}\ (u-u_{ij})$ at low energies up to the energy $u^* > u_{ij}$, we may write

$$C_{ij} = \sqrt{\frac{8k_BT_e}{\pi m}}\, 2k_BT_e\, c_{ij}\left\{\left(1+\frac{u_{ij}}{2k_BT_e}\right)\exp\left(-\frac{u_{ij}}{k_BT_e}\right)-\left(1+\frac{u^*}{2k_BT_e}\right)\right.$$

$$\exp\left(-\frac{u^*}{k_B T_e}\right)\Bigg\} + \sqrt{\frac{8}{\pi m}}\,\frac{a_{ij}}{(k_B T_e)^{3/2}}\int_{u^*}^{\infty}\ln\left(b_{ij}\,\frac{u}{u_{ij}}\right)\exp\left(-\frac{u}{k_B T_e}\right)\,du.$$

Making the substitution $\eta = b_{ij}\,u/u_{ij}$, we still obtain for the high-energy term

$$C_{ij} = \ldots\ldots\ldots + \sqrt{\frac{8}{\pi m}}\,\frac{a_{ij}}{(k_B T_e)^{3/2}}\,\frac{u_{ij}}{b_{ij}}\int_{\eta^*}^{\infty}\ln\eta\,\exp\left(-\frac{u_{ij}}{b_{ij}\,k_B T_e}\,\eta\right)\,d\eta,$$

which presents the following primitive for the integrand function with $a = u_{ij}/(b_{ij}\,k_B T_e)$

$$\int \ln\eta\,\exp(-\,a\eta)\,d\eta = -\frac{1}{a}\,\ln\eta\,\exp(-\,a\eta) + \frac{1}{a}\int\frac{1}{\eta}\,\exp(-\,a\eta)\,d\eta$$

$$= -\frac{1}{a}\,\ln\eta\,\exp(-\,a\eta) + \frac{1}{a}\left(\ln\eta - a\eta + \frac{(a\eta)^2}{2.\,2!} - \frac{(a\eta)^3}{3.\,3!} + \ldots\ldots\right).$$

Exercise 2.4. In a plasma column of radius R and infinite length the gas temperature presents a parabolic radial profile

$$T_o(r) = (T_o(0) - T_W)\left(1 - \frac{r^2}{R^2}\right) + T_W,$$

being $T_o(0)$ and $T_W = T_o(R)$ the gas temperatures at the axis and at the wall, respectively. Neglecting the radial dependence of the thermal conductivity coefficient through the gas temperature, determine the power loss by thermal conduction, per volume unity.

Resolution: Using an analogous expression for molecules as indicated in Sect. 2.4.3, the power loss by thermal conduction by volume unity is $P_{cond} = -\,k_T\,\nabla^2 T_o$, being k_T the thermal conductivity coefficient. We obtain therefore

$$P_{cond} = -\,k_T\,\frac{1}{r}\,\frac{d}{dr}\left(r\,\frac{dT_o}{dr}\right) = \frac{4}{R^2}\,k_T\,(T_o(0) - T_W)\,.$$

On the other hand, the radially averaged temperature is

$$\overline{T} = \frac{1}{R^2}\int_0^R T_o(r)\,2r\,dr = \frac{1}{2}\,(T_o(0) + T_W)\,,$$

so that the gas temperature at the axis of the column is $T_o(0) = 2\overline{T} - T_W$. Using this latter relation in the expression above, we obtain at the end

$$P_{cond} = \frac{8}{R^2}\,k_T\,\left(\overline{T} - T_W\right).$$

References

W.P. Allis, Motions of ions and electrons, in *Handbuch der Physik*, vol. 21. ed. by S. Flügge (Springer, Berlin, 1956), pp. 383–444

J.A. Bittencourt, *Fundamentals of Plasma Physics* (Springer, New York, 2004)

S. Chapman, T.G. Cowling, *The Mathematical Theory of Non-uniform Gases* (Cambridge University Press, Cambridge, 1939)

B.E. Cherrington, *Gaseous Electronics and Gas Lasers* (Pergamon Press, Oxford, 1980)

J.-L. Delcroix, *Physique des Plasmas: Volumes 1 and 2*. Monographies Dunod (Dunod, Paris, 1963, 1966) (in French)

J.-L. Delcroix, A. Bers, *Physique des Plasmas: Volumes 1 and 2*. Savoirs Actuels (InterÉditions/CNRS Éditions, Paris, 1994) (in French)

H. Goldstein, *Classical Mechanics*, 2nd edn. (Addison-Wesley, Reading, 1980)

K. Huang, *Statistical Mechanics* (Wiley, New York, 1963)

R.L. Liboff, *Introduction to the Theory of Kinetic Equations* (Wiley, New York, 1969)

H.S.W. Massey, E.H.S. Burhop, *Electronic and Ionic Impact Phenomena: Volume 1 – Collisions of Electrons with Atoms* (Oxford at the Clarendon Press, Oxford, 1969)

D.R. Nicholson, *Introduction to Plasma Theory* (Wiley, New York, 1983)

Y. Sone, *Molecular Gas Dynamics: Theory, Techniques and Applications* (Birkhäuser, Boston, 2007)

Chapter 3
Boltzmann Equation in Velocity Space

This chapter is devoted to the analysis of the electron Boltzmann transport equation in the velocity space for the case of an applied external direct-current (DC) electric field. The analysis is restricted to low and intermediate field values in order the two-term expansion approximation of the electron velocity distribution around the direction of electron drift is sufficiently accurate. The terms for the electron energy gain from the field and electron energy losses by elastic and inelastic collisions, as well as the corresponding energy-averaged power balance terms, are derived in detail. At the end of the chapter some solutions to the Boltzmann equation are shown and discussed for H_2 and N_2.

3.1 First-Order Expansion of Collision Integral

3.1.1 *Velocity Anisotropic Components*

Let us start by considering a weakly ionized plasma $n_e \ll n_0$, with $n_i \simeq n_e$, and where n_e, n_i and n_o denote the number densities of electrons, ions and molecules, respectively. We further assume the molecules in the ground-state only and the mass of electrons much smaller than that of molecules, $m/M \ll 1$.

Under the assumption of weakly ionized gases, the interactions between charged species are vanished when compared with the electron-molecule and ion-molecule interactions, so that we have in equation (2.72) for electrons

$$\left(\frac{\partial f_e}{\partial t}\right)_{coll} \simeq \left(\frac{\partial f_e}{\partial t}\right)_{e-o} \tag{3.1}$$

The original version of this chapter was revised. An erratum to this chapter can be found at DOI 10.1007/ 978-3-319-09253-9_12

J.M.A.H. Loureiro, J. de Amorim Filho, *Kinetics and Spectroscopy of Low Temperature Plasmas*, Graduate Texts in Physics,
DOI 10.1007/978-3-319-09253-9_3

and in the case of ions

$$\left(\frac{\partial f_i}{\partial t}\right)_{coll} \simeq \left(\frac{\partial f_i}{\partial t}\right)_{i-o} , \tag{3.2}$$

with $f_e(\mathbf{r}, \mathbf{v_e}, t)$ and $f_i(\mathbf{r}, \mathbf{v_i}, t)$ denoting the particle distribution functions of the two species.

Contrary to the charged species, the gas of molecules is homogeneous, i.e. it does not suffer the effects of diffusion gradients and external forces. The Boltzmann equation for molecules takes its simplest form (2.73) and the corresponding velocity distribution function is hence Maxwellian

$$F(v_o) = n_o \left(\frac{M}{2\pi k_B T_o}\right)^{3/2} \exp\left(-\frac{M v_o^2}{2k_B T_o}\right) , \tag{3.3}$$

with $v_o = |\mathbf{v_o}|$ denoting the velocity absolute value, and where $< \mathbf{v_o} >= 0$.

The interactions resulting from space-charge fields in the Boltzmann equation for electrons and ions, when they exist, are included in the terms for the applied external forces, by considering $\mathbf{F} + \mathbf{F}'$ instead of $\mathbf{F}$ in the Boltzmann equation (2.8) as we have seen in equation (2.79) for the case of Vlasov equation.

We still further assume that the external forces acting on the electrons and ions are due to the presence of an electromagnetic field (**E**,**B**), so that the electron acceleration due to Lorentz force is

$$\frac{\mathbf{F_e}}{m} = -\frac{e}{m}\mathbf{E} + [\vec{\omega}_{ce} \times \mathbf{v_e}] , \tag{3.4}$$

with $\vec{\omega}_{ce} = (e/m)\,\mathbf{B}$ denoting the electron cyclotron frequency vector, and where e denotes the absolute value of the electron charge, while for the ions we have

$$\frac{\mathbf{F_i}}{M} = \frac{e}{M}\mathbf{E} + [\vec{\omega}_{ci} \times \mathbf{v_i}] , \tag{3.5}$$

with $\vec{\omega}_{ci} = -\,(e/M)\,\mathbf{B}$, and where e is a single ion charge.

With equations (3.1) and (3.4), the Boltzmann equation for electrons (2.8), the only we will consider here, takes the following form

$$\frac{\partial f_e}{\partial t} + \left(\mathbf{v_e} \cdot \frac{\partial f_e}{\partial \mathbf{r}}\right) + \left(\left\{-\frac{e}{m}\mathbf{E} + [\vec{\omega}_{ce} \times \mathbf{v_e}]\right\} \cdot \frac{\partial f_e}{\partial \mathbf{v_e}}\right) = \left(\frac{\partial f_e}{\partial t}\right)_{e-o} , \tag{3.6}$$

with the collision term, assuming by now elastic collisions only, expressed as in equation (2.35)

$$\left(\frac{\partial f_e}{\partial t}\right)_{e-o} = \int_\Omega \int_{\mathbf{v_o}} (f_e' F_o' - f_e F_o)\, v\, \sigma(v, \Omega)\, d\Omega\, \mathbf{dv_o} \tag{3.7}$$

and where $v = |\mathbf{v}|$ denotes the absolute value of the relative velocity $\mathbf{v} = \mathbf{v_e} - \mathbf{v_o}$. The interactions electron-molecule are assumed as given by an isotropic central force depending only of the distance between the two particles, and in which the scattering in the solid-angle $d\Omega$ is $\sigma(v, \Omega)\ d\Omega = \sigma(v, \chi)\ 2\pi\ \sin\chi\ d\chi$, being $\sigma(v, \Omega)$ the differential cross section and χ the scattering angle in the centre-of-mass system.

By now we further assume, for simplification, the so-called Lorentz gas model for electron-molecule interactions, in which due to the hard inequality $m \ll M$, we assume approximately $|\mathbf{v_o'}| \simeq |\mathbf{v_o}| \simeq 0$ and $|\mathbf{v_e'}| \simeq |\mathbf{v_e}| \simeq |\mathbf{v}|$. This approximation corresponds to consider $m/M \simeq 0$. Under these approximations, the collision term (3.7) writes as follows

$$\left(\frac{\partial f_e}{\partial t}\right)_{e-o} = n_0\ v_e \int_0^{\pi} (f_e' - f_e)\ \sigma(v_e, \chi_0)\ 2\pi\ \sin\chi_0\ d\chi_0\ , \tag{3.8}$$

where f_e and f_e' denotes $f_e(\mathbf{v_e})$ and $f_e(\mathbf{v_e'})$, respectively, and χ_0 is the scattering angle in the laboratory system. Since f_e' may be expressed in terms of f_e and χ_0, the collisional term (3.8) is a linear operator represented as $I(f_e)$ from now on.

The electron transport due to the effects of external applied forces, resulting from electric and magnetic fields, as well as due to the presence of density gradients, is an anisotropic perturbation with respect to the situation of isotropic equilibrium, in which $f_e(v_e)$ depends only of the absolute value of the electron velocity, and not of its orientation. The drift or average vector velocity gained by the electrons from these perturbations is, in most cases, much smaller than the random or thermal velocity, so that the changes in the velocity distribution function are small perturbations relatively to the equilibrium distribution function. This allows the normal method for solution of the electron Boltzmann equation to be the expansion of the actual distribution function as a sum of the isotropic (equilibrium) component with small anisotropic corrective terms.

In the case of molecules and if the departures from equilibrium are caused by some agent α, the velocity distribution around the equilibrium can be expanded in powers of α. This procedure is known as the Enskog method for solving the Boltzmann equation and it was highly developed by Chapman and Cowling (1939). However, this method does not converge well for charged particles in a field. The technique most applicable in this case is the expansion of the velocity distribution function in terms of spherical harmonics. This approach has been presented by Allis (1956) and has the great advantage that very often one corrective term in the expansion is sufficient for obtaining an accurate approximation for the distribution function.

The coordinate system is selected so that the preferred direction of orientation (i.e. the polar axis of spherical coordinates) is the direction of the anisotropy produced either by the applied electric or magnetic fields, or the density gradients. Then we can expand the distribution function $f_e(\mathbf{r}, \mathbf{v_e}, t)$ in spherical harmonics, that is on spherical functions depending on the spatial orientation of the velocity vector $\mathbf{v_e}$ with respect to the direction of the anisotropy (z axis)

$$f_e(\mathbf{r}, \mathbf{v_e}, t) = \sum_{l=0}^{\infty} f_e^l(\mathbf{r}, v_e, t)\, P_l(\cos\theta), \tag{3.9}$$

where $P_l(\cos\theta)$ are the Legendre polynomials and the functions f_e^l depend only on the absolute value of velocity. The first Legendre polynomials are

$$P_0(\cos\theta) = 1\,; \quad P_1(\cos\theta) = \cos\theta\,; \quad P_2(\cos\theta) = \frac{1}{2}\,(3\cos^2\theta - 1)\,;$$
$$P_3(\cos\theta) = \frac{1}{2}\,\cos\theta\,(5\cos^2\theta - 3)\,;\ \text{etc}\,. \tag{3.10}$$

If the anisotropies are small the series (3.9) converges sufficiently rapidly for the first two terms to suffice, but we shall not assume that $f_e^0(\mathbf{r}, v_e, t)$ is Maxwellian, as this is generally far from true even when a relatively weak electric field is present

$$f_e(\mathbf{r}, \mathbf{v_e}, t) \simeq f_e^0(\mathbf{r}, v_e, t) + f_e^1(\mathbf{r}, v_e, t)\,\cos\theta\,. \tag{3.11}$$

It is worth noting here that directing the polar axis of spherical coordinates along the direction of anisotropy, the component $\mathbf{f_e^1}(\mathbf{r}, v_e, t)$ may be regarded as a vector oriented towards this direction (see Fig. 3.1)

$$f_e(\mathbf{r}, \mathbf{v_e}, t) \simeq f_e^0(\mathbf{r}, v_e, t) + \left(\mathbf{f_e^1}(\mathbf{r}, v_e, t)\,.\,\frac{\mathbf{v_e}}{v_e}\right)\,. \tag{3.12}$$

The equation (3.12) shows that at a given instant t, space position $\mathbf{r}$, and same absolute value of velocity but with different orientations, the electron velocity distribution function is equal to $f_e^0 + f_e^1$, f_e^0 and $f_e^0 - f_e^1$, for electrons with instantaneous velocities parallel, perpendicular and opposite to the anisotropy direction.

Because of the orthogonality of spherical harmonics

$$\int_\Omega P_l\, P_{l'}\, d\Omega = \frac{4\pi}{2l+1}\,\delta_{ll'}\,, \tag{3.13}$$

in which $\delta_{ll'} = 1$ for $l = l'$ and $\delta_{ll'} = 0$ for $l \neq l'$, we have for any scalar function $h(v_e)$ depending only on the absolute value of velocity

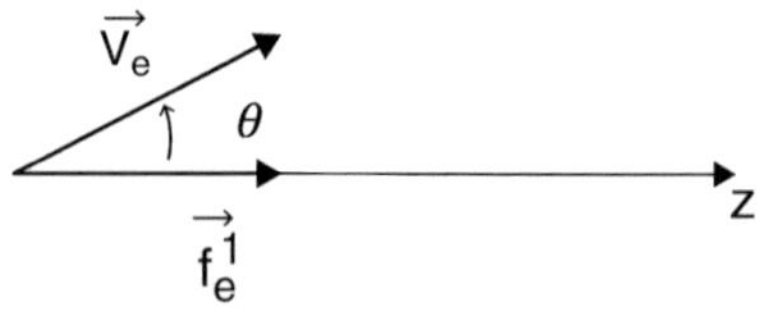

Fig. 3.1 Anisotropic component directed along the z axis. θ is the angle between the instantaneous velocity $\mathbf{v_e}$ and the anisotropy direction

$$\int_\Omega h(v_e)\, f_e(\mathbf{r}, \mathbf{v_e}, t)\; d\Omega \;=\; h(v_e) f_e^0(\mathbf{r}, v_e, t)\; 4\pi \; . \tag{3.14}$$

Thus, the two scalars of great interest here such as the electron number density $n_e(\mathbf{r}, t)$ and the mean energy $<u>(\mathbf{r}, t)$ depend on f_e^0 only

$$n_e(\mathbf{r}, t) \;=\; \int_{\mathbf{v_e}} f_e(\mathbf{r}, \mathbf{v_e}, t)\; \mathbf{dv_e} \;=\; \int_0^\infty f_e^0(\mathbf{r}, v_e, t)\; 4\pi v_e^{\,2}\; dv_e \tag{3.15}$$

$$<u>(\mathbf{r}, t) \;=\; \frac{1}{n_e} \int_{\mathbf{v_e}} u\, f_e(\mathbf{r}, \mathbf{v_e}, t)\; \mathbf{dv_e} \;=\; \frac{1}{n_e} \int_0^\infty u\, f_e^0(\mathbf{r}, v_e, t)\; 4\pi v_e^{\,2}\; dv_e \; , \tag{3.16}$$

with $u = \frac{1}{2}\, m v_e^{\,2}$ denoting the electron energy.

On the other hand, for the velocity vector $\mathbf{v_e}$, whose components are $v_{ex} = v_e \;\sin\theta\; \cos\phi$, $v_{ey} = v_e \;\sin\theta\; \sin\phi$, and $v_{ez} = v_e\; \cos\theta$ (see Fig. 3.2), in which v_{ez} is directed along the direction of the anisotropy $\mathbf{f_e^1}$ (polar axis), we have

$$\int_\Omega \mathbf{v_e}\, f_e(\mathbf{r}, \mathbf{v_e}, t)\; d\Omega \;=\; v_e \int_\Omega f_e(\mathbf{r}, \mathbf{v_e}, t)\; P_1\; d\Omega\; \mathbf{e_z} \;=\; \frac{v_e}{3}\, \mathbf{f_e^1}(\mathbf{r}, v_e, t)\; 4\pi \; , \tag{3.17}$$

so that the electron drift, or average vector velocity, $\mathbf{v_{ed}} = <\mathbf{v_e}>$ is given by

$$\begin{aligned} \mathbf{v_{ed}}(\mathbf{r}, t) &= \frac{1}{n_e} \int_{\mathbf{v_e}} \mathbf{v_e}\, f_e(\mathbf{r}, \mathbf{v_e}, t)\; \mathbf{dv_e} \;=\; \frac{1}{n_e} \int_0^\infty \int_\Omega \mathbf{v_e}\, f_e(\mathbf{r}, \mathbf{v_e}, t)\; v_e^{\,2}\; dv_e\; d\Omega \\ &= \frac{1}{n_e} \int_0^\infty \frac{v_e}{3}\, \mathbf{f_e^1}(\mathbf{r}, v_e, t)\; 4\pi v_e^{\,2}\; dv_e \; . \end{aligned} \tag{3.18}$$

Only the anisotropy of first order contributes for the average vector velocity. In a gas of electrons under isotropic equilibrium we have obviously $\mathbf{v_{ed}} = 0$. It is worth noting that when higher anisotropies exist, the electron drift velocity is always oriented along the direction of the first anisotropy, $\mathbf{f_e^1}$.

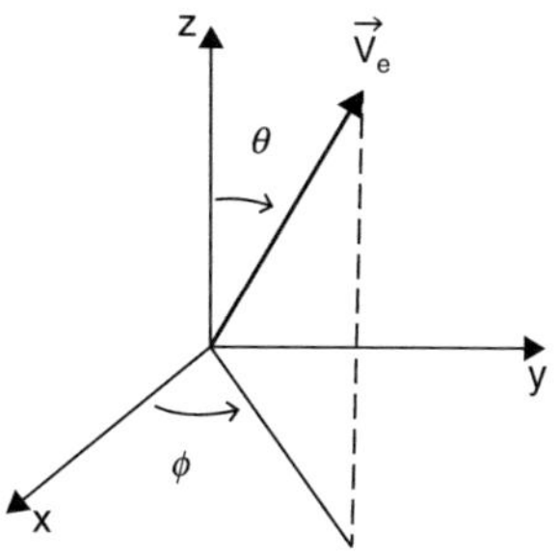

Fig. 3.2 Decomposition of instantaneous velocity $\mathbf{v_e}$

Let us consider now the situation where the symmetry around the polar axis does not exist, as it happens in the presence of a magnetic field. In this case the dependence on the azimuthal coordinate ϕ needs to be considered as well and we need to use the associated Legendre functions $P_l^m(\cos\theta)$. The expansion (3.9) should be then replaced with

$$f_e(\mathbf{r}, \mathbf{v_e}, t) = f_e^0(\mathbf{r}, v_e, t) + \sum_{l=1}^{\infty}\sum_{m=0}^{l} [p_{lm}(\mathbf{r}, v_e, t)\, C_{lm}(\theta, \phi) + q_{lm}(\mathbf{r}, v_e, t)\, S_{lm}(\theta, \phi)] , \tag{3.19}$$

with the functions C_{lm} and S_{lm} given by

$$C_{lm}(\theta, \phi) = P_l^m(\cos\theta)\, \cos(m\phi) , \quad S_{lm}(\theta, \phi) = P_l^m(\cos\theta)\, \sin(m\phi) \tag{3.20}$$

and where $P_l^m(\cos\theta)$ are the associated Legendre functions, in which $P_l^0 \equiv P_l$. The expansion (3.19) limited to first order anisotropies writes as follows

$$\begin{aligned} f_e(\mathbf{r}, \mathbf{v_e}, t) &\simeq f_e^0 + p_{11}\, \sin\theta\, \cos\phi + q_{11}\, \sin\theta\, \sin\phi + p_{10}\, \cos\theta \\ &\simeq f_e^0 + p_{11}\, \frac{v_{ex}}{v_e} + q_{11}\, \frac{v_{ey}}{v_e} + p_{10}\, \frac{v_{ez}}{v_e} . \end{aligned} \tag{3.21}$$

Then, the drift velocity (3.18) is given by

$$\begin{aligned} \mathbf{v_{ed}}(\mathbf{r}, t) &= \frac{1}{n_e}\int_0^{\infty}\int_{\Omega} v_e\, (p_{11}\, C_{11}^2\, \mathbf{e_x} + q_{11}\, S_{11}^2\, \mathbf{e_y} + p_{10}\, C_{10}^2\, \mathbf{e_z})\, v_e^2\, dv_e\, d\Omega \\ &= \frac{1}{n_e}\int_0^{\infty} \frac{v_e}{3}\, (p_{11}\, \mathbf{e_x} + q_{11}\, \mathbf{e_y} + p_{10}\, \mathbf{e_z})\, 4\pi v_e^2\, dv_e . \end{aligned} \tag{3.22}$$

In the absence of the dependence on ϕ, we obtain again equations (3.11) and (3.18) with $p_{10} \equiv f_e^1$.

3.1.2 *Relaxation of the Anisotropies*

Let us consider now the collision operator $I(f_e)$ for the interactions electron-molecule under the Lorentz gas approximation (3.8), in the case of expansion (3.9) as the electron velocity distribution $f_e \equiv f_e(\mathbf{r}, \mathbf{v_e}, t)$ does not exhibit the azimuthal dependence ϕ. The velocity distribution function after the collision $f_e' \equiv f_e(\mathbf{r}, \mathbf{v_e'}, t)$, with $|\mathbf{v_e'}| = |\mathbf{v_e}|$ under this approximation, can be written as

$$f_e(\mathbf{r}, \mathbf{v_e'}, t) = \sum_{l=0}^{\infty} f_e^l(\mathbf{r}, v_e, t)\, P_l(\cos\theta') , \tag{3.23}$$

where θ' is the angle of the instantaneous velocity $\mathbf{v_e'}$ with respect to the polar axis. From Fig. 3.3 and using the addition theorem for spherical harmonics, the Legendre polynomials $P_l(\cos\theta')$ are given by

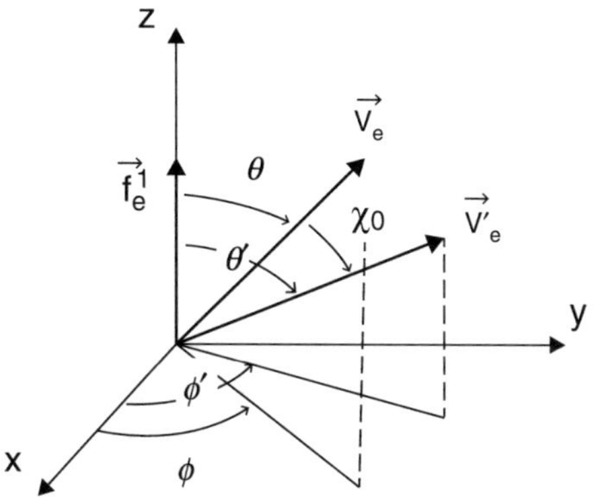

Fig. 3.3 Decomposition of $\mathbf{v_e}$ and $\mathbf{v'_e}$ vectors, with χ_0 denoting the scattering angle in the laboratory system

$$P_l(\cos\theta') = P_l(\cos\theta)\,P_l(\cos\chi_0) + 2\sum_{m=1}^{l}\frac{(l-m)!}{(l+m)!}\,P_l^m(\cos\theta)\,P_l^m(\cos\chi_0)\,\cos(m(\phi'-\phi))\,. \tag{3.24}$$

As $\sigma(v_e, \chi_0)$ is assumed independent of the azimuthal angle, the term in $\cos(m(\phi'-\phi))$ goes out in integrating equation (3.8), so that the Boltzmann collision integral writes as follows

$$I(f_e) = n_o\,v_e\int_0^{\pi}\sum_{l=0}^{\infty} f_e^l\,P_l(\cos\theta)\,[P_l(\cos\chi_0)-1]\,\sigma(v_e,\chi_0)\,2\pi\,\sin\chi_0\,d\chi_0\,. \tag{3.25}$$

Then the collision term can be expressed as

$$I(f_e) = \sum_{l=0}^{\infty} I^l(f_e^l)\,P_l(\cos\theta)\,, \tag{3.26}$$

introducing the collision integral terms of different order

$$I^l(f_e^l) = -\,\nu_{cl}\,f_e^l\,, \tag{3.27}$$

in which $\nu_{cl}(v_e)$ represents a set of collision frequencies in the laboratory system defined as follows

$$\nu_{cl}(v_e) = n_o\,v_e\int_0^{\pi}[1-P_l(\cos\chi_0)]\,\sigma(v_e,\chi_0)\,2\pi\,\sin\chi_0\,d\chi_0\,. \tag{3.28}$$

The first frequency is $\nu_{c0} = 0$ and accordingly $I^0(f_e^0) = 0$. This is not satisfactory, so that we will have to calculate later on this term using the next order of approximation. In fact, this is a consequence of the assumption $|\mathbf{v_e'}| = |\mathbf{v_e}|$, which neglects the modifications operated by the collisions into the isotropic part of the electron distribution function. The next two frequencies are respectively

$$\nu_{c1}(v_e) = n_o\, v_e \int_0^{\pi} (1 - \cos\chi_0)\, \sigma(v_e, \chi_0)\, 2\pi\ \sin\chi_0\, d\chi_0 \tag{3.29}$$

$$\nu_{c2}(v_e) = n_o\, v_e \int_0^{\pi} \frac{3}{2}\ \sin^2\chi_0\, \sigma(v_e, \chi_0)\, 2\pi\ \sin\chi_0\, d\chi_0. \tag{3.30}$$

The frequency ν_{c1} is the collision frequency for momentum transfer and is of such importance in transport theory that the simpler symbol ν_m will be used for it. It is a weighted collision frequency in which the backward scattering counts double, the right-angle scattering has the weight one, and the forward scattering is not counted. From such frequency we may define the collision cross section for momentum transfer

$$\sigma_m(v_e) = \frac{\nu_m}{n_o v_e} = \int_0^{\pi} (1 - \cos\chi_0)\, \sigma(v_e, \chi_0)\, 2\pi\ \sin\chi_0\, d\chi_0. \tag{3.31}$$

On the other hand, the frequency ν_{c2} may be seen as a scattering frequency in which neither forward nor backward scattering is accounted for and the right-angle scattering has the weight 3/2.

Obviously, another important collision frequency we may also define is the total collision frequency

$$\nu_c(v_e) = n_o\, v_e \int_0^{\pi} \sigma(v_e, \chi_0)\, 2\pi\ \sin\chi_0\, d\chi_0. \tag{3.32}$$

The integral in (3.32) is generally improper as $\sigma(v_e, \chi_0)$ has a singularity in the forward direction. However, the apertures of experimental apparatus prevent observations from covering this singularity so that this frequency is usually expressed in terms of an incomplete integral. As the scattering is isotropic the differential cross section is independent of the polar coordinate, $\sigma(v_e, \chi_0) \equiv \sigma(v_e)$, and due to the orthogonality of spherical harmonics ($\int_\Omega P_1\ P_0\ d\Omega = \int_\Omega \cos\chi_0\ d\Omega = 0$), the collision frequency for momentum transfer (3.29) transforms into the total frequency (3.32).

The equation (3.27) shows that the frequencies (3.28) are the relaxation frequencies for the components of the electron velocity distribution function expressed by expansion (3.9). When the external forces or other effects responsible for creation of anisotropies, such as density gradients, are suppressed, the electron distribution function initially out of equilibrium is restored to a situation of local equilibrium through the collisions. From equations (3.6), (3.9), (3.26) and (3.27), we find for each component f_e^l

$$\frac{\partial f_e^l}{\partial t} = -\, \nu_{cl}\, f_e^l, \tag{3.33}$$

from which we obtain

$$f_e^l(t) = f_e^l(0)\ \exp^{-\nu_{cl} t}\ . \tag{3.34}$$

Since $\nu_{c0} = 0$, we find $f_e^0(t) = f_e^0(0)$. This is a consequence of the assumption $m/M \simeq 0$ made at the beginning of this section, which means that with molecules infinitely heavy the collisions do not change the isotropic component of the velocity distribution function. If the isotropic part of the velocity distribution function is not Maxwellian, the collisions are unable to produce any evolution towards the equilibrium. On the contrary, the anisotropic components f_e^l, with $l \geq 1$, are rapidly smoothed with the frequencies ν_{cl}. If the deviation relatively to the isotropic distribution is not strong the first correction f_e^1 is sufficient and this component evolves at the frequency for momentum transfer, ν_m. The collisions are hence much more efficient to make the electron velocity distribution function becomes isotropic than to make the isotropic component becomes Maxwellian. Due to the small m/M ratio we need many collisions before the distribution component f_e^0 appreciably changes. This will be realized later on by introducing higher order terms into equation (3.7).

3.1.3 Coupling Between the Anisotropies

Let us consider in this section the situation where the anisotropic deviations, relatively to the equilibrium isotropic electron distribution function, are produced by an applied electric field directed along the negative z axis, $\mathbf{E} = -E\,\mathbf{e_z}$. Further only elastic collisions are considered by now. Due to the negative sign of the electron charge the first anisotropy $\mathbf{f_e^1}$ is directed along the positive z axis (see Fig. 3.4).

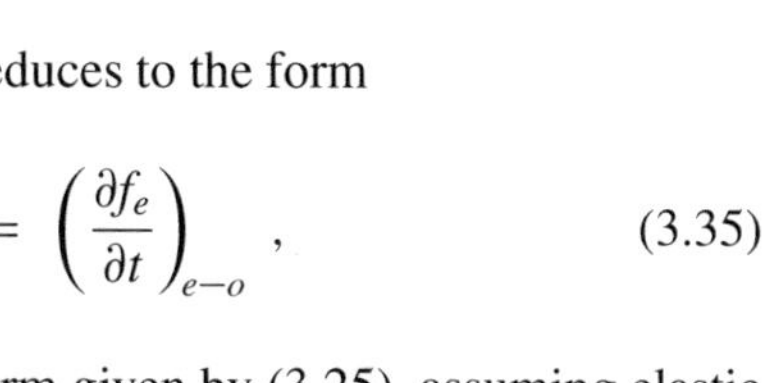

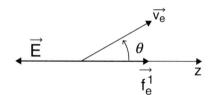

Fig. 3.4 Orientations of the electric field, anisotropic component $\mathbf{f_e^1}$ of the velocity distribution function, and instantaneous velocity $\mathbf{v_e}$

In this case the Boltzmann equation (3.6) reduces to the form

$$\frac{\partial f_e}{\partial t} - \frac{e}{m}\left(\mathbf{E}\cdot\frac{\partial f_e}{\partial \mathbf{v_e}}\right) = \left(\frac{\partial f_e}{\partial t}\right)_{e-o}, \tag{3.35}$$

in which $(\partial f_e/\partial t)_{e-o} = I(f_e)$ is the collision term given by (3.25), assuming elastic collisions only, under the Lorentz gas model approximation in which $m/M \simeq 0$.

Assuming the independence on the azimuthal angle, the equation (3.35) can be written under the following form using the limited expansion (3.11)

$$\frac{\partial f_e}{\partial t} - \frac{e}{m}\left(\mathbf{E}\cdot\left(\frac{\partial f_e}{\partial v_e}\,\mathbf{e_r} + \frac{1}{v_e}\,\frac{\partial f_e}{\partial \theta}\,\mathbf{e_\theta}\right)\right) = I(f_e)\,, \tag{3.36}$$

in which

$$\frac{\partial f_e}{\partial v_e} = \frac{\partial f_e^0}{\partial v_e} + \frac{\partial f_e^1}{\partial v_e} \cos\theta \,, \tag{3.37}$$

$$\frac{1}{v_e} \frac{\partial f_e}{\partial \theta} = -\frac{f_e^1}{v_e} \sin\theta \,. \tag{3.38}$$

Since $(\mathbf{e_z} \,.\, \mathbf{e_r}) = \cos\theta$ and $(\mathbf{e_z} \,.\, \mathbf{e_\theta}) = -\sin\theta$, and using the expansion for the collision term (3.26), we may write

$$\frac{\partial f_e^0}{\partial t} + \frac{\partial f_e^1}{\partial t} \cos\theta + \frac{eE}{m}\left[\left(\frac{\partial f_e^0}{\partial v_e} + \frac{\partial f_e^1}{\partial v_e}\cos\theta\right)\cos\theta + \frac{f_e^1}{v_e}\sin^2\theta\right]$$
$$= I^0(f_e^0) + I^1(f_e^1)\cos\theta \,. \tag{3.39}$$

Writing this equation in terms of the Legendre polynomials $P_1(\cos\theta) = \cos\theta$ and $P_2(\cos\theta) = \frac{1}{2}(3\cos^2\theta - 1)$, we obtain

$$\frac{\partial f_e^0}{\partial t} + \frac{\partial f_e^1}{\partial t} P_1(\cos\theta)$$
$$+ \frac{eE}{m}\left[\frac{\partial f_e^0}{\partial v_e} P_1(\cos\theta) + \frac{2}{3}\left(\frac{\partial f_e^1}{\partial v_e} - \frac{f_e^1}{v_e}\right) P_2(\cos\theta) + \frac{1}{3v_e^2}\frac{\partial}{\partial v_e}\left(v_e^2 f_e^1\right)\right]$$
$$= I^0(f_e^0) + I^1(f_e^1)\, P_1(\cos\theta) \,. \tag{3.40}$$

Separating now in equations for $P_0(\cos\theta) = 1$ and $P_1(\cos\theta) = \cos\theta$, and using equation (3.27) for the components of the collisional term, we finally obtain

$$\frac{\partial f_e^0}{\partial t} + \frac{eE}{m}\frac{1}{3v_e^2}\frac{\partial}{\partial v_e}\left(v_e^2 f_e^1\right) = -\nu_{c0} f_e^0 \tag{3.41}$$

$$\frac{\partial f_e^1}{\partial t} + \frac{eE}{m}\frac{\partial f_e^0}{\partial v_e} = -\nu_{c1} f_e^1 \,. \tag{3.42}$$

If instead of the limited expansion (3.11) we had inserted into equation (3.35) the full expansion (3.9), the equation (3.35) with an infinite number of terms would be converted into an infinite set of coupled equations with a finite number of terms. By using the orthogonality properties of the Legendre polynomials, as well as appropriate recursion relations between them, it is possible to convert equation (3.35) in an infinite set of coupled equations, in which the equation for the anisotropic component f_e^l is coupled to the equations for f_e^{l-1} and f_e^{l+1} (Delcroix 1963, 1966; Cherrington 1980; Delcroix and Bers 1994). In this system the equation for f_e^0 remains as equation (3.41), while the complete expressions for the equations of the first two anisotropies are as follows (see Appendix A.3.1)

$$\frac{\partial f_e^1}{\partial t} + \frac{eE}{m}\left[\frac{\partial f_e^0}{\partial v_e} + \frac{2}{5v_e^3}\frac{\partial}{\partial v_e}\left(v_e^3 f_e^2\right)\right] = -\nu_{c1} f_e^1 \qquad (3.43)$$

$$\frac{\partial f_e^2}{\partial t} + \frac{eE}{m}\left[\frac{2v_e}{3}\frac{\partial}{\partial v_e}\left(\frac{f_e^1}{v_e}\right) + \frac{3}{7v_e^4}\frac{\partial}{\partial v_e}\left(v_e^4 f_e^3\right)\right] = -\nu_{c2} f_e^2 \,. \qquad (3.44)$$

The electric field has hence the property of coupling the nearest anisotropies. The full set of equations can be solved using only approximated methods, in which the simplest solution is obtained assuming the rapid convergence (3.11) and the system limited to equations (3.41) and (3.42). The two-term approximation will suffice when the change of velocity due to the action of the applied electric field during the relaxation time of f_e^1, i.e. ν_m^{-1}, is much smaller than the velocity of electrons

$$\frac{eE}{m}\,\nu_m^{-1} \ll v_e. \qquad (3.45)$$

Considering the collision cross section for momentum transfer (3.31), we obtain

$$\frac{eE}{m}\,\lambda \ll v_e^2. \qquad (3.46)$$

being $\lambda = 1/(n_o\sigma_m)$ the collision mean free path of electrons. Thus an important criterion for the validity of the two-term approximation is the magnitude of the field to be sufficiently small in order the work realized by the field along the free path does not change appreciably the energy of electrons

$$eE\,\lambda \ll u\,, \qquad (3.47)$$

with $u = \frac{1}{2}\,mv_e^2$. At low E/n_o values the higher anisotropies become small corrections.

3.1.4 Electron Conductivity

For a constant electric field and steady-state conditions, which are attained after a time $\sim \nu_m^{-1}$, we have $\partial f_e^1/\partial t = 0$ in equation (3.42), so that we obtain

$$\mathbf{f_e^1} = -\frac{eE}{m\nu_m}\frac{df_e^0}{dv_e}\,\mathbf{e_z}\,. \qquad (3.48)$$

Introducing now this equation in (3.18), the drift or average vector velocity is

$$\mathbf{v_{ed}} = -\frac{eE}{n_e m}\int_0^\infty \frac{1}{\nu_m}\frac{df_e^0}{dv_e}\frac{4\pi v_e^3}{3}\,dv_e\,\mathbf{e_z}\,. \qquad (3.49)$$

Since $\mathbf{v_{ed}} = -\mu_e \mathbf{E}$, we immediately obtain the electron mobility μ_e from (3.49), whereas from the electron current density $\mathbf{J_e} = -en_e\mathbf{v_{ed}} = \sigma_{ce}\mathbf{E}$, we obtain the electron conductivity

$$\sigma_{ce} = -\frac{e^2}{m}\int_0^\infty \frac{1}{\nu_m}\frac{df_e^0}{dv_e}\frac{4\pi v_e^3}{3}\,dv_e\,. \tag{3.50}$$

Because $f_e^0 \to 0$ when $v_e \to \infty$, we may integrate (3.50) by parts and alternatively to write

$$\sigma_{ce} = \frac{e^2}{m}\int_0^\infty \frac{1}{v_e^2}\frac{d}{dv_e}\left(\frac{v_e^3}{3\nu_m}\right) f_e^0\, 4\pi v_e^2\, dv_e\,, \tag{3.51}$$

which can also be expressed as follows

$$\sigma_{ce} = \frac{e^2 n_e}{m} < \frac{1}{v_e^2}\frac{d}{dv_e}\left(\frac{v_e^3}{3\nu_m}\right) > . \tag{3.52}$$

Finally, as ν_m is independent of the velocity of electrons, we simply obtain

$$\sigma_{ce} = \frac{e^2 n_e}{m\nu_m}\,. \tag{3.53}$$

3.1.5 Equation for Momentum Conservation

In equations (3.26) and (3.27) we have seen that the electron neutral collision operator $(\partial f_e/\partial t)_{e-o} = I(f_e)$ may be expanded in spherical harmonics, with each term given by a collision frequency of different order

$$\left(\frac{\partial f_e}{\partial t}\right)_{e-o} = -\sum_{l=0}^{\infty} \nu_{cl}\, f_e^l\, P_l(\cos\theta)\,, \tag{3.54}$$

in which the frequency of 1st order ν_{c1} is the collision frequency for momentum transfer ν_m. We are now in conditions to determine the collision operator $\mathbf{I_{1(coll)}}$ (2.160) and (2.162) in equation for momentum conservation (2.167). In fact, using the orthogonality of the Legendre polynomials, we obtain the following expression for the collision source term $\mathbf{I_{1(coll)}}$ using equation (3.17)

$$\mathbf{I_{1(coll)}} = -\sum_{l=0}^{\infty}\int_{\mathbf{v_e}} m\,\mathbf{v_e}\,\nu_{cl}\, f_e^l\, P_l(\cos\theta)\,\mathbf{dv_e}$$

$$= - \int_0^\infty m \, \frac{v_e}{3} \, \nu_m \, \mathbf{f_e^1} \, 4\pi v_e^2 \, dv_e \, , \tag{3.55}$$

where $\mathbf{f_e^1}$ is the first anisotropic component. Assuming this anisotropy expressed as $\mathbf{f_e^1} = p_{11} \, \mathbf{e_x} + q_{11} \, \mathbf{e_y} + p_{10} \, \mathbf{e_z}$ and the drift velocity (3.18) given by

$$\mathbf{v_{ed}} = \frac{1}{n_e} \int_0^\infty \frac{v_e}{3} \, (p_{11} \, \mathbf{e_x} + q_{11} \, \mathbf{e_y} + p_{10} \, \mathbf{e_z}) \, 4\pi v_e^2 \, dv_e \, , \tag{3.56}$$

the equation (3.55) may be expressed as

$$\mathbf{I_{1(coll)}} = - \, n_e \, m \, (\widehat{\nu_m} \, . \, \mathbf{v_{ed}}) \, , \tag{3.57}$$

being $\widehat{\nu_m}$ a diagonal tensor of elements $\nu_{m(xx)}$, $\nu_{m(yy)}$, and $\nu_{m(zz)}$, such as

$$\nu_{m(xx)} = \frac{\int_0^\infty v_e \, \nu_m \, p_{11} \, v_e^2 \, dv_e}{\int_0^\infty v_e \, p_{11} \, v_e^2 \, dv_e} \tag{3.58}$$

and likewise expressions for the other two elements.

In the absence of a magnetic field and for a homogeneous plasma, in which the anisotropy exists along a sole direction, created e.g. by an applied electric field, we may write equation (3.57) as follows

$$\mathbf{I_{1(coll)}} = - \, n_e \, m \, \nu'_m \, \mathbf{v_{ed}} \, , \tag{3.59}$$

being ν'_m a scalar quantity. Substituting equation (3.59) in the equation for momentum conservation (2.167), we obtain

$$n_e \, m \left(\frac{\partial}{\partial t} + (\mathbf{v_{ed}} \, .\nabla) \right) \mathbf{v_{ed}} + \nabla p_e - n_e \, (\mathbf{F_e} + \mathbf{F_s}) = - \, n_e \, m \, (\nu'_m + \langle \nu_{ion} \rangle) \, \mathbf{v_{ed}} \, , \tag{3.60}$$

where $\mathbf{F_e} + \mathbf{F_s} = - \, e \, (\mathbf{E} + \mathbf{E_s})$ in case of electric fields are only present, and where e and $\mathbf{E_s}$ denote the absolute value of the electron charge and the space-charge field. In most cases we have $\nu'_m \gg \langle \nu_{ion} \rangle$.

Some particular derivations may now be obtained from equation (3.60). First of all, the inertia term $n_e \, m \, (\mathbf{v_{ed}} \, .\nabla)\mathbf{v_{ed}}$ can usually be neglected for electrons, which associated with the neglecting of $\langle \nu_{ion} \rangle$ as compared to ν'_m, and of the space-charge field, allows to write under stationary conditions

$$\nabla(n_e k_B T_e) + n_e \, e \, \mathbf{E} = - \, n_e \, m \, \nu'_m \, \mathbf{v_{ed}} \, . \tag{3.61}$$

If further T_e is spatially constant, we still have

$$\begin{aligned}\mathbf{\Gamma_e} &= n_e\,\mathbf{v_{ed}} = -\,\frac{k_B T_e}{m\,\nu'_m}\,\nabla n_e \;-\; \frac{n_e\,e}{m\,\nu'_m}\,\mathbf{E} \\ &= -\,D_e\,\nabla n_e \;-\; n_e\,\mu_e\,\mathbf{E},\end{aligned} \tag{3.62}$$

where $D_e = k_B T_e/(m\nu'_m)$ and $\mu_e = e/(m\nu'_m)$ are the electron free diffusion coefficient and the electron mobility, respectively. As shown in Exercise 3.6, the frequency ν'_m obtained here is related with the microscopic effective collision frequency $\nu^e_m(v_e)$ through the expression (the term effective results from the inclusion of inelastic collisions in parallel with elastic collisions; see Sect. 3.3.1)

$$\frac{1}{\nu'_m} = <\frac{1}{v_e^2}\,\frac{d}{dv_e}\left(\frac{v_e^3}{3\,\nu^e_m}\right)> . \tag{3.63}$$

On the other hand, near a cathode the electrons are distributed in the potential sheath $\mathbf{E} = -\,\nabla V(\mathbf{r})$, with a vanishingly small electron particle current towards the cathode, $\mathbf{\Gamma_e} \simeq 0$, which allows to write from equation (3.62)

$$0 \simeq -\,D_e\,\nabla n_e \;+\; n_e\,\mu_e\,\nabla V(\mathbf{r}). \tag{3.64}$$

Integrating we obtain the Maxwell-Boltzmann equilibrium relation

$$n_e(\mathbf{r}) = n_e^0\,\exp\left(\frac{eV(\mathbf{r})}{u_k}\right), \tag{3.65}$$

assuming $V(0) = 0$ and $n_e(0) = n_e^0$ at the plasma boundary, $V(\mathbf{r}) < 0$ in the cathode region, and where $u_k = eD_e/\mu_e$ is a parameter termed characteristic energy. When the velocity distribution is Maxwellian and D_e and μ_e are given by the expressions above, we obtain $u_k = k_B T_e$, otherwise u_k should be considered in equation (3.65) instead of $k_B T_e$ (Sect. 5.1.1).

Finally, the inertia term cannot be usually neglected for ions, so that as only the electric field exists together with the inertia term (which is valid for low densities and consequently low ion collision frequencies, $\nu_i \simeq 0$), and vanished ion temperatures $T_i \simeq 0$, we obtain from an equivalent equation (3.60) for ions

$$n_i\,M\,(\mathbf{v_{id}}\,.\nabla)\mathbf{v_{id}} \;+\; n_i\,e\,\nabla V = 0, \tag{3.66}$$

with v_{id} denoting the ion drift velocity. Then, in slab geometry we find

$$M\,v_{id}\,\frac{dv_{id}}{dx} = -\,e\,\frac{dV}{dx} \tag{3.67}$$

and integrating this equation with $v_{id} = 0$ and $V = 0$ at $x = 0$, and $V(x) < 0$ for $x > 0$, we obtain the equation for ion energy conservation

$$\frac{1}{2}\,M\,v_{id}^2 = -\,e\,V\,. \tag{3.68}$$

3.2 Cooling of Electron Gas

3.2.1 Elastic Collision Term

Up to now we have assumed the molecules infinitely heavy (Lorentz gas model) which has led to find $\nu_{c0} = 0$ in equation (3.28). Due to this fact, the elastic collisions, the only we are considering by now, do not act upon the isotropic component of the electron velocity distribution function, $f_e^0(v_e)$, and the thermodynamic equilibrium, in which the isotropic component goes towards a Maxwellian at temperature $T_e = T_o$, is never achieved. In fact, due to the small ratio m/M, it may take many thousand collisions to change the energy of an electron appreciably, whereas one collision defined by the frequency ν_m is enough to cancel its momentum. However, if we assume $m/M = 0$ we are not considering the small change of energy between the electrons and the molecules. In the absence of any mechanism for electron cooling, the electric field would produce an infinity heating. On the contrary, the first anisotropic component $f_e^1(v_e)$ changes with the frequency for momentum transfer ν_m, so that the isotropic equilibrium $\mathbf{v_{ed}} = <\mathbf{v_e}> = 0$ is achieved for times of few ν_m^{-1}.

Let us now consider the situation where the electrons transfer to the molecules a non-null energy as a result of elastic collisions ($v_e \rightarrow v_e' < v_e$ and $v_o \rightarrow v_o' > v_o$), but we will firstly assume the molecules at rest before the collision ($v_o = 0$). Under these conditions we have

$$v_e^2 - v_e'^2 = \alpha \, v_e^2 \,, \tag{3.69}$$

in which as shown in equation (2.89)

$$\alpha = \frac{2mM}{(m+M)^2} \, (1 - \cos\chi) \,, \tag{3.70}$$

being χ the scattering angle in the centre-of-mass system.

In order to consider the possibility of $f_e^0(v_e)$ tends toward an equilibrium distribution after many thousand collisions occur, higher order terms need to be included in the collision term (3.7). The corresponding collision term for the isotropic component, as given by (3.7) and (3.26), is as follows

$$I^0(f_e^0) = \int_\Omega \int_{\mathbf{v_o}} \left(f_e^0(v') \, F_o' \, v' \, \sigma(v', \Omega) - f_e^0(v) \, F_o \, v \, \sigma(v, \Omega) \right) \, d\Omega \, \mathbf{dv_o} \,, \tag{3.71}$$

with $v = |\mathbf{v_e} - \mathbf{v_o}|$ and $v' = |\mathbf{v_e'} - \mathbf{v_o'}|$. Since $\mathbf{v_o} = 0$ and $\mathbf{v_o'}$ is small, we have $v = v_e$ and $v' \simeq v_e'$, whereas for the two velocity-volume elements ($\mathbf{dv_e} = d^3v_e$) we may write

$$\frac{d^3 v'_e}{d^3 v_e} \simeq \frac{v'^3_e}{v_e^3} , \tag{3.72}$$

so that using Liouville's theorem

$$d^3 v_e \, d^3 v_0 = d^3 v'_e \, d^3 v'_o , \tag{3.73}$$

we find

$$d^3 v_0 = \frac{v'^3_e}{v_e^3} \, d^3 v'_o . \tag{3.74}$$

Substituting (3.74) into equation (3.71), we obtain

$$I^0(f_e^0) = \int_\Omega \int_{\mathbf{v_o}} f_e^0(v'_e) \, F'_o \, \frac{v'^4_e}{v_e^3} \, \sigma(v'_e, \Omega) \, d\Omega \, \mathbf{dv'_o} - \int_\Omega \int_{\mathbf{v_o}} f_e^0(v_e) \, F_o \, v_e \, \sigma(v_e, \Omega) \, d\Omega \, \mathbf{dv_o} . \tag{3.75}$$

Let us expand now $f_e^0(v'_e) \, v'^4_e \, \sigma(v'_e, \Omega)$ around $f_e^0(v_e) \, v_e^4 \, \sigma(v_e, \Omega)$ (see Allis 1956)

$$f_e^0(v'_e) \, v'^4_e \, \sigma(v'_e, \Omega) = f_e^0(v_e) \, v_e^4 \, \sigma(v_e, \Omega) + \frac{\partial}{\partial(v_e^2)} \left(f_e^0(v_e) \, v_e^4 \, \sigma(v_e, \Omega) \right) \, \Delta v_e^2 , \tag{3.76}$$

in which $\Delta v_e^2 = v_e^2 - v'^2_e = \alpha \, v_e^2$ as given by (3.69).

The first term of (3.76) cancels with the second term of (3.75) and since

$$\int_{\mathbf{v_o}} F'_o \, \mathbf{dv'_o} = n_o ,$$

we obtain for the collision term of the isotropic component

$$I^0(f_e^0) = n_o \int_\Omega \frac{\partial}{\partial(v_e^2)} \left(f_e^0(v_e) \, v_e^4 \, \sigma(v_e, \Omega) \right) \frac{\alpha}{v_e} \, d\Omega . \tag{3.77}$$

Substituting the expression (3.70) for α, we still have

$$I^0(f_e^0) = n_o \frac{2mM}{(m+M)^2} \frac{1}{v_e} \frac{\partial}{\partial(v_e^2)} \left(f_e^0(v_e) \, v_e^4 \int_0^\pi \sigma(v_e, \chi) \, (1 - \cos \chi) \, 2\pi \, \sin \chi \, d\chi \right) . \tag{3.78}$$

We may use now the frequency for momentum transfer in the centre-of-mass system similar to that defined in the laboratory system (3.29)

$$\nu_g(v_e) = n_o \, v_e \int_0^\pi (1 - \cos \chi) \, \sigma(v_e, \chi) \, 2\pi \, \sin \chi \, d\chi , \tag{3.79}$$

so that (3.78) takes the form

$$I^0(f_e^0) = \frac{2mM}{(m+M)^2} \frac{1}{v_e} \frac{\partial}{\partial(v_e^2)} \left(f_e^0\, v_e^3\, \nu_g\right) . \tag{3.80}$$

The relation between the frequencies ν_g and ν_m depends how the cross section varies with χ. In first order this is given by the relation between the electron velocities in the laboratory and centre-of-mass systems (2.84)

$$\nu_m(v_e) = \frac{M}{m+M}\, \nu_g(v_e) , \tag{3.81}$$

so that we may write (3.80) as

$$I^0(f_e^0) = \frac{2m}{m+M} \frac{1}{v_e} \frac{\partial}{\partial(v_e^2)} \left(f_e^0\, v_e^3\, \nu_m\right) . \tag{3.82}$$

The equation (3.82) is obtained assuming the molecules at rest before the collision. Nevertheless, the final expression for $I^0(f_e^0)$ should vanish as $f_e^0(v_e)$ tends toward the equilibrium with a Maxwellian at the same temperature as the molecules, $T_e = T_o$,

$$f_e^0(v_e) = n_e \left(\frac{m}{2\pi k_B T_o}\right)^{3/2} \exp\left(-\frac{m v_e^2}{2k_B T_o}\right) . \tag{3.83}$$

Since in equilibrium we have

$$f_e^0(v_e) + \frac{2k_B T_o}{m} \frac{df_e^0}{d(v_e^2)} = 0 ,$$

we should introduce this second term as a correction in (3.82) writing the collision term as follows (Allis 1956)

$$I^0(f_e^0) = \frac{2m}{m+M} \frac{1}{v_e} \frac{\partial}{\partial(v_e^2)} \left[v_e^3\, \nu_m \left(f_e^0 + \frac{2k_B T_o}{m} \frac{\partial f_e^0}{\partial(v_e^2)}\right)\right] . \tag{3.84}$$

This assures to be $I^0(f_e^0) = 0$ as the electron distribution function tends toward the equilibrium with the molecules at the temperature T_o. Replacing the derivative on v_e^2 with v_e and taking into account the strong inequality $m \ll M$, we may still write

$$I^0(f_e^0) = \frac{m}{M} \frac{1}{v_e^2} \frac{\partial}{\partial v_e} \left[v_e^3\, \nu_m \left(f_e^0 + \frac{k_B T_o}{m v_e} \frac{\partial f_e^0}{\partial v_e}\right)\right] . \tag{3.85}$$

This expression was firstly obtained by Chapman and Cowling (1939). It is worth noting here that as the electrons are generally hotter than the molecules in a discharge out of equilibrium, the corrective term in (3.85) represents the small heating for the electrons due to collisions with non-frozen molecules. The equation (3.85) is nonvanishing as long as the isotropic component of the electron velocity distribution function does not attain a Maxwellian with the same temperature of the molecules.

From equation (3.85) we obtain a characteristic relaxation frequency for energy transfer ν_e to replace our first estimate $\nu_{c0} = 0$, obtained in (3.28) for the case of molecules of infinite mass. Since we should have $I^0(f_e^0) = -\, \nu_e\, f_e^0$, we may write approximately

$$\nu_e(v_e) \simeq \frac{m}{M}\, \nu_m(v_e)\, . \tag{3.86}$$

This frequency represents the relaxation frequency for the isotropic component of the electron velocity distribution function as only elastic collisions are taken into account. This frequency is significantly smaller than the frequency ν_m for relaxation of the first anisotropy f_e^1, which means that the effect of elastic collisions upon the isotropic component is M/m times longer than the effect upon the anisotropic part of the distribution. The collisions are much more efficient to achieve the isotropic equilibrium than to establish the energy equilibration between the electrons and molecules (thermodynamic equilibrium).

For high and intermediate values of the ionization degree n_e/n_o, it may occur the frequency of electron-electron collisions to be larger than the frequency of energy transfer between the electrons and the molecules, $\nu_{e-e} \gg \nu_e$. In this case, the isotropic distribution firstly tends to a Maxwellian at a higher temperature $T_e > T_o$, in a time scale $\sim \nu_{e-e}^{-1}$, and later on the temperature T_e evolves toward T_o in a longer time scale $\sim \nu_e^{-1}$. During this time interval the corrective term in equation (3.85) is $\sim -\,(T_o/T_e)\, f_e^0$.

3.2.2 Flux in Velocity Space

It was seen through equations (3.43) and (3.44) that the neglect of f_e^2 and higher terms is valid for low and moderate values of the applied electric field E. Under such conditions, these equations reduce then to (3.41) and (3.42), with the collision operator in the first equation given by equation (3.85)

$$\frac{\partial f_e^0}{\partial t} + \frac{eE}{m}\frac{1}{3v_e^2}\frac{\partial}{\partial v_e}\left(v_e^2 f_e^1\right) = \frac{m}{M}\frac{1}{v_e^2}\frac{\partial}{\partial v_e}\left[v_e^3\, \nu_m \left(f_e^0 + \frac{k_B T_o}{m v_e}\frac{\partial f_e^0}{\partial v_e}\right)\right] \tag{3.87}$$

$$\frac{\partial f_e^1}{\partial t} + \frac{eE}{m}\frac{\partial f_e^0}{\partial v_e} = -\, \nu_m\, f_e^1\, . \tag{3.88}$$

Since the collisions are much more efficient to produce a steady-state for f_e^1 than for f_e^0, we may start by assuming $\partial f_e^1/\partial t = 0$ in equation (3.88) and therefore

$$f_e^1 = -\frac{eE}{m\nu_m}\frac{\partial f_e^0}{\partial v_e}. \tag{3.89}$$

Inserting now f_e^1 in (3.87), we obtain

$$\frac{\partial f_e^0}{\partial t} - \left(\frac{eE}{m}\right)^2 \frac{1}{v_e^2}\frac{\partial}{\partial v_e}\left(\frac{v_e^2}{3\nu_m}\frac{\partial f_e^0}{\partial v_e}\right) = \frac{m}{M}\frac{1}{v_e^2}\frac{\partial}{\partial v_e}\left[v_e^3\,\nu_m\left(f_e^0 + \frac{k_B T_o}{m v_e}\frac{\partial f_e^0}{\partial v_e}\right)\right]. \tag{3.90}$$

Let us consider by now steady-state conditions also for the isotropic component of the velocity distribution function, $\partial f_e^0/\partial t = 0$. By inspection of equation (3.90) we can see that it presents the form of a divergence of a radial vector in velocity space. In fact, the steady-state equation can be rewritten under the form

$$\frac{1}{v_e^2}\frac{d}{dv_e}\left(v_e^2(g_E^0 + g_{el}^0)\right) = 0\,, \tag{3.91}$$

with g_E^0 and g_{el}^0 denoting the components of a radial flow vector in velocity space, positive and negative, respectively, taking into account the acceleration of electrons by the electric field and the electron stopping due to recoil collisions

$$g_E^0 = -\frac{1}{3\nu_m}\left(\frac{eE}{m}\right)^2\frac{df_e^0}{dv_e} > 0 \tag{3.92}$$

$$g_{el}^0 = -\frac{m}{M}\,v_e\,\nu_m\left(f_e^0 + \frac{k_B T_o}{m v_e}\frac{df_e^0}{dv_e}\right) < 0. \tag{3.93}$$

Allis and Haus (1974) presented an analytic interpretation based upon the energy-flow processes in the electron gas, by now limited to acceleration by the electric field and deceleration by elastic collisions. In this approach the total radial flow through a sphere of radius v_e in velocity space is

$$G = \int_{\mathbf{v_e}} (\nabla.\,\mathbf{g})\,\mathbf{dv_e} = \int_{\Omega} (\mathbf{g}\,.\,\mathbf{e_v})\;v_e^2\,d\Omega\,, \tag{3.94}$$

with $\mathbf{g} = (g_E^0 + g_{el}^0)\,\mathbf{e_v}$, and where $\mathbf{e_v} = \mathbf{v_e}/v_e$ denotes the radial unitary vector in velocity space (see Fig. 3.5). The gain G represents the rate of flow of electrons across a sphere of velocity magnitude v_e in the direction of increasing velocity. Since $\mathbf{g}$ is radial, we simply obtain

$$G = G_E + G_{el} = (g_E^0 + g_{el}^0)\,4\pi\,v_e^2\,. \tag{3.95}$$

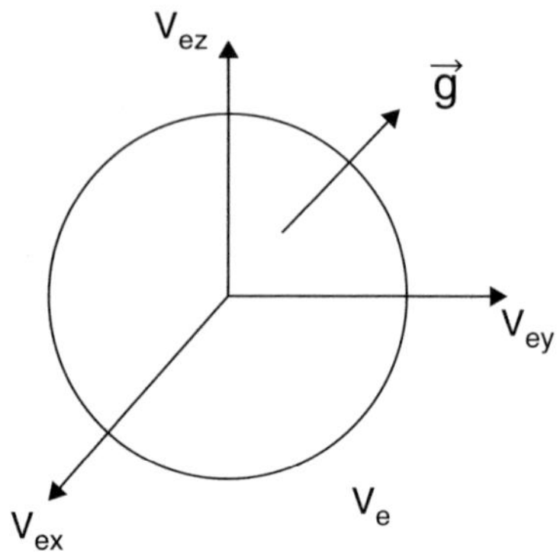

Fig. 3.5 Flux of electrons in velocity space

The interpretation due to Allis is clear. Since the electric field accelerates the electrons, the gain G_E is positive and it represents the upflux of electrons across the sphere of radius v_e, whereas the gain $G_{el} < 0$ represents the downflux across the same sphere due to elastic collisions. Multiplying the Boltzmann equation (3.91) by $4\pi\ v_e^2$, we obtain

$$\frac{d}{dv_e}(G_E + G_{el}) \ = \ 0\ , \tag{3.96}$$

which shows that the total gain is constant across any sphere of radius v_e (that is the divergence of **g** is null). Since $G = 0$ at $v_e = 0$ and $v_e = \infty$, we must also have $G(v_e) = 0$ over any sphere of radius v_e. The equation (3.96) transforms hence into

$$G_E + G_{el} \ = \ 0 \tag{3.97}$$

and because of (3.95), we still have

$$g_E^0 + g_{el}^0 \ = \ 0\ . \tag{3.98}$$

The equation (3.98) allows us to find a general solution to the Boltzmann equation. In fact, using (3.92) and (3.93) we obtain

$$\left(\frac{1}{3v_m^2}\left(\frac{eE}{m}\right)^2\frac{M}{m} + \frac{k_B T_o}{m}\right)\frac{df_e^0}{dv_e} \ = \ -\ v_e\, f_e^0\ , \tag{3.99}$$

whose solution is (Margenau 1946)

$$f_e^0(v_e) \ = \ f_e^0(0)\ \exp\left(-\int_0^{v_e}\frac{v_e'}{\frac{1}{3v_m^2}\left(\frac{eE}{m}\right)^2\frac{M}{m} + \frac{k_B T_o}{m}}\ dv_e'\right)\ . \tag{3.100}$$

If the electric field is zero, then the electron distribution function is a Maxwellian at the gas temperature T_o, as expected. If E is not zero, then the specific form of the distribution function will depend on the velocity dependence of the collision frequency $\nu_m(v_e)$. The function $f_e^0(v_e)$ is obviously a decreasing function of v_e being obtained by the normalization condition (3.15)

$$n_e = \int_0^\infty f_e^0(v_e)\, 4\pi v_e^2\, dv_e\,. \tag{3.101}$$

The simplest case to consider is the constant collision frequency case $\nu_m = \text{const}$. Defining an electron temperature by

$$k_B T_e = \frac{M}{3\nu_m^2}\left(\frac{eE}{m}\right)^2 + k_B T_o\,, \tag{3.102}$$

we obtain from equation (3.100)

$$f_e^0(v_e) = f_e^0(0)\,\exp\left(-\frac{m v_e^2}{2 k_B T_e}\right)\,. \tag{3.103}$$

In this case the solution is a Maxwellian, at temperature $T_e > T_0$, and the difference between the two temperatures increases as the magnitude of the electric field increases. On the other hand, if ν_m relies on v_e we may establish a criterion to define a critical electric field, E_c, above which the distribution function $f_e^0(v_e)$ significantly deviates from a Maxwellian. Such criterion may be expressed as

$$\frac{M}{3\nu_m^2}\left(\frac{eE_c}{m}\right)^2 \simeq k_B T_o\,, \tag{3.104}$$

which allows to obtain

$$E_c(v_e) \simeq \frac{m}{e}\sqrt{\frac{3 k_B T_o}{M}}\,\nu_m(v_e)\,. \tag{3.105}$$

E_c is hence function of v_e through ν_m. Thus, for a given electric field, f_e^0 is less perturbed at the velocities where the collision frequency is high, and significantly deviated from a Maxwellian at the velocities where the frequency ν_m is low.

The second simplest case to consider is the collision mean free path, λ, independent of the electron velocity. Since $\lambda = v_e/\nu_m$, with $\nu_m = n_o v_e \sigma_m$, we have $\lambda = 1/(n_o\sigma_m) = \text{const}$, which is known as hard sphere model. Assuming also a high electric field such that $E \gg E_c$ in equation (3.100), we obtain the following solution

$$f_e^0(v_e) = f_e^0(0)\,\exp\left(-\int_0^{v_e} \frac{4\, v_e'^3}{v_{e1}^4}\, dv_e'\right)\,. \tag{3.106}$$

with v_{e1} denoting a constant characteristic electron velocity defined as

$$v_{e1}^4 = \frac{4M}{3m}\left(\frac{eE}{m}\lambda\right)^2 . \tag{3.107}$$

Equation (3.106) is easily integrated given

$$f_e^0(v_e) = f_e^0(0) \exp\left(-\frac{v_e^4}{v_{e1}^4}\right) . \tag{3.108}$$

This solution is known in the literature as Druyvesteyn distribution (Druyvesteyn and Penning 1940) and it presents the particularity of varying with $\exp(-v_e^4)$ rather than $\exp(-v_e^2)$ as in a Maxwellian. The Druyvesteyn distribution drops off much more rapidly than the Maxwellian distribution at higher energies. In electrical discharge modelling the accurate determination of the high-energy tail of the electron velocity distribution function is mandatory, since this energy range determines the ionization rate and hence the steady-sate operating conditions of the discharge.

3.2.3 *Power Balance*

Let us consider the expression (3.16) for the mean electron energy $<u>= \frac{1}{2}\, m < v_e^2>$, by now assuming that the electron density remains constant, since we are not considering the creation of new electrons by ionization and the loss of electrons by diffusion or electron-ion recombination. In this case, the rate of change of the mean electron energy is

$$\frac{d<u>}{dt} = \frac{1}{n_e}\int_0^\infty \frac{1}{2}\, m\, v_e^2\, \frac{\partial f_e^0}{\partial t}\, 4\pi v_e^2\, dv_e . \tag{3.109}$$

Substituting now the equation for the rate of change of the isotropic component of the electron velocity distribution function (3.87)

$$\frac{\partial f_e^0}{\partial t} = -\frac{eE}{m}\frac{1}{3v_e^2}\frac{\partial}{\partial v_e}(v_e^2 f_e^1) + I(f_e^0) , \tag{3.110}$$

in which the collision term is given by (3.85), we may write the equation (3.109) under the form

$$n_e\frac{d<u>}{dt} = P_E - P_{el} , \tag{3.111}$$

where

$$P_E = -\frac{eE}{6}\int_0^\infty \frac{\partial}{\partial v_e}(v_e^2 f_e^1)\, 4\pi v_e^2\, dv_e \tag{3.112}$$

and

$$P_{el} = -\frac{m^2}{2M}\int_0^\infty \frac{\partial}{\partial v_e}\left[v_e^3\, \nu_m \left(f_e^0 + \frac{k_B T_o}{m v_e}\frac{\partial f_e^0}{\partial v_e}\right)\right] 4\pi v_e^2\, dv_e, \tag{3.113}$$

represent the power gain by the electrons from the field and the power lost in elastic collisions, per volume unity, respectively.

Integrating by parts, we obtain

$$P_E = eE\int_0^\infty \frac{v_e}{3} f_e^1\, 4\pi v_e^2\, dv_e \tag{3.114}$$

$$P_{el} = \frac{m^2}{M}\int_0^\infty v_e^2\, \nu_m \left(f_e^0 + \frac{k_B T_o}{m v_e}\frac{\partial f_e^0}{\partial v_e}\right) 4\pi v_e^2\, dv_e\,, \tag{3.115}$$

The equation (3.114) can still be expressed as

$$P_E = (\mathbf{J_e}\,.\,\mathbf{E}) = \sigma_{ce}\, E^2, \tag{3.116}$$

being $\mathbf{J_e} = -\, e n_e \mathbf{v_{ed}}$ the electron current density, with $\mathbf{v_{ed}}$ denoting the drift velocity (3.18), and σ_{ce} the electron conductivity (3.50). On the other hand, the equation (3.115), neglecting the corrective term due to the small heating of electrons in collisions with non-frozen molecules, may be written as follows

$$P_{el} = n_e \frac{2m}{M} <u\, \nu_m> \ . \tag{3.117}$$

In the case of electrons given by a Maxwellian distribution at temperature T_e, the neglected corrective term is of the order $\sim -\ (T_o/T_e)$, so that when such term is included equation (3.117) may be written as follows

$$P_{el} = n_e \frac{2m}{M} <u\, \nu_m> \left(1 - \frac{T_o}{T_e}\right) . \tag{3.118}$$

It is worth noting here that the equation for power balance under steady-state conditions can be directly obtained from the conservation of the total flux in velocity space (3.96). In fact, multiplying both members by the electron energy u and integrating for all velocities, we obtain the energy conservation equation (see Sect. 2.4.3)

$$\int_0^\infty u \frac{d}{dv_e}(G_E + G_{el})\, dv_e = 0 \tag{3.119}$$

and because $G_E + G_{el} = 0$ (3.97), the equation (3.119) can be integrated by parts giving

$$\int_0^\infty (G_E + G_{el})\, du \;=\; 0. \tag{3.120}$$

Taking into account now the following expressions for the gains

$$G_E \;=\; g_E^0\, 4\pi v_e^2 \;=\; \frac{eE}{3m} f_e^1\, 4\pi v_e^2 \;>\; 0 \tag{3.121}$$

$$G_{el} \;=\; g_{el}^0\, 4\pi v_e^2 \;=\; -\,\frac{m}{M}\, v_e\, \nu_m \left(f_e^0 + \frac{k_B T_o}{m v_e} \frac{df_e^0}{dv_e} \right) 4\pi v_e^2 \;<\; 0, \tag{3.122}$$

with G_E obtained from (3.92) and (3.95), and G_{el} from (3.93) and (3.95), we immediately obtain the steady-state energy conservation equation $P_E - P_{el} = 0$ from equation (3.120).

3.3 Inclusion of Inelastic Collisions

3.3.1 Inelastic Collision Term

In a weakly ionized plasma, created e.g. in a low-pressure discharge, they are usually the inelastic collisions, primarily the excitation processes, that dominate over elastic collisions. As we will show below this occurs except at low values of the reduced electric field, E/n_o, that is for low values of the ratio of the electric field magnitude to the gas number density, case in which the elastic collisions dominate. In an inelastic collision, the electron gives up part of its energy into excitation of rotational, vibrational, or electronic degrees of freedom of the molecules, or in ionizing them. In the case of excitation from a lower to a upper state, $X_i \rightarrow X_j$, the electrons of energy $u = \frac{1}{2} m v_e^2$ lose an energy equal to the threshold energy u_{ij}

$$e(u) \;+\; X_i \;\leftrightarrows\; e(u - u_{ij}) \;+\; X_j\,. \tag{3.123}$$

The reverse superelastic process can also occur when the electrons gain energy in assisting an excited molecule in returning to the lower energy level. The inelastic and the superelastic processes are expressed by the right and left arrows in equation (3.123). The latter are also termed inelastic processes of second kind. The ionization can also be considered as any other inelastic process, if the introduction of secondary electrons produced with the energy u' are not considered in the distribution

$$e(u) \;+\; X_0 \;\rightarrow\; e(u - u_{ion} - u') \;+\; X^+ \;+\; e(u')\,. \tag{3.124}$$

That is, if only the energy loss of primary electrons is taken into account.

As we have seen in Sect. 2.3.3, the inelastic collision term, with the inclusion of both inelastic and superelastic processes, is given by equation (2.127), renamed here as the operator $J(f_e)$

$$J(f_e) = \sum_{i,j} \left\{ n_j \frac{v_e'^{\,2}}{v_e} \int_{\Omega} f_e(\mathbf{v_e'})\, \sigma_{ji}(v_e', \Omega)\ d\Omega \ - \ n_i\, v_e\, f_e(\mathbf{v_e})\, \sigma_{ij}(v_e) \right\} + \sum_{i,j} \left\{ n_i \frac{v_e''^{\,2}}{v_e} \int_{\Omega} f_e(\mathbf{v_e''})\, \sigma_{ij}(v_e'', \Omega)\ d\Omega \ - \ n_j\, v_e\, f_e(\mathbf{v_e})\, \sigma_{ji}(v_e) \right\} . \tag{3.125}$$

As also seen in Sect. 2.3.3, the velocities v_e, v_e' and v_e'' are such that $v_e' < v_e < v_e''$. Due to the nature of this operator, we may assume as for the elastic collisions in (3.26), that $J(f_e)$ is given by the following expansion in spherical functions

$$J(f_e) \ = \ \sum_{l=0}^{\infty} J^l(f_e^l)\ P_l(\cos\theta)\ , \tag{3.126}$$

with each term J^l representing an operator acting on the component f_e^l of the electron velocity distribution function and $P_l(\cos\theta)$ the Legendre polynomials. Considering as before that the cross sections are independent of the azimuthal angle, $\sigma_{ij}(v_e, \Omega) \equiv \sigma_{ij}(v_e, \chi_0)$, with χ_0 representing the scattering angle in the laboratory system, we may assume, as in (3.23) and (3.24), that the electron velocity distribution at velocity $\mathbf{v_e'}$ is expanded in terms of the spherical harmonics on the angle θ' and that this angle may be transformed to a product of spherical functions on θ and χ_0

$$f_e(\mathbf{v_e'}) \ = \ \sum_{l=0}^{\infty} f_e^l(v_e')\ P_l(\cos\theta') \ = \ \sum_{l=0}^{\infty} f_e^l(v_e')\ P_l(\cos\theta)\ P_l(\cos\chi_0)\ . \tag{3.127}$$

Substituting equations (3.126) and (3.127) and the expansion for $f_e(\mathbf{v_e})$ into equation (3.125), we obtain for each individual term of the inelastic collision operator

$$J(f_e^l) = \sum_{i,j} \left\{ n_j \frac{v_e'^{\,2}}{v_e} \int_{\Omega} f_e^l(v_e')\ P_l(\cos\chi_0)\ \sigma_{ji}(v_e', \Omega)\ d\Omega \ - \ n_i\, v_e\, f_e^l(v_e)\, \sigma_{ij}(v_e) \right\} + \sum_{i,j} \left\{ n_i \frac{v_e''^{\,2}}{v_e} \int_{\Omega} f_e^l(v_e'')\ P_l(\cos\chi_0)\ \sigma_{ij}(v_e'', \Omega)\ d\Omega \ - \ n_j\, v_e\, f_e^l(v_e)\, \sigma_{ji}(v_e) \right\} . \tag{3.128}$$

We may define now the inelastic frequencies of order l as

$$\nu_{ij}^{l}(v_e) = n_i\, v_e \int_{\Omega} P_l(\cos\chi_0)\, \sigma_{ij}(v_e,\Omega)\, d\Omega \tag{3.129}$$

$$\nu_{ji}^{l}(v_e) = n_j\, v_e \int_{\Omega} P_l(\cos\chi_0)\, \sigma_{ji}(v_e,\Omega)\, d\Omega\ , \tag{3.130}$$

which allows to write (3.128) with the form

$$J^l(f_e^l) = \sum_{i,j} \left\{ \frac{v_e'}{v_e}\, \nu_{ji}^{l}(v_e')\, f_e^l(v_e') \; - \; \nu_{ij}^{0}(v_e)\, f_e^l(v_e) \right\} + \sum_{i,j} \left\{ \frac{v_e''}{v_e}\, \nu_{ij}^{l}(v_e'')\, f_e^l(v_e'') \; - \; \nu_{ji}^{0}(v_e)\, f_e^l(v_e) \right\} . \tag{3.131}$$

When $l = 0$, the frequencies ν_{ij}^l and ν_{ji}^l transform into the total collision frequencies (3.32).

The expression (3.131) represents a generalization of the expression obtained before for elastic collisions in Lorentz gas model. In fact, assuming $v_e' = v_e'' = v_e$, we obtain from (3.131)

$$J^l(f_e^l) = -\sum_{i,j} \{(\nu_{ij}^0 - \nu_{ij}^l) + (\nu_{ji}^0 - \nu_{ji}^l)\} f_e^l(v_e), \tag{3.132}$$

where the frequencies $(\nu_{ij}^0 - \nu_{ij}^l)$ and $(\nu_{ji}^0 - \nu_{ji}^l)$ are equivalent to the frequencies ν_{cl} defined in (3.28), that is they are weighted by the factor $(1 - P_l(\cos\chi_0))$.

Let us consider now firstly the expression for $J^0(f_e^0)$ from equation (3.131), rewritten here with a slightly different form

$$J^0(f_e^0) = \sum_{i,j} \left\{ \frac{v_e + v_{ij}}{v_e}\, \nu_{ij}(v_e + v_{ij})\, f_e^0(v_e + v_{ij}) - \nu_{ij}(v_e)\, f_e^0(v_e) \right\} + \sum_{j,i} \left\{ \frac{v_e - v_{ij}}{v_e}\, \nu_{ji}(v_e - v_{ij})\, f_e^0(v_e - v_{ij}) \; - \; \nu_{ji}(v_e)\, f_e^0(v_e) \right\} . \tag{3.133}$$

Here, $\nu_{ij} \equiv \nu_{ij}^0$ and $\nu_{ji} \equiv \nu_{ji}^0$ are the total collision frequencies. The first two terms of (3.133) are for the inelastic collisions $i \to j$ and they take into account, respectively, the introduction of electrons at velocity v_e that primarily had the velocity $(v_e + v_{ij})$ and the removal of electrons from velocity v_e. While the third and fourth terms are for the superelastic collisions $j \to i$ and they take into account the introduction of electrons at velocity v_e that primarily had the velocity $(v_e - v_{ij})$ and the removal of electrons from velocity v_e. The velocities $(v_e + v_{ij})$ and $(v_e - v_{ij})$ are defined such that $(u + u_{ij}) = \frac{1}{2}\, m(v_e + v_{ij})^2$ and $(u - u_{ij}) = \frac{1}{2}\, m(v_e - v_{ij})^2$, being u_{ij} the threshold energy of the inelastic process.

The term J^0 may be rewritten in terms of the electron energy $u = \frac{1}{2}\, m v_e^2$. In fact, making the substitutions $\nu_{ij}(v_e) = n_i v_e \sigma_{ij}(v_e)$ and $\nu_{ji}(v_e) = n_j v_e \sigma_{ji}(v_e)$, and using identical substitutions for $\nu_{ij}(v_e + v_{ij})$ and $\nu_{ji}(v_e - v_{ij})$, the operator J^0 may be expressed in terms of the energy u as follows

$$J^0(f_e^0) = \sqrt{\frac{2}{mu}} \sum_{i,j} n_i \left\{(u + u_{ij})\, \sigma_{ij}(u + u_{ij})\, f_e^0(u + u_{ij}) - \; u\, \sigma_{ij}(u)\, f_e^0(u)\right\}$$
$$+ \sqrt{\frac{2}{mu}} \sum_{j,i} n_j \left\{(u - u_{ij})\, \sigma_{ji}(u - u_{ij})\, f_e^0(u - u_{ij}) - \; u\, \sigma_{ji}(u)\, f_e^0(u)\right\} . \tag{3.134}$$

The equation (3.134) is a non-local equation in energy space, since the knowledge of $f_e^0(u)$ depends of the values for this same function at $(u + u_{ij})$ and $(u - u_{ij})$, which constitutes a very complicated problem in general.

Let us derive now the expression for the collision term of the first anisotropy $J^1(f_e^1)$ from equation (3.131). Introducing the frequencies for momentum transfer associated with inelastic and superelastic collisions, as we have done in (3.29) for the elastic collisions,

$$\nu_{ij}^m(v_e) \;=\; \nu_{ij}^0(v_e) - \nu_{ij}^1(v_e) \;=\; n_i\, v_e \int_\Omega (1 - \cos\chi_0)\, \sigma_{ij}(v_e, \Omega)\, d\Omega \tag{3.135}$$

$$\nu_{ji}^m(v_e) \;=\; \nu_{ji}^0(v_e) - \nu_{ji}^1(v_e) \;=\; n_j\, v_e \int_\Omega (1 - \cos\chi_0)\, \sigma_{ji}(v_e, \Omega)\, d\Omega\ , \tag{3.136}$$

we may write equation (3.131) as follows

$$J^1(f_e^1) = -\sum_{i,j} \left\{\nu_{ij}^m(v_e) \;+\; \nu_{ji}^m(v_e)\right\} f_e^1(v_e)$$
$$+ \sum_{i,j} \left\{ \frac{v_e''}{v_e}\, \nu_{ij}^1(v_e'')\, f_e^1(v_e'') \;-\; \nu_{ij}^1(v_e)\, f_e^1(v_e) \right\}$$
$$+ \sum_{j,i} \left\{ \frac{v_e'}{v_e}\, \nu_{ji}^1(v_e')\, f_e^1(v_e') \;-\; \nu_{ji}^1(v_e)\, f_e^1(v_e) \right\} . \tag{3.137}$$

This equation is written in this manner to evidence the contribution of three terms. The first one is identical to the corresponding term for elastic collisions (3.27), in which the collision integral $I^1(f_e^1)$ is given by the negative product of the collision frequency for momentum transfer, $\nu_{c1}(v_e) \equiv \nu_m(v_e)$, by the first anisotropic component of the electron velocity distribution function, $f_e^1(v_e)$. The second and the third terms are for inelastic and superelastic processes, respectively, but in both only the first anisotropic frequencies $\nu^1_{ij}(v_e)$ and $\nu^1_{ji}(v_e)$ given by (3.129) and (3.130) appear and these are vanishingly small as the scattering is nearly isotropic. In fact, when $\sigma_{ij}(v_e, \Omega) \simeq \sigma_{ij}(v_e)$ we obtain from equation (3.129)

$$\nu^1_{ij}(v_e) \;\simeq\; n_i\; v_e \int_0^{2\pi} \cos\chi_0\; \sigma_{ij}(v_e)\; 2\pi \sin\chi_0\; d\chi_0 \;=\; 0\;. \tag{3.138}$$

In many cases even as the inelastic scattering cannot be assumed isotropic, the second and third terms of (3.137) are small corrections in comparison with the first term, so that the collision integral takes the same form as that for elastic collisions (3.27). Both may be then associated by defining an effective collision frequency for momentum transfer

$$\nu^e_m(v_e) \;=\; \nu_m(v_e) + \sum_{i,j} \left\{\nu^m_{ij}(v_e) \;+\; \nu^m_{ji}(v_e)\right\}\;, \tag{3.139}$$

being the total collision term in equation (3.42) able to describe the evolution of the first anisotropy $f^1_e(v_e)$ given by

$$I^1(f^1_e) + J^1(f^1_e) \;\simeq\; -\;\nu^e_m(v_e)\, f^1_e(v_e)\;. \tag{3.140}$$

The equation for the steady-state anisotropy (3.48) should hence be rewritten as

$$\mathbf{f^1_e} \;=\; -\;\frac{eE}{m\nu^e_m}\,\frac{df^0_e}{dv_e}\;\mathbf{e_z}\;. \tag{3.141}$$

The total inelastic collision frequency for momentum transfer $\sum_{ij}(\nu^m_{ij} + \nu^m_{ji})$ is larger than the elastic collision frequency ν_m as the electron energy increases and it is obviously null for energies below the first threshold energy. This depends obviously on the set of electron cross section for each specific gas. Figure 3.6 shows schematically the typical behaviour of the electron cross sections for elastic, excitation and ionization processes, as a function of the electron energy u, in the case of an atomic gas as argon.

Let us estimate now the relaxation frequency for the isotropic component of the electron velocity distribution function f^0_e, i.e. the collision frequency for energy

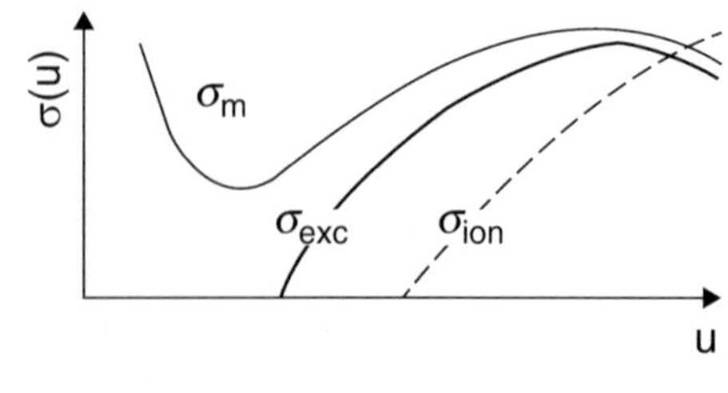

Fig. 3.6 Electron cross sections for elastic σ_m, excitation σ_{exc}, and ionization σ_{ion} collisions, in the case of a typical atomic gas as argon

transfer. We have previously found $\nu_{c0} = 0$ in equation (3.28) when we assumed $m/M = 0$, and it was necessary to include the first order corrective term on m/M to obtain the non-null value (3.86) from the collisional term $I^0(f_e^0)$ given by equation (3.85) (Chapman and Cowling 1939). Now when the inelastic and superelastic collisions are also taken into account by considering the term $J^0(f_e^0)$ given by (3.133), the collision frequency for energy transfer is approximately

$$\nu_e(v_e) \simeq \frac{m}{M}\,\nu_m(v_e) + \sum_{i,j}\{\nu_{ij}(v_e) + \nu_{ji}(v_e)\} . \tag{3.142}$$

This frequency ν_e is much smaller than ν_m^e given by (3.139) as the elastic collisions dominate, which occurs, as we will show below, for low values of the reduced electric field E/n_o. On the contrary, ν_e and ν_m^e tend to a common value as the inelastic processes prevail.

3.3.2 Analysis in Velocity Space and Power Balance

With the inclusion of inelastic processes the steady-sate Boltzmann equation (3.91) for determination of the isotropic component of the electron velocity distribution function f_e^0 should be replaced with

$$\frac{1}{v_e^2}\frac{d}{dv_e}\left(v_e^2\left(g_E^0 + g_{el}^0\right)\right) = J^0 , \tag{3.143}$$

where g_E^0 and g_{el}^0 are the components of a radial flow vector in velocity space, due to the acceleration of electrons by the electric field and to the recoil by elastic collisions, respectively, and J^0 is the inelastic collision term for the isotropic component of the velocity distribution (3.133). Multiplying both members by $4\pi v_e^2$, we have

$$\frac{d}{dv_e}\left(G_E + G_{el}\right) = J^0\, 4\pi v_e^2 , \tag{3.144}$$

where G_E and G_{el} are the gains associated with the upflux and downflux in velocity space. Integrating now from 0 to v_e, we obtain

$$G(v_e) = \int_0^{v_e} J^0\, 4\pi v_e^2\, dv_e , \tag{3.145}$$

with $G = G_E + G_{el}$, or under differential form

$$dG = J^0\, 4\pi v_e^2\, dv_e . \tag{3.146}$$

In the absence of inelastic collisions the total gain is null, since $G(0) = G(\infty) = 0$, i.e. the divergence of $\mathbf{g}$ is null. However, when the inelastic collisions are included these collisions introduce and remove electrons from each spherical shell of radius v_e and $v_e + dv_e$ in velocity space, in a way that the gain change dG in the shell is equal to the difference between the number of electrons introduced and removed from it, $J^0\, 4\pi v_e^2\, dv_e$. When the inelastic collisions exist, we have $G > 0$, because although they may also exist superelastic collisions, the energy supplied by these collisions is smaller than the energy spent in inelastic collisions. We have therefore

$$\int_0^{v_e} J^0\, 4\pi v_e^2\, dv_e \;>\; 0. \tag{3.147}$$

If the upper limit goes to $v_e \to \infty$, we have obviously a null equality in (3.147), since we are including only processes in which the number of electrons is kept constant. That is we are not considering by now the introduction of secondary electrons created by ionization and the loss of electrons by electron-ion recombination, electron attachment, or diffusion.

The power lost in inelastic collisions, per volume unit, can be obtained as in (3.119) by multiplying the right-hand side member of equation (3.144) by the electron energy and integrating over all velocity space

$$P_{inel} \;=\; -\int_0^{\infty} u\, J^0\, 4\pi v_e^2\, dv_e. \tag{3.148}$$

Then introducing equation (3.134), we obtain

$$\begin{aligned} P_{inel} = \frac{8\pi}{m^2} \sum_{i,j} n_i \int_0^{\infty} u \Big\{u\, \sigma_{ij}(u)\, f_e^0(u) \;-\; (u+u_{ij})\, \sigma_{ij}(u+u_{ij})\, f_e^0(u+u_{ij})\Big\}\, du \\ +\frac{8\pi}{m^2} \sum_{j,i} n_j \int_0^{\infty} u \Big\{u\, \sigma_{ji}(u)\, f_e^0(u) \;-\; (u-u_{ij})\, \sigma_{ji}(u-u_{ij})\, f_e^0(u-u_{ij})\Big\}\, du\,. \end{aligned} \tag{3.149}$$

Since $\sigma_{ij} = 0$ for $u < u_{ij}$, and $\sigma_{ji} \neq 0$ for any value of u, we still have

$$\begin{aligned} P_{inel} &= \frac{8\pi}{m^2} \sum_{i,j} n_i \int_{u_{ij}}^{\infty} \{u^2 \;-\; (u-u_{ij})\, u\}\; \sigma_{ij}(u)\, f_e^0(u)\, du \\ &\quad +\frac{8\pi}{m^2} \sum_{j,i} n_j \int_0^{\infty} \{u^2 \;-\; (u+u_{ij})\, u\}\; \sigma_{ji}(u)\, f_e^0(u)\, du\,. \\ &= \frac{8\pi}{m^2} \sum_{i,j} u_{ij} \left\{ n_i \int_{u_{ij}}^{\infty} u\, \sigma_{ij}(u)\, f_e^0(u)\, du \;-\; n_j \int_0^{\infty} u\, \sigma_{ji}(u)\, f_e^0(u)\, du \right\}\,. \end{aligned} \tag{3.150}$$

Using now the expression (2.136) for the electron rate coefficients of inelastic $C_{ij} = <v_e\ \sigma_{ij}>$ and superelastic $C_{ji} = <v_e\ \sigma_{ji}>$ processes, we finally obtain

$$P_{inel} = n_e \sum_{i,j} u_{ij}\ (n_i\ C_{ij}\ -\ n_j\ C_{ji}). \tag{3.151}$$

It is worth remembering here that the electron rate coefficients are linked to the average frequencies as $<\nu_{ij}>= n_i\ C_{ij}$ and $<\nu_{ji}>= n_j\ C_{ji}$. The equation (3.151) gives the difference between the power lost and the power gained by the electrons, per volume unity, in inelastic and superelastic collisions, respectively. The cross sections σ_{ij} and σ_{ji} are related one another by the Klein-Rosseland relation (2.140).

3.3.3 Continuous Approximation for Rotational Levels

Due to the small separation between the rotational levels of a molecule it is possible to use a continuous approximation to take into account the effects of electron inelastic and superelastic collisions with rotational levels in the Boltzmann equation (Frost and Phelps 1962).

Gerjuoy and Stein (1955) had pointed out that the homonuclear molecules such as H_2 and N_2 possesses an electric quadrupole moment and have calculated a cross section for rotational excitation, which increases very rapidly close to threshold and for which its magnitude can be calculated from measured electric quadrupole moments. For such molecules the energy levels of the rotating molecule are given to sufficient accuracy for present purposes by $E_J = J(J+1)\ B_0$, where B_0 is the rotational constant determined from spectroscopic data: $B_0 = 7.54 \times 10^{-3}$ eV for H_2 (Huber and Herzberg 1979) and $= 2.49 \times 10^{-4}$ eV for N_2 (Lofthus and Krupenie 1977). Using the selection rule for the rotational levels $\Delta J = \pm 2$, the energy lost by an electron in excitation is $u_{J,J+2} = E_{J+2} - E_J = (4J+6)\ B_0$, while that gained in de-excitation is $u_{J,J-2} = E_J - E_{J-2} = (4J-2)\ B_0$. According to Gerjuoy and Stein (1955) the cross sections for rotational excitation and de-excitation, as a function of the electron energy u, are given by

$$\sigma_{J,J+2}(u) = \frac{(J+2)(J+1)}{(2J+3)(2J+1)}\ \sigma_0 \sqrt{1 - \frac{(4J+6)\ B_0}{u}} \tag{3.152}$$

$$\sigma_{J,J-2}(u) = \frac{J(J-1)}{(2J-1)(2J+1)}\ \sigma_0 \sqrt{1 + \frac{(4J-2)\ B_0}{u}}, \tag{3.153}$$

where $\sigma_0 = 8\pi q^2 a_0^2/15$, q is the electric quadrupole moment in units of ea_0^2 ($q = 0.62$ for H_2 and $= 1.01$ for N_2, Frost and Phelps 1962), and a_0 is the Bohr radius.

Frost and Phelps (1962) had pointed out that the cross sections (3.152) and (3.153) may be assumed independent of u over the important energy range, so that the right-hand side member of equation (3.144) for rotational levels may be rewritten as follows using (3.134)

$$\begin{aligned} & J^0_{rot}(f^0_e)\; 4\pi v_e^2 \\ & = \frac{8\pi v_e}{m}\, n_o \sum_J \delta_J \left\{(u+u_{J,J+2})\;\sigma_{J,J+2}\; f^0_e(u+u_{J,J+2}) - u\;\sigma_{J,J+2}\; f^0_e(u)\right\} \\ & \quad + \frac{8\pi v_e}{m}\, n_o \sum_J \delta_J \left\{(u-u_{J,J-2})\;\sigma_{J,J-2}\; f^0_e(u-u_{J,J-2}) \;-\; u\;\sigma_{J,J-2}\; f^0_e(u)\right\}, \end{aligned} \tag{3.154}$$

with $\delta_J = n_J/n_o$ denoting the fractional population in the J-th level. Assuming also as in Frost and Phelps (1962) that the energy exchanges in rotational excitation and de-excitation are small enough to expand $f^0_e(u+u_{J,J+2})$ and $f^0_e(u-u_{J,J+2})$ in first-order Taylor series

$$f^0_e(u+u_{J,J+2}) \;\simeq\; f^0_e(u) \;+\; \frac{df^0_e}{du}\; u_{J,J+2} \tag{3.155}$$

$$f^0_e(u-u_{J,J-2}) \;\simeq\; f^0_e(u) \;-\; \frac{df^0_e}{du}\; u_{J,J-2}\,, \tag{3.156}$$

we obtain replacing (3.155) and (3.156) into equation (3.154) and neglecting terms in $u^2_{J,J+2}$ and $u^2_{J,J-2}$

$$J^0_{rot}(f^0_e)\; 4\pi v_e^2 \;=\; \frac{8\pi v_e}{m}\, n_o \sum_J \delta_J \left(u_{J,J+2}\;\sigma_{J,J+2} \;-\; u_{J,J-2}\;\sigma_{J,J-2}\right) \frac{d}{du}(u\, f^0_e)\,. \tag{3.157}$$

Substituting now (3.152) and (3.153) discarding the dependence on u in these expressions, and using the energy differences $u_{J,J+2} = (4J+6)\; B_0$ and $u_{J,J-2} = (4J-2)\; B_0$, we finally obtain considering $\sum_J \delta_J = 1$,

$$J^0_{rot}(f^0_e)\; 4\pi v_e^2 \;=\; \frac{8\pi v_e}{m}\, n_o\, 4B_0\sigma_0\, \frac{d}{du}(u\, f^0_e)\,. \tag{3.158}$$

As in recoil collisions, the effect of electron collisions with rotational levels, assuming the continuous approximation, can be represented in the form of a negative gain of the type considered in (3.122). In fact, assuming as in (3.144)

$$-\frac{dG_{rot}}{dv_e} \;=\; J^0_{rot}(f^0_e)\; 4\pi v_e^2\,, \tag{3.159}$$

we obtain

$$G_{rot} \;=\; -\, n_o\, \frac{4B_0\sigma_0}{m}\, f^0_e\; 4\pi v_e^2 \;<\; 0\,. \tag{3.160}$$

On the other hand, using (3.120) the power lost by the electrons in total excitation and de-excitation of rotational levels is given by

$$P_{rot} = -\int_0^{\infty} G_{rot}\, du = n_o\, 4B_0\, \sigma_0\, n_e <v_e> , \tag{3.161}$$

or $P_{rot} = 4B_0\ n_e < \nu_{rot} >$ introducing an average frequency for net rotational excitation as $<\nu_{rot}>= n_o\ \sigma_0\ <v_e>$.

Finally, we must notice that for very low values of the reduced electric field, typically for $E/n_o < 1\,\text{Td}$ ($1\,\text{Td} = 1 \times 10^{-21}\,\text{V m}^2$), the continuous approximation for the rotational levels is no more valid and the discrete collisional operator (3.154) needs to be used (Ridenti et al. 2015).

3.4 Resolution of the Boltzmann Equation

3.4.1 Independent Variables

Let us consider now the stationary electron Boltzmann transport equation (3.144) for the isotropic component of the electron velocity distribution function $f_e^0(v_e)$, with the inclusion of the negative gain due to the rotational losses

$$\frac{d}{du}(G_E + G_{el} + G_{rot})\, mv_e = J^0\, 4\pi v_e^{\,2} . \tag{3.162}$$

Here, G_E is given by (3.121), but with ν_m replaced by ν_m^e as explained in (3.140), and G_{el}, G_{rot} and J^0 are given by (3.122), (3.160) and (3.134), respectively. Substituting these expressions into (3.162), making the replacements $\nu_m^e(v_e) = n_o v_e \sigma_m^e(v_e)$ and $\nu_m(v_e) = n_o v_e \sigma_m(v_e)$, and introducing the fractional number populations $\delta_i = n_i/n_o$ and $\delta_j = n_j/n_o$, we obtain after division of both members by $-\ 8\pi n_o v_e/m$

$$\begin{aligned}
&\frac{d}{du}\left[\left(\frac{eE}{n_o}\right)^2 \frac{u}{3\sigma_m^e}\frac{df_e^0}{du} + \frac{2m}{M}\, u^2\, \sigma_m \left(f_e^0 + k_B T_o\, \frac{df_e^o}{du}\right) + 4B_0\sigma_0\, u\, f_e^0\right] \\
&\quad + \sum_{i,j} \delta_i \left\{(u + u_{ij})\, \sigma_{ij}(u + u_{ij})\, f_e^0(u + u_{ij}) - \ u\, \sigma_{ij}(u)\, f_e^0(u)\right\} \\
&\quad + \sum_{j,i} \delta_j \left\{(u - u_{ij})\, \sigma_{ji}(u - u_{ij})\, f_e^0(u - u_{ij}) - \ u\, \sigma_{ji}(u)\, f_e^0(u)\right\} = 0 .
\end{aligned} \tag{3.163}$$

The equation (3.163) is the so-called homogeneous electron Boltzmann transport equation and the isotropic component of the electron velocity distribution function, $f_e^0(v_e)$, can be expressed in terms of the electron energy $u = \frac{1}{2} mv_e^{\,2}$, being then termed electron energy distribution function (EEDF). Equation (3.163) shows that the homogeneous Boltzmann equation is a continuity equation for $f_e^0(u)$ along the energy axis only, expressing the fact that the change in the total electron flux (i.e., the net flux resulting from the upflux due to the applied electric field and the downfluxes due to the elastic recoil and rotational excitation), in an energy interval du, is equal to the difference between the rates for removal and re-introduction of electrons by inelastic and superelastic collisions.

In an inelastic collision the electrons can give up some of their energy into excitation of vibrational or electronic states, while in a superelastic collision the electrons receive energy in assisting an excited state to return to the lower energy state. The contribution of rotational levels is taken into account through a continuous approximation. The second term in equation (3.163) accounts for the inelastic loss processes by considering, respectively, electrons of energy $u + u_{ij}$ undergoing a collision in which they lose the energy u_{ij} and appear as electrons of energy u, and electrons of energy u undergoing a collision in which they also lose the energy u_{ij} and appear as electrons of energy $u - u_{ij}$. The third term accounts for the reverse processes since it takes into account the gain of energy by electrons due to superelastic collisions (see Fig. 3.7).

For a given particular gas, in which the cross sections for momentum transfer σ_m and for excitation of vibrational and electronic states σ_{ij} are known, the equation (3.163) is solved as a function of the reduced electric field E/n_o, the gas temperature T_o, and the fractional population concentrations δ_i and δ_j of all excited states. The cross sections for de-excitation processes σ_{ji} may be obtained from the Klein-Rosseland relation (2.140). The ionization process is taken into account as any other inelastic process by considering, by now, only the energy loss of primary electrons. Finally, the effective collision cross section for momentum transfer is according to (3.139) defined as follows

$$\sigma_m^e(u) = \sigma_m(u) + \sum_{i,j} \{\delta_i \, \sigma_{ij}^m(u) + \delta_j \, \sigma_{ji}^m(u)\} . \tag{3.164}$$

In the case of nearly isotropic scattering, the cross sections for momentum transfer in inelastic and superelastic collisions, σ_{ij}^m and σ_{ji}^m, are close to the total cross sections; see equations (3.135) and (3.136).

In an atomic gas, the fractional populations of the excited electronic states are usually very small as compared with that of the electronic ground-state, so that the superelastic processes ($j \to i$) may be neglected, while for the inelastic ones ($i \to j$)

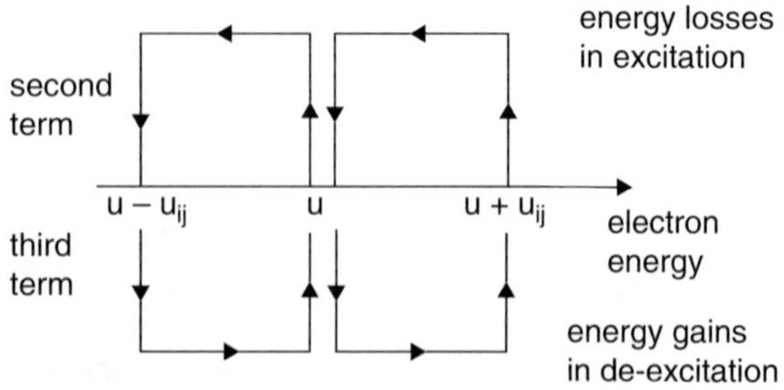

Fig. 3.7 Scheme with the energy losses in excitation and energy gains in de-excitation due to inelastic and superelastic collisions

only processes from the ground-state usually need to be taken into consideration, by making in equation (3.163) $\delta_0 = 1$, $\delta_i = 0$ for $i > 0$, and $\delta_j = 0$. In this case, the proper variables reduce to E/n_o and T_o. Furthermore, if the mean electron energy $<u>$ is much larger than $k_B T_o$, the correction due to the influence of non-frozen molecules may also be neglected and the sole variable is E/n_o

$$\frac{d}{du}\left[\left(\frac{eE}{n_o}\right)^2 \frac{u}{3\sigma_m^e} \frac{df_e^0}{du} + \frac{2m}{M} u^2 \sigma_m f_e^0\right] + \sum_{j>0} \{(u+u_j)\,\sigma_j(u+u_j)\,f_e^0(u+u_j) - u\,\sigma_j(u)\,f_e^0(u)\} = 0\,, \tag{3.165}$$

with $\sigma_j \equiv \sigma_{0j}$. On the contrary, in a molecular gas, such as H_2, N_2, O_2, CO_2, etc, both inelastic and superelastic electron collisions upon the excited vibrational levels need to be considered in the determination of the EEDF, $f_e^0(u)$, except at conditions for which the excitation of vibrational levels cannot occur. Whereas for the electronic states only inelastic collisions of electrons with electronic ground-state molecules need, in general, to be considered due to vanishingly small populations of the electronic excited states.

Under the present approximations of neglecting the space gradients and conservation of the electron number density, the EEDF can be re-normalized for practical purposes such that

$$\int_0^\infty f(u)\,\sqrt{u}\,du = 1\,. \tag{3.166}$$

Taking into account the normalization condition (3.15), $f_e^0(u)$ and $f(u)$ are linked one another as follows

$$f(u) = \frac{1}{n_e}\frac{4\pi}{m}\sqrt{\frac{2}{m}}\,f_e^0(u)\,. \tag{3.167}$$

With such normalization, the Maxwellian electron energy distribution at temperature T_e writes

$$f(u) = \frac{2}{\sqrt{\pi}}\,(k_B T_e)^{-3/2}\exp\left(-\frac{u}{k_B T_e}\right). \tag{3.168}$$

In terms of the EEDF, $f(u)$, the average energy is expressed as

$$<u> = \int_0^\infty f(u)\, u^{3/2}\, du \,, \tag{3.169}$$

with $<u>= \frac{3}{2}\, k_B T_e$ in the case of a Maxwellian, while the electron mobility $\mu_e = \sigma_{ce}/(e n_e)$, being σ_{ce} the electron conductivity given by equation (3.50), is

$$\mu_e \;=\; -\,\frac{2}{3}\,\frac{e}{m}\int_0^\infty \frac{u^{3/2}}{\nu_m^e(u)}\,\frac{df}{du}\,du \,. \tag{3.170}$$

Making the substitution $\nu_m^e = n_o\, \sigma_m^e\, v_e$, we obtain the following expression for the reduced mobility

$$n_o\,\mu_e \;=\; -\,\frac{e}{3}\sqrt{\frac{2}{m}}\int_0^\infty \frac{u}{\sigma_m^e(u)}\,\frac{df}{du}\,du \,, \tag{3.171}$$

which depends only of the proper variables mentioned above for the EEDF. On the other hand, the electron rate coefficient (2.136) associated with the transition $i \to j$ takes the form

$$C_{ij} \;=\; \sqrt{\frac{2}{m}}\int_{u_{ij}}^\infty u\, \sigma_{ij}(u)\, f(u)\, du \tag{3.172}$$

and this depends on the same independent variables.

Let us consider now the stationary power balance, per volume unity,

$$P_E \;=\; P_{el} \;+\; P_{rot} \;+\; P_{inel} \,. \tag{3.173}$$

Using equations (3.116), (3.118), (3.151), and (3.161), we obtain in the case of frozen molecules ($T_o = 0$)

$$\begin{aligned} e\, n_e\, \mu_e\, E^2 = n_e\, \frac{2m}{M}\, <u\, \nu_m> + 4B_0\, n_e\, <\nu_{rot}> \\ + n_e \sum_{i,j} u_{ij}\, (n_i\, C_{ij} - n_j\, C_{ji}) \,. \end{aligned} \tag{3.174}$$

Dividing both members by $n_e n_o$, we obtain the following equation for the power balance per electron at unit gas density (i.e. expressed in eV s^{-1} m^3)

$$\begin{aligned} e\, n_o \mu_e \left(\frac{E}{n_o}\right)^2 = \frac{2m}{M}\, <u\, \sigma_m\, v_e> + 4B_0\, \sigma_0\, <v_e> \\ + \sum_{ij} u_{ij}\, (\delta_i\, C_{ij} - \delta_j\, C_{ji}) \,, \end{aligned} \tag{3.175}$$

in which the first and the second terms, on the right-hand side member, have the following expressions in terms of the EEDF, $f(u)$,

$$\frac{P_{el}}{n_e n_o} = \frac{2m}{M}\sqrt{\frac{2}{m}}\int_0^\infty u^2\,\sigma_m(u)\,f(u)\,du \tag{3.176}$$

$$\frac{P_{rot}}{n_e n_o} = 4B_0\,\sigma_0\sqrt{\frac{2}{m}}\int_0^\infty u\,f(u)\,du\,. \tag{3.177}$$

Once again equation (3.175) shows that, besides the fractional population concentrations δ_i and δ_j, the only variable from which the power balance depends is the reduced electric field E/n_o. Finally, the drift velocity (3.49) is also function of the same variable, since $\mathbf{v_{ed}} = n_o\mu_e\,(E/n_o)\,\mathbf{e_z}$, with $n_o\mu_e$ given by (3.171). In conclusion: $<u>$, $n_o\mu_e$, C_{ij}, C_{ji}, $P_E/(n_e n_o)$, $P_{el}/(n_e n_o)$, $P_{rot}/(n_e n_o)$, $P_{inel}/(n_e n_o)$, and $\mathbf{v_{ed}}$ are function of E/n_o (or of E/n_o, δ_i, δ_j, and T_o, in case of a molecular gas with non-frozen molecules).

3.4.2 Numerical Procedure

The steady-state equation (3.163) can be converted to a set of n-coupled ordinary algebraic equations by finite differencing the electron energy axis into n cells of width Δu. By setting, $u_k^+ = k\,\Delta u$ and $u_k^- = u_{k-1}^+$ to the upper and lower boundary energy limits of a cell k, while for a given quantity X we have $X_k^- = X_{k-1}^+$, $(dX/du)_k = (X_k^+ - X_{k-1}^+)/\Delta u$, and $(dX/du)_k^+ = (X_{k+1} - X_k)/\Delta u$, being X_k and $(dX/du)_k$ the values at the middle of the cell k, with $X_k^+ = (X_k + X_{k+1})/2$, we obtain using a procedure close to Rockwood (1973) the following set of algebraic coupled equations

$$A_{k-1}\,f_{k-1} - (A_k + B_k)\,f_k + B_{k+1}\,f_{k+1} + J_k = 0\,, \tag{3.178}$$

with the matrix elements given by

$$\begin{aligned} A_k &= A_{1k} + A_{2k} + A_{3k} \\ &= \frac{k\,(eE/n_o)^2}{3\,\Delta u\,\sigma_{m(k)}^{e(+)}} + \frac{m}{M}\,\sigma_{m(k)}^{(+)}\,k^2\,(2k_BT_o - \Delta u) - 2k\,B_0\,\sigma_0 \end{aligned} \tag{3.179}$$

$$\begin{aligned} B_k &= B_{1k} + B_{2k} + B_{3k} \\ &= \frac{(k-1)\,(eE/n_o)^2}{3\,\Delta u\,\sigma_{m(k-1)}^{e(+)}} + \frac{m}{M}\,\sigma_{m(k-1)}^{(+)}\,(k-1)^2\,(2k_BT_o + \Delta u) \\ &\quad + 2(k-1)\,B_0\,\sigma_0\,. \end{aligned} \tag{3.180}$$

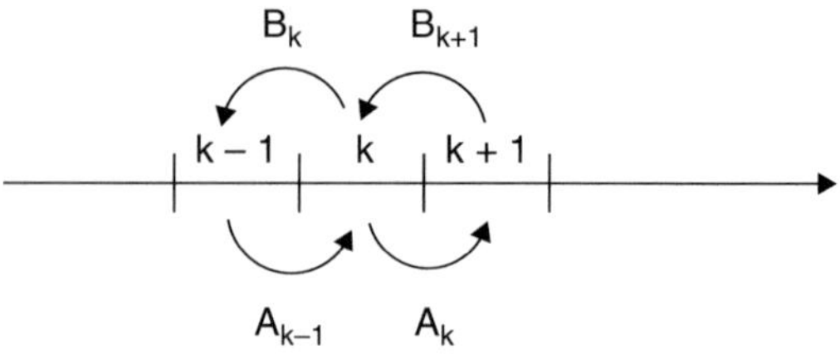

Fig. 3.8 Scheme showing the promotion and demotion of electrons in the energy cell k and neighbouring cells

$$J_k = \sum_{i,j} \delta_i \{u_{k+m_{ij}}\, \sigma_{ij(k+m_{ij})}\, f_{k+m_{ij}} \; - \; u_k\, \sigma_{ij(k)}\, f_k\} \\ + \sum_{i,j} \delta_j \{u_{k-m_{ij}}\, \sigma_{ji(k-m_{ij})}\, f_{k-m_{ij}} \; - \; u_k\, \sigma_{ji(k)}\, f_k\} \; . \qquad (3.181)$$

A_k may be interpreted as the rate at which electrons with energy u_k are promoted to energy u_{k+1} as a result of the action of the continuous terms in equation (3.163), while B_k is the rate for demotion from u_k to u_{k-1} (see Fig. 3.8). The vector f_k represents the EEDF in every cell k. As a matter of fact, it represents either $f_e^0(u)$ or $f(u)$, since the two functions are linked one another through the constant factor (3.167).

In concerning now the terms in equation (3.181), $m_{ij} = u_{ij}/\Delta u$ represents the number of cells jumped by an electron along the energy grid in result of an inelastic or a superelastic collision. It occurs an entrance into energy u_k of electrons originally at energies $u_{k+m_{ij}}$ and $u_{k-m_{ij}}$ (first and third terms) and an exit from energy u_k to energies $u_{k-m_{ij}}$ and $u_{k+m_{ij}}$ (second and fourth terms). Making use of the Klein-Rosseland relation (2.140), the term for the superelastic collisions (second term) can also be written using the cross sections σ_{ij} for the excitation processes as follows

$$\sum_{i,j} \delta_j \; (g_i/g_j)\{u_k\, \sigma_{ij(k)}\, f_{k-m_{ij}} \; - \; u_{k+m_{ij}}\, \sigma_{ij(k+m_{ij})}\, f_k\} \; , \qquad (3.182)$$

being g_i and g_j the statistical weights of the two states.

The set of n algebraic coupled equations (3.178) is linearly dependent, so that it can be solved by matrix inversion using $n - 1$ equations together with the normalization condition (3.166) written now as

$$\sum_k f_k\, u_k^{1/2}\, \Delta u \; = \; 1 \; . \qquad (3.183)$$

Once the vector of the EEDF is known, we may determine $<u>$, C_{ij}, C_{ji}, $P_E/(n_e n_o)$, $P_{el}/(n_e n_o)$, and $P_{rot}/(n_e n_o)$, using the following expressions

$$<u> = \sum_k u_k^{3/2}\, f_k\, \Delta u \qquad (3.184)$$

$$C_{ij} = \sqrt{\frac{2}{m}} \sum_k \sigma_{ij(k)} f_k \, u_k \, \Delta u \tag{3.185}$$

$$C_{ji} = \sqrt{\frac{2}{m}} \frac{g_i}{g_j} \sum_k \sigma_{ij(k+m_{ij})} f_k \, u_{k+m_{ij}} \, \Delta u \tag{3.186}$$

$$\frac{P_E}{n_e n_o} = \sqrt{\frac{2}{m}} \sum_k (A_{1k} - B_{1k}) f_k \, \Delta u^2 \tag{3.187}$$

$$\frac{P_{el}}{n_e n_o} = \sqrt{\frac{2}{m}} \sum_k (B_{2k} - A_{2k}) f_k \, \Delta u^2 \tag{3.188}$$

$$\frac{P_{rot}}{n_e n_o} = \sqrt{\frac{2}{m}} \sum_k (B_{3k} - A_{3k}) f_k \, \Delta u^2 \, , \tag{3.189}$$

with $u_k = (k - 1/2)\, \Delta u$ and where the coefficients A_k and B_k are given by (3.179) and (3.180), while for $n_o \mu_e$ and $P_{inel}/(n_e n_o)$ we may use the relations

$$n_o \mu_e = \frac{P_E/(n_e n_o)}{e\,(E/n_o)^2} \tag{3.190}$$

$$\frac{P_{inel}}{n_e n_o} = \sum_{i,j} m_{ij} \, \Delta u \, (\delta_i \, C_{ij} - \delta_j \, C_{ji}) \, . \tag{3.191}$$

3.5 Electron Cross Sections

In the study of low temperature plasmas, with low and moderate degrees of ionization, they are usually the inelastic and superelastic collisions that dominate in determining the EEDF. Specifically, the inelastic and the superelastic collisions upon the vibrational levels of the electronic ground-state and the inelastic collisions for excitation of upper electronic states, in the case of a molecular gas, and the inelastic collisions for excitation of electronic states from the ground-state, in the case of an atomic gas, being the effects due to superelastic collisions negligibly small in this latter case.

The knowledge of reliable sets of electron cross sections is then mandatory to determine accurate EEDFs for each specific gas, as well as to determine accurate electron transport parameters and excitation and ionization rate coefficients using the EEDFs so determined. Accordingly, since the sixties of last century a systematic work has been realized by many authors in determining reliable electron cross sections' sets for many gases, such as He, Ne, Ar, Kr, H_2, N_2, O_2, CO_2, etc. Among them we must evidence Phelps and his co-authors for their tremendous work in determining cross sections for the most common gases (Pack and Phelps 1961; Frost and Phelps 1962; Engelhardt and Phelps 1963, 1964; Engelhardt et al. 1964; Frost and Phelps 1964, etc).

The method used is basically an iterative procedure on initial estimates for the cross sections until a good agreement is obtained between calculated and experimental data for different quantities. The electron Boltzmann equation is numerically solved to determine the EEDF taking into account both elastic and inelastic collisions in the case of only a DC electric field is present. Furthermore, the experiments used are in conditions of null vibrational excitation, in order the superelastic collisions do not need to be taken into consideration. These experiments are usually termed as electron swarm experiments or tube drift experiments (see Gilardini 1972). In the simplest derivations conducted in the sixties, the quantities under observation were the electron diffusion coefficient D_e, the electron mobility μ_e, and the first Townsend ionization coefficient α.

The electron diffusion coefficient is determined, as it will be shown later on in Chap. 5, taking an appropriate average over the EEDF, $D_e = < v_e^2/(3\ \nu_m^e) >$. We present now the expression for this parameter to evidence its dependence on the isotropic component of the electron velocity distribution function $f_e^0(v_e)$

$$D_e = \frac{1}{n_e} \int_0^\infty \frac{v_e^2}{3\ \nu_m^e(v_e)} f_e^0\, 4\pi v_e^2\, dv_e. \tag{3.192}$$

In terms of the EEDF, $f(u)$, normalized through equation (3.166), we obtain the following expression for the reduced diffusion coefficient

$$n_o D_e = \frac{1}{3} \sqrt{\frac{2}{m}} \int_0^\infty \frac{u}{\sigma_m^e(u)} f\, du\,, \tag{3.193}$$

with n_o denoting the gas number density, expressing the fact that the reduced diffusion coefficient is a function of the independent variable E/n_o only (in the case of negligibly small concentrations of the excited states and low gas temperature). In what concerns the reduced electron mobility, this quantity is also function of E/n_o only, having the corresponding expression already been indicated in equation (3.171). The first Townsend ionization coefficient has been introduced in Chap. 1 and it gives the number of electrons produced by each primary electron, per length unit of its path in the field direction, $\alpha = < \nu_{ion} > / v_{ed}$ (von Engel 1965), being $< \nu_{ion} >$ and v_{ed} the average electron ionization frequency and the absolute value of the drift velocity, respectively. Since $< \nu_{ion} > = n_o\ C_{ion}$, with C_{ion} denoting the ionization rate coefficient

$$C_{ion} = \sqrt{\frac{2}{m}} \int_{u_{ion}}^\infty u\, \sigma_{ion}(u)\, f\, du\,, \tag{3.194}$$

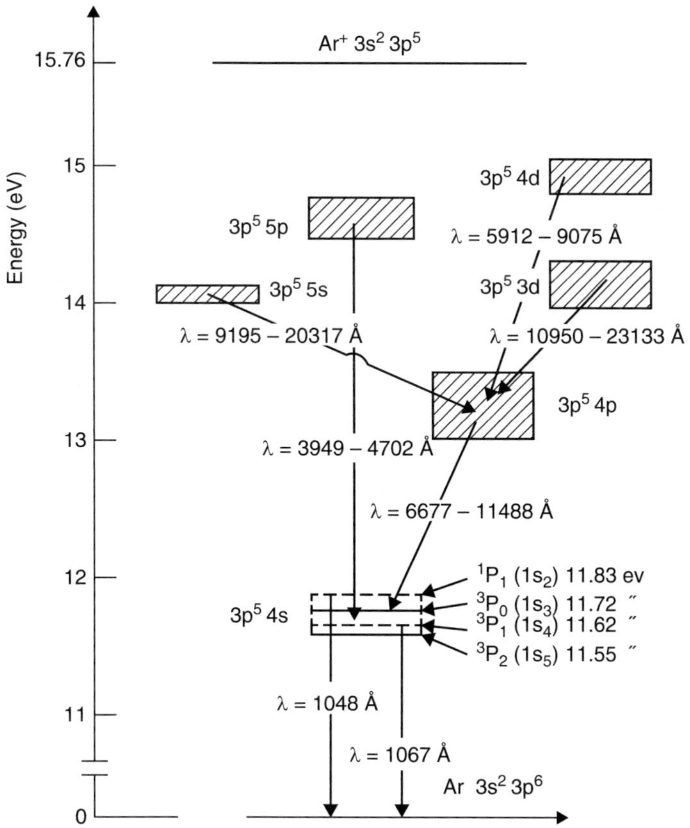

Fig. 3.9 Schematic energy level diagram of Ar atom (Ferreira et al. 1985)

the reduced Townsend ionization coefficient α/n_o, in m^2, is also function of E/n_o

$$\frac{\alpha}{n_o} = \frac{C_{ion}}{v_{ed}} . \tag{3.195}$$

Figure 3.9 shows the schematic energy level diagram of the different states of argon atom with the corresponding transition wavelengths, while Fig. 3.10 shows the electron excitation cross sections obtained for this gas in Ferreira and Loureiro (1983a).

The different curves in Fig. 3.10 represent the electron cross sections for direct excitation of the two metastable states $^3P_2$ and $^3P_0$ altogether (curve A), of the resonant states $^3P_1$ and $^1P_1$ (curves B and C), of the forbidden states 3p^{5}4p with the threshold energy of 12.9 eV (curve D), and of the higher-lying optically allowed states with the threshold of 14.0 eV (curve E). The momentum transfer cross sections of Frost and Phelps (1964) (with its Ramsauer minimum close to the

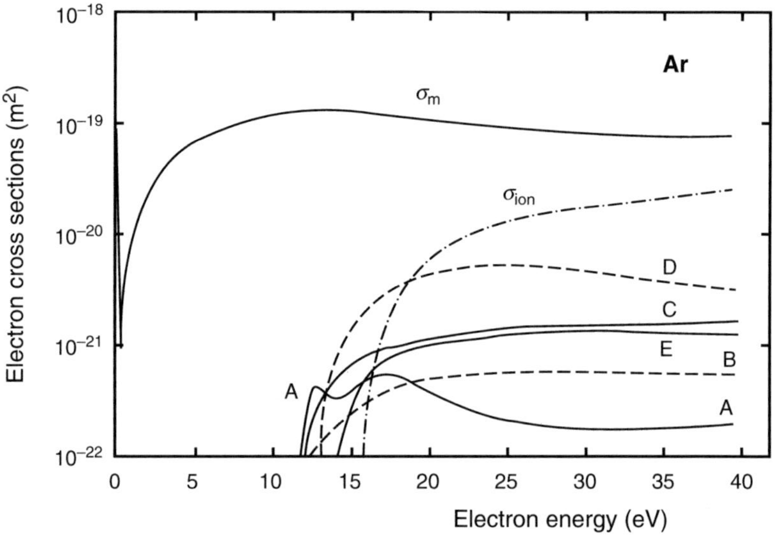

Fig. 3.10 Electron cross sections for excitation of electronic states of Ar in m^2, as a function of the electron energy in eV. The curves are for excitation of the following states or group of states: (*A*) $^3P_2 + {}^3P_0$; (*B*) $^3P_1$; (*C*) $^1P_1$; (*D*) forbidden states (mostly $3p^54p$ multiple); (*E*) higher-lying optically allowed states. For comparison the electron cross section for momentum transfer (σ_m) and the ionization cross section (σ_{ion}) are also plotted (Ferreira and Loureiro 1983a)

origin) and the ionization cross section (with a threshold of 15.76 eV) of Rapp and Englander-Golden (1965) have been adopted with no modifications, as they seemed to be well established.

With the purpose of providing a useful estimate to be used in modelling of electrical discharges, analytic fitting laws have been obtained in Ferreira and Loureiro (1983b). It has been seen that the cross section for momentum transfer is well fitted by the expression $\sigma_m(u) = \alpha_m(u/u_X)$, for $u \leq u_X$, and $\sigma_m(u) = \alpha_m \sqrt{u_X/u}$, for $u \geq u_X$ (that is, $\nu_m \propto u^{3/2}$ for $u \leq u_X$, and ν_m = const for $u \geq u_X$), being α_m a constant equal to $1.59 \times 10^{-19}\,m^2$ and $u_X = 11.55\,eV$ the lowest threshold energy (that for excitation of $^3P_2$ state). Accordingly the total cross section for excitation plus ionization has been seen to be given approximately by $\sigma_X(u) = \alpha_X(u/u_X - 1)\sqrt{u_X/u}$, with $\alpha_X = 1.56 \times 10^{-20} m^2$.

In molecular gases the determination of accurate electron cross sections sets is performed using the same iterative procedure but the situation is more complicated due to the larger number of states involved. The initial estimative is constructed using experimental and theoretical electron cross sections for excitation of individual states but in most cases only the dependence of the cross sections on energy is kept, i.e. the shapes of the cross sections, being their magnitudes modified by successive adjustments until theoretical and experimental values of electron transport and rate coefficients are brought into good agreement. The results are by no means unique, but they certainly do represent a consistent and realistic set of elastic and inelastic electron cross sections (Engelhardt and Phelps 1963).

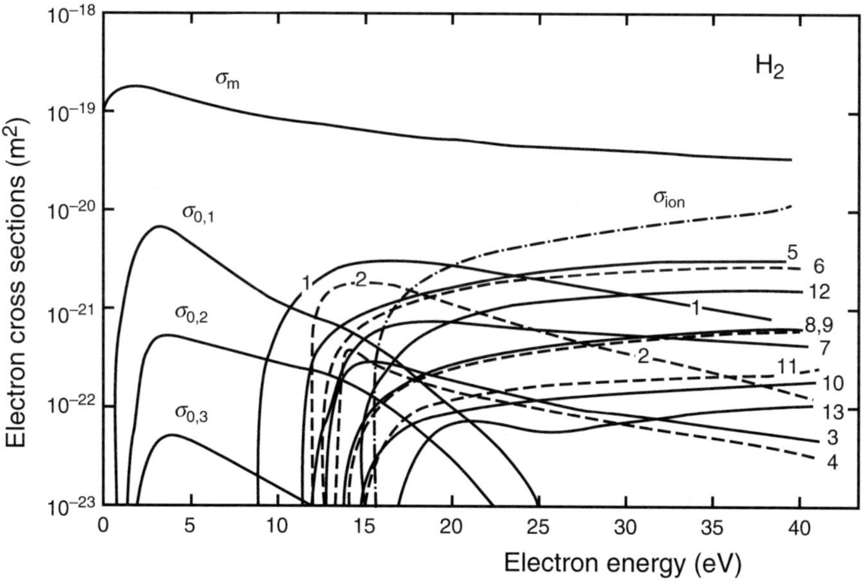

Fig. 3.11 Initial set of electron cross sections for H_2 used in Loureiro and Ferreira (1989b) in m^2, as a function of the electron energy in eV. The various curves represent the cross section for momentum transfer σ_m, excitation of $v = 1, 2,$ and3 levels of $H_2(X,v)$, and excitation of electronic states with the following notation: (*1*) b; (*2*) c; (*3*) a; (*4*) e; (*5*) B; (*6*) C; (*7*) E + F; (*8*) B′; (*9*) D; (*10*) B″; (*11*) D′. It is also shown the ionization cross section, σ_{ion}, and the cross section for dissociation into H(1*s*) + H($n = 2$) (*12*) and H(1*s*) + H($n = 3$) (*13*)

Figure 3.11 shows the initial set of electron cross sections for H_2 found in the literature and from which a final self-consistent set has been constructed in Loureiro and Ferreira (1989b). The different curves represent the cross section for momentum transfer, the cross sections for excitation of the vibrational levels $v = 1, 2,$ and 3 belonging to the electronic ground-state X $^1\Sigma_g^+$ state (with the threshold energies of 0.5, 1.0 and 1.5 eV, respectively), the ionization cross section (15.4 eV), and the cross sections for excitation of the following electronic states with the corresponding threshold energies in eV: b $^3\Sigma_u^+$ (8.9 eV); c $^3\Pi_u$ (11.9 eV); a $^3\Sigma_g^+$ (11.9 eV); e $^3\Sigma_u^+$ (13.4 eV); B $^1\Sigma_u^+$ (11.4 eV); C $^1\Pi_u$ (12.4 eV); E $^1\Sigma_g^+$ + F $^1\Sigma_g^+$ (12.4 eV); B′ $^1\Sigma_u^+$ (13.8 eV); D $^1\Pi_u$ (14.1 eV); B″ $^1\Sigma_u^+$ (14.6 eV); and D′ $^1\Pi_u$ (14.7 eV). Also plotted are the cross sections for dissociation: $e + H_2(X\ ^1\Sigma_g^+) \rightarrow e + H(1s) + H(n = 2)$ (14.7 eV); and $e + H_2(X\ ^1\Sigma_g^+) \rightarrow e + H(1s) + H(n = 3)$ (16.6 eV).

The electron cross section set shown in Fig. 3.11 is the original set for H_2 found in the literature and from which a final set was established in Loureiro and Ferreira (1989b). This final set was obtained using the requirement that when inserted into the electron Boltzmann equation, in conditions of null-vibrational excitation (in order the superelastic collisions do not need to be taken into account), should result in calculated transport, excitation and ionization rate coefficients in agreement with available measurements. The comparison has been realised for the characteristic energy u_k (i.e. the ratio between the electron diffusion coefficient and the electron mobility, eD_e/μ_e, see Chap. 5), the drift velocity v_{ed}, the reduced first ionization coefficient $\alpha/n_o = C_{ion}/v_{ed}$, and similar reduced coefficients for total dissociation $\alpha_{diss}/n_o = \sum_j C_{0j}/v_{ed}$, and total emission of radiation $\alpha_{rad}/n_o = \sum_k C_{0k}/v_{ed}$, with j and k denoting the various states contributing for dissociation and radiative emission.

The momentum transfer and the ionization cross sections were taken from Engelhardt and Phelps (1963) and Rapp and Englander-Golden (1965). For excitation of the levels $v = 1-3$ from $v = 0$ the cross sections of Ehrhardt et al. (1968) obtained from beam experiments have been multiplied by a factor of 1.3. Multiplication by this factor was necessary in order to fit transport data and, simultaneously, the ionization rate coefficient at the lower E/n_o values. In so doing, the sum of the cross sections for the excitation of $v = 1-3$ levels is practically coincident with the total vibrational excitation cross section obtained in Engelhardt and Phelps (1963).

The total cross section for electronic excitation includes the following contributions: *(i)* excitation of the triplet states b $^3\Sigma_u^+$, c $^3\Pi_u$, a $^3\Sigma_g^+$, and e $^3\Sigma_u^+$ (note that excitation to these states leads to dissociation into H(1s) + H(1s)); *(ii)* excitation of the singlet states B $^1\Sigma_u^+$, C $^1\Pi_u$, E $^1\Sigma_g^+$, F $^1\Sigma_g^+$, B′ $^1\Sigma_u^+$, D $^1\Pi_u$, B″ $^1\Sigma_u^+$, and D′ $^1\Pi_u$; *(iii)* dissociation into H(1s) + H($n = 2$) and H(1s) + H($n = 3$).

The cross sections for the triplet states were taken from the close-coupling calculations of Chung and Lin (1978) but these had to be corrected by scaling factors in order to fit to the experimental ionization and dissociation rate coefficients. The cross section for the b $^3\Sigma_u^+$ state was raised by a factor of 1.1 whereas all the other triplet cross sections were reduced by a factor of 0.33. The resulting total cross section for the sum of the triplet states is close to that proposed by Buckman and Phelps (1985). It approximately agrees also with the total cross section for dissociation into H(1s) + H(1s) obtained by Chung et al. (1975) from the experimental dissociation cross section of Corrigan (1965).

The cross sections for the singlet states were taken from Ajello et al. (1984) multiplied by a factor of 0.6. According to Ajello et al. (1984) the excitation of the sates D, B″ and D′ leads either to radiative cascades or to dissociation into H(1s) + H($n = 2$), with branching ratios for radiation of 0.702, 0.033 and 0.421, respectively, at $u = 100$ eV. For this reason, the branching ratios of Ajello et al. (1984) have been considered here being only the radiative part of these cross sections included in group *(ii)*. The other part leading to dissociation is included in group *(iii)*. However, the errors introduced by this procedure have little effect on the results since these cross sections are small as compared with those for excitation of B and C states. Finally, the cross sections of Lavrov (1977) have been considered for dissociation into H(1s) + H($n = 2$) and H(1s) + H($n = 3$).

Figure 3.12 shows the initial total electron cross section for excitation of electronic states of H_2 found in the literature, that is the sum of curves (1) to (13) plotted in Fig. 3.11, and from which the final set used in Loureiro and Ferreira (1989b) has been constructed. The modified total excitation cross section of electronic states so

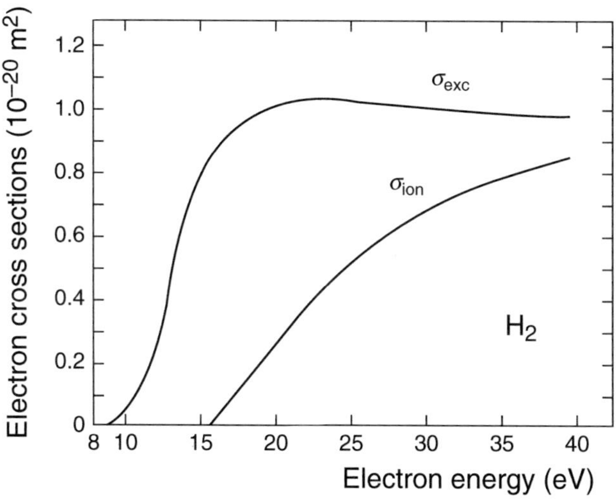

Fig. 3.12 Initial total electron cross section for excitation of the manifold of triplet and singlet states of H_2, including dissociation into H(1*s*) + H(1*s*), H(1*s*) + H($n = 2$) and H(1*s*) + H(n=3), used in Loureiro and Ferreira (1989b), together with the ionization cross section of Rapp and Englander-Golden (1965)

obtained is close to that proposed in Engelhardt and Phelps (1963). The ionization cross section of Rapp and Englander-Golden (1965) is also plotted in Fig. 3.12 for comparison.

However, this procedure of successive adjustments does not lead to a unique solution. The method permits to obtain only a reliable set of elastic and inelastic cross sections which when inserted into the Boltzmann equation results in realistic values for the electron transport parameters and the excitation, dissociation and ionization rates. For instance, Buckman and Phelps (1985) have proposed a slightly different solution: the momentum transfer cross section is larger for electron energies above 4 eV; the total vibrational excitation cross section is smaller by a factor of 1.3; the excitation cross sections for the B $^1\Sigma_u^+$ and C $^1\Pi_u$ states agree with the results of Ajello et al. (1984) and thus are larger than those obtained in Loureiro and Ferreira (1989b) by a factor of 1/0.6. The other differences between the two sets have little effect on the calculated results.

The cross sections for N_2 must also be briefly referred here since they constitute a very particular case due to the large magnitudes of the cross sections for momentum transfer and total vibrational excitation (Engelhardt et al. 1964). This latter has a very peaked maximum at ~2 eV. Figure 3.13 shows the cross sections set reported in Pitchford and Phelps (1982) and used in Loureiro and Ferreira (1986). These cross sections are assumed to result in isotropic scattering, so that the effective momentum transfer cross section σ_m^e (see discussion in Sect. 3.3.1) equals the sum of the total elastic and inelastic cross sections. The rotational excitation cross sections are plotted for one example only, σ_{46}, along with the corresponding cross

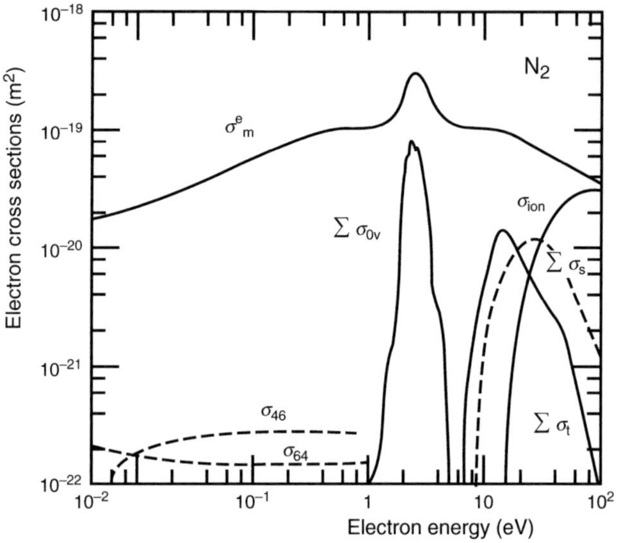

Fig. 3.13 Electron cross sections set for N_2 reported in Pitchford and Phelps (1982) as follows: σ_m^e effective cross section for momentum transfer; σ_{46} and σ_{64} example of rotational cross sections; $\sum \sigma_{0v}$ total cross section for excitation of vibrational levels; $\sum \sigma_t$ total cross section for excitation of triplet electronic states (*full*); $\sum \sigma_s$ idem for total excitation of singlet electronic states (*broken*); σ_{ion} ionization cross section

section for the reverse superelastic process, σ_{64}. The cross sections for excitation of the vibrational and electronic states are shown as sums of the individual cross sections. It is represented a total cross section for excitation of the manifold of vibrational levels $\sum_v \sigma_{0v}$, a total cross sections for excitation of the triplet electronic states (A $^3\Sigma_u^+$, B $^3\Pi_g$, W $^3\Delta_u$, B′ $^3\Sigma_u^-$, C $^3\Pi_u$, E $^3\Sigma_g^+$, F $^3\Pi_u$, G $^3\Pi_u$, etc), another for total excitation of the singlet electronic states (a′ $^1\Sigma_u^-$, a $^1\Pi_g$, w $^1\Delta_u$, a″ $^1\Sigma_g^+$, c′ $^1\Sigma_u^+$, b′ $^1\Sigma_u^+$, etc), and the ionization cross section. However, in the numerical calculations of the Boltzmann equation the cross sections have been used individually for every state (Loureiro and Ferreira 1986).

The individual transitions between the v–th levels of the electronic ground-state N_2(X $^1\Sigma_g^+$,v) and the v'–th levels of upper electronic states N_2(Y,v'), with $Y \equiv$ A $^3\Sigma_u^+$, B $^3\Pi_g$, etc, were considered in Loureiro and Ferreira (1989a) using the Franck-Condon approximation through an expression of the type

$$\sigma_{X,v}^{Y,v'}(u) \;=\; q_{X,v}^{Y,v'}\, R_e^2\, \sigma_X^Y\left(u/u_{X,v}^{Y,v'}\right)\;, \tag{3.196}$$

where $q_{X,v}^{Y,v'}$ is the Franck-Condon factor, R_e^2 is the square of the matrix element for the electronic transition, and σ_X^Y is a universal function for the X–Y transition which reflects the dependence of the cross section on the electron energy in units of the threshold energy $u_{X,v}^{Y,v'}$ (Bauer and Bartky 1965; Massey and Burhop 1969).

Nowadays, the determination of complete and consistent sets of electron cross sections is a topic of great interest in the modelling of electrical discharges leading to the appearing of a large number of publications on the subject. For example, in Hayashi (1987) the elastic and inelastic electron cross sections for the most common molecules with interest for plasma processes, gas lasers, and gaseous electronics are given. These cross sections are determined from electron beam and electron swarm data utilizing the Boltzmann equation and Monte Carlo simulation. Due to the importance of determining reliable electron cross sections sets to be used in fluid models of gas discharges, user-friendly Boltzmann equation solvers have been developed especially for this purpose, such as it is the case of the freely available solver BOLSIG+ (Hagelaar and Pitchford 2005), which is more general and easier to use than most other solvers available. Recently, an open-access website (http://www.lxcat.net) has been created, in the framework of LXcat project, with the aim of data interchanging, making possible the download of electron and ion cross sections. Finally, the studies of multiple-term solutions to the Boltzmann equation and their application to swarm systems are intrinsically coupled to the availability of the integral and differential cross sections as shown e.g. in Pitchford and Phelps (1982) and Phelps and Pitchford (1985). The generalization of the traditional two-term Legendre expansion of the Boltzmann equation in the context of electron swarm experimental analysis has been conducted in many works, e.g. in Pitchford et al. (1981) and Pitchford and Phelps (1982). However, such discussion is beyond the scope of this book.

3.6 Results from the Boltzmann Equation

The most important aspect to notice at this point is the fact that, in most cases, the EEDFs are far from Maxwellian. Obviously, the shape of the EEDFs depends on the electron cross sections set but globally the EEDFs break sharply just above the lowest threshold energy of the inelastic processes. In the case of atomic gas, where only excitation of electronic sates exists, the EEDFs are essentially Maxwellian (given by straight lines on logarithmic scale) below the lowest excitation threshold energy and they may also be considered almost Maxwellian above this energy, but with lower temperatures. Therefore, the calculation of excitation and ionization rates, even in the case of atomic gases, cannot be made with any degree of accuracy if a simple Maxwellian distribution is used. Also the approximation of a two temperature Maxwellian distribution can only be used in atomic gases where only one inelastic process dominates the shape of the EEDFs (Cherrington 1980). When the inelastic processes are more complicated, computer coding of the Boltzmann equation needs to be used, e.g. in the way described in Sect. 3.4.2.

Because of the variety and complexity of the inelastic processes occurring in molecular gases, analytic solutions to the Boltzmann equation are usually not attempted, rather computer calculations are used. In most cases the very rapid depletion of the high-energy tail of the EEDFs by inelastic collisions is very

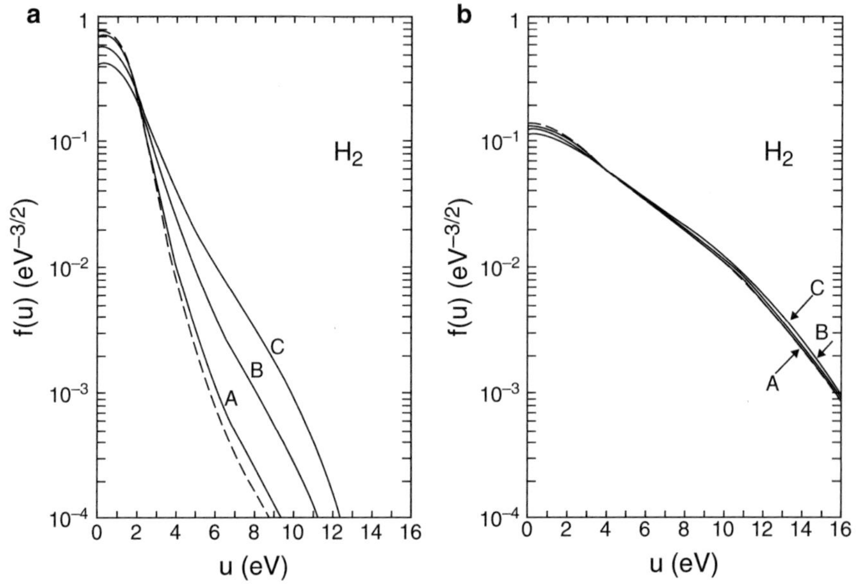

Fig. 3.14 (**a**) Electron energy distribution functions in H_2 for $E/n_o = 30\,\text{Td}$ and the following values of the vibrational temperature of $H_2(X\,^1\Sigma_g^+, v)$ distribution (Loureiro and Ferreira 1989b): (A) 2000 K; (B) 3000 K; (C) 4000 K. The *broken curve* is for $T_v = T_o = 400\,\text{K}$. (**b**) As in figure (**a**) but for $E/n_o = 100\,\text{Td}$ with the same values of T_v

evident. However, the EEDFs calculated in H_2 do not reflect the dramatic deviations relatively to Maxwellian distributions as it occurs with other molecules, such as e.g. in N_2 (see Loureiro and Ferreira 1986).

Figure 3.14a, b show the EEDFs calculated in H_2 for $E/n_o = 30$ Td (figure a) and 100 Td (b) (1 Td $= 1 \times 10^{-21}\,\text{V m}^2$), respectively, when the effects of electron superelastic collisions on the excited vibrational levels $H_2(X\,^1\Sigma_g^+, v > 0)$ are taken into account assuming the levels populated by vibrational distributions characterized by the temperatures $T_v = 2000\,\text{K}$, 3000 K, and 4000 K (see Loureiro and Ferreira 1989b). The gas temperature is assumed constant equal to $T_o = 400\,\text{K}$. In Fig. 3.14a, b the EEDFs calculated in the absence of appreciable vibrational excitation (i.e. for $T_v = T_o = 400\,\text{K}$) are also plotted for comparison. These results illustrate the effects of the electron superelastic collisions upon the excited vibrational levels on the tail of the EEDFs. The effects are significant at $E/n_o = 30\,\text{Td}$, allowing that more electrons may reach the high-energy tail of the EEDFs, but they produce only small modifications at $E/n_o = 100$ Td.

The effects of electron-vibration (e—V) superelastic collisions are dramatically reinforced in N_2 due to the large peak of the total electron cross section for vibrational excitation at $u \sim 2\,\text{eV}$ as shown in Fig. 3.13. Figure 3.15 shows the EEDFs calculated in N_2 for $E/n_o = 30\,\text{Td}$, 80 Td, and 211 Td, either for the cases of a vibrational temperature of $N_2(X\,^1\Sigma_g^+, v)$ molecules equal to $T_v = 4000\,\text{K}$, or for

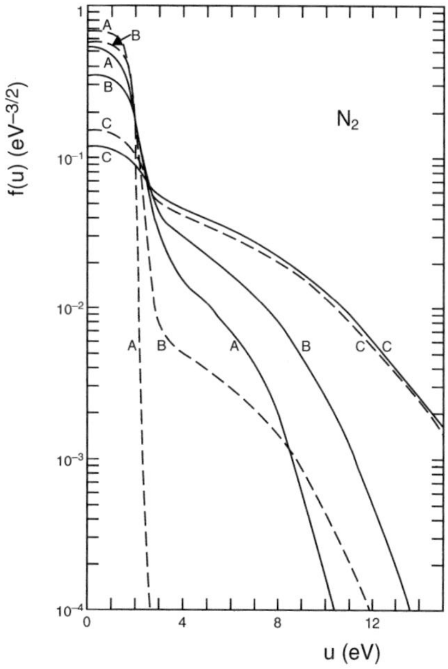

Fig. 3.15 Electron energy distribution functions in N_2 for $T_v = 4000$ K (*full curves*) and $T_v = T_o = 400$ K (*broken curves*) and the following values of E/n_o: (*A*) 30 Td; (*B*) 80 Td; (*C*) 211 Td (Loureiro and Ferreira 1986)

$T_v = T_o = 400$ K (Loureiro and Ferreira 1986). Figure 3.15 illustrates the important effects of the e–V superelastic collisions in enhancing the high-energy tail of the distributions as Nighan (1970) firstly pointed out. It can be seen that these effects become relatively small in N_2 only at sufficiently high E/n_o values, while they cause a strong increase of the high-energy tails (thus, of the electron rate coefficients) at lower E/n_o values.

Figures 3.16 and 3.17 show, as a function of E/n_o and for $T_v = 400$ K, 3000 K and 4000 K, with $T_o = 400$ K (see Loureiro and Ferreira 1989b), the electron rate coefficients for excitation of $H_2(b\ ^3\Sigma_u^+)$ state, excitation of the group of triplet states $H_2(b\ ^3\Sigma_u^+,\ c\ ^3\Pi_u,\ a\ ^3\Sigma_g^+,\ e\ ^3\Sigma_u^+)$ which dissociate to yield two H(1s) atoms, and dissociation into $H(1s) + H(n = 2)$, in the case of Fig. 3.16, and for excitation of the radiative singlet states $H_2(B\ ^1\Sigma_u^+)$ and $H_2(C\ ^1\Pi_u)$, in the case of Fig. 3.17. An increase in T_v produces an increase of all the rates, especially at lower E/n_o values, as a result of the enhancement of the high-energy tail of the EEDFs.

The fractional power transferred by the electrons to H_2 molecules through the various collisional mechanisms is presented in Fig. 3.18, as a function of E/n_o and for $T_v = 4000$ K and $T_v = T_o = 400$ K. The various curves represent the power lost by the electrons through elastic collisions, rotational excitation, vibrational excitation (the net power loss corresponding to the difference between inelastic and superelastic collisions), dissociation, excitation of radiative states, and ionization. In the presence of vibrationally excited molecules the superelastic e–V collisions cause a significant decrease in the net power spent in vibrational excitation and, consequently, an increase in the power transferred to electronic excited states. As seen from Fig. 3.18, at $T_v = 4000$ K dissociation by electron impact constitutes the major energy loss channel among all the electronic mechanisms in the range $E/n_o = 50 - 300$ Td.

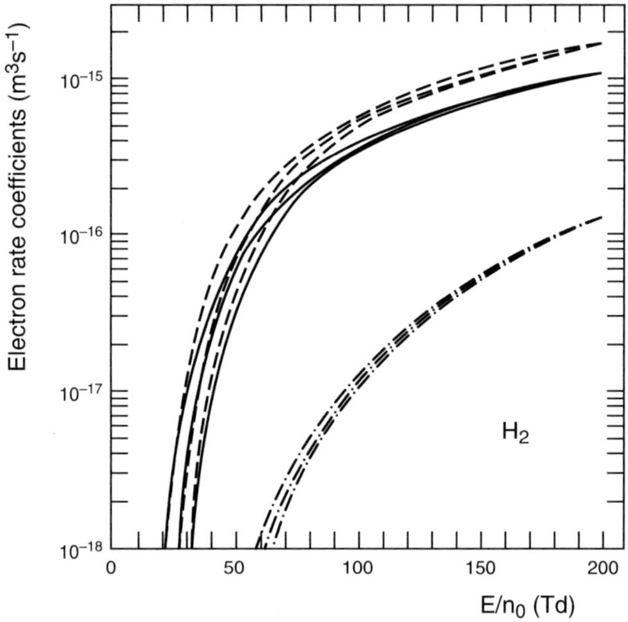

Fig. 3.16 Electron rate coefficients, against E/n_o, for excitation of $H_2(b^3\Sigma_u^+)$ state (*full curves*), total dissociation into H(1s) + H(1s) (*broken curves*), and dissociation into H(1*s*) + H(n = 2) (*chain curves*). In each case, the lower, intermediate and upper curves correspond to the values T_v = 400 K, 3000 K, 4000 K, respectively (Loureiro and Ferreira 1989b)

Appendices

A.3.1 Expansion of the Boltzmann Equation in Spherical Harmonics

In this appendix we present the derivation of the complete expansion of the Boltzmann equation (3.35) in spherical harmonics (Cherrington 1980; Delcroix 1963, 1966). The electron velocity distribution function is expressed in terms of a sum of spherical harmonics (3.9)

$$f_e(\mathbf{r}, \mathbf{v_e}, t) = \sum_{l=0}^{\infty} f_e^l(\mathbf{r}, v_e, t)\, P_l(\cos\theta)\,, \tag{3.197}$$

where θ is the angle between the instantaneous velocity vector $\mathbf{v_e}$ and the direction of the anisotropy created by the applied electric field $\mathbf{E} = -E\,\mathbf{e_z}$, $P_l(\cos\theta)$ are the Legendre polynomials, and $f_e^l(\mathbf{r}, v_e, t)$ depends only on the absolute value of the velocity. The Boltzmann equation (3.35) writes as

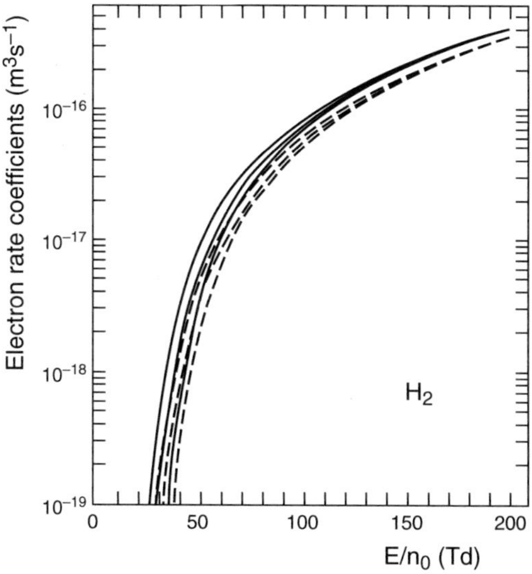

Fig. 3.17 As in Fig. 3.16 but for excitation of the singlet states $H_2(B\ ^1\Sigma_u^+)$ (*full curves*) and $H_2(C\ ^1\Pi_u)$ (*broken curves*) (Loureiro and Ferreira 1989b)

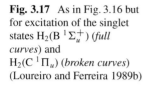

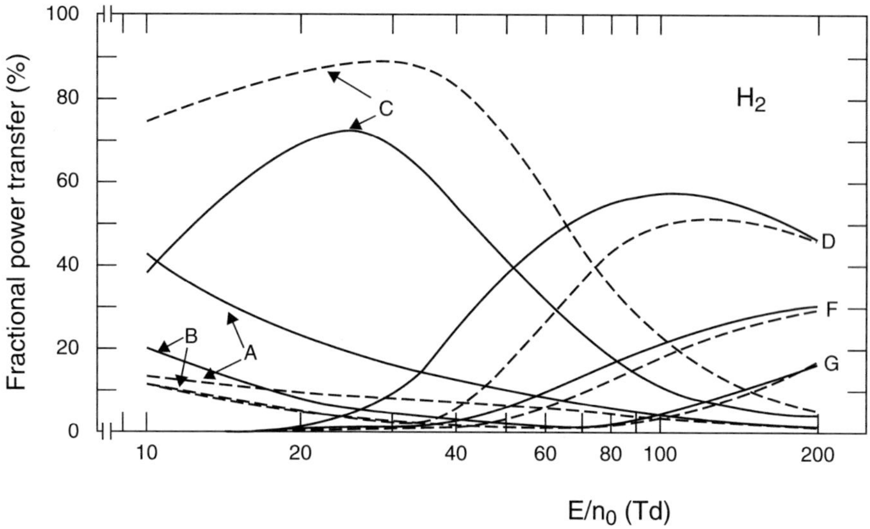

Fig. 3.18 Fractional power transferred by the electrons to H_2 molecules, against E/n_o and for $T_v = 4000$ K (*full curves*) and $T_v = T_o = 400$ K (*broken curves*), associated with the following mechanisms: (*A*) elastic collisions; (*B*) rotational excitation; (*C*) vibrational excitation; (*D*) dissociation; (*F*) excitation of radiative states; (*G*) ionization (Loureiro and Ferreira 1989b)

$$\sum_l \frac{\partial f_e^l}{\partial t} P_l + \frac{e}{m}\left(E\,\mathbf{e_z}\cdot\frac{\partial}{\partial \mathbf{v_e}}\left(\sum_l f_e^l P_l\right)\right) = \sum_l I^l(f_e^l)\,P_l\,, \tag{3.198}$$

using an expansion similar to (3.197) for the collision term.

Since the electric field has only the component in z, we obtain

$$\frac{\partial}{\partial v_{ez}}\left(\sum_l f_e^l\, P_l\right) = \sum_l \left(\frac{\partial f_e^l}{\partial v_{ez}}\, P_l + f_e^l\, \frac{\partial P_l}{\partial v_{ez}}\right) , \tag{3.199}$$

with

$$\frac{\partial f_e^l}{\partial v_{ez}} = \frac{\partial f_e^l}{\partial v_e}\, \frac{\partial v_e}{\partial v_{ez}} \tag{3.200}$$

and

$$\frac{\partial P_l}{\partial v_{ez}} = \frac{dP_l}{d(\cos\theta)}\, \frac{\partial(\cos\theta)}{\partial v_{ez}} . \tag{3.201}$$

Using the relations

$$v_e^2 = v_{ex}^2 + v_{ey}^2 + v_{ez}^2 \tag{3.202}$$

$$v_{ez} = v_e\, \cos\theta \tag{3.203}$$

$$\frac{\partial v_e}{\partial v_{ez}} = \frac{v_{ez}}{v_e} = \cos\theta \tag{3.204}$$

$$\frac{\partial(\cos\theta)}{\partial v_{ez}} = \frac{\partial}{\partial v_{ez}}\left(\frac{v_{ez}}{v_e}\right) = \frac{v_e - v_{ez}\, v_{ez}/v_e}{v_e^2} = \frac{1}{v_e}\left(1-\cos^2\theta\right), \tag{3.205}$$

we obtain for the right-hand side terms of equation (3.199)

$$\frac{\partial f_e^l}{\partial v_{ez}}\, P_l = \frac{\partial f_e^l}{\partial v_e}\, \cos\theta\, P_l \tag{3.206}$$

$$f_e^l\, \frac{\partial P_l}{\partial v_{ez}} = f_e^l\, \frac{dP_l}{d(\cos\theta)}\, \frac{(1-\cos^2\theta)}{v_e} . \tag{3.207}$$

Making use now of the recursion relation for the Legendre polynomials

$$(l+1)\, P_{l+1} = (2l+1)\, \cos\theta\, P_l - l\, P_{l-1} , \tag{3.208}$$

we obtain

$$\cos\theta\, P_l = \frac{(l+1)\, P_{l+1} + l\, P_{l-1}}{2l+1} \tag{3.209}$$

and equation (3.206) writes as follows

$$\frac{\partial f_e^l}{\partial v_{ez}}\, P_l \;=\; \frac{\partial f_e^l}{\partial v_e}\; \frac{(l+1)\, P_{l+1} + l\, P_{l-1}}{2l+1}\;. \qquad (3.210)$$

On the other hand, using the recursion relation

$$(\cos^2\theta - 1)\, \frac{dP_l}{d(\cos\theta)} \;=\; l\, \cos\theta\, P_l \;-\; l\, P_{l-1} \qquad (3.211)$$

which may be written as follows using (3.209)

$$(\cos^2\theta - 1)\, \frac{dP_l}{d(\cos\theta)} \;=\; \frac{l\,(l+1)}{2l+1}\, (P_{l+1} - P_{l-1})\;, \qquad (3.212)$$

we obtain for equation (3.207)

$$f_e^l\, \frac{\partial P_l}{\partial v_{ez}} \;=\; \frac{f_e^l}{v_e}\; \frac{l\,(l+1)}{2l+1}\, (P_{l-1} - P_{l+1})\;. \qquad (3.213)$$

Inserting now equations (3.199), (3.210) and (3.213) into equation (3.198), we find

$$\sum_l \frac{\partial f_e^l}{\partial t}\, P_l \;+\; \frac{eE}{m} \sum_l \left[\frac{l+1}{2l+1} \left(\frac{\partial f_e^l}{\partial v_e} \;-\; l\, \frac{f_e^l}{v_e} \right) P_{l+1} \right.$$
$$\left. +\; \frac{l}{2l+1} \left(\frac{\partial f_e^l}{\partial v_e} \;+\; (l+1)\, \frac{f_e^l}{v_e} \right) P_{l-1} \right] \;=\; \sum_l I^l(f_e^l)\, P_l\;. \quad (3.214)$$

The second term on the left-hand side member may be rewritten as follows

$$\sum_l \frac{\partial f_e^l}{\partial t}\, P_l \;+\; \frac{eE}{m} \sum_l \left[\frac{l}{2l-1} \left(\frac{\partial f_e^{l-1}}{\partial v_e} \;-\; (l-1)\, \frac{f_e^{l-1}}{v_e} \right) \right.$$
$$\left. +\; \frac{l+1}{2l+3} \left(\frac{\partial f_e^{l+1}}{\partial v_e} \;+\; (l+2)\, \frac{f_e^{l+1}}{v_e} \right) \right] P_l \;=\; \sum_l I^l(f_e^l)\, P_l\;. \quad (3.215)$$

Then, we obtain an infinite set of coupled equations, each one with a finite number of terms, in which the equation of order l is

$$\frac{\partial f_e^l}{\partial t} \;+\; \frac{eE}{m} \left[\frac{l}{2l-1} \left(\frac{\partial f_e^{l-1}}{\partial v_e} - (l-1)\, \frac{f_e^{l-1}}{v_e} \right) + \frac{l+1}{2l+3} \left(\frac{\partial f_e^{l+1}}{\partial v_e} + (l+2)\, \frac{f_e^{l+1}}{v_e} \right) \right]$$
$$=\; I^l(f_e^l)\;. \qquad (3.216)$$

Using now the relations

$$v_e^{l-1}\frac{\partial}{\partial v_e}\left(\frac{f_e^{l-1}}{v_e^{l-1}}\right) = \frac{\partial f_e^{l-1}}{\partial v_e} - (l-1)\,\frac{f_e^{l-1}}{v_e} \tag{3.217}$$

$$\frac{1}{v_e^{l+2}}\frac{\partial}{\partial v_e}\left(v_e^{l+2} f_e^{l+1}\right) = \frac{\partial f_e^{l+1}}{\partial v_e} + (l+2)\,\frac{f_e^{l+1}}{v_e}\,, \tag{3.218}$$

we obtain the final expression

$$\frac{\partial f_e^l}{\partial t} + \frac{eE}{m}\left[\frac{l}{2l-1}\,v_e^{l-1}\,\frac{\partial}{\partial v_e}\left(\frac{f_e^{l-1}}{v_e^{l-1}}\right) + \frac{l+1}{2l+3}\,\frac{1}{v_e^{l+2}}\,\frac{\partial}{\partial v_e}\left(v_e^{l+2} f_e^{l+1}\right)\right] = I^l(f_e^l). \tag{3.219}$$

From this we find the first order equations expressed through equations (3.41), (3.43) and (3.44).

Exercises

Exercise 3.1. In a given gas the effective collision frequency for momentum transfer $\nu_m^e(u)$ is independent of the energy of electrons. Write the expressions in units for the reduced mobility and the power gain from the field per electron at unit gas density.

Resolution: The electron conductivity is given by equation (3.53) as only elastic collisions are taken into account with a collision frequency $\nu_m(u) = \text{const}$, so that as both elastic and inelastic collisions are considered an effective collision frequency for momentum transfer (3.139) must be considered and the electron mobility is $\mu_e = \sigma_{ce}/(en_e) = e/(m\nu_m^e)$. Since we may write

$$n_o\mu_e = \frac{e}{m\,(\nu_m^e/n_o)}\,,$$

the reduced mobility is independent of the gas number density n_o. Expressing this formula with the corresponding SI units, we obtain

$$n_o\mu_e\;[10^{24}\ \mathrm{V^{-1}\ m^{-1}\ s^{-1}}] = \frac{0.1759}{\nu_m^e/n_o\;[10^{-12}\ \mathrm{m^3\ s^{-1}}]}\,.$$

On the other hand, the power gain from the field per electron at unit gas density is from (3.116) given by

$$\frac{P_E}{n_e n_o} = e\,(n_o\mu_e)\left(\frac{E}{n_o}\right)^2,$$

so that we obtain

$$\frac{P_E}{n_e n_o}\,[10^{-16}\ \mathrm{eV\ m^3\ s^{-1}}] \;=\; n_o\mu_e\,[10^{24}\ \mathrm{V^{-1}m^{-1}s^{-1}}]\,\left(\frac{E}{n_o}\,[10\ \mathrm{Td}]\right)^2,$$

with 1 Td (Townsend) = 1×10^{-21} V m^2.

Exercise 3.2. Write the expressions in units for the power lost in elastic collisions and the power lost in rotational excitation, per electron at unit gas density, in the case of a gas of homonuclear diatomic molecules, with the collision frequency for momentum transfer independent of the energy, and an electron energy distribution function Maxwellian at temperature T_e.

Resolution: The power lost in elastic collisions, per electron at unit gas density, in the case of a collision frequency for momentum transfer $\nu_m(u)$ = const and a Maxwellian EEDF is from (3.117) given by

$$\frac{P_{el}}{n_e n_o} \;=\; \frac{3m}{M}\,k_B T_e\,\frac{\nu_m}{n_o},$$

so that we obtain

$$\frac{P_{el}}{n_e n_o}\,[10^{-16}\ \mathrm{eV\ m^3\ s^{-1}}] \;=\; \frac{3m}{M} \times 10^4\, k_B T_e\,[\mathrm{eV}]\,\frac{\nu_m}{n_o}\,[10^{-12}\ \mathrm{m^3\ s^{-1}}].$$

On the other hand, the power lost in rotational excitation is from (3.161) given by

$$\frac{P_{rot}}{n_e n_o} \;=\; 4B_o\,\sigma_o\,\sqrt{\frac{8k_B T_e}{\pi m}},$$

so that we find

$$\frac{P_{rot}}{n_e n_o}\,[10^{-16}\ \mathrm{eV\ m^3\ s^{-1}}] \;=\; 2.677 \times 10^2\, B_o\,[\mathrm{eV}]\,\sigma_o\,[10^{-20}\ \mathrm{m^2}]\,\sqrt{k_B T_e\,[\mathrm{eV}]},$$

with

$$\sigma_o\,[10^{-20}\ \mathrm{m^2}] \;=\; 0.4692\, q^2$$

and where q is the quadrupole moment of the homonuclear diatomic molecule in units of ea_o^2, with a_o denoting the Bohr radius.

Exercise 3.3. Obtain the expression in units for the electron rate coefficient of a given inelastic process of cross section σ_{ij} and threshold energy u_{ij}.

Resolution: The expression (3.172) writes as follows

$$C_{ij}\,[10^{-16}\ \mathrm{m^3\ s^{-1}}] \;=\; 59.309\int_{u_{ij}}^{\infty} u\,\sigma_{ij}\,[10^{-20}\ \mathrm{m^2}]\,f(u)\,du\,,$$

with u in eV and $f(u)$ in eV$^{-3/2}$ according to the normalization (3.166).

Exercise 3.4. An inelastic process for excitation of a given state j from state i with the threshold energy u_{ij} has a constant cross section $\sigma_{ij} = a_{ij}$ (see Exercise 2.1 *(a)* at the end of Chap. 2). Determine the expressions in units for the electron rate coefficient and the power lost, per electron at unit gas density, through this process.

Resolution: This electron rate coefficient has been obtained in Exercise 2.1 *(a)* and it presents the following expression in units

$$C_{ij}\,[10^{-16}\,\mathrm{m}^3\,\mathrm{s}^{-1}] = 66.923\,\sqrt{k_BT_e\,[\mathrm{eV}]}\,a_{ij}\,[10^{-20}\,\mathrm{m}^2]\left(1+\frac{u_{ij}}{k_BT_e}\right)\exp\left(-\frac{u_{ij}}{k_BT_e}\right),$$

so that the corresponding power lost is

$$\frac{P_{inel}}{n_e n_o}\,[10^{-16}\,\mathrm{eV}\,\mathrm{m}^3\mathrm{s}^{-1}] = u_{ij}\,[\mathrm{eV}]\,C_{ij}\,[10^{-16}\,\mathrm{m}^3\,\mathrm{s}^{-1}]\,.$$

Exercise 3.5. Determine the expression of the power absorbed from the field, per electron at unit gas density, for the case of an electric field of amplitude E and a Maxwellian EEDF at temperature T_e. Specify for the two situations: *(i)* the effective collision frequency for momentum transfer, ν_m^e, is independent of the velocity of electrons: *(ii)* the effective collision cross section for momentum transfer, σ_m^e, is independent of the velocity of electrons.

Resolution: Since in a Maxwellian distribution $df_e^0/dv_e \propto (-\,mv_e/(k_BT_e))\,f_e^0$, we obtain the following expression for the electron conductivity from equation (3.50)

$$\sigma_{ce} = \frac{e^2 n_e}{3k_BT_e} < \frac{v_e^2}{\nu_m^e} > ,$$

so that the power absorbed from the field, per electron and at unit gas density, is

$$\frac{P_E}{n_e n_o} = \frac{e^2}{3k_BT_e} < \frac{v_e^2}{\nu_m^e/n_o} > \left(\frac{E}{n_o}\right)^2 .$$

When $\nu_m^e(v_e) = \mathrm{const}$, we obtain

$$\frac{P_E}{n_e n_o} = \frac{e^2}{m\,(\nu_m^e/n_o)}\left(\frac{E}{n_o}\right)^2 ,$$

while when $\sigma_m^e(v_e) = \mathrm{const}$, we have

$$\frac{P_E}{n_e n_o} = \frac{2e^2}{3\sigma_m^e}\,\frac{\sqrt{2}}{\sqrt{\pi m\,k_BT_e}}\left(\frac{E}{n_o}\right)^2 .$$

Exercise 3.6. Determine the expression of the velocity-averaged effective collision frequency for momentum transfer that allows to keep the electron mobility as $\mu_e = e/(m\,\nu_m')$, when the microscopic effective collision frequency depends on the electron velocity: $\nu_m^e(v_e)$.

Resolution: From equation (3.52), we obtain the following expression for the required velocity-averaged collision frequency

$$\frac{1}{\nu'_m} = < \frac{1}{v_e^2} \frac{d}{dv_e} \left(\frac{v_e^3}{3\,\nu_m^e(v_e)} \right) > .$$

Integrating by parts and using the normalization condition (3.166), this expression writes as follows in accordance with (3.170)

$$\frac{1}{\nu'_m} = -\frac{2}{3} \int_0^\infty \frac{u^{3/2}}{\nu_m^e(u)} \frac{df}{du}\, du .$$

In a gas with $\nu_m^e(v_e) \neq$ const, the present expression for ν'_m should be used in writing $\mu_e = e/(m\,\nu'_m)$.

Exercise 3.7. Write the Boltzmann equation under the form expressed by equation (3.144), in which the Boltzmann equation is given by the variation of gains in velocity space, using the EEDF normalized through equation (3.166).

Resolution: Using the energy of electrons $u = \frac{1}{2}\,mv_e^2$ as independent variable, the equation (3.144) transforms to

$$\frac{d}{du}(G_E + G_{el}) = \frac{4\pi}{m}\sqrt{\frac{2u}{m}}\, J^0 ,$$

where G_E is the positive gain associated with the upflux produced by the electric field in velocity space (3.92) and (3.95)

$$G_E = -\,n_e\, \frac{2}{3} \frac{u^{3/2}}{\nu_m^e} \frac{(eE)^2}{m} \frac{df}{du}$$

and G_{el} is the negative gain associated with the downflux produced by the elastic collisions (3.93) and (3.95)

$$G_{el} = -\,n_e\, \frac{2m}{M}\, u^{3/2} \nu_m \left(f + k_B T_o \frac{df}{du} \right) .$$

On the other hand, the term on the right-hand side member taking into account the introduction and removal of electrons in velocity space by inelastic and superelastic collisions (3.134) writes as

$$\frac{4\pi}{m}\sqrt{\frac{2u}{m}}\, J^0 = n_e \sum_{i,j} \left\{ \sqrt{u + u_{ij}}\; \nu_{ij}(u + u_{ij})\, f(u + u_{ij}) - \sqrt{u}\; \nu_{ij}(u)\, f(u) \right\}$$
$$+ n_e \sum_{j,i} \left\{ \sqrt{u - u_{ij}}\; \nu_{ji}(u - u_{ij})\, f(u - u_{ij}) - \sqrt{u}\; \nu_{ji}(u)\, f(u) \right\} .$$

Exercise 3.8. Write the expression equivalent to those of Exercise 3.7 for the negative gain due to excitation and de-excitation of rotational levels (3.160) using the continuous approximation (3.158).

Resolution: As in the previous exercise, we obtain

$$G_{rot} \;=\; -\, n_e\, n_o\, 4B_0\, \sigma_0 \sqrt{\frac{2}{m}}\, u f\,,$$

using the EEDF normalized through condition (3.166).

Exercise 3.9. Determine the condition at which the effective cross section for momentum transfer should obey in order the electron mobility may be negative.

Resolution: By inspection of equation (3.171) for the reduced electron mobility

$$n_o\, \mu_e \;=\; -\,\frac{e}{3}\sqrt{\frac{2}{m}} \int_0^{\infty} \frac{u}{\sigma_m^e(u)}\, \frac{df}{du}\, du$$

in which $f(u)$, with $u = \frac{1}{2}\, m v_e^2$, is the EEDF normalized through equation (3.166), we find that it exists the possibility of the electron mobility to be negative if $df/du > 0$ in a certain energy range. Moreover, integrating the above equation by parts we obtain

$$n_o\, \mu_e \;=\; \frac{e}{3}\sqrt{\frac{2}{m}} \int_0^{\infty} \frac{d}{du}\left(\frac{u}{\sigma_m^e(u)}\right) f\, du\,,$$

so that another conditions is

$$\frac{d}{du}\left(\frac{u}{\sigma_m^e(u)}\right) \;<\; 0,$$

which means that $\sigma_m^e(u)$ should be a super-linear function of energy in a certain energy range. This latter condition can also be expressed as (Dyatko 2007)

$$\frac{d(\ln(\sigma_m^e(u)))}{d(\ln u)} \;>\; 1\,.$$

The fulfillment of these two conditions is necessary but not sufficient for the electron mobility to be negative. For each case, the integrals should result negative.

The condition $df/du > 0$, known in the literature as inverse EEDF, can form in quite different situations such as during the EEDF relaxation in a plasma of heavy rare gas (Ar, Kr, Xe); in steady or decaying plasmas of heavy rare gas with admixture of electronegative gases (due to the elimination of low energy electrons by attachment processes); or in optically excited plasmas of heavy rare gases with admixture of metal atoms (due to superelastic collisions of electrons with excited metal atoms). On the other hand, the condition $d(u/\sigma_m^e(u))/du < 0$ is satisfied in the case of the electron cross sections for momentum transfer of Ar, Kr, and Xe in the energy range above the Ramsauer minimum ($\sim$ 0.2, 0.5 and 0.6 eV, respectively). Experimentally the negative mobility has only been observed in a transient Xe plasma (Warman et al. 1985).

References

J.M. Ajello, D. Shemansky, T.L. Kwok, Y.L. Yung, Studies of extreme-ultraviolet emission from Rydberg series of H_2 by electron impact. Phys. Rev. A **29**(2), 636–653 (1984)

W.P. Allis, Motions of ions and electrons, in *Handbuch der Physik*, vol. 21, ed. by S. Flügge (Springer, Berlin, 1956), pp. 383–444

W.P. Allis, H.A. Haus, Electron distributions in gas lasers. J. Appl. Phys. **45**(2), 781–791 (1974)

E. Bauer, C.D. Bartky, Calculation of inelastic electron-molecule collision cross sections by classical methods. J. Chem. Phys. **43**(7), 2466–2476 (1965)

S.J. Buckman, A.V. Phelps, Vibrational excitation of D_2 by low energy electrons. J. Chem. Phys. **82**(11), 4999–5011 (1985)

S. Chapman, T.G. Cowling, *The Mathematical Theory of Non-Uniform Gases* (Cambridge University Press, Cambridge, 1939)

B.E. Cherrington, *Gaseous Electronics and Gas Lasers* (Pergamon Press, Oxford, 1980)

S. Chung, C.C. Lin, Application of the close-coupling method to excitation of electronic states and dissociation of H_2 by electron impact. Phys. Rev. A **17**(6), 1874–1891 (1978)

S. Chung, C.C. Lin, E.T.P. Lee, Dissociation of the hydrogen molecule by electron impact. Phys. Rev. A **12**(4), 1340–1349 (1975)

S.J.B. Corrigan, Dissociation of molecular hydrogen by electron impact. J. Chem. Phys. **43**(12), 4381–4386 (1965)

J.-L. Delcroix, *Physique des Plasmas: Volumes 1 and 2*. Monographies Dunod (Dunod, Paris, 1963, 1966) (in French)

J.-L. Delcroix, A. Bers, *Physique des Plasmas: Volumes 1 and 2*. Savoirs Actuels (InterÉditions/CNRS Éditions, Paris, 1994) (in French)

M.J. Druyvesteyn, F.M. Penning, The mechanism of electrical discharges in gases of low pressure. Rev. Mod. Phys. **12**(2), 87–174 (1940)

N.A. Dyatko, Negative electron conductivity in gases and semiconductors. J. Phys.: Conf. Ser. **71**, 012005 (2007)

H. Ehrhardt, L. Langhans, F. Linder, H.S. Taylor, Resonance scattering of slow electrons from H_2 and CO angular distributions. Phys. Rev. **173**(1), 222–230 (1968)

A.G. Engelhardt, A.V. Phelps, Elastic and inelastic cross sections in hidrogen and deuterium from transport coefficients. Phys. Rev. **131**(5), 2115–2128 (1963)

A.G. Engelhardt, A.V. Phelps, Transport coefficients and cross sections in argon and hydrogen-argon mixtures. Phys. Rev. **133**(2A), A375–A380 (1964)

A.G. Engelhardt, A.V. Phelps, C.G. Risk, Determination of momentum transfer and inelastic collision cross sections for electrons in nitrogen using transport coefficients. Phys. Rev. **135**(6A), A1566– A1574 (1964)

C.M. Ferreira, J. Loureiro, Electron transport parameters and excitation rates in argon. J. Phys. D: Appl. Phys. **16**, 1611–1621 (1983a)

C.M. Ferreira, J. Loureiro, Electron energy distributions and excitation rates in high-frequency argon discharges. J. Phys. D: Appl. Phys. **16**, 2471–2483 (1983b)

C.M. Ferreira, J. Loureiro, A. Ricard, Populations in the metastable and the resonance levels of argon and setpwise ionization effects in a low-pressure argon positive column. J. Appl. Phys. **57**(1), 82–90 (1985)

L.S. Frost, A.V. Phelps, Rotational excitation and momentum transfer cross sections for electrons in H_2 and N_2 from transport coefficients. Phys. Rev. **127**(5), 1621–1633 (1962)
L.S. Frost, A.V. Phelps, Momentum-transfer cross sections for slow electrons in He, Ar, Kr, and Xe from transport coefficients. Phys. Rev. **136**(6A), A1538–A1545 (1964)
E. Gerjuoy, S. Stein, Rotational excitation by slow electrons. Phys. Rev. **97**(6), 1671–1679 (1955)
A.L. Gilardini, *Low Energy Electron Collisions in Gases: Swarm and Plasma Methods Applied to Their Study* (Wiley, New York, 1972)
G.J.M. Hagelaar, L.C. Pitchford, Solving the Boltzmann equation to obtain electron transport coefficients and rate coefficients for fluid models. Plasma Sources Sci. Technol. **14**, 722–733 (2005)
M. Hayashi, Electron collision cross-sections for molecules determined from beam and swarm data, in *Swarm Studies and Inelastic Electron-Molecule Collisions*. ed. by L.C. Pitchford, B.V. McKoy, A. Chutjian, S. Trajmar (Springer, New York, 1987), pp. 167–187
K.P. Huber, G. Herzberg, Molecular Spectra and Molecular Structure: Vol. IV – Constants of Diatomic Molecules (Van Nostrand Reinhold, New York, 1979)
B.P. Lavrov, Role of dissociative excitation in populating the atomic levels in a hydrogen plasma with a small degree of dissociation. Opt. Spectrosc. **42**(3), 250–253 (1977)
A. Lofthus, P.H. Krupenie, The spectrum of molecular nitrogen. J. Phys. Chem. Ref. Data **6**(1), 113–307 (1977)
J. Loureiro, C.M. Ferreira, Coupled electron energy and vibrational distribution functions in stationary N_2 discharges. J. Phys. D: Appl. Phys. **19**, 17–35 (1986)
J. Loureiro, C.M. Ferreira, Electron excitation rates and transport parameters in direct-current N_2 discharges. J. Phys. D: Appl. Phys. **22**, 67–75 (1989a)
J. Loureiro, C.M. Ferreira, Electron and vibrational kinetics in the hydrogen positive column. J. Phys. D: Appl. Phys. **22**, 1680–1691 (1989b)
H. Margenau, Conduction and dispersion of ionized gases at high frequencies. Phys. Rev. **69** (9–10), 508–513 (1946)
H.S.W. Massey, E.H.S. Burhop, *Electronic and Ionic Impact Phenomena: Volume 1 – Collisions of Electrons with Atoms* (Oxford at the Clarendon Press, Oxford, 1969)
W.L. Nighan, Electron energy distributions and collision rates in electrically excited N_2, CO, and CO_2. Phys. Rev. A **2**(5), 1989–2000 (1970)
J.L. Pack, A.V. Phelps, Drift velocities of slow electrons in helium, neon, argon, hydrogen and nitrogen. Phys. Rev. **121**(3), 798–806 (1961)
A.V. Phelps, L.C. Pitchford, Anisotropic scattering of electrons by N_2 and its effect on electron transport. Phys. Rev. A **31**(5), 2932–2949 (1985)
L.C. Pitchford, S.V. O'Neil, J.R. Rumble Jr., Extended Boltzmann analysis of electron swarm experiments. Phys. Rev. A **23**(1), 294–304 (1981)
L.C. Pitchford, A.V. Phelps, Comparative calculation of electron-swarm properties in N_2 at moderate E/N values. Phys. Rev. A **25**(1), 540–554 (1982)
D. Rapp, P. Englander-Golden, Total cross sections for ionization and attachment in gases by electron impact. I. Positive ionization. J. Chem. Phys. **43**(5), 1464–1479 (1965)
M.A. Ridenti, L. L. Alves, V. Guerra, J. Amorim, The role of rotational mechanisms in electron swarm parameters at low reduced electric field in N_2, O_2 and H_2. Plasma Sources Sci. Technol. **24**, 035002 (2015)
S.D. Rockwood, Elastic and inelastic cross sections for electron-Hg scattering from Hg transport data. Phys. Rev. A **8**(5), 2348–2358 (1973)
J.M. Warman, U. Sowada, M.P. De Haas, Transient negative mobility of hot electrons in gaseous xenon. Phys. Rev. A **31**(3), 1974–1976 (1985)
A. von Engel, *Ionized Gases* 2nd edn. (Clarendon Press, Oxford, 1965)

Chapter 4
Boltzmann Equation with Time-Varying Fields

This chapter analyses the effects produced by a time-varying electric field in the electron kinetics. The behaviour exhibited by the electron velocity distribution function is controlled by two characteristic relaxation frequencies, one for energy and another for momentum transfer, when compared with the field frequency. The cases of high-frequency (HF) and radio-frequency (RF) fields are analysed separately, since they correspond to situations in which no time-modulation and large time-modulation exist, respectively, in the isotropic part of the electron velocity distribution. This chapter also analyses the electron kinetics under the simultaneous effects of a HF electric field and a stationary external magnetic field, with leads to electron cyclotron resonance (ECR) when the electron cyclotron frequency equals the field-frequency.

4.1 High-Frequency Electric Fields

4.1.1 Characteristic Frequencies

The situation to be analysed here is that of electron kinetics when a uniform time-varying electric field $E(t) = E_0 \cos(\omega t)$ is applied to a plasma. A qualitative analysis of the time-dependence of the isotropic and the anisotropic components of the electron velocity distribution function under the effects of time-varying fields can be realised using the characteristic relaxation frequencies for energy and momentum transfer, ν_e and ν_m^e respectively, previously introduced in Chap. 3 through equations (3.142) and (3.139)

J.M.A.H. Loureiro, J. de Amorim Filho, *Kinetics and Spectroscopy of Low Temperature Plasmas*, Graduate Texts in Physics,
DOI 10.1007/978-3-319-09253-9_4

$$\nu_e(u) = \frac{2m}{M}\, \nu_m(u) + \sum_j \nu_j(u) \tag{4.1}$$

$$\nu_m^e(u) = \nu_m(u) + \sum_j \nu_j^m(u), \tag{4.2}$$

where ν_m and $\sum_j \nu_j^m$ are the elastic and the inelastic electron collision frequencies for momentum transfer, the latter assumed here with transitions from the ground-state only, and in which the frequencies ν_j^m equal the total frequencies ν_j in the case of isotropic scattering (see discussion in Sect. 3.1.2). Further, m/M is the electron-molecule mass ratio and $u = \frac{1}{2}\, mv_e^2$ is the electron energy. The factor 2 in the first term of equation (4.1) holds as the Chapman and Cowling expression (3.84) is written in terms of the electron energy.

As we will show below, it is possible to distinguish the following situations in this analysis (see e.g. Loureiro 1993):

(i) At low field frequencies such that $\omega \ll \nu_e$, that is at radio-frequency (RF) fields, the isotropic component of the electron velocity distribution, i.e. the electron energy distribution function (EEDF), follows in a quasistationary way the RF field, presenting consequently a very large time-modulation in those parts of the relevant range of the electron energy where the above inequality is satisfied. Therefore, in this low-frequency limit, the EEDF can be obtained by solving the Boltzmann equation for a direct-current (DC) field for each time-varying value of the instantaneous RF field strength;

(ii) When the field frequency increases up to $\nu_e \simeq \omega \ll \nu_m^e$, in most parts of the relevant electron-energy range, the time-modulation of the EEDF is significantly reduced and a time-resolved solution of the Boltzmann equation is required instead of a quasi-stationary one. However, in this range of ω values, since the inequality $\omega \ll \nu_m^e$ holds in most of the significant electron-energy range, the anisotropic component of the electron velocity distribution is not significantly modified by the field frequency and, consequently, the magnitude of the EEDF is only slightly dependent on the field frequency;

(iii) Finally, for higher values of ω, such that $\nu_e \ll \nu_m^e \simeq \omega$, the inequality $\nu_e \ll \omega$ determines that no time-modulation of the EEDF can occur. However, the proximity of frequencies $\omega \simeq \nu_m^e$ produces a time-delay of the anisotropic component of the velocity distribution relatively to the field, which reaches $\phi = -\pi/2$ when $\omega \gg \nu_m^e$, producing as we will show below a strong reduction in the magnitude of the EEDF.

It follows from the present discussion that we can expect an important time-modulation of the EEDF for angular field frequencies $\omega < \nu_e$, in the whole significant electron-energy range. In an atomic gas, at field frequencies not too low ($\omega > 10^8\, s^{-1}$) and at gas pressures typical of RF plasma processing (typically $p \sim 100\, Pa$), the inequality $\omega < \nu_e$ holds only in the high-energy tail of the EEDF, so that there is no time-modulation on the body of the distribution.

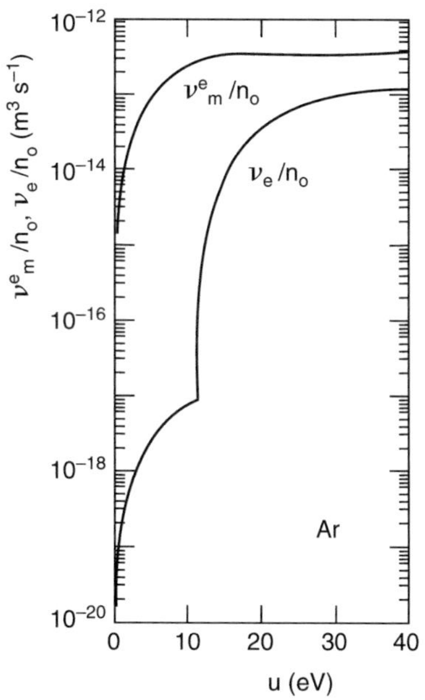

Fig. 4.1 Ratios of the characteristic relaxation frequencies to the gas number density for electron-neutral energy transfer, ν_e/n_o, and momentum transfer, ν_m^e/n_o, as a function of the electron energy, in argon (Sá et al. 1994)

We note that the characteristic frequency for energy transfer ν_e is a monotonously increasing function of the electron energy. Figure 4.1 shows the ratios ν_e/n_o and ν_m^e/n_o for argon as a function of the electron energy (Sá et al. 1994). The ratio ν_m^e/n_o largely exceeds ν_e/n_o in the energy range 0–40 eV, which is a consequence of the important contribution of elastic collisions to the total (elastic + inelastic) electron cross section for momentum transfer in Ar. On the other hand, $\nu_e(u) \simeq \nu_m^e(u)$ when $u \to \infty$, mainly due to the predominance of the ionization cross section in equations (4.1) and (4.2).

However, this is no longer true if we deal with a molecular gas, where the characteristic frequency for energy transfer ν_e presents important values even at electron energies as low as a few eV (e.g. in N_2, at $u \simeq 2\,\text{eV}$), as a result of the dissipation of electron energy in vibrational excitation. Here, the time-modulation of the body of the EEDF also occurs. Figure 4.2 shows the ratios of the characteristic relaxation frequencies to the gas number density ν_e/n_o and ν_m^e/n_o, as a function of the electron energy, in N_2 and H_2, in the case of absence of appreciable vibrational excitation, i.e. for $T_v = T_o$, in order the effects of superelastic electron-vibration (e-V) collisions do not need to be taken into account (Loureiro 1993). The ratio ν_e/n_o presents a sharp maximum in N_2 at about 2 eV due to vibrational excitation and a monotonic growth at higher energies in accordance with the cross sections shown in Fig. 3.13 for this gas. On the contrary, in H_2 only a smooth maximum exists associated with vibrational excitation (see Fig. 3.11).

In molecular gases the characteristic relaxation frequencies ν_e and ν_m^e must include the contributions due to excitation of rotational and vibrational levels. Thus, even in the case of vanishingly small vibrational excitation, both frequencies (4.1)

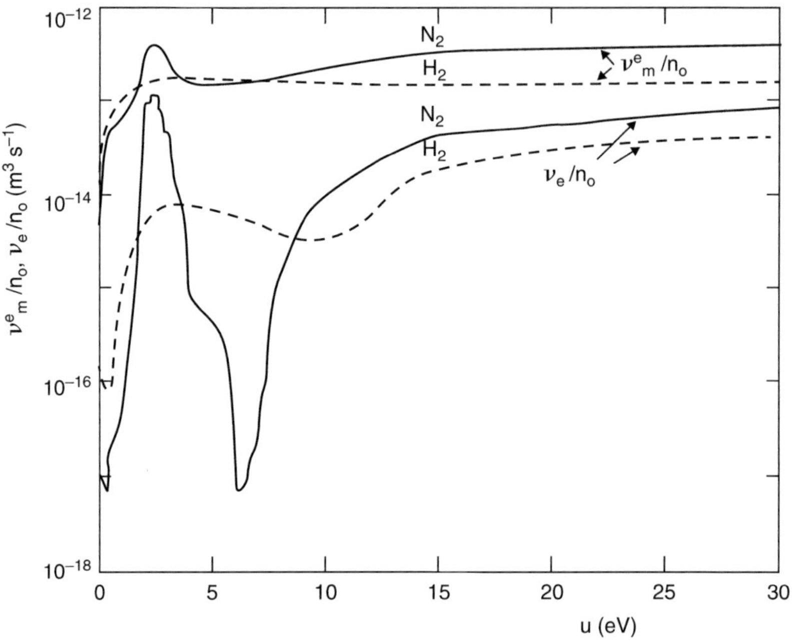

Fig. 4.2 Ratios of the characteristic relaxation frequencies to the gas number density for electron-neutral energy transfer, ν_e/n_o, and momentum transfer, ν_m^e/n_o, as a function of the electron energy, in N_2 (*full curves*) and in H_2 (*broken curves*) (Loureiro 1993)

and (4.2) must be corrected to include inelastic and superelastic rotational exchanges (see Sect. 3.3.3) and inelastic vibrational exchanges, by writing (see also Makabe and Goto 1988 and Goto and Makabe 1990)

$$\nu_e(u) \;=\; \frac{2m}{M}\,\nu_m(u) \,+\, 8B_0\,\frac{d\nu_0(u)}{du} \,+\, \sum_v \nu_v(u) \,+\, \sum_j \nu_j(u) \tag{4.3}$$

$$\nu_m^e(u) \;=\; \nu_m(u) \,+\, 2\,\nu_0(u) \,+\, \sum_v \nu_v(u) \,+\, \sum_j \nu_j(u)\,, \tag{4.4}$$

with $\nu_0(u) = n_o\sigma_0\sqrt{2u/m}$, and where it is assumed $\nu_{J,J+2} + \nu_{J,J-2} \simeq 2\nu_0$ and $(u_{J,J+2} - u_{J,J-2})/2 = 4B_0$, with J denoting the rotational quantum number. On the other hand, ν_v is the collision frequency for vibrational excitation of v-th levels from $v = 0$, within the electronic ground state $N_2(X\,^1\Sigma_g^+)$ or $H_2(X\,^1\Sigma_g^+)$.

4.1.2 Power Absorbed at High-Frequency Fields

Let us consider now in equation (3.35) a time-varying electric field of frequency ω and amplitude E_0, directed along the negative z axis: $\mathbf{E}(t) = - E(t)\ \mathbf{e_z}$. Assuming the only anisotropies present on the electron velocity distribution are those caused by the field and that they are sufficiently weak for a two-term expansion suffices (3.11), we may write from (3.41) and (3.42) the following equations for the evolution of the isotropic and the anisotropic components of the electron velocity distribution function

$$\frac{\partial f_e^0}{\partial t} + \frac{eE(t)}{m} \frac{1}{3v_e^2} \frac{\partial}{\partial v_e} \left(v_e^2 f_e^1\right) = I^0(f_e^0) + J^0(f_e^0) \tag{4.5}$$

$$\frac{\partial f_e^1}{\partial t} + \frac{eE(t)}{m} \frac{\partial f_e^0}{\partial v_e} = - \nu_m^e f_e^1, \tag{4.6}$$

where the elastic and the inelastic collision terms in the equation for f_e^0 are given by equations (3.85) and (3.134), respectively, and the effective collision frequency for momentum transfer is given by (3.139).

In principle, f_e^0 and f_e^1 are now functions of v_e and t, so that we may expand them in Fourier series in ωt

$$f_e^l(v_e, t) = \sum_{k=0}^{\infty} \mathrm{Re}\{\overline{f_{e,k}^l}(v_e)\ e^{jk\omega t}\}, \tag{4.7}$$

where Re{ } means "the real part of". The situation to be considered here firstly is that of a stationary plasma created by a high-frequency (HF) field. Thus, if the field frequency is significantly larger than the characteristic frequency for energy transfer, the isotropic component of the velocity distribution does not change appreciably during a cycle of the field oscillation and hence $\partial f_e^0/\partial t = 0$ in equation (4.5), while in equation (4.6) we have a dependence in ωt through the field. We may write then

$$f_e^1(v_e, t) = \mathrm{Re}\{\overline{f_e^1}\ e^{j\omega t}\} \tag{4.8}$$

and to obtain from (4.6)

$$\overline{f_e^1}(v_e) = - \frac{1}{\nu_m^e + j\omega} \frac{eE_0}{m} \frac{df_e^0}{dv_e}. \tag{4.9}$$

The drift or mean vector velocity $\mathbf{v_{ed}} = <\mathbf{v_e}>$ given by equation (3.18) also oscillates with the frequency ω and takes the form

$$\begin{aligned} \mathbf{v_{ed}}(t) &= \frac{1}{n_e} \int_0^{\infty} \frac{v_e}{3}\ \mathbf{f_e^1}(v_e, t)\ 4\pi v_e^2\ dv_e \\ &= \mathrm{Re}\{\overline{V_{ed}}\ e^{j\omega t}\}\ \mathbf{e_z}\ , \end{aligned} \tag{4.10}$$

being the complex amplitude $\overline{V_{ed}}$ after the substitution of (4.9) given by

$$\overline{V_{ed}} = -\frac{eE_0}{n_e m}\int_0^\infty \frac{1}{\nu_m^e + j\omega}\frac{df_e^0}{dv_e}\frac{4\pi v_e^3}{3}\,dv_e. \tag{4.11}$$

Since $\mathbf{v_{ed}} = \text{Re}\{\overline{\mu_e}\, E_0\, e^{j\omega t}\}\, \mathbf{e_z}$, we obtain the complex electron mobility

$$\overline{\mu_e} = -\frac{e}{n_e m}\int_0^\infty \frac{1}{\nu_m^e + j\omega}\frac{df_e^0}{dv_e}\frac{4\pi v_e^3}{3}\,dv_e\,. \tag{4.12}$$

On the other hand, the power absorbed by the electrons from the field, per volume unit, is

$$P_E(t) = (\mathbf{J_e}\,.\,\mathbf{E}) = \text{Re}\{\overline{\sigma_{ce}}\, E_0\, e^{j\omega t}\}\,\text{Re}\{E_0\, e^{j\omega t}\}\,, \tag{4.13}$$

being $\overline{\sigma_{ce}} = en_e\overline{\mu_e}$ the complex conductivity. Making the product of complex quantities, we obtain

$$P_E(t) = \frac{1}{2}\,\text{Re}\{\overline{\sigma_{ce}}\}\, E_0^2 + \frac{1}{2}\,\text{Re}\{\overline{\sigma_{ce}}\, e^{j2\omega t}\}\, E_0^2\,, \tag{4.14}$$

and because the second term vanishes as the time-average is taken, the time-averaged absorbed power is

$$\overline{P_E(t)} = \text{Re}\{\overline{\sigma_{ce}}\}\, E_{rms}^2\,, \tag{4.15}$$

with $E_{rms} = E_0/\sqrt{2}$ denoting the root mean square field and

$$\text{Re}\{\overline{\sigma_{ce}}\} = -\frac{e^2}{m}\int_0^\infty \frac{\nu_m^e}{\nu_m^{e\,2} + \omega^2}\frac{df_e^0}{dv_e}\frac{4\pi v_e^3}{3}\,dv_e \tag{4.16}$$

the real part of the complex electron conductivity. The mean power absorbed per electron in eV s^{-1}, $\overline{P_E(t)}/n_e$, is usually represented in literature by Θ (see e.g. Ferreira and Loureiro 1984, 1989).

The time-averaged absorbed power rapidly decreases as ω increases beyond ν_m^e, as a result of a progressive shift delay between the electron current density and the field which reaches $-\pi/2$ as $\omega \gg \nu_m^e$. In this limit of extremely high frequencies, we obtain from equations (4.10) and (4.11)

$$\mathbf{v_{ed}}(t) = \frac{eE_0}{m\omega}\,\cos(\omega t - \pi/2)\,\mathbf{e_z}\,, \tag{4.17}$$

and consequently

$$P_E(t) = \frac{e^2 n_e}{2m\omega}\,E_0^2\,\sin(2\omega t)\,, \tag{4.18}$$

with $\overline{P_E(t)} = 0$. As a matter of fact, they are the collisions that are responsible for the energy transfer from the field to the electron motion so that as they exist the phase shift progressively departs from $\pi/2$ and $\overline{P_E(t)}$ becomes non-null.

On the other hand, when the effective collision frequency for momentum transfer is independent of the electron energy, $\nu_m^e(v_e) = \text{const}$, we obtain from equations (4.15) and (4.16) a similar expression to that of a DC field

$$\overline{P_E(t)} = \frac{e^2 n_e}{m\nu_m^e} E_e^2 , \tag{4.19}$$

having defined the effective electric field E_e given by

$$E_e = \frac{1}{\sqrt{1+(\omega/\nu_m^e)^2}} E_{rms} , \tag{4.20}$$

which represents the equivalent field magnitude capable to produce the same energy absorption as a DC field. This effective field is $E_e = E_{rms}$, for $\omega \ll \nu_m^e$, and $E_e = (\nu_m^e/\omega)\, E_{rms}$, for $\omega \gg \nu_m^e$, so that in this latter case we have $E_e \to 0$ as $\omega \to \infty$, due to the above mentioned phase shift delay of $\pi/2$ between the electron current density and the HF field.

4.1.3 Stationary Electron Energy Distribution Functions

As the field frequency is sufficiently high so that the isotropic component of the electron velocity distribution remains time constant, we may assume $\partial f_e^0/\partial t = 0$ in equation (4.5) and write

$$-\frac{e}{m}\frac{1}{3v_e^2}\frac{d}{dv_e}\left(\frac{v_e^2}{2}\,\mathrm{Re}\{(\mathbf{f_e^1}\,.\,\mathbf{E}^*)\}\right) = I^0(f_e^0) + J^0(f_e^0) , \tag{4.21}$$

with $\mathbf{E}^*$ denoting the complex conjugate of the electric field. Inserting now equation (4.9) in (4.21) and making the product of complexes quantities, we easily obtain

$$-\frac{1}{2}\left(\frac{eE_0}{m}\right)^2 \frac{1}{v_e^2}\frac{d}{dv_e}\left(\frac{v_e^2}{3\nu_m^e}\frac{1}{1+(\omega/\nu_m^e)^2}\frac{df_e^0}{dv_e}\right) = I^0(f_e^0) + J^0(f_e^0) . \tag{4.22}$$

From equations (4.19) and (4.20) we see that when $\nu_m^e(v_e)$ depends on the electron velocity, $\overline{P_E(v_e,t)}/n_e$ may be seen as representing the time-averaged energy transferred from the field to an electron of velocity v_e, while $u_c(v_e) = \overline{P_E(v_e,t)}/(n_e\nu_m^e)$ is the energy transferred per collision. Comparing u_c with the term under brackets in (4.22), this equation may be written under the form

$$-\frac{1}{3m}\frac{1}{v_e^2}\frac{d}{dv_e}\left(v_e^2\,\nu_m^e\,u_c\,\frac{df_e^0}{dv_e}\right) = I^0(f_e^0) + J^0(f_e^0)\,, \tag{4.23}$$

or still under the following more explicit form using (3.85), (3.134), and (3.158) and multiplying both members by $\sqrt{u}$

$$\begin{aligned}-\frac{d}{du}&\left[\frac{2}{3}u^{3/2}\,\nu_m^e\,u_c\,\frac{df_e^0}{du} + \frac{2m}{M}u^{3/2}\,\nu_m\left(f_e^0 + k_BT_o\,\frac{df_e^o}{du}\right) + 4B_0\,\nu_0\,\sqrt{u}\,f_e^0\right]\\ &= \sum_{i,j}\left\{\sqrt{u+u_{ij}}\;\nu_{ij}(u+u_{ij})\,f_e^0(u+u_{ij}) - \sqrt{u}\;\nu_{ij}(u)\,f_e^0(u)\right\}\\ &\quad+\sum_{j,i}\left\{\sqrt{u-u_{ij}}\;\nu_{ji}(u-u_{ij})\,f_e^0(u-u_{ij}) - \sqrt{u}\;\nu_{ji}(u)\,f_e^0(u)\right\}\,. \end{aligned}\tag{4.24}$$

This equation can now be written in terms of the electron energy distribution function (EEDF), $f(u)$, normalized through condition (3.166). Both equations are identical since the two distributions are related one another through a constant factor.

Inspection of equation (4.24) reveals that the dependence of the EEDF on the parameters E_0, ω, and n_o arises through the time-averaged energy gain per collision

$$u_c(u) = \frac{e^2E_0^2}{2m\,(\nu_m^e(u)^2 + \omega^2)}\,, \tag{4.25}$$

with $\nu_m^e(u)/n_o = \sqrt{2u/m}\;\sigma_m^e(u)$, so that we can express all results in terms of the two independent reduced parameters E_0/n_o and ω/n_o as follows

$$u_c(u) = \frac{e^2}{2m}\,\frac{(E_0/n_o)^2}{(\nu_m^e(u)/n_o)^2 + (\omega/n_o)^2}\,. \tag{4.26}$$

When $\omega \gg \nu_m^e$ throughout the significant electron energy range, we have

$$u_c \sim \frac{e^2}{2m}\left(\frac{E_0}{\omega}\right)^2 \tag{4.27}$$

and the EEDF is function of E_0/ω only.

Since both members of equation (4.24) can be divided by the gas number density n_o, the other independent variables are the fractional population concentrations $\delta_i = n_i/n_o$ and $\delta_j = n_j/n_o$ in the inelastic and superelastic collision terms, respectively, and the gas temperature T_o due the small heating of electrons in collisions with non-frozen molecules (3.85) . Generally, the populations of excited states are important only for the vibrational levels (the effects of rotational levels are already included through a continuous approximation, see Sect. 3.3.3) and the effects of v-th levels can be taken into consideration through a given vibrational distribution function (VDF) characterized by a vibrational temperature T_v. The determination of the

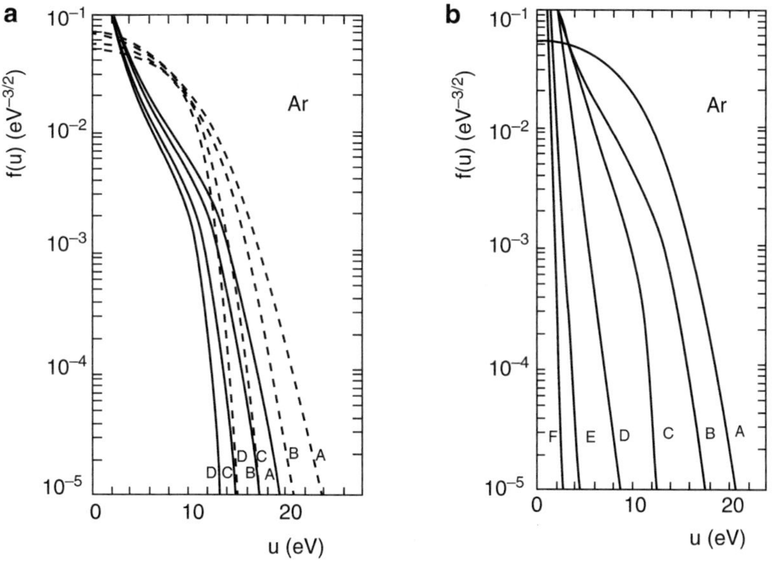

Fig. 4.3 Electron energy distribution functions in argon. (**a**) *Full curves*: $\omega/n_o = 4\times10^{-13}\,\mathrm{m^3\,s^{-1}}$; *Broken curves*: DC case. The different curves are for the following values of E_{rms}/n_o *(full)* or E/n_o *(broken)* in Td: *(A)* 100; *(B)* 65; *(C)* 30; *(D)* 15. (**b**) $E_{rms}/n_o = 65$ Td and the following values of ω/n_o in $10^{-12}\,\mathrm{m^3\,s^{-1}}$: *(A)* DC case; *(B)* 0.4; *(C)* 3; *(D)* 5; *(E)* 7.5; *(F)* 10 (Ferreira and Loureiro 1983)

VDFs is beyond the scope of this book, but the reader is invited to search this subject in Loureiro and Ferreira (1989, 1986) for H_2 and N_2 cases, respectively. In those papers it was shown that the coupling between the EEDF and VDF needs to be taken into consideration for a self-consistent determination of both distributions.

Figure 4.3a, b show the calculated EEDFs in argon obtained in Ferreira and Loureiro (1983) for various combinations of the reduced parameters E_0/n_o and ω/n_o. The full curves in Fig. 4.3a are for constant $\omega/n_o = 4\times10^{-13}\,\mathrm{m^3\,s^{-1}}$ and various values of E_{rms}/n_o, with $E_{rms} = E_0/\sqrt{2}$, ranging from 15 to 100 Td ($1\,\mathrm{Td} = 1\times10^{-21}\,\mathrm{V\,m^2}$). These distributions are remarkably different in shape from those obtained in a DC field at the same E/n_o values, which are also shown in Fig. 4.3a for comparison (broken curves). The distributions shown in Fig. 4.3b are for constant $E_{rms}/n_o = 65$ Td and various values of ω/n_o ranging from zero (DC case) to $\omega/n_o = 1\times10^{-11}\,\mathrm{m^3\,s^{-1}}$. From Fig. 4.3b it is seen that the EEDFs are strongly depleted for $\omega/n_o > 5\times10^{-12}\,\mathrm{m^3\,s^{-1}}$, because beyond this value we have $\omega \gg \nu_m^e$ in the whole energy range (see Fig. 4.1) and, consequently, the absorption of energy is significantly reduced due to the electron current density becomes $\sim \pi/2$ out of phase to the field. It is also assumed here $T_o = 300$ K for the gas temperature, so that the small heating of the electrons due to collisions with non-frozen atoms is negligible.

Moreover, it is also seen in Fig. 4.3b that the EEDFs are Maxwellian at the higher field frequencies $\omega/n_o > 5\times10^{-12}\,\mathrm{m^3\,s^{-1}}$ (Margenau 1946). This behaviour results from the fact that in Ar the EEDFs are strongly depleted at these ω/n_o values and, consequently, the electrons do not gain enough energy to produce an inelastic collision. The collisions are of elastic type only and the effective collision frequency for momentum transfer is equal to the elastic collision frequency, $\nu_m^e \equiv \nu_m$. Then equation (4.24) reduces to

$$\frac{d}{du}\left\{u^{3/2}\,\nu_m\left[\left(\frac{e^2}{3m}\left(\frac{E_0}{\omega}\right)^2+\frac{2m}{M}\,k_BT_o\right)\frac{df_e^0}{du}+\frac{2m}{M}\,f_e^0\right]\right\}=0, \qquad (4.28)$$

whose solution is a Maxwellian distribution at temperature

$$T_e = T_o + \frac{M}{6k_B}\left(\frac{eE_0}{m\omega}\right)^2. \qquad (4.29)$$

For $E_0/\omega = \sqrt{2}\times10^{-8}\,\mathrm{V\,m^{-1}\,s}$ and $T_o = 300\,\mathrm{K}$ in argon, we obtain $k_BT_e = 0.43\,\mathrm{eV}$.

As seen in Fig. 4.3a the EEDFs in argon at high field frequencies sharply increase near the origin of low electron energies. They are also depleted relatively to DC case at intermediate energies and have larger slopes at high energies. This behaviour can be well understood through Fig. 4.4, in which the EEDFs having the same mean

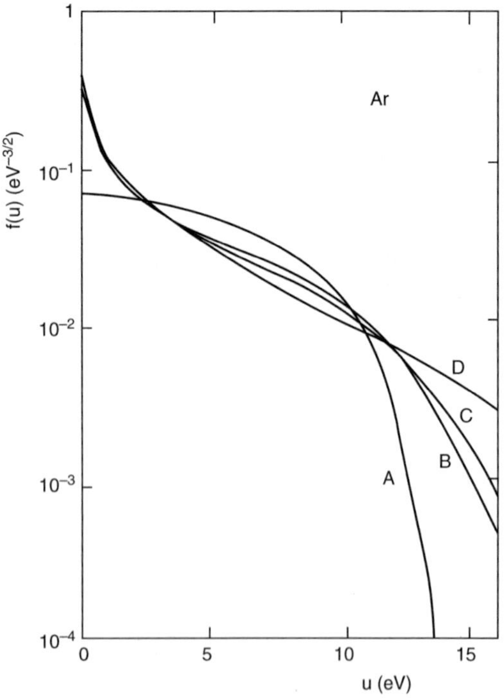

Fig. 4.4 Electron energy distribution functions in argon with the same mean energy of 3.5 eV for the following cases: (*A*) DC; (*B*) $\omega/n_o = 1\times10^{-13}\,\mathrm{m^3\,s^{-1}}$;(*C*) $\omega/n_o = 1.6\times10^{-13}\,\mathrm{m^3\,s^{-1}}$; (*D*) $\omega\to\infty$ (Ferreira et al. 1987)

energy $<u>= 3.5\,\text{eV}$ are plotted for various values of ω/n_o. We can see that the relative number of high-energy electrons increases with ω/n_o, although $<u>$ is kept unchanged. This is a direct consequence of the increasing efficiency of the power transfer from the field to the electrons of energy well above the average. Indeed, the power transfer per electron $u_c(u)\ \nu_m^e(u)$ (with $u_c(u)$ given by equation (4.25)) is maximum when $\omega = \nu_m^e(u)$ and, in argon, the effective collision frequency for momentum transfer is an increasing function of energy, up to $u \sim 12\,\text{eV}$ (see Fig. 4.1). Therefore, as ω increases the electron energy u^* for which the transfer of energy is maximum, $\nu_m^e(u^*) = \omega$, also increases, which produces the strong enhancement of the high-energy tail of the EEDFs shown in Fig. 4.4 (Winkler et al. 1984; Ferreira et al. 1987; Karoulina and Lebedev 1988).

The percentage electron energy losses by elastic, excitation, and ionization collisions are shown in Fig. 4.5 for argon, as a function of the time-averaged input power per electron at unit gas density, $\overline{P_E(t)}/(n_e n_o)$, for the high-frequency (HF) limit ($\omega \gg \nu_m^e(u)$ in the whole significant electron energy range) and for the stationary (DC) limit (Ferreira and Loureiro 1984). It may be concluded that, for similar $\overline{P_E(t)}/(n_e n_o)$ values, the input energy transferred to ionization

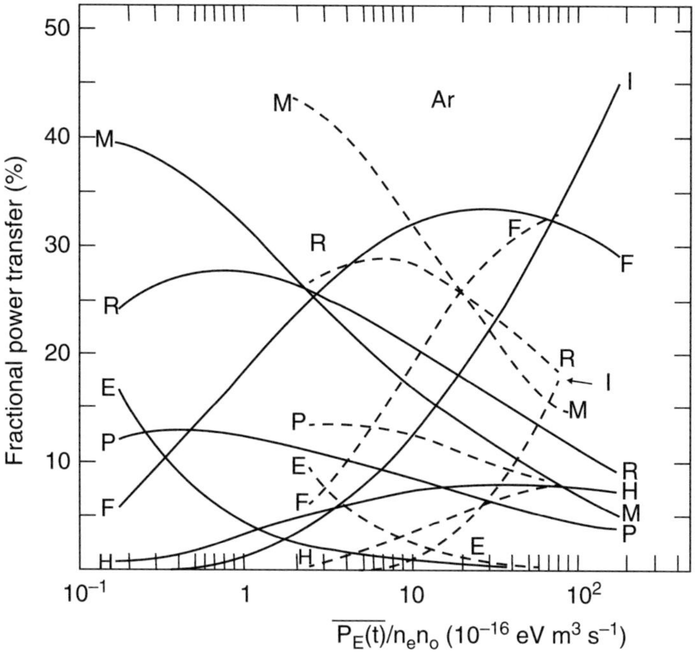

Fig. 4.5 Percentage electron energy losses in argon as a function of the average input power per electron at unit gas density, $\overline{P_E(t)}/(n_e n_o)$, in HF (*full curves*) and DC (*broken curves*). The labels of the curves correspond to the following energy loss channels (see Fig. 3.9): (*E*) elastic; (*M*) metastable states $^3P_2 + ^3P_0$; (*P*) resonant state $^3P_1$; (*R*) resonant $^1P_1$; (*F*) forbidden states; (*H*) higher-lying allowed states; (*I*) ionization (Ferreira and Loureiro 1984)

increases with increasing ω, i.e. when one goes from the DC case to HF. This is explained by an enhancement of the high-energy tail of the EEDF. The power transfer has a maximum at $\omega = \nu_m^e(u)$ and in argon $\nu_m^e(u)$ increases with u in the range $0.2 < u < 12\,\text{eV}$ and becomes approximately constant at energies $u > 12\,\text{eV}$, which corresponds to the high-energy tail of the distribution. Thus, as ω increases, the power transfer to the electrons in the tail of the distribution also increases. Besides elastic collisions and ionization, the excitation channels shown in Fig. 4.5 correspond to the cross sections represented in Fig. 3.10 as follows: conjoint excitation of the metastable states $^3P_2+^3P_0$; resonant state $^3P_1$; resonant state $^1P_1$; forbidden states $3p^54p$ with the threshold energy of 12.9 eV; and higher-lying optically allowed states with the threshold energy of 14.0 eV.

Figure 4.6a, b show the EEDFs calculated in nitrogen for $E_{rms}/n_o = 100\,\text{Td}$, various values of ω/n_o, including the DC case, and for the cases of appreciable vibrational excitation of $N_2(X\ ^1\Sigma_g^+, v)$ levels (figure a) and null vibrational excitation (figure b). As ω/n_o increases up to $4 \times 10^{-13}\,\text{m}^3\,\text{s}^{-1}$ the relative number of high-energy electrons decreases due to small energy absorption, as a result of the increase of the phase shift between the electron current density and the electric field. However, the new effect shown here that did not exist in argon is the strong enhancement of the high-energy tail of the EEDF as the electron superelastic collisions are also taken into account. More precisely, the electron superelastic collisions with vibrationally excited molecules $N_2(X\ ^1\Sigma_g^+, v > 0)$, i.e. the effect of the so-called electron-vibration (e–V) superelastic processes. Due to the strong peak at $u \sim 2\,\text{eV}$ in the electron cross section for vibrational excitation (see Fig. 3.13), the EEDFs exhibit a sharp decrease at this value of energy when the

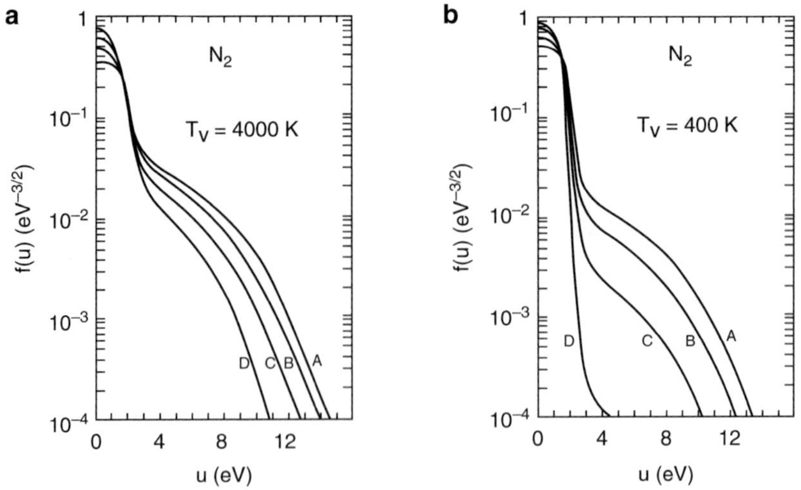

Fig. 4.6 Electron energy distribution functions in N_2 for $E_{rms}/n_o = 100\,\text{Td}$ and the following values of ω/n_o in $10^{-13}\,\text{m}^3\,\text{s}^{-1}$: (*A*) DC case; (*B*) 1; (*C*) 2; (*D*) 4. Figure (**a**): $T_v = 4000\,\text{K}$ and $T_o = 400\,\text{K}$. Figure (**b**): $T_v = T_o = 400\,\text{K}$ (Ferreira and Loureiro 1989)

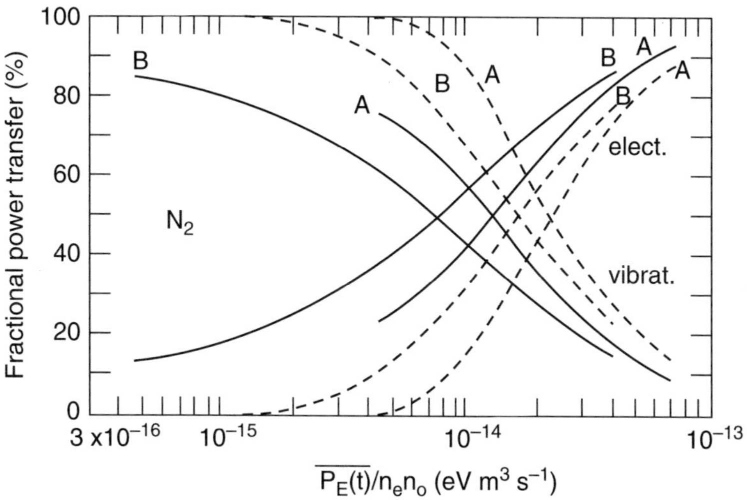

Fig. 4.7 Fractional power transfer into vibrational excitation and electronic excitation plus ionization in N_2, as a function of the mean power absorbed per electron at unit gas density, for (*A*) DC and (*B*) $\omega/n_o = 4 \times 10^{-13}\,\mathrm{m^3\,s^{-1}}$, and for $T_v = 4000\,\mathrm{K}$ (*full curves*) and $T_v = T_o = 400\,\mathrm{K}$ (*broken curves*) (Ferreira et al. 1987)

electron superelastic collisions are not taken into account, because even in DC case the E/n_o values are not high enough in order an appreciable amount of electrons may cross over the energy barrier at 2 eV (figure b). The situation dramatically changes as the e-V superelastic collisions are included (figure a) being now the high-energy portion of the EEDF populated with electrons that receive energy from the de-excitation of v-th levels. Figure 4.6a is for $T_v = 4000\,\mathrm{K}$ and $T_o = 400\,\mathrm{K}$, while Fig. 4.6b is for $T_v = T_o = 400\,\mathrm{K}$ (see Loureiro and Ferreira 1986 for DC and Ferreira and Loureiro 1989 for HF cases).

Figure 4.7 shows the percentage electron energy losses in nitrogen through vibrational excitation and electronic excitation plus ionization, as a function of the time-averaged absorbed power per electron at unity gas density, $\overline{P_E(t)}/(n_en_o)$. An increase in ω/n_o or in T_v results in a decrease of the power transferred to the vibrational mode and, consequently, in an increase of the power transferred to electronic excitation plus ionization. Both effects are consequence of the enhancement of the high-energy tail of the EEDF which occurs with increasing either ω/n_o or T_v. The effects of the changes in ω/n_o are qualitatively the same in nitrogen as those found before in Fig. 4.5 for argon.

Finally, Fig. 4.8 shows the percentage electron energy losses into excitation of the most important triplet states of N_2 (A $^3\Sigma_u^+$, B $^3\Pi_g$, C $^3\Pi_u$) and ionization, for $T_v = 4000\,\mathrm{K}$, and for the cases DC and $\omega/n_o = 4 \times 10^{-13}\,\mathrm{m^3\,s^{-1}}$. Due to the enhancement of the high-energy tail of the EEDF in HF, the excitation of the triplet states of N_2 is obtained at lower $\overline{P_E(t)}/n_en_o$ values than in DC case (although as we have seen before the E_0/n_o values are much larger).

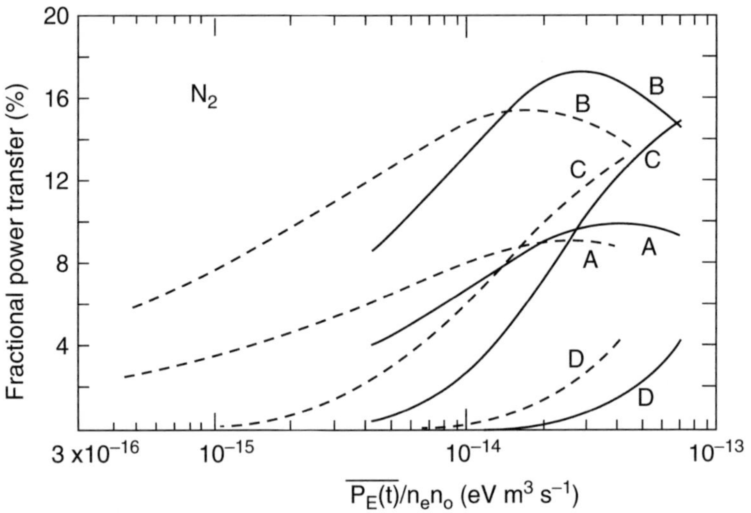

Fig. 4.8 Fractional power transfer into excitation of the triplet states in N_2: (*A*) A $^3\Sigma_u^+$; (*B*) B $^3\Pi_g$; (*C*) C $^3\Pi_u$; (*D*) ionization, as a function of the mean power absorbed per electron at unit gas density, for $T_v = 4000$ K, and for DC (*full curves*) and $\omega/n_o = 4 \times 10^{-13}$ m^3 s^{-1} (*broken curves*) (Ferreira and Loureiro 1989)

4.2 Electron Cyclotron Resonance

4.2.1 Hydrodynamic Description

As seen in Sect. 4.1.2, when the electrons of a plasma are submitted to a HF electric field they oscillate at the same frequency of the field. However, in the collisionless limit $\omega \gg \nu_m^e$, they oscillate with a phase shift of $\pi/2$ and the average energy acquired over a field period by the electrons is zero. They are the collisions that temporarily interrupt the movement of the electrons and make the phase shift deviates from $\pi/2$. The electrons may absorb then energy from the field.

Due to this particular aspect of HF discharges, the electrons in HF fields may absorb significantly less energy than in DC fields. So, to avoid this inconvenience, the superimposing of a static magnetic field on the HF system is of great interest (Margot et al. 1992). Besides the fact that a discharge with this configuration significantly reduces the electron losses to the wall, the electrons when submitted to a static **B** field rotate circularly around its field lines with an angular electron cyclotron frequency $\omega_{ce} = eB/m$. If a HF wave with frequency ω also exists, with the electric field perpendicular to the direction of **B** field, rotating in the same direction, the electrons in their reference system see a DC field when both frequencies are adjusted, $\omega_{ce} = \omega$. The electrons are then continuously accelerated until their movement is interrupted by a collision.

As the electron frequency for momentum transfer is independent of velocity, $\nu_m^e(v_e) = \text{const}$, the electron movement is described by the hydrodynamic equation for local momentum conservation (see equation (3.60) in Sect. 3.1.5)

$$m \, \frac{\mathbf{d v_{ed}}}{dt} \;=\; -\, e\, \mathbf{E} \;-\; e\, [\mathbf{v_{ed}} \times \mathbf{B}] \;-\; m\, \nu_m^e \, \mathbf{v_{ed}} \, . \tag{4.30}$$

Here, $\mathbf{v_{ed}} = <\mathbf{v_e}>$ is the electron drift or average vector velocity, $\mathbf{E} = \mathbf{E_0}\, e^{j\omega t}$ is the HF electric field with amplitude $\mathbf{E_0}$ and frequency ω, and $\mathbf{B} = B\, \mathbf{e_z}$ is the static magnetic field oriented along the z axis. Since $\mathbf{v_{ed}} \propto e^{j\omega t}$, we may still write using complex quantities

$$(\nu_m^e \, + \, j\omega)\, \mathbf{v_{ed}} \;=\; -\, \frac{e}{m}\, \mathbf{E} \, + \, \omega_{ce}\, [\mathbf{e_z} \times \mathbf{v_{ed}}] \, . \tag{4.31}$$

This equation can be then decomposed through the three axis as follows

$$(\nu_m^e \, + \, j\, \omega)\, \overline{v}_x \, + \, \omega_{ce}\, \overline{v}_y \;=\; -\, \frac{e}{m}\, \overline{E}_{0x} \tag{4.32}$$

$$(\nu_m^e \, + \, j\, \omega)\, \overline{v}_y \, - \, \omega_{ce}\, \overline{v}_x \;=\; -\, \frac{e}{m}\, \overline{E}_{0y} \tag{4.33}$$

$$(\nu_m^e \, + \, j\, \omega)\, \overline{v}_z \;=\; -\, \frac{e}{m}\, \overline{E}_{0z} \, , \tag{4.34}$$

where the velocity amplitudes are complex quantities to include origin phase shifts.

The vector equation is better written using tensor notation

$$\mathbf{v_{ed}} \;=\; -\widehat{\boldsymbol{\mu}_\mathbf{e}}\, \mathbf{E} \, . \tag{4.35}$$

Introducing the quantities as in Allis (1956)

$$\overline{r} \;=\; \frac{1}{\nu_m^e \, + \, j\, (\omega - \omega_{ce})} \, , \quad \overline{l} \;=\; \frac{1}{\nu_m^e \, + \, j\, (\omega + \omega_{ce})} \, , \quad \overline{p} \;=\; \frac{1}{\nu_m^e \, + \, j\, \omega} \, , \tag{4.36}$$

to correspond to right and left circular polarization of the **E** field normal to **B**, and **E** parallel to **B**, the mobility tensor writes as

$$\widehat{\boldsymbol{\mu}_\mathbf{e}} \;=\; \begin{pmatrix} \overline{\mu}_{xx} & \overline{\mu}_{xy} & 0 \\ -\overline{\mu}_{xy} & \overline{\mu}_{yy} & 0 \\ 0 & 0 & \overline{\mu}_{zz} \end{pmatrix} , \tag{4.37}$$

in which the different components are

$$\overline{\mu}_{xx} \;=\; \overline{\mu}_{yy} \;=\; \frac{e}{m}\, \frac{(\overline{r} + \overline{l})}{2} \tag{4.38}$$

$$\overline{\mu}_{xy} = j\,\frac{e}{m}\,\frac{(\overline{r}-\overline{l})}{2} \tag{4.39}$$

$$\overline{\mu}_{zz} = \frac{e}{m}\,\overline{p}\,. \tag{4.40}$$

The electron current density is then $\mathbf{J_e} = -en_e\mathbf{v_{ed}} = \widehat{\boldsymbol{\sigma}_{\mathbf{ce}}}\,\mathbf{E}$, being $\widehat{\boldsymbol{\sigma}_{\mathbf{ce}}} = en_e\widehat{\boldsymbol{\mu}_{\mathbf{e}}}$ the electron conductivity tensor.

As seen in equation (4.13), the power absorbed per volume unit is

$$P_E(t) = (\mathbf{J_e}\,.\,\mathbf{E}) = (\widehat{\boldsymbol{\sigma}_{\mathbf{ce}}}\,\mathbf{E}\,.\,\mathbf{E})\,, \tag{4.41}$$

being the time-averaged power given by

$$\overline{P_E(t)} = \frac{1}{2}\,\mathrm{Re}\{(\widehat{\boldsymbol{\sigma}_{\mathbf{ce}}}\,\mathbf{E}\,.\,\mathbf{E}^*)\}\,, \tag{4.42}$$

with $\mathbf{E}^*$ denoting the complex conjugate. Defining now the components of the field as $\overline{E}_x = E_{0x}\,e^{j\omega t}$, $\overline{E}_y = E_{0y}\,e^{j(\omega t+\theta)}$ and $\overline{E}_z = E_{0z}\,e^{j\omega t}$, in order θ may represent the phase shift between $\overline{E}_y$ and $\overline{E}_x$, we obtain

$$\begin{aligned}\overline{P_E(t)} = \frac{1}{2}\,&\left[\mathrm{Re}\{\overline{\sigma}_{xx}\}\,E_{0x}^2 + \mathrm{Re}\{\overline{\sigma}_{yy}\}\,E_{0y}^2 + \mathrm{Re}\{\overline{\sigma}_{zz}\}\,E_{0z}^2\right]\\ &- \mathrm{Im}\{\overline{\sigma}_{xy}\}\,E_{0x}\,E_{0y}\,\sin\theta\,,\end{aligned} \tag{4.43}$$

in which

$$\mathrm{Re}\{\overline{\sigma}_{xx}\} = \mathrm{Re}\{\overline{\sigma}_{yy}\} = \frac{e^2n_e}{m}\,\frac{\nu_m^e\,(\nu_m^{e\,2}+\omega^2+\omega_{ce}^2)}{(\nu_m^{e\,2}+(\omega+\omega_{ce})^2)\,(\nu_m^{e\,2}+(\omega-\omega_{ce})^2)} \tag{4.44}$$

$$\mathrm{Im}\{\overline{\sigma}_{xy}\} = \frac{e^2n_e}{m}\,\frac{2\,\nu_m^e\,\omega\,\omega_{ce}}{(\nu_m^{e\,2}+(\omega+\omega_{ce})^2)\,(\nu_m^{e\,2}+(\omega-\omega_{ce})^2)} \tag{4.45}$$

$$\mathrm{Re}\{\overline{\sigma}_{zz}\} = \frac{e^2n_e}{m}\,\frac{\nu_m^e}{\nu_m^{e\,2}+\omega^2}\,. \tag{4.46}$$

The angle θ is the polarization angle of the electric field in the direction perpendicular to the $\mathbf{B}$ field. For a right-hand circularly polarized wave, we have $E_{0x} = E_{0y} = E_0$, $E_{0z} = 0$, and $\theta = -\pi/2$, so that equation (4.43) becomes

$$\begin{aligned}\overline{P_{E(R)}} &= \left(\mathrm{Re}\{\overline{\sigma}_{xx}\} + \mathrm{Im}\{\overline{\sigma}_{xy}\}\right)\,E_0^2\\ &= \frac{e^2n_e}{m}\,\frac{\nu_m^e}{\nu_m^{e\,2}+(\omega-\omega_{ce})^2}\,E_0^2\,,\end{aligned} \tag{4.47}$$

while for a left-hand circularly polarized wave, we have $\theta = \pi/2$ and we obtain

$$\overline{P_{E(L)}} = \left(\mathrm{Re}\{\overline{\sigma}_{xx}\} - \mathrm{Im}\{\overline{\sigma}_{xy}\}\right) E_0^2 = \frac{e^2 n_e}{m} \frac{\nu_m^e}{\nu_m^{e\,2} + (\omega + \omega_{ce})^2} E_0^2 . \tag{4.48}$$

Equation (4.47) shows that for $\omega_{ce} = \omega$, the power absorbed by the electrons from a right-hand circularly polarized wave has a maximum as a result of the exact match between the rotating field and the cyclotron motion of electrons. The power becomes equal to that absorbed from a DC field of amplitude E_0. On the other hand, in the case of a left-hand wave the power absorbed is significantly reduced, being the ratio of the power absorbed by the two waves at $\omega_{ce} = \omega$ equal to $\nu_m^{e\,2}/(\nu_m^{e\,2} + 4\omega^2)$.

It is still worth noting that in the case of a linearly polarized wave in the plane perpendicular to the **B** field of amplitude $E_{0(l)}$, which may be decomposed into two circularly polarized waves of amplitudes $E_{0(l)}/2$, we obtain

$$\overline{P_E(t)} = \frac{e^2 n_e}{m} \left(\frac{\nu_m^e}{\nu_m^{e\,2} + (\omega - \omega_{ce})^2} + \frac{\nu_m^e}{\nu_m^{e\,2} + (\omega + \omega_{ce})^2} \right) \left(\frac{E_{0(l)}}{2} \right)^2 . \tag{4.49}$$

Thus, when $\mathbf{B} = 0$ this equation transforms into the power absorbed by a HF field of amplitude $E_{0(l)}$ given by equation (4.19)

$$\overline{P_E(t)} = \frac{e^2 n_e}{m} \frac{\nu_m^e}{\nu_m^{e\,2} + \omega^2} \frac{E_{0(l)}^2}{2} . \tag{4.50}$$

4.2.2 Boltzmann Equation

In the presence of the **E** and **B** fields, and in the absence of space gradients, the electron Boltzmann equation takes the form given by equation (3.6)

$$\frac{\partial f_e}{\partial t} + \left(\left\{ -\frac{e}{m} \mathbf{E} + [\vec{\boldsymbol{\omega}}_{\mathbf{ce}} \times \mathbf{v_e}] \right\} \cdot \frac{\partial f_e}{\partial \mathbf{v_e}} \right) = \left(\frac{\partial f_e}{\partial t} \right)_{e-o} , \tag{4.51}$$

with $\vec{\boldsymbol{\omega}}_{\mathbf{ce}} = eB/m\, \mathbf{e_z}$. In this case due to the magnetic field we need to consider the dependence on the azimuthal coordinate ϕ, so that the expansion of $f_e(\mathbf{v_e}, t)$ limited to the first order anisotropies takes the form (3.21)

$$f(\mathbf{v_e}, t) = f_e^0(v_e, t) + \left(\mathbf{f_e^1}(v_e, t) \cdot \frac{\mathbf{v_e}}{v_e} \right) = f_e^0 + p_{11} \frac{v_{ex}}{v_e} + q_{11} \frac{v_{ey}}{v_e} + p_{10} \frac{v_{ez}}{v_e} . \tag{4.52}$$

Then, the term with the magnetic field in equation (4.51) is

$$\left([\vec{\boldsymbol{\omega}}_{\mathbf{ce}} \times \mathbf{v_e}] \cdot \frac{\partial f_e}{\partial \mathbf{v_e}} \right) = \omega_{ce}\, q_{11}\, \frac{v_{ex}}{v_e} - \omega_{ce}\, p_{11}\, \frac{v_{ey}}{v_e} . \tag{4.53}$$

Since $v_{ex}/v_e = \sin\theta \ \cos\phi$ and $v_{ey}/v_e = \sin\theta \ \sin\phi$, equation (4.53) leads to the appearing of two new terms in the equations for the anisotropic components p_{11} and q_{11}, being equation (3.42) replaced with the following system of equations

$$\frac{\partial p_{11}}{\partial t} - \frac{eE_x}{m} \frac{\partial f_e^0}{\partial v_e} + \omega_{ce}\, q_{11} = -\, \nu_m^e\, p_{11} \tag{4.54}$$

$$\frac{\partial q_{11}}{\partial t} - \frac{eE_y}{m} \frac{\partial f_e^0}{\partial v_e} - \omega_{ce}\, p_{11} = -\, \nu_m^e\, q_{11} \tag{4.55}$$

$$\frac{\partial p_{10}}{\partial t} - \frac{eE_z}{m} \frac{\partial f_e^0}{\partial v_e} = -\, \nu_m^e\, p_{10} , \tag{4.56}$$

while for the isotropic component we have in place of (3.41)

$$\frac{\partial f_e^0}{\partial t} - \frac{e}{m} \frac{1}{3v_e^2} \frac{\partial}{\partial v_e} \left(v_e^2 \left(p_{11}\, E_x + q_{11}\, E_y + p_{10}\, E_z \right) \right) = I^0(f_e^0) + J^0(f_e^0) , \tag{4.57}$$

with the elastic and the inelastic collision terms given by equations (3.85) and (3.134).

Considering now the field frequency much larger than the characteristic relaxation frequency for energy transfer ν_e, in order we may assume the time invariance of f_e^0, the equations (4.54), (4.55) and (4.56) take the following form using complex quantities

$$(\nu_m^e + j\omega)\, \overline{p}_{11} + \omega_{ce}\, \overline{q}_{11} = \frac{e}{m}\, \overline{E}_x\, \frac{df_e^0}{dv_e} \tag{4.58}$$

$$(\nu_m^e + j\omega)\, \overline{q}_{11} - \omega_{ce}\, \overline{p}_{11} = \frac{e}{m}\, \overline{E}_y\, \frac{df_e^0}{dv_e} \tag{4.59}$$

$$(\nu_m^e + j\omega)\, \overline{p}_{10} = \frac{e}{m}\, \overline{E}_z\, \frac{df_e^0}{dv_e} \tag{4.60}$$

and as in (4.37) we may express under tensorial form

$$\begin{bmatrix} \overline{p}_{11} \\ \overline{q}_{11} \\ \overline{p}_{10} \end{bmatrix} = \frac{e}{m} \begin{bmatrix} \overline{E}_x\, (\overline{r} + \overline{l})/2 & j\overline{E}_y\, (\overline{r} - \overline{l})/2 & 0 \\ -\, j\overline{E}_x\, (\overline{r} - \overline{l})/2 & \overline{E}_y\, (\overline{r} + \overline{l})/2 & 0 \\ 0 & 0 & \overline{E}_x\, \overline{p} \end{bmatrix} \frac{df_e^0}{dv_e} . \tag{4.61}$$

The drift or average vector velocity (4.10)

$$\mathbf{v_{ed}} = \frac{1}{n_e}\int_0^\infty (\overline{p}_{11}\,\mathbf{e_x} + \overline{q}_{11}\,\mathbf{e_y} + \overline{p}_{10}\,\mathbf{e_z})\,\frac{4\pi v_e^3}{3}\,dv_e \tag{4.62}$$

may be written in the form (4.35) being the components of the tensor mobility as follows

$$\overline{\mu}_{xx} = -\frac{e}{mn_e}\int_0^\infty \left(\frac{\overline{r}+\overline{l}}{2}\right)\frac{df_e^0}{dv_e}\,\frac{4\pi v_e^3}{3}\,dv_e \tag{4.63}$$

$$\overline{\mu}_{xy} = -j\,\frac{e}{mn_e}\int_0^\infty \left(\frac{\overline{r}-\overline{l}}{2}\right)\frac{df_e^0}{dv_e}\,\frac{4\pi v_e^3}{3}\,dv_e \tag{4.64}$$

$$\overline{\mu}_{zz} = -\frac{e}{mn_e}\int_0^\infty \overline{p}\,\frac{df_e^0}{dv_e}\,\frac{4\pi v_e^3}{3}\,dv_e \tag{4.65}$$

and $\overline{\mu}_{yy} = \overline{\mu}_{xx}$. Obviously when $\nu_m^e(v_e) = \text{const}$ and because of

$$\int_0^\infty \frac{df_e^0}{dv_e}\,\frac{4\pi v_e^3}{3}\,dv_e = -\,n_e\,, \tag{4.66}$$

we return back to equations (4.38), (4.39) and (4.40).

Let us consider now equation (4.57) for the isotropic component assuming $\omega \gg \nu_e$, and hence $\partial f_e^0/\partial t = 0$. Substituting the equation (4.61) for the anisotropic component in equation (4.57), written as in equation (4.21), we obtain

$$-\left(\frac{e}{m}\right)^2\frac{1}{3v_e^2}\,\frac{d}{dv_e}\left[\frac{v_e^2}{2}\left(\mathrm{Re}\left\{\frac{\overline{r}+\overline{l}}{2}\right\}(E_{0x}^2+E_{0y}^2) + \mathrm{Re}\{\overline{p}\}\,E_{0z}^2 - \mathrm{Im}\left\{j\left(\frac{\overline{r}-\overline{l}}{2}\right)\right\}2\,E_{0x}\,E_{0y}\,\sin\theta\right)\frac{df_e^0}{dv_e}\right] = I^0(f_e^0) + J^0(f_e^0), \tag{4.67}$$

in which

$$\mathrm{Re}\left\{\frac{\overline{r}+\overline{l}}{2}\right\} = \frac{\nu_m^e\,(\nu_m^{e\,2}+\omega^2+\omega_{ce}^2)}{(\nu_m^{e\,2}+(\omega+\omega_{ce})^2)\,(\nu_m^{e\,2}+(\omega-\omega_{ce})^2)} \tag{4.68}$$

$$\mathrm{Im}\left\{j\left(\frac{\overline{r}-\overline{l}}{2}\right)\right\} = \frac{2\,\nu_m^e\,\omega\,\omega_{ce}}{(\nu_m^{e\,2}+(\omega+\omega_{ce})^2)\,(\nu_m^{e\,2}+(\omega-\omega_{ce})^2)} \tag{4.69}$$

$$\mathrm{Re}\,\{\overline{p}\} = \frac{\nu_m^e}{\nu_m^{e\,2}+\omega^2}\,. \tag{4.70}$$

For a right-hand circularly polarized wave $E_{0x} = E_{0y} = E_0$, $E_{0z} = 0$, and $\theta = -\pi/2$, we obtain

$$-\left(\frac{e}{m}\right)^2 \frac{1}{3v_e^2} \frac{d}{dv_e} \left[v_e^2 \left(\mathrm{Re}\left\{ \frac{\bar{r}+\bar{l}}{2} \right\} + \mathrm{Im}\left\{ j\left(\frac{\bar{r}-\bar{l}}{2} \right) \right\} \right) E_0^2 \frac{df_e^0}{dv_e} \right]$$
$$= I^0(f_e^0) + J^0(f_e^0)$$

and therefore

$$-\left(\frac{e}{m}\right)^2 \frac{1}{3v_e^2} \frac{d}{dv_e} \left[v_e^2 \frac{\nu_m^e}{\nu_m^{e\,2} + (\omega - \omega_{ce})^2} E_0^2 \frac{df_e^0}{dv_e} \right] = I^0(f_e^0) + J^0(f_e^0)\,, \tag{4.71}$$

while for a left-hand circularly polarized wave $\theta = \pi/2$, we find

$$-\left(\frac{e}{m}\right)^2 \frac{1}{3v_e^2} \frac{d}{dv_e} \left[v_e^2 \frac{\nu_m^e}{\nu_m^{e\,2} + (\omega + \omega_{ce})^2} E_0^2 \frac{df_e^0}{dv_e} \right] = I^0(f_e^0) + J^0(f_e^0)\,. \tag{4.72}$$

We may check that in case of a linearly polarized wave in the plane perpendicular to the **B** field of amplitude $E_{0(l)}$, decomposed into two circularly polarized waves of amplitudes $E_{0(l)}/2$, we have

$$-\left(\frac{e}{m}\right)^2 \frac{1}{3v_e^2} \frac{d}{dv_e} \left[v_e^2 \left(\frac{\nu_m^e}{\nu_m^{e\,2} + (\omega - \omega_{ce})^2} + \frac{\nu_m^e}{\nu_m^{e\,2} + (\omega + \omega_{ce})^2} \right) \frac{E_{0l}^2}{4} \frac{df_e^0}{dv_e} \right]$$
$$= I^0(f_e^0) + J^0(f_e^0)\,, \tag{4.73}$$

so that as $\omega_{ce} = 0$, equation (4.73) transforms into the previous expression (4.22) derived for a linearly polarized HF field.

Finally, let us consider the most general case of a right-hand circularly polarized wave of amplitude $E_{0(R)}$ simultaneously with a left-hand circularly polarized wave of amplitude $E_{0(L)}$. Multiplying both members of equations (4.71) and (4.72) by $\sqrt{u}$, we obtain the following expression for the Boltzmann equation in terms of the EEDF $f_e^0(u)$ equivalent to equation (4.24) as **B** is discarded

$$-\frac{d}{du}\left[\frac{2}{3} u^{3/2} \nu_m^e \frac{e^2}{m} \left(\frac{E_{0(R)}^2}{\nu_m^{e\,2} + (\omega - \omega_{ce})^2} + \frac{E_{0(L)}^2}{\nu_m^{e\,2} + (\omega + \omega_{ce})^2} \right) \frac{df_e^0}{du} \right]$$
$$= \frac{d}{du}\left[\frac{2m}{M} u^{3/2} \nu_m \left(f_e^0 + k_B T_o \frac{df_e^o}{du} \right) + 4B_0\, \nu_0 \sqrt{u} f_e^0 \right]$$
$$+ \sum_{i,j} \left\{ \sqrt{u + u_{ij}}\ \nu_{ij}(u + u_{ij})\, f_e^0(u + u_{ij}) - \sqrt{u}\ \nu_{ij}(u)\, f_e^0(u) \right\}$$
$$+ \sum_{j,i} \left\{ \sqrt{u - u_{ij}}\ \nu_{ji}(u - u_{ij})\, f_e^0(u - u_{ij}) - \sqrt{u}\ \nu_{ji}(u)\, f_e^0(u) \right\}. \tag{4.74}$$

Equation (4.74) determines the EEDF in terms of $E_{0(R)}/n_0$ and $E_{0(L)}/n_0$ (or $E_{0(l)}/n_0$), ω/n_0 and ω_{ce}/ω. When $\omega_{ce}/\omega = 1$ the right-hand circularly polarized wave leads to electron cyclotron resonance.

4.2.3 Power Absorbed from the Field

Following a procedure identical to (4.43) we obtain the following expression for the mean power absorbed from the HF field, per volume unit, for the case of right-hand and left-hand circularly polarized waves

$$\overline{P_E(t)} = en_e \left(\mathrm{Re}\{\overline{\mu}_{xx}\} + \mathrm{Im}\{\overline{\mu}_{xy}\}\right) E_{0(R)}^2 + en_e \left(\mathrm{Re}\{\overline{\mu}_{xx}\} - \mathrm{Im}\{\overline{\mu}_{xy}\}\right) E_{0(L)}^2 . \tag{4.75}$$

Individually, the power absorbed from the right-hand and the left-hand circularly polarized waves are as follows using the EEDF with the normalization (3.166)

$$\overline{P_{E(R)}} = -\frac{2}{3}\frac{e^2 n_e}{m}\int_0^\infty u^{3/2}\,\nu_m^e\,\frac{E_{0(R)}^2}{{\nu_m^e}^2 + (\omega - \omega_{ce})^2}\,\frac{df}{du}\,du \tag{4.76}$$

$$\overline{P_{E(L)}} = -\frac{2}{3}\frac{e^2 n_e}{m}\int_0^\infty u^{3/2}\,\nu_m^e\,\frac{E_{0(L)}^2}{{\nu_m^e}^2 + (\omega + \omega_{ce})^2}\,\frac{df}{du}\,du\,. \tag{4.77}$$

Figure 4.9 presents the mean power per electron at unit gas density, $\overline{P_E(t)}/n_e n_o$, absorbed from the R and L waves in argon, as a function of ω_{ce}/ω, for the reduced *rms* field $E_{rms(l)}/n_o = 20$ Td of a linearly polarized wave decomposed into two circularly polarized waves, and for $\omega/n_o = 1.17 \times 10^{-13}\,\mathrm{m^3\,s^{-1}}$ and $4.78 \times 10^{-13}\,\mathrm{m^3\,s^{-1}}$ (Loureiro 1995). The full curves represent results obtained from equation (4.74) in the absence of electron-electron (*e-e*) collisions, while the broken curves show for comparison the mean power calculated as the EEDFs are assumed Maxwellian (3.168), with their temperatures obtained from the energy balance equation (3.174)

$$\overline{P_E(t)} = P_{el} + P_{inel}\,, \tag{4.78}$$

in which the average power loss terms are given by (3.151) and (3.176). In the case of a gas pressure $p = 133.3$ Pa (1 Torr) and temperature of neutrals $T_o = 300$ K, the values of ω/n_o used here correspond to $\omega/2\pi = 600$ MHz and 2.45 GHz. The equality $\omega_{ce} = \omega$ is achieved for $B = 214$ and 875 G, respectively. Figure 4.9 shows that maximum heating is obtained for a Maxwellian EEDF. Further, the power absorbed from the L wave decreases as ω/n_o increases from 1.17×10^{-13} to $4.78 \times 10^{-13}\,\mathrm{m^3\,s^{-1}}$, which signifies higher efficiency in ECR when a linearly polarized wave is launched. The small maximum in the power absorbed from the L wave as $\omega_{ce} = \omega$ results from the changes on the EEDFs for conditions close to ECR originated by the R wave.

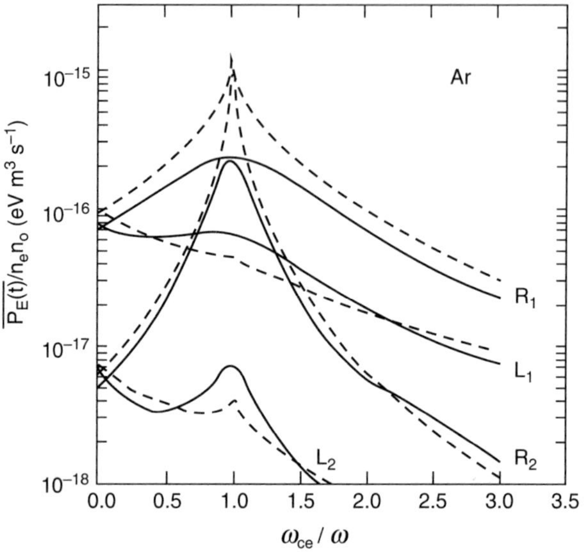

Fig. 4.9 Mean power absorbed per electron at unit gas density against ω_{ce}/ω in argon, for $E_{rms(l)}/n_o = 20\,\mathrm{Td}$ and $\omega/n_o = 1.17 \times 10^{-13}\,\mathrm{m^3\,s^{-1}}$ (R_1; L_1) and $4.78 \times 10^{-13}\,\mathrm{m^3\,s^{-1}}$ (R_2; L_2). R and L are for the R and L waves. *Full curves*: absence of *e-e* collisions; *Broken curves*: Maxwellian EEDFs (Loureiro 1995)

Figure 4.10 shows values of $\overline{P_E(t)}/n_e n_o$ calculated for $E_{rms(l)}/n_o = 120\,\mathrm{Td}$, keeping all other conditions as in Fig. 4.9. From inspection of Figs. 4.9 and 4.10 we conclude that the differences between the values of $\overline{P_E(t)}/n_e n_o$ calculated in the absence of *e-e* collisions (i.e. with non-Maxwellian EEDFs) and in the case of Maxwellian EEDFs significantly reduce as $E_{rms(l)}/n_o$ increases, in consequence of modifications on the shape of the EEDFs.

Figure 4.11 shows values of $\overline{P_E(t)}/n_e n_o$ in Ar, as a function of the reduced *rms* amplitude of a linearly polarized wave, $E_{rms(l)}/n_o$, and for $\omega_{ce} = \omega$. As in Fig. 4.9 we consider $\omega/n_o = 1.17 \times 10^{-13}$ and $4.78 \times 10^{-13}\,\mathrm{m^3\,s^{-1}}$ in the absence of *e-e* collisions (full curves) and assuming Maxwellian EEDFs (broken curves). We note that the results shown in Fig. 4.11 for the R wave are equivalent to those in a DC electric field, if the L wave was omitted and a R wave with $E_{0(R)} = E_{DC}$ was directly launched instead of the linearly polarized wave. The values for R waves are indistinguishable for the two values of ω/n_o.

Finally, Fig. 4.12 shows 2/3 of the average electron energy, $<u>$, as a function of the ratio of the *rms* amplitude of a linearly polarized wave to the gas number density, $E_{rms(l)}/n_o$, in argon at ECR conditions (i.e. for $\omega_{ce} = \omega$), and for $\omega/n_o = 1.17 \times 10^{-13}$ and $4.78 \times 10^{-13}\,\mathrm{m^3\,s^{-1}}$. As before the full curves are obtained in the absence of *e-e* collisions, while the broken ones are for Maxwellian EEDFs. In the case of these latter, $2/3\ <u>$ corresponds to the electron kinetic temperature $k_B T_e$ of the distributions.

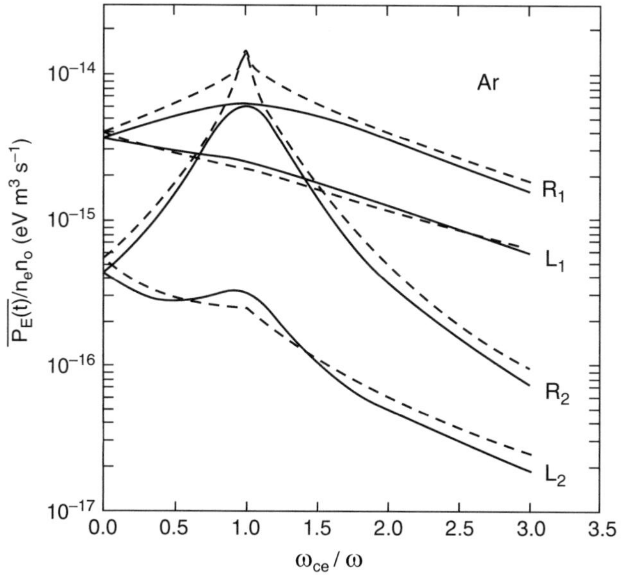

Fig. 4.10 As in Fig. 4.9 but for $E_{rms(l)}/n_o = 120$ Td (Loureiro 1995)

4.3 Radio-Frequency Electric Fields

4.3.1 Time-Dependent Boltzmann Equation

As the field frequency decreases up to values of the same order or smaller than the characteristic relaxation frequency for energy transfer, $\omega < \nu_e$, with ν_e given by equation (4.1), the collisional energy dissipation is faster than the radio-frequency (RF) field variation, $\nu_e^{-1} < T$, with T denoting the field period, and a large time-modulation occurs in the isotropic component of the electron velocity distribution function, i.e. in the EEDF (Margenau and Hartman 1948; Delcroix 1963,1966; Winkler et al. 1987; Capitelli et al. 1988; Loureiro 1993). The energy range where the time-modulation takes place depends on the form as ν_e varies with the electron energy. In argon, for example, ν_e sharply increases with the energy u (see Fig. 4.1), so that at field frequencies in RF range and for gas pressures typically $p \sim 100$ Pa, the inequality $\omega < \nu_e$ holds only in the high-energy tail of the EEDF (Sá et al. 1994). On the contrary, in a molecular gas such as H_2 or N_2 the above inequality holds at much lower electron energies, due to the contribution of vibrational excitation to the characteristic relaxation frequency ν_e, and a large time-modulation occurs also in the bulk of the distribution (Loureiro 1993). This is particularly evident in N_2 (see Fig. 4.2), where a sharp and pronounced maximum exists at $u \sim 2$ eV in the total cross section for vibrational excitation (see also Fig. 3.13).

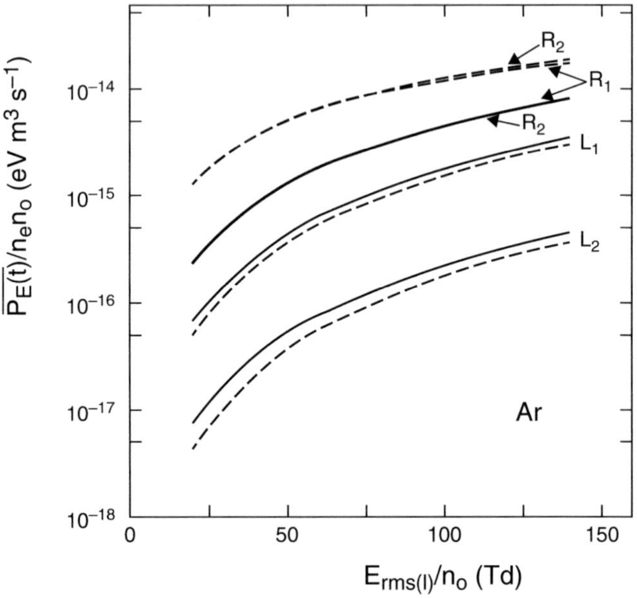

Fig. 4.11 Mean power absorbed per electron at unit gas density against $E_{rms(l)}/n_o$ in argon, for $\omega_{ce} = \omega$ and $\omega/n_o = 1.17 \times 10^{-13}$ m^3 s^{-1} (R_1; L_1) and 4.78×10^{-13} m^3 s^{-1} (R_2; L_2). R and L are for the right and left waves. *Full curves*: absence of *e-e* collisions; *Broken curves*: Maxwellian EEDFs (Loureiro 1995)

Fig. 4.12 2/3 of the average electron energy, <*u*>, against $E_{rms(l)}/n_o$ in argon, for $\omega_{ce} = \omega$, and for (*A*) $\omega/n_o = 1.17 \times 10^{-13}$ m^3 s^{-1} and (*B*) 4.78×10^{-13} m^3 s^{-1}. *Full curves*: absence of *e-e* collisions; *Broken curves*: Maxwellian EEDFs (Loureiro 1995)

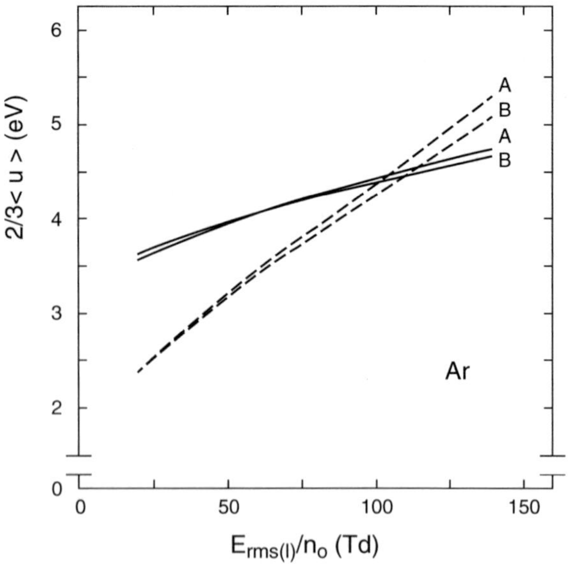

The EEDF in the collision dominated bulk plasma created by a RF field can be obtained by solving the time-dependent, spatially homogeneous, Boltzmann equation under the action of a time-varying sinusoidal electric field

$$\mathbf{E}(t) \ = \ \mathbf{E_0} \ \cos(\omega t) \ , \tag{4.79}$$

with $\mathbf{E_0} = - E_0 \, \mathbf{e_z}$, and where ω is the angular field frequency. Then, the electron velocity distribution function $f_e(\mathbf{v_e}, t)$ can be obtained by solving the time-dependent Boltzmann equation

$$\frac{\partial f_e}{\partial t} - \left(\frac{e}{m} \, \mathbf{E} \, . \, \frac{\partial f_e}{\partial \mathbf{v_e}} \right) \ = \ I(f_e) \ + \ J(f_e) \ , \tag{4.80}$$

where e and m are the electron absolute charge and mass, respectively, $\mathbf{v_e}$ is the electron velocity, and I and J denote the collision operator for elastic and for inelastic collisions, respectively. Here, as before, we will neglect processes which result in the production or loss of electrons, i.e., production of secondary electrons by ionization, electron-ion recombination, and electron attachment, so that $J(f_e)$ includes only the effects of energy-exchange processes, both for inelastic and superelastic collisions. The electron velocity distribution is normalized through the condition

$$n_e \ = \ \int_{\mathbf{v_e}} f_e(\mathbf{v_e}, t) \ \mathbf{dv_e} \ , \tag{4.81}$$

where n_e denotes the electron number density assumed here time-independent.

Equation (4.80) is solved by expanding f_e in Legendre polynomials in velocity space and Fourier series in time

$$f_e(\mathbf{v_e}, t) \ = \ \sum_{l=0} \sum_{k=0} \mathrm{Re}\{\overline{f_k^l}(v_e) \ e^{jk\omega t}\} \ P_l(\cos\theta) \ , \tag{4.82}$$

with θ denoting the angle between the instantaneous velocity $\mathbf{v_e}$ and the direction of the anisotropy along the z axis, Re{ } means "the real part of", $v_e = |\mathbf{v_e}|$, and $\overline{f_k^l}(v_e)$ is a complex function expressing the time delay of the electron transport with respect to the applied electric field (4.79). Here, we will assume that the anisotropies resulting from the field are sufficiently small, so that the first two terms in spherical functions suffice for the expansion

$$f_e(\mathbf{v_e}, t) \ \simeq \ f_e^0(v_e, t) \ + \ f_e^1(v_e, t) \ \cos\theta \ . \tag{4.83}$$

Under this assumption, the lowest-order approximation for the expansion in Fourier series, allowing a periodic time-variation in the isotropic component of the electron velocity distribution (that is, in the EEDF), is

$$f_e^0(v_e, t) \ \simeq \ f_0^0(v_e) \ + \ f_2^0(v_e) \ \cos(2\omega t + \phi_2^0(v_e)) \ , \tag{4.84}$$

while for the anisotropic component we have

$$f_e^1(v_e,t) \simeq f_1^1(v_e)\cos(\omega t + \phi_1^1(v_e))\,. \tag{4.85}$$

As shown below, the fact that the isotropic and the anisotropic components of the electron velocity distribution are functions of $E(t)^2$ and $E(t)$, respectively, originates that the isotropic velocity distribution, $f_e^0(v_e,t)$, only has even harmonics in the Fourier expansion, whereas $f_e^1(v_e,t)$ only has odd ones (Margenau and Hartman 1948; Delcroix 1963,1966). Then, the normalization condition (4.81) appropriate to the present simplification should be written as follows

$$\int_0^\infty f_0^0(v_e)\,4\pi v_e^2\,dv_e = n_e\,, \tag{4.86}$$

$$\int_0^\infty f_{2R}^0(v_e)\,4\pi v_e^2\,dv_e = 0\,, \tag{4.87}$$

$$\int_0^\infty f_{2I}^0(v_e)\,4\pi v_e^2\,dv_e = 0\,, \tag{4.88}$$

where f_{2R}^0 and f_{2I}^0 denote, respectively, the real and imaginary part of the complex amplitude $\overline{f_2^0} = f_2^0\,\exp(j\phi_2^0)$ of the time-varying isotropic component at frequency 2ω.

Introducing (4.82) into equation (4.80), one obtains the following system of nonlocal equations in velocity space for $f_0^0, \overline{f_2^0}$, and $\overline{f_1^1}$, respectively

$$\frac{1}{3v_e^2}\frac{d}{dv_e}\left(\frac{eE_0}{2m}\,v_e^2\,\mathrm{Re}\{\overline{f_1^1}\}\right) = I^0(f_0^0) + J^0(f_0^0)\,, \tag{4.89}$$

$$j2\omega\,\overline{f_2^0} + \frac{1}{3v_e^2}\frac{d}{dv_e}\left(\frac{eE_0}{2m}\,v_e^2\,\overline{f_1^1}\right) = I^0(\overline{f_2^0}) + J^0(\overline{f_2^0})\,, \tag{4.90}$$

$$j\omega\,\overline{f_1^1} + \frac{eE_0}{m}\frac{df_0^0}{dv_e} - \frac{eE_0}{2m}\frac{d\overline{f_2^0}}{dv_e} = I^1(\overline{f_1^1}) + J^1(\overline{f_1^1})\,. \tag{4.91}$$

The elastic and inelastic collision terms of the isotropic components, I^0 and J^0, are given by (3.85) and (3.133), whereas in equation (4.91) we have

$$I^1(\overline{f_1^1}) + J^1(\overline{f_1^1}) = -\,\nu_m^e\,\overline{f_1^1}\,, \tag{4.92}$$

with ν_m^e given by (4.2). As seen in equation (3.139) the use of an effective collision frequency in equation (4.92) is justifiable in gases for which the inelastic scattering is not negligible but is nearly isotropic. The evaluation of such an approximation in N_2 is treated in Phelps and Pitchford (1985).

Substituting (4.92) into equation (4.91), one obtains for $\overline{f_1^1}$

$$\overline{f_1^1} = -\frac{eE_0}{m(\nu_m^e + j\omega)}\frac{df_0^0}{dv_e} - \frac{eE_0}{2m(\nu_m^e + j\omega)}\frac{d\overline{f_2^0}}{dv_e} . \tag{4.93}$$

This expression when inserted in equations (4.89) and (4.90) allows us to write the following system of equations for f_0^0, f_{2R}^0, and f_{2I}^0 (Loureiro 1993)

$$-\frac{1}{6v_e^2}\left(\frac{eE_0}{m}\right)^2\frac{d}{dv_e}\left[v_e^2\left(A\frac{df_0^0}{dv_e} + \frac{A}{2}\frac{df_{2R}^0}{dv_e} + \frac{B}{2}\frac{df_{2I}^0}{dv_e}\right)\right]$$
$$= I^0(f_0^0) + J^0(f_0^0) , \tag{4.94}$$

$$-2\omega f_{2I}^0 - \frac{1}{6v_e^2}\left(\frac{eE_0}{m}\right)^2\frac{d}{dv_e}\left[v_e^2\left(A\frac{df_0^0}{dv_e} + \frac{A}{2}\frac{df_{2R}^0}{dv_e} + \frac{B}{2}\frac{df_{2I}^0}{dv_e}\right)\right]$$
$$= I^0(f_{2R}^0) + J^0(f_{2R}^0) , \tag{4.95}$$

$$2\omega f_{2R}^0 + \frac{1}{6v_e^2}\left(\frac{eE_0}{m}\right)^2\frac{d}{dv_e}\left[v_e^2\left(B\frac{df_0^0}{dv_e} + \frac{B}{2}\frac{df_{2R}^0}{dv_e} - \frac{A}{2}\frac{df_{2I}^0}{dv_e}\right)\right]$$
$$= I^0(f_{2I}^0) + J^0(f_{2I}^0) , \tag{4.96}$$

with

$$A = \frac{\nu_m^e}{\nu_m^{e\,2} + \omega^2} ; \qquad B = \frac{\omega}{\nu_m^{e\,2} + \omega^2} . \tag{4.97}$$

Once the functions $f_0^0(v_e)$, $f_{2R}^0(v_e)$, and $f_{2I}^0(v_e)$ are obtained, the anisotropic component $\overline{f_1^1}(v_e)$ is calculated from equation (4.93). Then, the drift velocity $\mathbf{v_{ed}}(t) = v_{ed}(t)\,\mathbf{e_z}$, the energy-averaged electron energy $<u>$ (t), or, for example, the electron rate coefficient $C_{ij}(t)$ for excitation of a j state from i sate can be readily obtained.

Transforming now as in (3.166) the isotropic component of the electron velocity distribution $f_e^0(v_e, t)$ to the electron energy distribution function $F^0(u, t)$ as follows

$$f_e^0(v_e, t)\, 4\pi v_e^2\, dv_e = n_e\, F^0(u, t)\, \sqrt{u}\, du , \tag{4.98}$$

and $f_0^0(v_e)$, $f_{2R}^0(v_e)$ and $f_{2I}^0(v_e)$ transformed to $F_0^0(u)$, $F_{2R}^0(u)$ and $F_{2I}^0(u)$, while the anisotropic component $\overline{f_1^1}(v_e)$ is transformed to $\overline{F_1^1}(u)$

$$\overline{f_1^1}(v_e)\, 4\pi v_e^2\, dv_e = n_e\, \overline{F_1^1}(u)\, \sqrt{u}\, du , \tag{4.99}$$

with $u = \frac{1}{2} m v_e^2$ denoting the electron energy, we obtain successively for the different quantities:

(i) Drift velocity

$$v_{ed}(t) = \mathrm{Re}\{\overline{V_{d0}}\, e^{j\omega t}\} , \tag{4.100}$$

with $\overline{V_{d0}}$ given by

$$\overline{V_{d0}} = \frac{1}{3}\sqrt{\frac{2}{m}} \int_0^\infty u\, \overline{F_1^1}(u)\, du ; \tag{4.101}$$

(ii) Energy-averaged electron energy

$$<u>(t) = \int_0^\infty u^{3/2} F^0(u,t)\, du , \tag{4.102}$$

which can also be written under the form

$$<u>(t) = <u>_0 + <u>_{2R} \cos(2\omega t) - <u>_{2I} \sin(2\omega t) , \tag{4.103}$$

with $<u>_0$ given by

$$<u>_0 = \int_0^\infty u^{3/2} F_0^0(u)\, du , \tag{4.104}$$

and where $<u>_{2R}$ and $<u>_{2I}$ are given by similar expressions but in which $F_0^0(u)$ is replaced with $F_{2R}^0(u)$ and $F_{2I}^0(u)$, respectively;

(iii) Electron rate coefficient for a $i \rightarrow j$ state transition

$$C_{ij}(t) = <v_e\, \sigma_{ij}> = \sqrt{\frac{2}{m}} \int_0^\infty u\, \sigma_{ij}(u)\, F^0(u,t)\, du , \tag{4.105}$$

or in the form

$$C_{ij}(t) = (C_{ij})_0 + (C_{ij})_{2R} \cos(2\omega t) - (C_{ij})_{2I} \sin(2\omega t) , \tag{4.106}$$

with $(C_{ij})_0$ given by

$$(C_{ij})_0 = \sqrt{\frac{2}{m}} \int_0^\infty u\, \sigma_{ij}(u)\, F_0^0(u)\, du , \tag{4.107}$$

and where $(C_{ij})_{2R}$ and $(C_{ij})_{2I}$ are given by similar expressions using $F_{2R}^0(u)$ and $F_{2I}^0(u)$.

The energy-averaged power balance equation may also be decoupled into three equations as follows

$$(P_E)_0 = (P_{el})_0 + (P_{rot})_0 + (P_{inel})_0 \tag{4.108}$$

$$(P_E)_{2R} = (P_{el})_{2R} + (P_{rot})_{2R} + (P_{inel})_{2R} + (P_\omega)_{2R} \tag{4.109}$$

$$(P_E)_{2I} = (P_{el})_{2I} + (P_{rot})_{2I} + (P_{inel})_{2I} + (P_\omega)_{2I} , \tag{4.110}$$

in which the terms in the left-hand side members are

$$(P_E)_0 = \frac{en_e E_0}{2} \text{Re}\{\overline{V_{d0}}\} \tag{4.111}$$

$$(P_E)_{2R} = (P_E)_0 \tag{4.112}$$

$$(P_E)_{2I} = \frac{en_e E_0}{2} \text{Im}\{\overline{V_{d0}}\} . \tag{4.113}$$

The terms for P_{el}, P_{rot} and P_{inel} are identical to equations (3.176), (3.177), and (3.151) replacing F_0^0, $F_{2R}^0(u)$ and $F_{2I}^0(u)$, whereas the terms $(P_\omega)_{2R}$ and $(P_\omega)_{2I}$ associated with the interchange in equations for $F_{2R}^0(u)$ and $F_{2I}^0(u)$ due to the first term in equation (4.90), taking into consideration the phase-shift ϕ_2^0 in equation (4.84), are

$$(P_\omega)_{2R} = -\ 2n_e\omega <u>_{2I} \tag{4.114}$$

$$(P_\omega)_{2I} = 2n_e\omega <u>_{2R} . \tag{4.115}$$

Finally, it follows from this formulation that the instantaneous energy-averaged power absorbed from the field is given by

$$P_E(t) = (\mathbf{J_e} \cdot \mathbf{E}) = (P_E)_0 \, [1 + \cos(2\omega t)] - (P_E)_{2I} \, \sin(2\omega t) , \tag{4.116}$$

with $\mathbf{J_e} = -\ en_e\mathbf{v_{ed}}$ denoting the electron current density.

4.3.2 Time-Dependent Velocity Distributions

The time-dependent electron Boltzmann transport equation written in the form (4.94), (4.95) and (4.96) can be solved to yield the EEDF, $F^0(u,t)$, as a function of the independent parameters: ratio of the electric field amplitude to the gas number density, E_0/n_o; ratio of the angular field frequency to the gas density, ω/n_o; gas temperature, T_o; and vibrational temperature, T_v. The latter used to take into account the effects produced by the inelastic e-V processes starting from a level $v > 0$, and the superelastic e-V processes (Loureiro 1993).

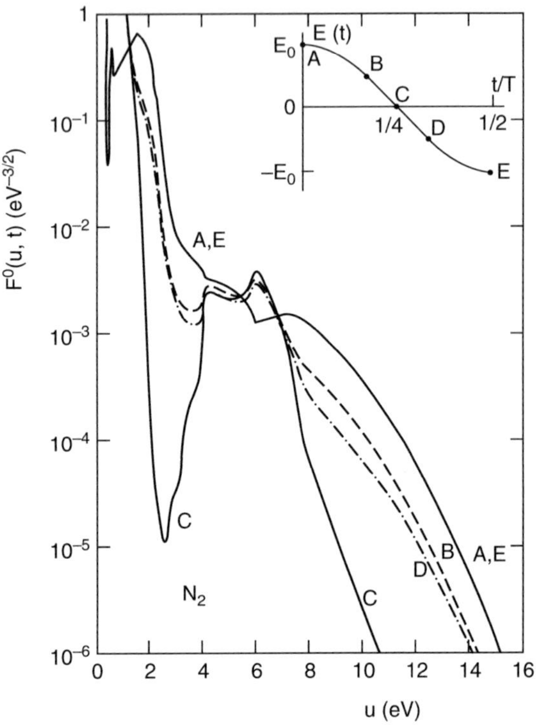

Fig. 4.13 EEDFs in N_2 for $E_0/n_o = 60\sqrt{2}$ Td, $\omega/n_o = 5 \times 10^{-16}\,\text{m}^3\,\text{s}^{-1}$, and $T_v = T_o = 400$ K. The various curves are the following instants with T denoting the field period: (*A*) 0; (*B*) *T*/6; (*C*) *T*/4; (*D*) *T*/3; (E) *T*/2 (Loureiro 1993)

Figure 4.13 shows the EEDFs calculated in nitrogen for $E_0/n_o = 60\sqrt{2}$ Td ($1\,\text{Td} = 1 \times 10^{-21}\,\text{V m}^2$), $\omega/n_o = 5 \times 10^{-16}\,\text{m}^3\,\text{s}^{-1}$, and $T_v = T_o = 400$ K, i.e. in the absence of appreciable vibrational excitation, and for the different times during the half period of the RF electric field shown in the inset.

For the considered RF field the EEDFs oscillate with twice the field frequency presenting a modulation of many orders of magnitude and a small phase delay, relative to the applied electric field, in those parts of the electron-energy range where ω is appreciably smaller than the relaxation frequency for energy transfer, ν_e, shown in Fig. 4.2. The sharp maximum of ν_e/n_o at $u \simeq 2$ eV, of about $10^{-13}\,\text{m}^3\,\text{s}^{-1}$, due to vibrational excitation is clearly larger than the value chosen here for ω/n_o, so that for electron energies around 2 eV the EEDFs follow the RF field in a quasi-stationary way excepting when the field goes through zero. We note that in this region of electron energies the EEDFs are maximum when the absolute value of the RF field passes through its maximum, decrease strongly as the absolute value of the field decreases, and reach a minimum when the field passes through zero. In particular, for zero field, the EEDFs are extremely reduced and most of the electrons have only very small energy.

The other parts of the EEDFs can be interpreted as well by looking at the dependence of the relaxation frequency ν_e on the electron energy shown in Fig. 4.2. There are two energy regions $u < 1.5$ eV and $4 < u < 8$ eV practically devoid of vibrational or electronic inelastic processes, which correspond to relatively small values of ν_e/n_o. In both regions the inequality $\omega \ll \nu_e$ is no longer valid so the EEDFs follow the RF field with a much smaller modulation and large phase delay. On the other hand, in the high-energy range $u > 8$ eV, dominated by the excitation of electronic states, the relaxation frequency ν_e has approximately the same amplitude as for $u \simeq 2$ eV, so that the EEDFs show a very large modulation as well. It is also interesting to note that, as a result of the nonequilibrium between the EEDFs and the applied RF field, the EEDFs are different at the instants t and $T/2 - t$, with T denoting the field period. The EEDFs are naturally larger when the RF is decreasing, $0 < t < T/4$, because of memory effect. Obviously, the differences between the EEDFs at both instants are decreasingly smaller as ν_e/n_o increases, which is the case as the electron energy increases from $\simeq$8 to $\simeq$14 eV (see curves B and D in Fig. 4.13).

Figure 4.14 shows the EEDFs calculated in nitrogen for the same values of $E_0/n_o = 60\sqrt{2}$ Td and $T_v = T_o = 400$ K as in Fig. 4.13, but for the higher value of $\omega/n_o = 1 \times 10^{-14}$ m^3 s^{-1}. In this case we have $\omega > \nu_e$ in large parts of the relevant range of the electron energies so that the amplitudes of the time modulation of the EEDFs are strongly diminished with respect to Fig. 4.13.

Figure 4.15 shows the EEDFs calculated in nitrogen, still for the same values of E_0/n_o and $T_v = T_o$ as in Figs. 4.13 and 4.14, but for the higher value of

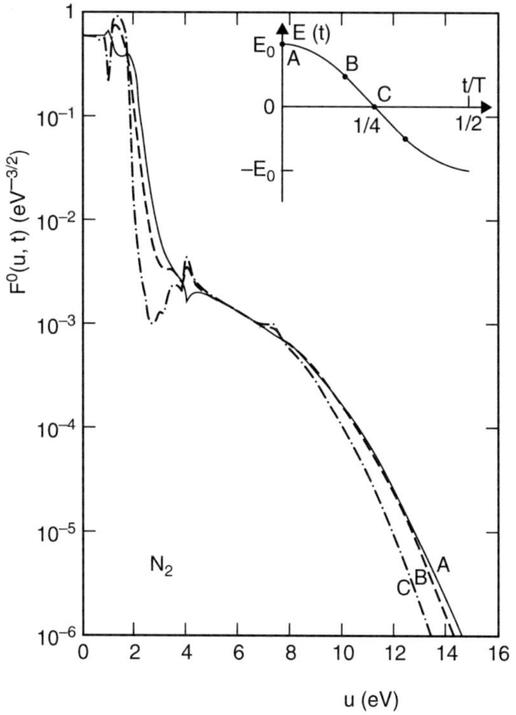

Fig. 4.14 EEDFs in N_2 for the same values of E_0/n_o and $T_v = T_o$ as in Fig. 4.13, but for $\omega/n_o = 1 \times 10^{-14}$ m^3 s^{-1}. The various curves are for the following instants with T denoting the field period: (*A*) 0; (*B*) *T*/6; (*C*) *T*/4 (Loureiro 1993)

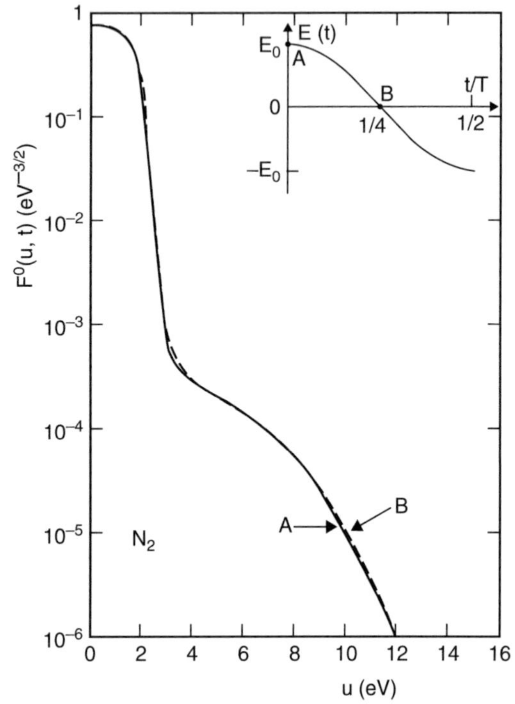

Fig. 4.15 EEDFs in N_2 for the same values of E_0/n_o and $T_v = T_o$ as in Figs. 4.13 and 4.14, but for $\omega/n_o = 1 \times 10^{-13}\,\mathrm{m^3\,s^{-1}}$. The curves are for the instants: (*A*) 0; (*B*) *T*/4 (Loureiro 1993)

$\omega/n_o = 1 \times 10^{-13}\,\mathrm{m^3\,s^{-1}}$. In this case we have no time-modulation at all because the inequality $\omega > \nu_e$ holds in the entire electron-energy range. On the other hand, as long as ω is smaller than the relaxation frequency for momentum transfer ν_m^e, the time-averaged value of the EEDF, $\overline{F^0(u,t)} = F_0^0(u)$, is only slightly dependent on ω. In the limit $\omega \ll \nu_m^e$, we have $A \simeq 1$ and $B \simeq 0$ in equations (4.97), and equation (4.94) becomes identical to the Boltzmann equation for a DC field of magnitude equal to the effective field strength $E_{eff} = E_0/\sqrt{2}$, excepting that here we must also keep in equation (4.94) the term with F_{2R}^0. However, as the field frequency increases beyond ν_m^e, the time-averaged value $\overline{F^0(u,t)} = F_0^0(u)$ is strongly reduced. As we have seen in Sects. 4.1.2 and 4.1.3, when $\omega \gg \nu_m^e$ the time delay of the electron current density approaches $T/4$ and the electrons cannot gain energy from the field on the average. In this limit the time average over one period of equation (4.116) yields only a vanishingly small value $\overline{P_E(t)} = \overline{(\mathbf{J_e} \cdot \mathbf{E})} \simeq 0$. For frequencies of this order, which correspond to microwave fields at gas pressures typically $p \sim 100\,\mathrm{Pa}$, the EEDFs exhibit a strong peak at zero energy. See, for example, the EEDFs in argon for the highest values of ω/n_o shown in Fig. 4.3.

This latter aspect is well understood through the evolution of the drift velocity with the ω/n_o values. Figure 4.16 shows the drift velocity $\mathbf{v_{ed}} = v_{ed}(t)\,\mathbf{e_z}$ calculated for $E_0/n_o = 60\sqrt{2}\,\mathrm{Td}$, $T_v = T_o = 400\,\mathrm{K}$, and various values of ω/n_o between 5×10^{-16} and $1 \times 10^{-12}\,\mathrm{m^3\,s^{-1}}$. This figure shows that for $\omega/n_o < 1 \times 10^{-14}\,\mathrm{m^3\,s^{-1}}$

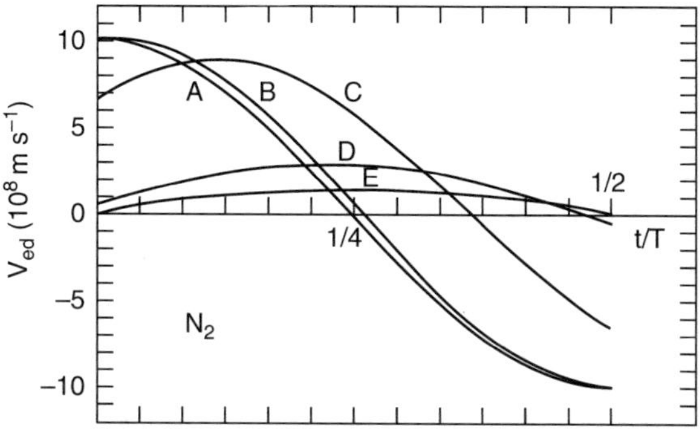

Fig. 4.16 Electron drift velocity in N_2 as a function of the reduced time t/T, with T denoting the field period, for $E_0/n_o = 60\sqrt{2}\,$Td, $T_v = T_o = 400\,$K, and the following values of ω/n_o in $\mathrm{m^3\,s^{-1}}$: (*A*) 5×10^{-16}; (*B*) 1×10^{-14}; (*C*) 1×10^{-13}; (*D*) 5×10^{-13}; (*E*) 1×10^{-12} (Loureiro 1993)

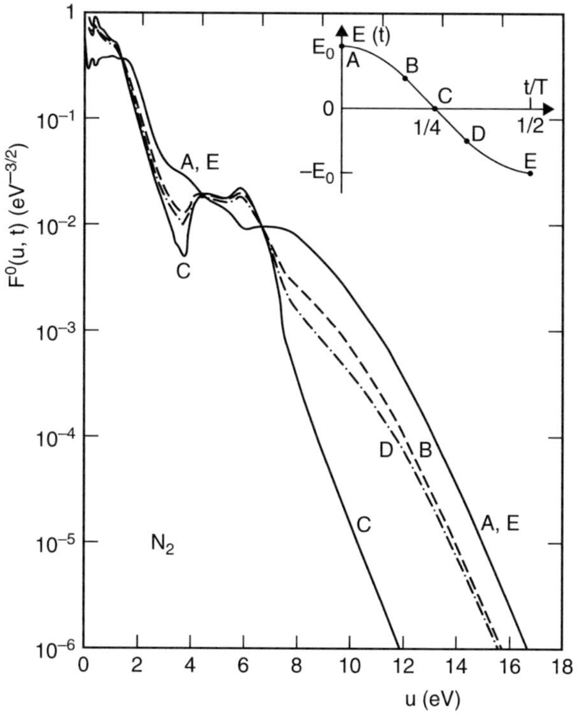

Fig. 4.17 EEDF in N_2 as in Fig. 4.13 but for $T_v = 4000\,$K (Loureiro 1993)

there is no phase delay with respect to the applied field $\mathbf{E}(t) = -E_0\cos(\omega t)\,\mathbf{e_z}$; for $\omega/n_o \simeq 1\times10^{-13}\,\mathrm{m^3\,s^{-1}}$, the delay is $\simeq -\pi/4$ in agreement with the fact of $\omega \sim \nu_m^e$ over most of the relevant electron-energy range; and for the highest values of ω/n_o the delay approaches $-\pi/2$. On the other hand, for the highest values of ω/n_o the drift velocity strongly reduces in magnitude.

The large modulation of the EEDFs in the electron-energy range $1.5 < u < 4\,$eV is strongly reduced when we consider the effects produced by the e-V superelastic collisions. Figure 4.17 shows the EEDFs for the same values of E/n_o, ω/n_o, and T_o as in Fig. 4.13, but for the case of vibrational excitation corresponding to a characteristic vibrational temperature $T_v = 4000\,\mathrm{K}$. The comparison between Figs. 4.13 and 4.17 allows to evaluate the effects caused by the *e-V* superelastic collisions in reducing the effectiveness of the characteristic relaxation frequency for energy transfer ν_e, shown in Fig. 4.2, in the electron-energy range under discussion. The *e-V* superelastic processes produce a decrease in the amplitude and an increase in phase delay of the EEDFs, for energies $1.5 < u < 4\,\mathrm{eV}$, as well as an enhancement of the high-energy tail of the EEDFs. This latter aspect has already been discussed for the EEDFs obtained in a DC field in Sect. 3.6. Finally, it is worth noting that the enhancement of the high-energy tail of the EEDFs from Figs. 4.13 to 4.17 is not very significant because we have chosen here a relatively high value of E_0/n_o. In the case of lower E_0/n_o values, e.g. as small as $30\sqrt{2}\,\mathrm{Td}$, the effects produced by the *e-V* superelastic collisions would become much larger (see Sect. 3.6).

The EEDFs in hydrogen are much lesser modulated than in nitrogen due to the relaxation frequency for energy transfer is about one order of magnitude smaller. Figure 4.18 shows the EEDFs in H_2 calculated for $E_0/n_o = 30\sqrt{2}\,\mathrm{Td}$, $\omega/n_o = 2 \times 10^{-15}\,\mathrm{m^3\,s^{-1}}$, $T_v = T_o = 400\,\mathrm{K}$, and for the different times shown in the inset. As seen from this figure, as compared with those presented before for N_2, the time modulation is clearly less pronounced here, which is a consequence of the relative magnitudes of the relaxation frequency for energy transfer in both gases plotted in

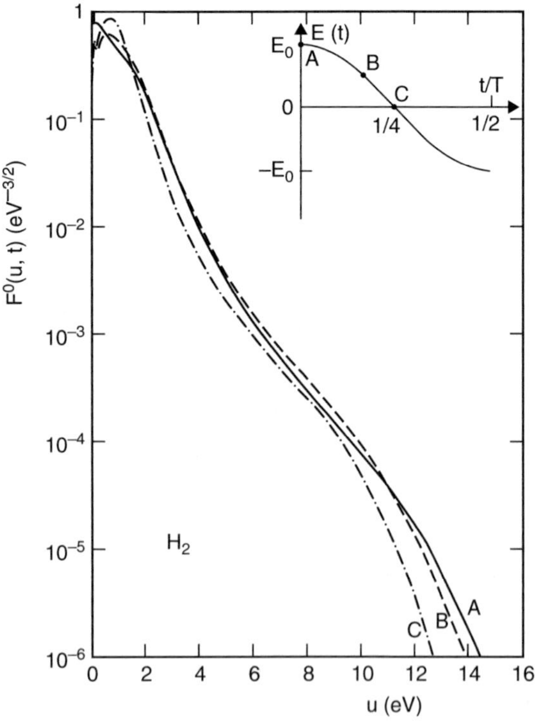

Fig. 4.18 EEDFs in H_2 for $E_0/n_o = 30\sqrt{2}\,\mathrm{Td}$, $\omega/n_o = 2 \times 10^{-15}\,\mathrm{m^3\,s^{-1}}$, $T_v = T_o = 400\,\mathrm{K}$, and for the following instants: (*A*) 0; (*B*) *T*/6; (*C*) *T*/4 (Loureiro 1993)

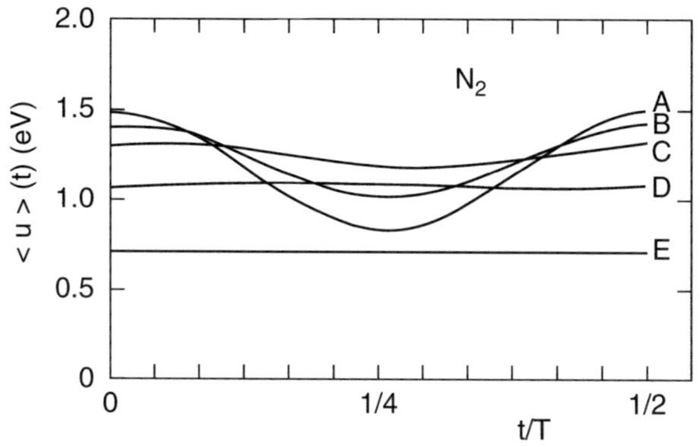

Fig. 4.19 Energy-averaged electron energy in N_2 as a function of the reduced time t/T, for $E_0/n_o = 60\sqrt{2}$ Td, $T_v = T_o = 400$ K, and the following values of ω/n_o in $m^3\,s^{-1}$: (*A*) 5×10^{-16}; (*B*) 1×10^{-15}; (*C*) 1×10^{-14}; (*D*) 1×10^{-13}; (E) 1×10^{-12} (Loureiro 1993)

Fig. 4.2. Besides the fact of ν_e/n_o is smaller by one order of magnitude in H_2, it does not present any pronounced maximum due to the dissipation of electron energy in vibrational excitation as it exists in N_2 at $\sim$2 eV. The absence of such a maximum in H_2 also signifies that the effects of *e*-*V* superelastic collisions produce only minor modifications in the EEDFs in H_2 as compared to those observed in N_2.

Once the EEDF are obtained, the various energy-averaged quantities with interest for plasma modelling can also be obtained as a function of time. Figure 4.19 shows the energy-averaged electron energy, $<u>(t)$, calculated in nitrogen for $E_0/n_o = 60\sqrt{2}$ Td, $T_v = T_o = 400$ K, and various values of ω/n_o between 5×10^{-16} and $1\times10^{-12}\,m^3\,s^{-1}$. This figure shows that there is a marked reduction in the time modulation (at the frequency 2ω) as ω/n_o increases together with a reduction in the amplitude of the time-averaged values. There is also an increasing phase delay which approaches $-\pi$, i.e. a time delay $-T/4$, as $\omega\to\infty$, but this latter aspect is not particularly visible in this figure.

Finally, Fig. 4.20 shows the electron rate coefficient, $C_X^B(t)$, for excitation of the triplet state $N_2(B\,^3\Pi_g)$ at the same conditions as the previous figure for $<u>(t)$. Both the time modulation and the reduction of amplitude for the highest values of ω are particularly visible now because the rate coefficient depends of the high-energy tail of the EEDF only, whereas $<u>(t)$ in Fig. 4.19 results from an integration over the entire energy range. Also the increasing phase delay approaching $-T/4$ is more clearly visible in this figure. It is worth noting here that instead of equation (4.106) the electron rate coefficient for excitation of a given $i\to j$ transition may also be written under the form

$$C_{ij}(t) = (C_{ij})_0 + (C_{ij})_2\cos(2\omega t + \phi_2^0)\,, \tag{4.117}$$

with $(C_{ij})_2 \simeq (C_{ij})_0$ and $\phi_2^0 \simeq 0$ when $\omega \ll \nu_e$, and $(C_{ij})_2 \ll (C_{ij})_0$ and $\phi_2^0 \simeq -\pi$ when $\omega \gg \nu_e$. Then, the term $(C_{ij})_0$ is strongly reduced when $\omega \gg \nu_m^e$.

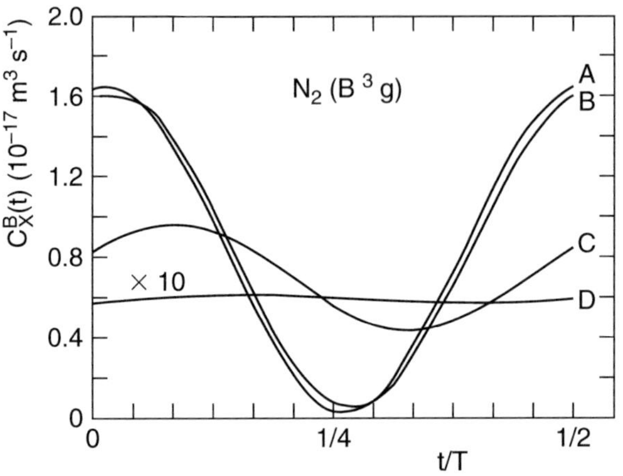

Fig. 4.20 Electron rate coefficient for excitation of $N_2(B\ ^3\Pi_g)$ state as a function of the reduced time t/T, for $E_0/n_o = 60\sqrt{2}$ Td, $T_v = T_o = 400$ K, and the following values of ω/n_o in $m^3\,s^{-1}$: (*A*) 5×10^{-16}; (*B*) 1×10^{-15}; (*C*) 1×10^{-14}; (*D*) 1×10^{-13} (Loureiro 1993)

It follows from the present analysis that the instantaneous energy-averaged power absorbed from the RF field and the instantaneous power lost in electron collisions are not in phase with each other. The time-modulation at frequency 2ω of the instantaneous power lost by electron collisions is strongly diminished and its phase-delay approaches $-\pi$ (i.e. $\Delta t = T/4$) as ω/n_o increases, such as it occurred with $<u>$ (t) and $C_X^B(t)$. On the contrary, the instantaneous power absorbed from the field shows no reduction of its modulation and has an increasing phase-delay going to $-\pi/2$ (i.e. $\Delta t = T/8$, see equation (4.116) and Exercise 4.6). We note that $P_E(t) \simeq -(P_E)_{2I} \sin(2\omega t)$ in equation (4.116) as $\omega \to \infty$. Obviously, the time-averaged values of the energy-loss and energy-gain terms exactly compensate each other and both go towards zero as ω/n_o increases beyond ν_m^e/n_o.

Appendices

A.4.1 Effective Collision Frequency and Electron Density in High-Frequency Discharges

When the effective collision frequency for momentum transfer ν_m^e, given by equation (3.139), is independent of electron velocity, the equation for electron momentum conservation (3.60) in high-frequency (HF) fields reduces to

$$n_e\, m\, \frac{d\mathbf{v}_{ed}}{dt} + n_e\, e\, \mathbf{E} = -\, n_e\, m\, \nu_m^e\, \mathbf{v}_{ed}, \tag{4.118}$$

with $\mathbf{v_{ed}}$ denoting the electron drift velocity. Here, we are assuming the field frequency much larger than the characteristic relaxation frequency for energy transfer $\omega \gg \nu_e$, in order the electron density does not change appreciably during a cycle of the field oscillation. Then, for an applied field of frequency ω, we have $\mathbf{dv_{ed}}/dt = j\omega\ \mathbf{v_{ed}}$, and the drift velocity is given by the well known formula

$$\mathbf{v_{ed}} = -\frac{e}{m\,(\nu_m^e + j\omega)}\,\mathbf{E}\,. \tag{4.119}$$

The electron current density $\mathbf{J_e} = -\,en_e\ \mathbf{v_{ed}}$, allows to obtain the complex conductivity from $\mathbf{J_e} = \overline{\sigma_{ce}}\ \mathbf{E}$, with the form

$$\overline{\sigma_{ce}} = \frac{e^2 n_e}{m\,(\nu_m^e + j\omega)}\,, \tag{4.120}$$

whereas the time-averaged power absorbed from the field (4.15) is

$$\overline{P_E(t)} = \frac{1}{2}\,\frac{e^2 n_e\ \nu_m^e}{m\,(\nu_m^{e\,2} + \omega^2)}\,E_0^2\,. \tag{4.121}$$

However, when the frequency ν_m^e is velocity-dependent, we must consider the expression (4.9) for the complex anisotropic component

$$\mathbf{f_e^1} = \overline{f_e^1}(v_e)\,e^{j\omega t}\,\mathbf{e_z} = -\,\frac{1}{\nu_m^e + j\omega}\,\frac{eE_0}{m}\,\frac{df_e^0}{dv_e}\,e^{j\omega t}\,\mathbf{e_z}\,, \tag{4.122}$$

in the collision term of equation for momentum conservation (3.55)

$$\mathbf{I_1} = -\int_0^\infty m\,\frac{v_e}{3}\,\nu_m^e\,\mathbf{f_e^1}\,4\pi v_e^2\,dv_e\,, \tag{4.123}$$

assuming here the interactions of collision type only and the applied field with the form $\mathbf{E} = -\,E_0\,e^{j\omega t}\,\mathbf{e_z}$. The complex drift velocity (4.10) and (4.11) is then

$$\mathbf{v_{ed}} = \overline{V_{ed}}\,e^{j\omega t}\,\mathbf{e_z} = -\,\frac{eE_0}{n_e m}\int_0^\infty \frac{1}{\nu_m^e + j\omega}\,\frac{df_e^0}{dv_e}\,\frac{4\pi v_e^3}{3}\,dv_e\,e^{j\omega t}\,\mathbf{e_z}\,, \tag{4.124}$$

so that substituting equation (4.123) in equation (4.118), we find

$$n_e m\,j\omega\,\overline{V_{ed}} - n_e\,eE_0 = eE_0\int_0^\infty \frac{\nu_m^e}{\nu_m^e + j\omega}\,\frac{df_e^0}{dv_e}\,\frac{4\pi v_e^3}{3}\,dv_e\,. \tag{4.125}$$

In the case of a velocity-independent frequency ν_m^e, we immediately obtain equation (4.119) from (4.125). In fact, integrating by parts

$$\int_0^\infty \frac{df_e^0}{dv_e} \frac{4\pi v_e^3}{3} dv_e = -\int_0^\infty f_e^0 \, 4\pi v_e^2 \, dv_e = -n_e \,, \tag{4.126}$$

we find

$$n_e m \, j\omega \, \overline{V}_{ed} - n_e \, eE_0 = -eE_0 \frac{\nu_m^e}{\nu_m^e + j\omega} n_e \tag{4.127}$$

and therefore

$$\overline{V_{ed}} = \frac{eE_0}{m\,(\nu_m^e + j\omega)} \,. \tag{4.128}$$

A different situation occurs as ν_m^e depends on the electron velocity. In this case in order to express equation (4.125) with the form (4.118) and to obtain an equivalent expression (4.120) for the complex conductivity, we should consider an effective complex collision frequency $\overline{\nu}_{eff}$ in equation (4.125), such as

$$n_e m \, j\omega \, \overline{V_{ed}} - n_e \, eE_0 = -n_e m \, \overline{\nu}_{eff} \, \overline{V_{ed}} \,. \tag{4.129}$$

Then, using equation (4.124) for $\overline{V_{ed}}$, we obtain

$$\overline{\nu}_{eff} = \int_0^\infty \frac{\nu_m^e}{\nu_m^e + j\omega} \frac{df_e^0}{dv_e} v_e^3 \, dv_e \times \left(\int_0^\infty \frac{1}{\nu_m^e + j\omega} \frac{df_e^0}{dv_e} v_e^3 \, dv_e \right)^{-1} \,. \tag{4.130}$$

Obviously $\overline{\nu}_{eff}$ allows to obtain ν_m^e again as this latter frequency is time-independent. When such is not the case, equation (4.129) allows to obtain an equivalent expression to equation (4.128) but with a complex frequency in place of ν_m^e

$$\overline{V_{ed}} = \frac{eE_0}{m\,(\overline{\nu}_{eff} + j\omega)} \,, \tag{4.131}$$

whereas the complex conductivity is

$$\overline{\sigma_{ce}} = \frac{e^2 n_e}{m\,(\overline{\nu}_{eff} + j\omega)} \,. \tag{4.132}$$

However, separating $\overline{\nu}_{eff}$ in real and imaginary parts, $\overline{\nu}_{eff} = \nu_R + j\,\nu_I$, we obtain an exact equivalent expression (4.120) as follows

$$\overline{\sigma_{ce}} = \frac{e^2 \, n_e^*}{m\,(\nu^* + j\omega)} \,, \tag{4.133}$$

with n_e^* and ν^* denoting an effective electron density and a new effective collision frequency given by

$$n_e^* = \frac{n_e}{1 + \nu_I/\omega} \tag{4.134}$$

$$\nu^* = \frac{\nu_R}{1 + \nu_I/\omega} . \tag{4.135}$$

Separating equation (4.130) in equations for the real and imaginary parts of complex quantities, we obtain from equation for the imaginary part

$$\nu_R \int_0^\infty \frac{1}{\nu_m^{e\,2} + \omega^2} \, v_e^{\,3} \, \frac{df_e^0}{dv_e} \, dv_e = \left(1 + \frac{\nu_I}{\omega}\right) \int_0^\infty \frac{\nu_m^e}{\nu_m^{e\,2} + \omega^2} \, v_e^{\,3} \, \frac{df_e^0}{dv_e} \, dv_e \, , \tag{4.136}$$

from which we may write

$$\nu^* = \int_0^\infty \frac{\nu_m^e}{\nu_m^{e\,2} + \omega^2} \, v_e^{\,3} \, \frac{df_e^0}{dv_e} \, dv_e \times \left(\int_0^\infty \frac{1}{\nu_m^{e\,2} + \omega^2} \, v_e^{\,3} \, \frac{df_e^0}{dv_e} \, dv_e \right)^{-1} . \tag{4.137}$$

On the other hand, from the comparison between the imaginary part of the complex conductivity (4.12)

$$\overline{\sigma_{ce}} = e n_e \, \overline{\mu_e} = - \frac{e^2}{m} \int_0^\infty \frac{1}{\nu_m^e + j\omega} \, \frac{df_e^0}{dv_e} \, \frac{4\pi v_e^{\,3}}{3} \, dv_e \tag{4.138}$$

and the imaginary part of equation (4.133), we obtain

$$n_e^* = - \left(\nu^{*\,2} + \omega^2\right) \int_0^\infty \frac{1}{\nu_m^{e\,2} + \omega^2} \, \frac{df_e^0}{dv_e} \, \frac{4\pi v_e^{\,3}}{3} \, dv_e \, . \tag{4.139}$$

When ν_m^e is velocity-independent, equations (4.137) and (4.139) reduce to $\nu^* = \nu_m^e$ and $n_e^* = n_e$. Finally, in terms of the EEDF normalized through equation (3.166), ν^* and n_e^* are written as follows

$$\nu^* = \int_0^\infty \frac{\nu_m^e}{\nu_m^{e\,2} + \omega^2} \, u^{3/2} \, \frac{df}{du} \, du \times \left(\int_0^\infty \frac{1}{\nu_m^{e\,2} + \omega^2} \, u^{3/2} \, \frac{df}{du} \, du \right)^{-1} \tag{4.140}$$

$$n_e^* = - \frac{2}{3} \, n_e \left(\nu^{*\,2} + \omega^2\right) \int_0^\infty \frac{1}{\nu_m^{e\,2} + \omega^2} \, u^{3/2} \, \frac{df}{du} \, du \, . \tag{4.141}$$

As firstly pointed out in Whitmer and Herrmann (1966), the two effective parameters ν^* and n_e^* have the advantage to permit the writing of a relation of the type (4.120) for the electron conductivity in HF fields, for gases in which the collision frequency for momentum transfer depends on the electron energy.

Exercises

Exercise 4.1. Determine the expression of the velocity-averaged collision frequency for momentum transfer valid as $\omega \gg \nu_m^e$, that allows to keep the time-averaged power absorbed from the field $\overline{P_E(t)}$ with the same form as the effective collision frequency for momentum transfer is independent of energy.

Resolution: When $\omega \gg \nu_m^e$ and $\nu_m^e(v_e) = \text{const}$, we obtain the following expression for the time-averaged power absorbed from the field from equations (4.15) and (4.16)

$$\overline{P_E(t)} = \frac{e^2 n_e}{m} \frac{\nu_m^e}{\omega^2} E_{rms}^2 ,$$

so that as $\nu_m^e(v_e) \neq \text{const}$, we must consider the energy-averaged frequency

$$\nu^* = -\frac{1}{n_e} \int_0^\infty \nu_m^e \frac{df_e^0}{dv_e} \frac{4\pi v_e^3}{3} dv_e ,$$

in order an equivalent expression for $\overline{P_E(t)}$ may be used

$$\overline{P_E(t)} = \frac{e^2 n_e}{m} \frac{\nu^*}{\omega^2} E_{rms}^2 .$$

Integrating by parts the frequency ν^* can also be written as

$$\nu^* = < \frac{1}{v_e^2} \frac{d}{dv_e} \left(\frac{\nu_m^e v_e^3}{3} \right) > .$$

Using the normalization (3.166) the first expression for ν^* writes as follows

$$\nu^* = -\frac{2}{3} \int_0^\infty \nu_m^e u^{3/2} \frac{df}{du} du .$$

This frequency corresponds to the effective collision frequency introduced in Appendix A.4.1, in the limit $\omega \gg \nu_m^e$, and it differs from the effective collision frequency ν_m' used with a DC field to write the electron mobility as $\mu_e = e/(m\, \nu_m')$ (see Exercise 3.6). Obviously, both result in ν_m^e when the frequency ν_m^e is velocity-independent.

Exercise 4.2. Obtain the time-averaged power absorbed from the field by the electrons, for the case of a high-frequency field $\omega \gg \nu_m^e$, directly from equation (4.22) for the isotropic component of the electron velocity distribution f_e^0.

Resolution: Multiplying the symmetric of the left-hand side member of equation (4.22) by the electron velocity $u = \frac{1}{2} m v_e^2$ and integrating in all velocity space, we obtain

$$\overline{P_E(t)} = \int_0^\infty u \, \frac{1}{2} \left(\frac{eE_0}{m} \right)^2 \frac{1}{v_e^2} \frac{d}{dv_e} \left(\frac{v_e^2}{3} \frac{\nu_m^e}{\nu_m^{e\,2} + \omega^2} \frac{df_e^0}{dv_e} \right) 4\pi v_e^2 \, dv_e \, .$$

Integrating now by parts, we find

$$\begin{aligned} \overline{P_E(t)} &= -\frac{1}{2} \left(\frac{eE_0}{m} \right)^2 \int_0^\infty \frac{d}{dv_e} \left(\frac{1}{2} m v_e^2 \right) \frac{v_e^2}{3} \frac{\nu_m^e}{\nu_m^{e\,2} + \omega^2} \frac{df_e^0}{dv_e} 4\pi \, dv_e \\ &= -\frac{1}{2} \frac{(eE_0)^2}{m} \int_0^\infty \frac{\nu_m^e}{\nu_m^{e\,2} + \omega^2} \frac{df_e^0}{dv_e} \frac{4\pi v_e^3}{3} \, dv_e \end{aligned}$$

and identifying $\text{Re}\{\overline{\sigma_{ce}}\}$ given by equation (4.16) in this expression, we can write $\overline{P_E(t)}$ under the form

$$\overline{P_E(t)} = \frac{1}{2} \, \text{Re}\{\overline{\sigma_{ce}}\} \, E_0^2 \, ,$$

in accordance with equation (4.15).

Exercise 4.3. Write the expression for the mean power absorbed from the field per electron, $\overline{P_E(t)}/n_e$, in terms of the mean energy absorbed per collision $u_c(u)$ by an electron of energy u given by equation (4.25).

Resolution: Replacing equation (4.25) into equations (4.15) and (4.16), we find

$$\frac{\overline{P_E(t)}}{n_e} = -\frac{1}{n_e} \int_0^\infty \nu_m^e \, u_c \, \frac{df_e^0}{dv_e} \frac{4\pi v_e^3}{3} \, dv_e \, .$$

Making the replacement of v_e with u and using the EEDF normalized such as

$$\int_0^\infty f \, \sqrt{u} \, du = 1 \, ,$$

with both distributions linked each other through equation (3.167), we obtain

$$\frac{\overline{P_E(t)}}{n_e} = -\frac{2}{3} \int_0^\infty u^{3/2} \, \nu_m^e \, u_c \, \frac{df}{du} \, du \, .$$

On the other hand, starting from the first expression in the answer to Exercise 4.2, we also obtain

$$\frac{\overline{P_E(t)}}{n_e} = \frac{2}{3}\int_0^\infty u\,\frac{d}{du}\left(u^{3/2}\,\nu_m^e\,u_c\,\frac{df}{du}\right)du$$

and integrating by parts we find the same expression.

Exercise 4.4. Write the expression for the complex mobility (4.12) and for the time-averaged absorbed power from the field per volume unit (4.15) and (4.16) in the case of an HF field, using the EEDF normalized through equation (3.166).

Resolution: The two expressions asked are:

$$\overline{\mu_e} = -\frac{2}{3}\frac{e}{m}\int_0^\infty \frac{u^{3/2}}{\nu_m^e + j\omega}\,\frac{df}{du}\,du$$

$$\overline{P_E(t)} = -\frac{2}{3}\frac{e^2 n_e}{m}\,E_{rms}^2\int_0^\infty u^{3/2}\,\frac{\nu_m^e}{\nu_m^{e\,2}+\omega^2}\,\frac{df}{du}\,du\,.$$

Although the notation seems to be identical the first expression represents a complex quantity and the second a time-averaged value of a real quantity.

Exercise 4.5. Write the expressions of the time-averaged gain produced by the field in velocity space, $\overline{G_E(t)}$, and of the power absorbed by the electrons in terms of this gain, for the case of a HF frequency electric field.

Resolution: The gain produced by a HF field in velocity space is obtained by inserting (4.20) into equations (3.92) and (3.95)

$$\overline{G_E(t)} = -\frac{1}{6}\left(\frac{eE_0}{m}\right)^2\frac{\nu_m^e}{\nu_m^{e\,2}+\omega^2}\,\frac{df_e^0}{dv_e}\,4\pi v_e^2\,,$$

so that using equation (4.25), we still have

$$\overline{G_E(t)} = -\frac{1}{3}\,\frac{\nu_m^e\,u_c}{m}\,\frac{df_e^0}{dv_e}\,4\pi v_e^2 > 0\,.$$

On the other hand using equation (3.119), the time-averaged power absorbed by the electrons is given by

$$\overline{P_E(t)} = -\int_0^\infty u\,\frac{d\overline{G_E}}{dv_e}\,dv_e = \int_0^\infty \overline{G_E}\,mv_e\,dv_e$$

$$= -\int_0^\infty \nu_m^e\,u_c\,\frac{df_e^0}{dv_e}\,\frac{4\pi v_e^3}{3}\,dv_e\,.$$

In terms of the EEDF normalized through equation (3.166), we find

$$\overline{G_E(t)} = -n_e \frac{2}{3} u^{3/2} \nu_m^e u_c \frac{df}{du}$$

and

$$\overline{P_E(t)} = \int_0^\infty \overline{G_E} du = -n_e \frac{2}{3} \int_0^\infty u^{3/2} \nu_m^e u_c \frac{df}{du} du .$$

Exercise 4.6. By writing the drift velocity of electrons under a RF field as $v_{ed}(t) = V_{d0} \cos(\omega t + \Phi)$ show that the power absorbed from the field (4.116) can be expressed as

$$P_E(t) = (P_E)_0 + \frac{en_e E_0}{2} V_{d0} \cos(2\omega t + \Phi) .$$

Obtain the limit values of $P_E(t)$ at low and high field frequencies.

Resolution: The power absorbed is given by

$$\begin{aligned} P_E(t) = (\mathbf{J_e} \,.\, \mathbf{E}) &= en_e V_{d0} \operatorname{Re}\{e^{j(\omega t+\Phi)}\} E_0 \operatorname{Re}\{e^{j\omega t}\} \\ &= \frac{en_e E_0}{2} V_{d0} [\cos(\Phi) + \cos(2\omega t + \Phi)] , \end{aligned}$$

so that using equation (4.111) we obtain

$$P_E(t) = (P_E)_0 + \frac{en_e E_0}{2} V_{d0} \cos(2\omega t + \Phi) .$$

When $\omega \ll \nu_m^e$, $v_{ed}(t)$ and $E(t)$ are in phase so that

$$P_E(t) = en_e E_0 V_{d0} \cos^2(\omega t) ,$$

with $V_{d0} = eE_0/(m\nu_m^e)$ and $\overline{P_E(t)} = en_e E_0 V_{d0}/2$. On the other hand, when $\omega \gg \nu_m^e$, the phase shift of the drift velocity approaches $-\pi/2$ and we obtain

$$P_E(t) = \frac{en_e E_0}{2} V_{d0} \sin(2\omega t) ,$$

with $V_{d0} = eE_0/(m\omega)$ in accordance with equation (4.18), and therefore $\overline{P_E(t)} = 0$.

References

W.P. Allis, Motions of ions and electrons, in *Handbuch der Physik*, vol. 21. ed. by S. Flügge (Springer, Berlin, 1956) pp. 383–444

M. Capitelli, R. Celiberto, C. Gorse, R. Winkler, J. Wilhelm, Electron energy distribution functions in radio-frequency collision dominated nitrogen discharges. J. Phys. D: Appl. Phys. **21**, 691–699 (1988)

J.-L. Delcroix, *Physique des Plasmas: Volumes 1 and 2*. Monographies Dunod (Dunod, Paris, 1963/1966) (in French)

C.M. Ferreira, J. Loureiro, Electron energy distributions and excitation rates in high-frequency argon discharges. J. Phys. D: Appl. Phys. **16**, 2471–2483 (1983)

C.M. Ferreira, J. Loureiro, Characteristics of high-frequency and direct-current argon discharges at low pressures: a comparative analysis. J. Phys. D: Appl. Phys. **17**, 1175–1188 (1984)

C.M. Ferreira, J. Loureiro, Electron excitation rates and transport parameters in high-frequency N_2 discharges. J. Phys. D: Appl. Phys. **22**, 76–82 (1989)

C.M. Ferreira, J. Loureiro, A.B. Sá, The modelling of high-frequency discharges at low and intermediate pressure, in *Proceedings 18th International Conference on Phenomena in Ionized Gases (ICPIG)*, Swansea, vol. Invited papers (1987), pp. 220–230

N. Goto, T. Makabe, Time-dependent electron swarm parameters in RF fields in CH_4 and H_2. J. Phys. D: Appl. Phys. **23**, 686–693 (1990)

E.V. Karoulina, Yu A, Lebedev, The influence of the electron transport cross sectional shape on electron energy distribution functions in DC and microwave plasmas. J. Phys. D: Appl. Phys. **21**, 411–417 (1988)

J. Loureiro, Time-dependent electron kinetics in N_2 and H_2 for a wide range of the field frequency including electron-vibration superelastic collisions. Phys. Rev. E **47**(2), 1262–1275 (1993)

J. Loureiro, Wave-to-plasma power transfer in magnetically assisted high-frequency discharges in argon. J. Appl. Phys. **78**(8), 4850–4854 (1995)

J. Loureiro, C.M. Ferreira, Coupled electron energy and vibrational distribution functions in stationary N_2 discharges. J. Phys. D: Appl. Phys. **19**, 17–35 (1986)

J. Loureiro, C.M. Ferreira, Electron and vibrational kinetics in the hydrogen positive column. J. Phys. D: Appl. Phys. **22**, 1680–1691 (1989)

T. Makabe, N. Goto, The time behaviour of electron transport in RF fields in gases. J. Phys. D: Appl. Phys. **21**, 887–895 (1988)

H. Margenau, Conduction and dispersion of ionized gases at high frequencies. Phys. Rev. **69** (9–10), 508–513 (1946)

H. Margenau, L.M. Hartman, Theory of high frequency gas discharges. II. Harmonic components of the distribution function. Phys. Rev. **73**(4), 309–315 (1948)

J. Margot, T.W. Johnston, J. Musil, Principles of magnetically assisted microwave discharges, in *Microwave Excited Plasmas*, ed. by M. Moisan, J. Pelletier. Plasma Technology, vol. 4 (Elsevier, Amsterdam, 1992), pp. 181–212

A.V. Phelps, L.C. Pitchford, Anisotropic scattering of electrons by N_2 and its effect on electron transport. Phys. Rev. A **31**(5), 2932–2949 (1985)

P.A. Sá, J. Loureiro, C.M. Ferreira, Time-dependent kinetics of electrons and $3p^5$ 4s levels of argon in high frequency plasmas. J. Phys. D: Appl. Phys. **27**, 1171–1183 (1994)

R.F. Whitmer, G.F. Herrmann, Effects of a velocity-dependent collision frequency on wave-plasma interactions. Phys. Fluids **9**(4), 768–773 (1966)

R. Winkler, H. Deutsch, J. Wilhelm, Ch. Wilke, Electron kinetics of weakly ionized HF plasmas. II. Decoupling in the Fourier hierarchy and simplified kinetics at higher frequencies. Beiträge aus der Plasmaphysik (Contributions on Plasma Physics) **24**(4), 303–316 (1984)

R. Winkler, M. Dilonardo, M. Capitelli, J. Wilhelm, Time-dependent solution of Boltzmann equation in RF plasmas: a comparison with the effective field approximation. Plasma Chem. Plasma Process. **7**(1), 125–137 (1987)

Chapter 5
Electron Transport, Ionization and Attachment

This chapter is devoted to the analysis of electron transport by using the Boltzmann equation. Contrary to the previous chapters where only an electric field exists, here the electron drift also results from a density gradient. The chapter initiates by considering the situation where only the density gradient exists, which leads to free diffusion of electrons. Since the diffusion to the walls leads to the disappearing of electrons from the swarm, the reintroduction of secondary electrons produced by ionization into the distribution needs to be properly taken into account to obtain the breakdown self-sustaining field. Here, only the breakdown produced by a high-frequency (HF) field is considered, since in the direct-current (DC) case the situation is much more complicated because the breakdown also depends on the nature of the cathode. The chapter ends by considering another mechanism for electron sink: the electron attachment. In conclusion: this chapter deals with mechanisms in which the electron density is not kept constant in the Boltzmann equation, such as free diffusion, ionization and attachment.

5.1 Electron Diffusion

5.1.1 Free Electron Diffusion

In Sect. 3.1.3 we have analysed the situation where the anisotropic deviations relatively to the isotropic equilibrium were due to the action of an external electric field **E** and the gas, as a whole, was in a homogeneous state. Here, on the contrary, we will start by assuming that $\mathbf{E} = 0$ but the electron distribution function varies

The original version of this chapter was revised. An erratum to this chapter can be found at DOI 10.1007/ 978-3-319-09253-9_12

J.M.A.H. Loureiro, J. de Amorim Filho, *Kinetics and Spectroscopy of Low Temperature Plasmas*, Graduate Texts in Physics,
DOI 10.1007/978-3-319-09253-9_5

from one point to another due to the presence of density gradients, ∇n_e. Then, the Boltzmann equation (3.6) takes the form

$$\frac{\partial f_e}{\partial t} + \left(\mathbf{v_e} \cdot \frac{\partial f_e}{\partial \mathbf{r}}\right) = \left(\frac{\partial f_e}{\partial t}\right)_{e-o}, \tag{5.1}$$

where the right-hand side member is the collision integral (3.7). Furthermore, we will assume that the diffusion of electrons takes place in a gas of molecules, which is itself homogeneous, and the molecules with the density n_o are in thermodynamic equilibrium at temperature T_o. The electrons diffuse then in the gas of molecules due to their thermal motion only and they are out from the action of electric fields of any sort. This phenomenon is called free diffusion.

As in the case of the anisotropies caused by an electric field, we may expand the electron distribution function in spherical harmonics in velocity space (3.9)

$$f_e(\mathbf{r}, \mathbf{v_e}, t) = \sum_{l=0}^{\infty} f_e^l(\mathbf{r}, v_e, t)\, P_l(\cos\theta), \tag{5.2}$$

where, if necessary, the dependence on the azimuthal coordinate ϕ may also be taken into account by considering the associated Legendre functions $P_l^m(\cos\theta)$ and equation (3.19). Taking the polar axis along the direction of the density gradient and assuming that it exists rotational symmetry about this axis, in order the associated Legendre functions are not required, we obtain the following expansion limited to the first two terms (see Appendix A.5.1 and Allis 1956)

$$\begin{aligned}\left(\mathbf{v_e} \cdot \frac{\partial f_e}{\partial \mathbf{r}}\right) &= \sum_{l=0}^{\infty} v_e \cos\theta\, \frac{\partial f_e^l}{\partial z}\, P_l \\ &= \frac{v_e}{3}\frac{\partial f_e^1}{\partial z} + v_e \cos\theta\, \frac{\partial}{\partial z}\left(f_e^0 + \frac{2}{5} f_e^2\right) + \dots\dots\dots,\end{aligned} \tag{5.3}$$

or still under the form

$$\left(\mathbf{v_e} \cdot \frac{\partial f_e}{\partial \mathbf{r}}\right) = \frac{v_e}{3}(\nabla . \mathbf{f_e^1}) + \left(\mathbf{v_e} . \nabla\left(f_e^0 + \frac{2}{5} f_e^2\right)\right) + \dots\dots\dots \tag{5.4}$$

In equation (5.4) the scalar function f_e^1 has been replaced by the vector $\mathbf{f_e^1}$ oriented along the anisotropy direction. This notation removes the initial restriction on the direction of the polar axis of spherical harmonics.

Since according to equations (3.26) and (3.126) the first term on the left-hand side member and the right-hand side member of equation (5.1) are

$$\frac{\partial f_e}{\partial t} = \frac{\partial f_e^0}{\partial t} + \frac{\partial f_e^1}{\partial t} \cos\theta + \dots\dots\dots \tag{5.5}$$

$$\left(\frac{\partial f_e}{\partial t}\right)_{e-o} = (I^0 + J^0) + (I^1 + J^1) \cos\theta + \dots\dots\dots, \tag{5.6}$$

with I^l and J^l denoting the collision terms of order l for elastic and inelastic collisions (3.85), (3.133), and (3.140), the equation (5.1) may be decomposed into a system of equations in which the first two are

$$\frac{\partial f_e^0}{\partial t} + \frac{v_e}{3}\left(\nabla . \mathbf{f_e^1}\right) = I^0 + J^0 \tag{5.7}$$

$$\frac{\partial \mathbf{f_e^1}}{\partial t} + v_e \, \nabla\!\left(f_e^0 + \frac{2}{5} f_e^2\right) = -\, \nu_m^e \, \mathbf{f_e^1} \, . \tag{5.8}$$

In the case of weak anisotropies, the spherical harmonic expansion converges rapidly and the term in f_e^2 is negligible. Further, for steady-state conditions of $\mathbf{f_e^1}$, which are attained for times $\sim (\nu_m^e)^{-1}$, the equation (5.8) takes the form

$$\mathbf{f_e^1} = -\, \frac{v_e}{\nu_m^e} \, \nabla f_e^0 \, . \tag{5.9}$$

Then, the equation (5.9) inserted into (5.7) allows to obtain the following equation for the isotropic component of the electron distribution function

$$\frac{\partial f_e^0}{\partial t} - \frac{v_e^2}{3\nu_m^e} \, \nabla^2 f_e^0 = I^0 + J^0 \, . \tag{5.10}$$

As in Sect. 3.1.4 we are assuming that the anisotropic component reaches the steady-state equilibrium in a time $(\nu_m^e)^{-1}$ much shorter than the time at which the isotropic component also attains its equilibrium, $(\nu_e)^{-1}$.

On the other hand, inserting equation (5.9) into the expression (3.18) for the electron drift velocity, $\mathbf{v_{ed}} = <\mathbf{v_e}>$, we obtain the following expression for the electron particle current density

$$\begin{aligned} \mathbf{\Gamma_e} = n_e <\mathbf{v_e}> &= -\int_0^\infty \frac{v_e^2}{3\nu_m^e} \, \nabla f_e^0 \, 4\pi v_e^2 \, dv_e \\ &= -\, \nabla (n_e \, D_e) \, , \end{aligned} \tag{5.11}$$

where

$$D_e = < \frac{v_e^2}{3\nu_m^e} > = \frac{1}{n_e} \int_0^\infty \frac{v_e^2}{3\nu_m^e} \, f_e^0 \, 4\pi v_e^2 \, dv_e \, , \tag{5.12}$$

is the electron free diffusion coefficient.

It is worth noting here that the ratio between the electron free diffusion coefficient and the electron mobility, $\mu_e = \sigma_{ce}/(en_e)$ being σ_{ce} the electron conductivity given by (3.52),

$$\mu_e = \frac{e}{m} < \frac{1}{v_e^2} \frac{d}{dv_e} \left(\frac{v_e^3}{3\nu_m^e} \right) > \tag{5.13}$$

is termed characteristic energy

$$u_k = \frac{e\, D_e}{\mu_e} . \tag{5.14}$$

This parameter has therefore the following expression

$$u_k = m \frac{< \frac{v_e^2}{\nu_m^e} >}{< \frac{1}{v_e^2} \frac{d}{dv_e} \left(\frac{v_e^3}{\nu_m^e} \right) >} \tag{5.15}$$

and when f_e^0 is a Maxwellian at temperature T_e, $f_e^0 \sim \exp(-mv_e^2/(2k_BT_e))$, we obtain the well-known Einstein relation

$$u_k = k_B T_e . \tag{5.16}$$

On the other hand, when the collision frequency for momentum transfer is independent of velocity, $\nu_m^e(v_e) = \text{const}$, we still obtain $D_e = k_BT_e/(m\nu_m^e)$ and $\mu_e = e/(m\nu_m^e)$ for the electron free diffusion coefficient and the electron mobility, respectively. However, the characteristic energy should replace the electron temperature in many situations of plasma physics where the isotropic component of the velocity distribution function is non-Maxwelllian.

5.1.2 Local Field Approximation

Let us assume now the more general situation where an electric field $\mathbf{E}$ exists in simultaneous with a density gradient ∇n_e. Using equations (3.35) and (5.1) the Boltzmann equations writes under the more complete form

$$\frac{\partial f_e}{\partial t} + \left(\mathbf{v_e} \cdot \frac{\partial f_e}{\partial \mathbf{r}} \right) - \frac{e}{m} \left(\mathbf{E} \cdot \frac{\partial f_e}{\partial \mathbf{v_e}} \right) = \left(\frac{\partial f_e}{\partial t} \right)_{e-o} . \tag{5.17}$$

The electron distribution function $f_e(\mathbf{r}, \mathbf{v_e}, t)$ is not in equilibrium with the instantaneous and local electric field if the frequency for energy relaxation, and/or momentum relaxation, is small compared to the rate for time change of the electric field

$$\nu_e \text{ (and/or } \nu_m^e) < \frac{1}{|\mathbf{E}|} \frac{\partial |\mathbf{E}|}{\partial t} \tag{5.18}$$

and if the frequency for space relaxation, λ^{-1}, being λ the mean free path, is small compared to the rate for space change of the electric field

$$\frac{1}{\lambda} < \frac{1}{|\mathbf{E}|} \left| \frac{\partial \mathbf{E}}{\partial \mathbf{r}} \right| . \tag{5.19}$$

The first condition has already been analysed in Sect. 4.3.2, for a sinusoidal electric field of angular frequency ω, and it corresponds to have ν_e (and/or ν_m^e) $< \omega$. This condition is usually found at high-frequency (typically > few MHz) and low-pressure (< few 10^2 Pa) discharges. The second condition says that the electron distribution function is not in equilibrium with the field if the distance over which the distribution function equilibrates, λ, is larger than the distance over which the field significantly varies. A situation of non-equilibrium for this latter condition occurs in the cathode fall of a glow discharge.

A suggestive figure showing the characteristic time and distance required for the electron distribution function to come into equilibrium with an applied electric field was presented in DiCarlo and Kushner (1989) for the case of electron swarms in Ar and N_2 (see Fig. 5.1). These results were obtained by applying a step function in the electric field to a Maxwellian velocity distribution (with $k_B T_e = 0.05\,\mathrm{eV}$) and observing the time and the distance required for the electron distribution to reach the steady-state. Both are normalized by the gas density, so that they represent $\tau_e n_o = \nu_e^{-1} n_o$ and λn_o ($n_o = 3.22 \times 10^{22}\,\mathrm{m}^{-3}$ at $p = 133.3\,\mathrm{Pa}$ and $T_o = 300\,\mathrm{K}$). In low-pressure argon plasmas (i.e. $p <$ few 10^2 Pa) and moderate reduced electric fields ($E/n_o \simeq 10 - 100\,\mathrm{Td}$, with $1\ \mathrm{Td} = 1 \times 10^{-21}\ \mathrm{V\,m^2}$) the equilibration time and distance are $10 - 100\,\mathrm{ns}$ and $0.1 - 1.0\,\mathrm{cm}$. Below these values temporal and/or spatial developments of the electron distribution function need to be considered.

Let us consider now the conjoint action of diffusion in the space of positions and diffusion in the space of velocity, the latter due to the presence of a homogeneous electric field $\mathbf{E} = -E\ \mathbf{e_z}$. Considering weak anisotropies and steady-state conditions for the anisotropic component $\mathbf{f_e^1}$, we may write from equations (3.89) and (5.9)

$$\mathbf{f_e^1} = -\frac{eE}{m\nu_m^e} \frac{\partial f_e^0}{\partial v_e} \mathbf{e_z} - \frac{v_e}{\nu_m^e} \nabla f_e^0 , \tag{5.20}$$

while for the equation describing the evolution of the isotropic component, f_e^0, we obtain using equations (3.41) and (5.7)

$$\frac{\partial f_e^0}{\partial t} - \frac{e}{m} \frac{1}{3 v_e^2} \frac{\partial}{\partial v_e} \left(v_e^2 \left(\mathbf{E} \,.\, \mathbf{f_e^1} \right) \right) + \frac{v_e}{3} \left(\nabla .\, \mathbf{f_e^1} \right) = I^0 + J^0 . \tag{5.21}$$

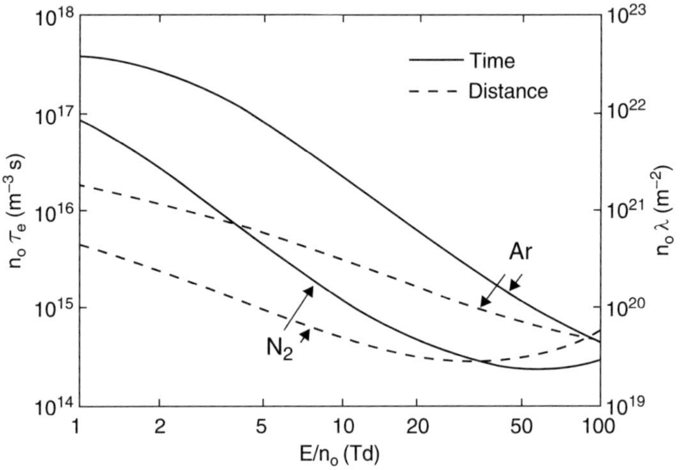

Fig. 5.1 Characteristic times (*full curves*) and distances (*broken*), as a function of E/n_o, required for the electron distribution function to come into equilibrium with a step function in electric field for electron swarms in Ar and N_2. The quantities are normalized by the gas density (DiCarlo and Kushner 1989)

Substituting (5.20) into (5.21), we find

$$\frac{\partial f_e^0}{\partial t} - \left(\frac{eE}{m}\right)^2 \frac{1}{3v_e^2} \frac{\partial}{\partial v_e}\left(\frac{v_e^2}{\nu_m^e} \frac{\partial f_e^0}{\partial v_e}\right) - \frac{eE}{m} \frac{1}{3v_e^2} \frac{\partial}{\partial v_e}\left(\frac{v_e^3}{\nu_m^e} (\mathbf{e_z} . \nabla f_e^0)\right)$$
$$- \frac{eE}{m} \frac{v_e}{3\nu_m^e} \nabla . \left(\frac{\partial f_e^0}{\partial v_e} \mathbf{e_z}\right) - \frac{v_e^2}{3\nu_m^e} \nabla^2 f_e^0 = I_0 + J_0. \tag{5.22}$$

The third term on the left-hand side member of (5.22) may be discarded as long as the diffusion occurs perpendicularly to the direction of the applied electric field. However, the presence of a crossed space and velocity dependence in the fourth term indicates that an exact separable solution does not exist for the electron distribution function, unless the distance required for the distribution to come into equilibrium with the field is smaller than the diffusion length. When such condition exists the electron distribution function is factorizable into a function of the velocity and a function of the space only. The ratio of the electron distribution function to the electron density (3.15), f_e^0/n_e, is then independent of space. This assumption is known as *local field approximation* (LFA).

The first consequence of LFA is the diffusion coefficient (5.12) becomes independent of space and we may write the electron particle current density (5.11) as follows

$$\mathbf{\Gamma_e} = n_e <\mathbf{v_e}> = - D_e \nabla n_e. \tag{5.23}$$

Also the Boltzmann equation for the isotropic component of the electron distribution function can be written now as

$$\frac{\partial f_e^0}{\partial t} - \left(\frac{eE}{m}\right)^2 \frac{1}{3v_e^2}\frac{\partial}{\partial v_e}\left(\frac{v_e^2}{\nu_m^e}\frac{\partial f_e^0}{\partial v_e}\right) - \frac{eE}{m}\frac{v_e}{3\nu_m^e}\frac{\partial}{\partial v_e}\left(\frac{f_e^0}{n_e}\right)\frac{\partial n_e}{\partial z}$$
$$-\frac{v_e^2}{3\nu_m^e}\frac{f_e^0}{n_e}\nabla^2 n_e = I_0 + J_0 \qquad (5.24)$$

and once again if the diffusion occurs perpendicularly to the electric field we have $\partial n_e/\partial z = 0$ in the third term of the left-hand side member of equation (5.24). The form achieved for the Boltzmann equation after this last simplification corresponds essentially to the equation valid for breakdown, where the electron density is very low. On the contrary, in a glow-discharge produced either by a DC or an HF electric field, the electron density is large and a space-charge field emerges in the plasma along the direction of diffusion, with opposite direction relatively to the electron density gradient. In this latter case an equivalent term to the third term of equation (5.24) needs to be considered. We will treat this situation later on in Chap. 6.

5.1.3 Continuity Equation and Diffusion Length

Let us multiply now both members of equation (5.22) by $4\pi v_e^2$, after having neglected the third and fourth terms on the left-hand side member due to the reasons pointed before, and integrating between $0 \to \infty$. We obtain the continuity equation (see Sect. 2.4.1)

$$\frac{\partial n_e}{\partial t} + \int_0^\infty \frac{\partial G_E}{\partial v_e}\, dv_e - D_e\nabla^2 n_e = -\int_0^\infty \frac{\partial G_{el}}{\partial v_e}\, dv_e + \int_0^\infty J^0\, 4\pi v_e^2\, dv_e, \qquad (5.25)$$

in which

$$G_E = g_E^0\, 4\pi v_e^2 = -\frac{1}{3\nu_m^e}\left(\frac{eE}{m}\right)^2 \frac{\partial f_e^0}{\partial v_e}\, 4\pi v_e^2 \qquad (5.26)$$

is the positive gain in velocity space due to the field (3.92) and (3.95), G_{el} is the negative gain due to the losses by elastic collisions (3.122) defined such as

$$-\int_0^\infty \frac{\partial G_{el}}{\partial v_e}\, dv_e = \int_0^\infty I^0\, 4\pi v_e^2\, dv_e \qquad (5.27)$$

and the second term on the right-hand side member of equation (5.25) represents the equivalent term for the inelastic collisions.

The terms with G_E and G_{el} vanish both as we integrate over the whole velocity space, since both keep constant the number of electrons. They simply redistribute a fixed number of electrons along the energy axis. The same situation occurs with the term for inelastic collisions (3.134) because so far we are simply considering in this term the promotion and the demotion of a fixed number of electrons in the energy cells. Therefore, it exists an inconsistency in the form as equation (5.25) is written. On the left-hand side we include a term for diffusion which leads to the disappearing of electrons from the swarm, while on the right-hand side we are not including the counterpart associated with the electron production by ionization. The continuity equation resulting from the electron Boltzmann equation, with the terms we have considered by now, is hence

$$\frac{\partial n_e}{\partial t} - D_e \nabla^2 n_e = 0. \tag{5.28}$$

This corresponds to the equation for a decaying plasma starting from a given initial density at $t = 0$, such as it is the case of a post-discharge.

The form of the solution to equation (5.28) will depend upon the boundary conditions applied. In the case of a infinite medium the solution of the three dimensional equation

$$\frac{\partial n_e}{\partial t} - D_e \frac{1}{r^2} \frac{\partial}{\partial r} \left(r^2 \frac{\partial n_e}{\partial r} \right) = 0 \tag{5.29}$$

for diffusion away from a point source is a Gaussian function

$$n_e(r, t) = \frac{N_0}{(4\pi D_e t)^{3/2}} \exp\left(- \frac{r^2}{4D_e t} \right), \tag{5.30}$$

with N_0 denoting the total constant population of electrons created at the point source, which allows the normalization

$$N_0 = \int_0^\infty n_e(r, t)\, 4\pi r^2\, dr. \tag{5.31}$$

If instead of the population N_0, we have at time $t = 0$ an initial space density $n_{e0}(r)$, the density as a function of the position and time is

$$n_e(r, t) = \frac{1}{(4\pi D_e t)^{3/2}} \int_0^\infty n_{e0}(r') \exp\left(- \frac{(r - r')^2}{4D_e t} \right) 4\pi r'^2\, dr' . \tag{5.32}$$

When we deal with a gas discharge, we need to obtain the solution to the diffusion equation in a bounded, rather than an infinite medium. The solution of equation (5.28), with $n_e = 0$ at the boundary of a plasma container by assuming total electron recombination at the wall, may be written under the form of a sum of orthogonal functions of the type

$$n_e(\mathbf{r},t) = \sum_j a_j(t)\, n_j(\mathbf{r}) \,. \tag{5.33}$$

The equation (5.28) may be then transformed into a set of equations

$$\frac{\partial a_j}{\partial t}\, n_j - a_j\, D_e\, \nabla^2 n_j = 0 \,. \tag{5.34}$$

Defining the diffusion length of mode j as

$$\nabla^2 n_j = -\frac{n_j}{\Lambda_j^2} \,, \tag{5.35}$$

we obtain

$$\frac{\partial a_j}{\partial t} + \frac{D_e}{\Lambda_j^2}\, a_j = 0 \,, \tag{5.36}$$

whose solution is

$$a_j(t) = a_j(0)\, e^{-\nu_j t} \,, \tag{5.37}$$

being $\nu_j = D_e/\Lambda_j^2$ the relaxation frequency of mode j. In the case of the diffusion mode lengths (5.35) for a rectangular (a,b,c) or a cylindrical (R,h) cavity, we obtain

$$\frac{1}{\Lambda_{lmn}^2} = \left(\frac{l\pi}{a}\right)^2 + \left(\frac{m\pi}{b}\right)^2 + \left(\frac{n\pi}{c}\right)^2 \tag{5.38}$$

$$\frac{1}{\Lambda_{lmn}^2} = \left(\frac{\alpha_{lm}}{R}\right)^2 + \left(\frac{n\pi}{h}\right)^2 \,, \tag{5.39}$$

being α_{lm} the l–th root of the Bessel function of first kind and order m, in which $\alpha_{10} = 2.405$.

The density $n_j(\mathbf{r})$ of the fundamental mode has a maximum at the centre of the cavity and decreases towards the wall. The corresponding fundamental diffusion length (Λ_{111} for rectangular and Λ_{101} for cylindrical cavities) is of the order of the dimensions of the container. When one of the dimensions is significantly smaller than the others (e.g. $R \ll h$ in the case of a long cylindrical tube), the diffusion length is of the order of the smallest dimension ($\Lambda_{lmn} = R/\alpha_{lm}$). The higher order modes have diffusion lengths smaller and relaxation frequencies higher than the fundamental one. The diffusion of higher modes may be then considered ended ($a_j(t) \simeq 0$, for $j > 1$) when the relaxation of the slowly fundamental mode still occurs.

5.1.4 Free Diffusion in Presence of a Magnetic Field

The electron free diffusion is naturally modified as a static magnetic field is put together. Assuming the magnetic field directed along the z axis, we may write using equations (3.6) and (5.1)

$$\frac{\partial f_e}{\partial t} + \left(\mathbf{v_e} \cdot \frac{\partial f_e}{\partial \mathbf{r}}\right) + \left([\vec{\boldsymbol{\omega}}_{\mathbf{ce}} \times \mathbf{v_e}] \cdot \frac{\partial f_e}{\partial \mathbf{v_e}}\right) = \left(\frac{\partial f_e}{\partial t}\right)_{e-o}, \tag{5.40}$$

where $\vec{\boldsymbol{\omega}}_{\mathbf{ce}} = eB/m\ \mathbf{e_z}$ is the electron cyclotron frequency vector, with e denoting the absolute value of the electron charge. As seen in Sect. 4.2.2, we need to consider now the associated Legendre functions in the expansion of $f_e(\mathbf{r}, \mathbf{v_e}, t)$

$$\begin{aligned} f(\mathbf{r}, \mathbf{v_e}, t) &= f_e^0(\mathbf{r}, v_e, t) + \left(\mathbf{f_e^1}(\mathbf{r}, v_e, t) \cdot \frac{\mathbf{v_e}}{v_e}\right) \\ &= f_e^0 + p_{11}\ \sin\theta\ \cos\phi + q_{11}\ \sin\theta\ \sin\phi + p_{10}\ \cos\theta. \end{aligned} \tag{5.41}$$

Using the expansions (4.53) and (5.4) and neglecting the term on f_e^2, we obtain the following set of equations for the isotropic and anisotropic components of the electron velocity distribution function

$$\frac{\partial f_e^0}{\partial t} + \frac{v_e}{3}(\nabla . \mathbf{f_e^1}) = I^0(f_e^0) + J^0(f_e^0) \tag{5.42}$$

$$\frac{\partial p_{11}}{\partial t} + v_e \frac{\partial f_e^0}{\partial x} + \omega_{ce}\, q_{11} = -\nu_m^e\, p_{11} \tag{5.43}$$

$$\frac{\partial q_{11}}{\partial t} + v_e \frac{\partial f_e^0}{\partial y} - \omega_{ce}\, p_{11} = -\nu_m^e\, q_{11} \tag{5.44}$$

$$\frac{\partial p_{10}}{\partial t} + v_e \frac{\partial f_e^0}{\partial z} = -\nu_m^e\, p_{10}. \tag{5.45}$$

The anisotropies decay with $(\nu_m^e)^{-1}$ more rapidly than f_e^0, so that we may assume $\boldsymbol{\partial}\mathbf{f_e^1}/\partial t = 0$ in equations (5.43), (5.44), and (5.45). Solving the system so obtained, we can write under tensorial form

$$\begin{bmatrix} p_{11} \\ q_{11} \\ p_{10} \end{bmatrix} = -v_e \begin{bmatrix} \frac{\nu_m^e}{\nu_m^{e\,2}+\omega_{ce}^2} & -\frac{\omega_{ce}}{\nu_m^{e\,2}+\omega_{ce}^2} & 0 \\ \frac{\omega_{ce}}{\nu_m^{e\,2}+\omega_{ce}^2} & \frac{\nu_m^e}{\nu_m^{e\,2}+\omega_{ce}^2} & 0 \\ 0 & 0 & \frac{1}{\nu_m^e} \end{bmatrix} \begin{bmatrix} \frac{\partial f_e^0}{\partial x} \\ \frac{\partial f_e^0}{\partial y} \\ \frac{\partial f_e^0}{\partial z} \end{bmatrix}, \tag{5.46}$$

or simply writing

$$\mathbf{f_e^1} = -\, v_e \left(\hat{\mathbf{d}} \,.\nabla f_e^0 \right), \tag{5.47}$$

with $\hat{\mathbf{d}}$ denoting the tensor 3×3 of equation (5.46). Inserting now (5.47) into (5.42), we obtain the following equation for the isotropic component of the electron velocity distribution

$$\frac{\partial f_e^0}{\partial t} - \frac{v_e^2}{3} \left(\nabla . \left(\hat{\mathbf{d}} \,.\nabla f_e^0 \right) \right) = I^0(f_e^0) \ + J^0(f_e^0) \tag{5.48}$$

and making the calculations, we obtain

$$\frac{\partial f_e^0}{\partial t} - \frac{v_e^2}{3} \left(\frac{\nu_m^e}{\nu_m^{e\,2} + \omega_{ce}^2} \left(\frac{\partial^2 f_e^0}{\partial x^2} + \frac{\partial^2 f_e^0}{\partial y^2} \right) + \frac{1}{\nu_m^e} \frac{\partial^2 f_e^0}{\partial z^2} \right) = I^0(f_e^0) \ + J^0(f_e^0) \,. \tag{5.49}$$

As in (5.35) we may define the diffusion lengths

$$\frac{\partial^2 f_e^0}{\partial x^2} = -\, \frac{f_e^0}{\Lambda_x^2} \,, \quad \text{etc.} \tag{5.50}$$

so that at the end the equation describing the evolution of the isotropic component writes as follows

$$\frac{\partial f_e^0}{\partial t} + \frac{v_e^2}{3\nu_m^e} \frac{f_e^0}{\Lambda_m^2} = I^0(f_e^0) \ + J^0(f_e^0) \,, \tag{5.51}$$

being Λ_m an effective diffusion length taking into account the effect of the magnetic field

$$\frac{1}{\Lambda_m^2} = \frac{\nu_m^{e\,2}}{\nu_m^{e\,2} + \omega_{ce}^2} \left(\frac{1}{\Lambda_x^2} + \frac{1}{\Lambda_y^2} \right) + \frac{1}{\Lambda_z^2} \,. \tag{5.52}$$

Equation (5.52) shows that due to the electron trapping caused by the magnetic field the effective diffusion lengths in the directions perpendicularly to the field are larger than the diffusion length in the axial direction.

On the other hand, the electron particle current density (5.11) using (5.47) takes the form

$$\begin{aligned} \mathbf{\Gamma_e} &= n_e <\mathbf{v_e}> = \int_0^\infty \frac{v_e}{3} \, \mathbf{f_e^1} \, 4\pi v_e^2 \, dv_e \\ &= -\int_0^\infty \frac{v_e^2}{3} \left(\hat{\mathbf{d}} \,.\nabla f_e^0 \right) 4\pi v_e^2 \, dv_e \,. \end{aligned} \tag{5.53}$$

When the dependence of $f_e^0(\mathbf{r}, v_e, t)$ on the space coordinate is the same as $n_e(\mathbf{r}, t)$, i.e. in conditions of the local field approximation (LFH), the ratio f_e^0/n_e is independent of $\mathbf{r}$, and we may write (5.53) under the form

$$\boldsymbol{\Gamma}_\mathbf{e} = -\frac{1}{3}\left(< v_e^2\, \hat{\mathbf{d}} > . \nabla n_e\right) = -\left(\widehat{D}_\mathbf{e} . \nabla n_e\right) \tag{5.54}$$

having defined the electron diffusion tensor

$$\widehat{D}_\mathbf{e} = \frac{1}{3} < v_e^2\, \hat{\mathbf{d}} > , \tag{5.55}$$

in which $\hat{\mathbf{d}}$ is the tensor 3×3 of equation (5.46).

When ν_m^e is independent of velocity, we find

$$\widehat{D}_\mathbf{e} = \frac{< v_e^2 >}{3}\, \hat{\mathbf{d}} \tag{5.56}$$

and if further f_e^0 is Maxwellian, $< v_e^2 >= 3k_B T_e/m$, we obtain

$$\widehat{D}_\mathbf{e} = \frac{k_B T_e}{m}\, \hat{\mathbf{d}} . \tag{5.57}$$

The electron diffusion coefficient is then replaced with the antisymmetric tensor

$$\widehat{D}_\mathbf{e} = \begin{bmatrix} D_1 & -D_2 & 0 \\ D_2 & D_1 & 0 \\ 0 & 0 & D_0 \end{bmatrix} , \tag{5.58}$$

in which

$$D_1 = \frac{1}{1 + (\omega_{ce}/\nu_m^e)^2}\, D_0 \tag{5.59}$$

$$D_2 = \frac{\omega_{ce}/\nu_m^e}{1 + (\omega_{ce}/\nu_m^e)^2}\, D_0 \tag{5.60}$$

and

$$D_0 = \frac{k_B T_e}{m \nu_m^e} . \tag{5.61}$$

These expressions naturally result in $D_1 = D_0$ and $D_2 = 0$ when $\mathbf{B} = 0$. Equations (5.59), (5.60), and (5.61) show that the diffusion is smaller perpendicularly to the field, being the coefficients D_1 and D_2 as $\omega_{ce}/\nu_m^e \gg 1$ given by

$$D_1 \simeq \left(\frac{\nu_m^e}{\omega_{ce}}\right)^2 D_0 \ll D_0 \;, \qquad D_2 \simeq \left(\frac{\nu_m^e}{\omega_{ce}}\right) D_0 \ll D_0 \;. \tag{5.62}$$

On the other hand, when ν_m^e depends on v_e, we obtain the following expressions for the components of equation (5.55)

$$D_1 = \frac{1}{3} < \frac{v_e^2}{\nu_m^e} \frac{1}{1 + (\omega_{ce}/\nu_m^e)^2} > \tag{5.63}$$

$$D_2 = \frac{1}{3} < \frac{v_e^2}{\nu_m^e} \frac{\omega_{ce}/\nu_m^e}{1 + (\omega_{ce}/\nu_m^e)^2} > \tag{5.64}$$

$$D_0 = \frac{1}{3} < \frac{v_e^2}{\nu_m^e} > \;. \tag{5.65}$$

Finally, it is still worth noting that when $\nu_m^e(v_e) = \text{const}$, the electron mobility is $\mu_e = e/(m\nu_m^e)$, so that using $\omega_{ce} = eB/m$, we find

$$\frac{\omega_{ce}}{\nu_m^e} = \mu_e \, B \;, \tag{5.66}$$

which leads to the coefficients (5.59) and (5.60) may be still written with the form

$$D_1 = \frac{1}{1 + (\mu_e B)^2} D_0 \quad \text{and} \quad D_2 = \frac{\mu_e B}{1 + (\mu_e B)^2} D_0 \;. \tag{5.67}$$

The free diffusion of ions under the effect of a magnetic field takes place through the same expressions being necessary only to replace the electron mass with the ion mass M, and the electron collision frequency ν_m^e with the ion-neutral collision frequency ν_{io}, in the ion mobility $\mu_i = e/(M\nu_{io})$. So, due to the fact of the ion mobility is smaller than the electron mobility, it may occur that for **B** fields high enough the ions may freely diffuse perpendicularly faster than the electrons. In this case, the diffusion coefficient D_1 for ions is

$$D_1 \simeq \frac{k_B T_i}{eB^2} \frac{M\nu_{io}}{e} \;. \tag{5.68}$$

5.2 Electron Breakdown

5.2.1 Electron Ionization

We have seen in Sect. 5.1.3 that the Boltzmann equation as written in (5.24), with the term for inelastic collisions J^0 given by equation (3.134), leads to an inconsistency. From one side we admit electron losses by diffusion and from another we do not include any term for production of secondary electrons by ionization. This equation is surely valid to describe an afterglow plasma, i.e. a post-discharge, but it is invalid for the case of a sustaining plasma. Holstein (1946) has shown that the term for ionization may be included into the Boltzmann equation in a way that allows to separate formally the contributions of the secondary electrons produced and the primary electrons scattered. This description was later on used in Yoshida et al. (1983) to study the effects of the electrons produced by ionization on the electron energy distribution function (EEDF) and transport coefficients, at high values of the ratio of the electric field to the gas number density, E/n_o.

Considering here the formulation used by Holstein (1946) and Yoshida et al. (1983), the collision term for inelastic collisions (3.134) should be modified to included the following terms for ionization

$$J^0(f_e^0) = \ldots\ldots + \frac{2}{mv_e}\, n_o \left\{ \int_{2u+u_I}^{\infty} u'\, \sigma_{ion}^{sec}(u',u)\, f_e^0(u')\, du' \right.$$
$$\left. + \int_{u+u_I}^{2u+u_I} u'\, \sigma_{ion}^{sca}(u',u)\, f_e^0(u')\, du' \; - u\, \sigma_{ion}(u)\, f_e^0(u) \right\} , \quad (5.69)$$

where $\sigma_{ion}^{sec}(u',u)$ is the differential ionization cross section for the process in which a primary electron of energy u' produces a secondary electron of energy u, $\sigma_{ion}^{sca}(u',u)$ is the differential cross section for production of a scattered electron of energy u by a primary electron of energy u', and $\sigma_{ion}(u)$ is the total ionization cross section. Further, u_I is the ionization threshold energy, and n_o is the gas density. Here, we assume that the ionization occurs due to electron collisions upon the electronic ground state only, i.e. we neglect stepwise ionization. The first term in equation (5.69) accounts for the secondary electrons which enter the distribution at energy u as the result of ionizing collisions by primary electrons with energies between $u' = 2u + u_I$ and ∞. Similarly, the second term accounts for scattered primary electrons which reenter the distribution at the energy u as the result of ionizing collisions by electrons with energies between $u' = u + u_I$ and $u' = 2u + u_I$. The relationships among these energies are illustrated in the energy map shown in Fig. 5.2 taken from Yoshida et al. (1983). Finally, the third term in equation (5.69) accounts for electrons leaving the distribution at energy u in result of ionization collisions.

Although the two electrons produced by ionization are indistinguishable, we use the conventional terms of secondary and scattered primary for helping the visualization. The abscissa and ordinate in Fig. 5.2 are the energy u' of the primary electrons before ionization and the energies u of the electrons produced by ionization. The secondary electrons are defined as the low-energy product electrons with energies below $u = (u' - u_I)/2$, whereas the scattered primary electrons are those with energies between $u = (u' - u_I)/2$ and $u = (u' - u_I)$. The indistinguishability of the electrons and the conservation of energy lead to the requirement that the distribution

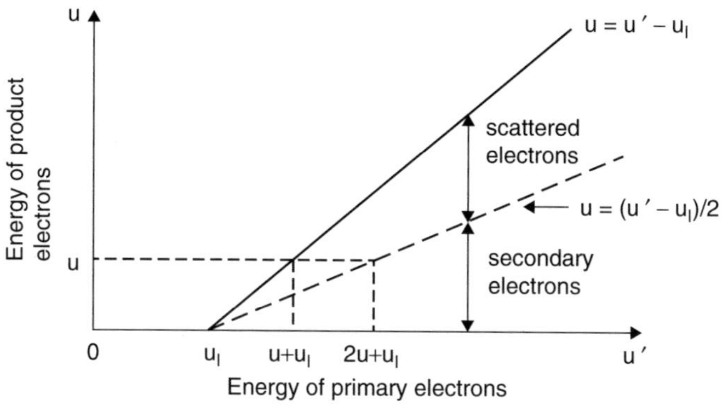

Fig. 5.2 Energy of product electrons resulting from electron impact ionization against the energy of primary electrons. Secondary electrons are defined as those below the *dashed line*, while scattered primary electrons are those between the *dashed line* and the *solid line* (Yoshida et al. 1983)

in energy u of electrons produced by ionization is symmetrical about the energy $u = (u' - u_I)/2$. The limits of integration of equation (5.69) can be recognized in Fig. 5.2 as the limits of the energies for secondary and scattered primary electrons for a fixed value of the electron energy u.

An extreme case for the sharing of energy among the electrons produced by the ionization event is when the secondary electrons appear with zero energy, i.e. along the horizontal axis of Fig. 5.2. In this case the scattered electrons are produced with energy $(u' - u_I)$, i.e. along the solid sloping line. These differential cross sections may be expressed in the form given by Thomas (1969), using Dirac $\delta(u)$ functions at $u = 0$ and $u = (u' - u_I)$.

In general, for a secondary electron appeared with the energy u_X, the scattered electron has the energy $u' - (u_I + u_X)$, so that we may write

$$\sigma_{ion}^{sec}(u', u) = \sigma_{ion}(u')\, \delta(u - u_X) \tag{5.70}$$

$$\sigma_{ion}^{sac}(u', u) = \sigma_{ion}(u')\, \delta(u - (u' - u_I - u_X)) \,, \tag{5.71}$$

where u_X is a function of $(u' - u_I)$. When $u_X = 0$ we minimize the energy of the secondary electrons and maximize the energy of the scattered electrons. In this case, the secondary electrons enter the distribution at $u = 0$, while the primary electrons lose only an energy equal to the ionization threshold energy as any other inelastic process. This secondary-electron distribution is equivalent to that obtained by adjusting the flux of electrons entering the distribution at $u = 0$ to equal the energy-averaged ionization frequency, $<\nu_{ion}>$. With such a flux the collision term for inelastic collisions with the inclusion of ionization (5.69) may be written in the following form, considering here equation (3.133) instead of (3.134)

$$J^0(f_e^0) = \ldots\ldots + n_e <\nu_{ion}> \frac{\delta(v_e)}{4\pi v_e^2} + \frac{v_e + v_I}{v_e} \nu_{ion}(v_e + v_I) f_e^0(v_e + v_I)$$
$$- \nu_{ion}(v_e) f_e^0(v_e) . \tag{5.72}$$

Here, $\nu_{ion}(v_e) = n_o v_e \sigma_{ion}(v_e)$ is the ionization collision frequency, $v_e = \sqrt{2u/m}$ is the electron velocity, $(v_e + v_I) = \sqrt{2(u + u_I)/m}$, and $< \nu_{ion} >$ is the energy-averaged ionization frequency

$$<\nu_{ion}> = \frac{1}{n_e} \int_{u_I}^{\infty} \nu_{ion}(v_e) f_e^0(v_e)\, 4\pi v_e^2\, dv_e. \tag{5.73}$$

For an ionization process in which both electrons produced appear at the centre of the available energy range, i.e. at the energy $u = (u' - u_I)/2$, the differential cross sections of the secondary and scattered primary electrons are

$$\sigma_{ion}^{sec}(u', u) = \sigma_{ion}^{sac}(u', u) = \sigma_{ion}(u')\, \delta(u - (u' - u_I)/2). \tag{5.74}$$

This case gives the maximum available energy to the secondary electrons and the highest possible energy loss to the scattered electrons.

Let us consider now the more general case in which the share of energy between the secondary and scattered electrons is specified by a parameter Δ, with $0 \le \Delta \le 1/2$. For a primary electron of energy u', the energy received by the two electrons is $\Delta\ (u' - u_I)$ and $(1 - \Delta)\ (u' - u_I)$, respectively. Since the energy u' that makes to appear a secondary electron with the energy u is $u' = (u/\Delta) + u_I$, while the energy u' that makes to appear a scattered electron with the energy u is $u' = (u/(1 - \Delta)) + u_I$, the collision term (5.69) writes under the following form after integration of δ functions with $du' = du/\Delta$ and $du' = du/(1 - \Delta)$ in the first and second terms (Itoh et al. 1988; Pinheiro and Loureiro 2002)

$$J^0(f_e^0) = \ldots\ldots + \frac{2}{m v_e} n_o \left\{ \left(\frac{u}{\Delta} + u_I\right) \sigma_{ion}\left(\frac{u}{\Delta} + u_I\right) f_e^0\left(\frac{u}{\Delta} + u_I\right) \frac{1}{\Delta} \right.$$
$$+ \left(\frac{u}{1-\Delta} + u_I\right) \sigma_{ion}\left(\frac{u}{1-\Delta} + u_I\right) f_e^0\left(\frac{u}{1-\Delta} + u_I\right) \frac{1}{1-\Delta}$$
$$\left. - u\, \sigma_{ion}(u) f_e^0(u) \right\}. \tag{5.75}$$

5.2.2 *Equations for Electron Breakdown*

As in Sect. 5.1.3., if we multiply both members of equation (5.22) by $4\pi v_e^2$, after having neglected as before the third and fourth terms on the left-hand side member in that equation, but including now the terms for ionization as given by equation (5.72), we obtain the continuity equation

$$\frac{\partial n_e}{\partial t} - D_e \nabla^2 n_e = n_e <\nu_{ion}> . \tag{5.76}$$

All other terms vanish because they simply redistribute a constant number of electrons along the energy axis. The only two non-conservative terms for the electron density are the diffusion loss term and the flux gain term associated with the production of secondary electrons. It is worth noting here that the equation (5.76) is also obtained by integrating in velocity space the equation (5.7) for the isotropic component of the velocity distribution function

$$\frac{\partial n_e}{\partial t} + (\nabla \,.\, \mathbf{\Gamma_e}) = n_e <\nu_{ion}> \tag{5.77}$$

and using after the equation (5.11), under LFA conditions.

The equation (5.22), without the third and fourth terms on the left-hand side member due to the reasons appointed before, may be written now as follows

$$\frac{\partial f_e^0}{\partial t} - \left(\frac{eE}{m}\right)^2 \frac{1}{3v_e^2} \frac{\partial}{\partial v_e}\left(\frac{v_e^2}{\nu_m^e} \frac{\partial f_e^0}{\partial v_e}\right) + \frac{v_e^2}{3\nu_m^e} \frac{f_e^0}{\Lambda^2}$$
$$= I_0 + J_0^* + n_e <\nu_{ion}> \frac{\delta(v_e)}{4\pi v_e^2}, \tag{5.78}$$

in which we have introduced the diffusion length Λ defined in (5.35) characteristic of the dimensions of the plasma container (i.e. of the discharge vessel), and where J_0^* represents the inelastic term with the inclusion of the last two terms only of equation (5.72), that is of the terms for reintroduction and removal of primary electrons. These two terms are equivalent to any other terms of an inelastic process of excitation. The only non-conservative terms are those for electron diffusion and for the flux of secondary electrons appeared into the distribution. If we multiply then both members of equation (5.78) firstly by the electron energy, $u = \frac{1}{2} m v_e^2$, and then by $4\pi v_e^2$, and integrate in all velocity space, we obtain the following updated equation for energy conservation, which replaces equation (3.111)

$$\frac{\partial (n_e <u>)}{\partial t} = P_E - P_{dif} - P_{el} - P_{inel} - P_{rot} + P_{ion}. \tag{5.79}$$

The terms P_E, P_{el}, P_{inel} and P_{rot} are the power gained from the field per volume unit (3.116), and the power lost in elastic collisions (3.118), inelastic collisions (3.151), and rotational excitation (3.177). The new terms P_{dif} and P_{ion} are the power lost in diffusion and the power brought to the distribution by the secondary electrons. This last term is obviously null since the secondary electrons appear at $u = 0$

$$P_{ion} = n_e <\nu_{ion}> \int_0^\infty u \;\delta(v_e)\; dv_e = 0, \tag{5.80}$$

while for the power lost by diffusion we obtain

$$P_{dif} = \int_0^\infty u \; \frac{v_e^2}{3\nu_m^e} \, \frac{f_e^0}{\Lambda^2} \; 4\pi v_e^2 \, dv_e \; = \; \frac{n_e}{\Lambda^2} \; < u \; \frac{v_e^2}{3\nu_m^e} > . \tag{5.81}$$

This term may be written approximately as follows

$$P_{dif} \simeq \frac{n_e}{\Lambda^2} \; <u> < \frac{v_e^2}{3\nu_m^e} > = \frac{n_e}{\Lambda^2} \; <u> D_e. \tag{5.82}$$

5.2.3 Breakdown Produced by a High-Frequency Field

The breakdown of a gas occurs when the gain in electron density resulting from ionization of the gas becomes equal to the loss of electrons by diffusion, recombination, and attachment. When the loss is only by diffusion the problem becomes relatively simple.

The breakdown of a low pressure gas under the action of a DC electric field has two sources of electrons. Most of the electrons are generated in the volume of the gas through ionization by electron collisions, but the original source of electrons results from secondary emission from the cathode caused by positive ion or photon bombardment (see Sect. 1.1.4). Thus the prediction of breakdown voltage requires the knowledge of the efficiency of these latter processes, which leads to some difficulties on the predictions. On the contrary, the breakdown caused by a HF electric field is determined by the primary ionization process only, i.e. by volume ionization collisions between electrons and molecules, having the electron production at the walls a negligible effect. It is therefore possible to predict the electric field for breakdown from the knowledge of the electron ionization frequency (Brown 1956; MacDonald et al. 1963).

Let us assume then a gas discharge bounded by walls that absorb electrons. The continuity equation for electrons (5.77), with the electron particle current density given by equation (5.23), allows to write equation (5.76). Introducing the diffusion length characteristic of the plasma container as defined in (5.35), we obtain the following equation for the time dependent electron density

$$\frac{\partial n_e}{\partial t} = n_e <\nu_{ion}> - D_e \frac{n_e}{\Lambda^2}. \tag{5.83}$$

The steady-state condition for breakdown occurs as the energy-averaged ionization frequency increases up to a value that equilibrates the frequency of electron loss by diffusion, in order an electron lost by diffusion may be always replaced by another created by ionization

$$<\nu_{ion}> = \frac{D_e}{\Lambda^2} . \tag{5.84}$$

Although an accurate description of the breakdown can be given theoretically only by taking into account the electron energy distribution function (EEDF) obtained by solving the electron Boltzmann equation, a physical picture of the mechanisms involved can be given qualitatively for the case of helium containing small admixtures of mercury vapour (Brown 1956). This mixture has the advantage of acting as a gas of atoms without excitation levels. The first excited state of helium is a metastable state with the energy of 19.8 eV, and transitions from this state to the He ground-state by radiation are hence forbidden. Since the mestastable state has a mean life of the order of ~1 ms, practically every helium metastable with the energy of 19.8 eV will collide with a mercury atom and loses its energy by ionizing the mercury. Therefore, each inelastic collision of helium will produce the ionization of mercury, so that the effective ionization potential is $u_I = 19.8$ eV. Furthermore, for the mixture He:Hg the collision frequency may be considered independent of velocity: $\nu_m^e(v_e) = \text{const}$.

Let us consider first the limit of high pressures $\nu_m^e \gg \omega$, but where the field frequency is not however so small that it still permits to consider the EEDF quasi-stationary (i.e. we have $\omega \gg \nu_e$, being ν_e the characteristic frequency for energy relaxation, see discussion in Sect. 4.1.1). In this case, the time-averaged absorbed power from the field (4.19) is

$$\overline{P_E(t)} = \frac{e^2 n_e}{m} \frac{\nu_m^e}{\nu_m^2 + \omega^2} E_{rms}^2 \simeq \frac{e^2 n_e}{m \nu_m^e} E_{rms}^2, \tag{5.85}$$

being $E_{rms} = E_0/\sqrt{2}$ the root-mean square electric field. In this limit the power is totally dissipated through elastic collisions (3.118)

$$\sum P_{loss} \simeq P_{el} = n_e \frac{2m}{M} \nu_m^e \langle u \rangle \tag{5.86}$$

and $\nu_e/\nu_m^e \sim m/M$ (see equations (4.1) and (4.2)), which confirms the validity of the assumption $\nu_e < \omega < \nu_m^e$. Equalling equations (5.85) and (5.86), we obtain the following expression for the *rms* electric field

$$E_{rms} \simeq \frac{m}{e} \sqrt{\frac{2 \langle u \rangle}{M}} \nu_m^e. \tag{5.87}$$

Since $\nu_m^e \propto n_o$ and $p = k_B T_o n_o$, we may still write

$$E_{rms} \simeq \frac{m}{e k_B T_o} \sqrt{\frac{2 u_I}{3M}} \frac{\nu_m^e}{n_o} p, \tag{5.88}$$

where we have assumed following Brown (1956) that $\langle u \rangle = u_I/3$ in the He:Hg mixture. If we further assume $T_o = \text{const}$, we obtain a linear dependence of E_{rms} with p, independent of the diffusion length Λ.

Let us consider now the limit of low pressures $\nu_m^e \ll \omega$ in the expression for the time-averaged absorbed power from the field

$$\overline{P_E(t)} = \frac{e^2 n_e}{m} \frac{\nu_m^e}{\nu_m^2 + \omega^2} E_{rms}^2 \simeq \frac{e^2 n_e \nu_m^e}{m\,\omega^2} E_{rms}^2. \tag{5.89}$$

In this low pressure limit, the electric field is high, and since we are analysing the He:Hg mixture, practically all collisions are inelastic, and then practically all ionizing ones; see equation (3.151)

$$\sum P_{loss} \simeq P_{ion} = n_e\, u_I <\nu_{ion}> . \tag{5.90}$$

This is obviously the power lost by the primary electrons since the power brought to the distribution by the secondary electrons (5.80) is vanishingly small. In the low pressure regime, the electrons make many oscillations per collision and gain appreciable energy between two collisions. Equating equations (5.89) and (5.90), we obtain

$$E_{rms} \simeq \frac{\omega}{e} \sqrt{m\, u_I \frac{<\nu_{ion}>}{\nu_m^e}}. \tag{5.91}$$

The energy-averaged ionization frequency may be now determined by equating the number of collisions to ionize to the number of collisions to diffuse out from the tube, that is from the breakdown condition (5.84)

$$<\nu_{ion}> = \frac{D_e}{\Lambda^2}, \tag{5.92}$$

where the free diffusion coefficient (5.12), in this case where $\nu_m^e(v_e) = \text{const}$, is

$$D_e = \frac{<v_e^2>}{3\,\nu_m^e} = \frac{2\,<u>}{3\,m\,\nu_m^e}. \tag{5.93}$$

Substituting equations (5.92) and (5.93) into equation (5.91), we obtain

$$E_{rms} \simeq \frac{\omega}{e\,\Lambda\,\nu_m^e} \sqrt{\frac{2\,u_I\,<u>}{3}}. \tag{5.94}$$

The ionization threshold energy is 19.8 eV, so that if we assume a very low pressure, in which all of the input energy is channelled to ionization, the average energy is of the order of u_I, and we obtain

$$E_{rms} \simeq \sqrt{\frac{2}{3}} \frac{\omega\, u_I\, k_B T_o}{e\,(\nu_m^e/n_o)} \frac{1}{p\,\Lambda} . \tag{5.95}$$

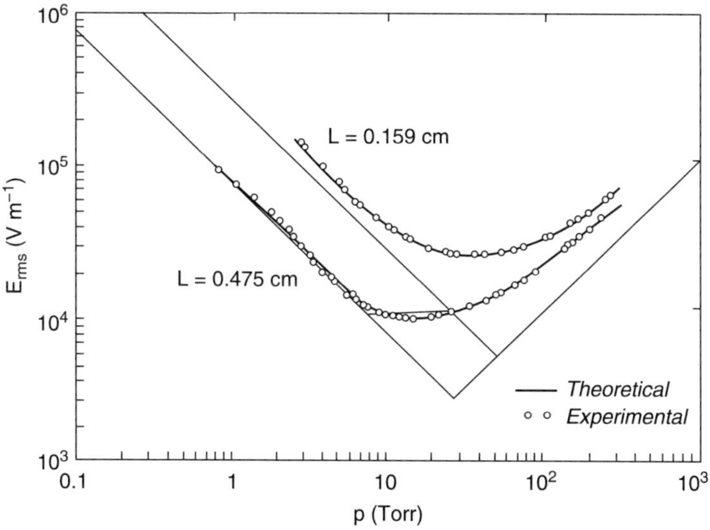

Fig. 5.3 Experimental breakdown *rms* electric field in HF discharges in He:Hg mixture, against the pressure (1 Torr = 133.3 Pa), compared with predicted values obtained from a simplified theory (Brown 1956)

In this case E_{rms} is inversely proportional to the product of pressure by the diffusion length, which agrees fairly well with the low pressure measurements shown in Fig. 5.3 taken from Brown (1956).

Typical values of the behaviour of the breakdown field at high frequency with changes in gas pressure are the curves shown in Fig. 5.3 taken from Brown (1956). At first sight these curves look very similar to corresponding data taken from DC fields, that is, as the pressure decreases the breakdown field first decreases and then increases again at low pressures (Paschen's curves). However, in the low pressure region, the rising breakdown field with decreasing pressure in HF discharges results from the increasing loss of efficiency in the transfer of energy from the field to the electrons. As we have seen in Sect. 4.1.2, in the low pressure limit the electrons oscillate out of phase with the field and gain no energy from it. Thus, as the pressure decreases, one must increases the field to make up for the loss of efficiency by just the factor of the effective field given by equation (4.20).

On the contrary, in the high pressure limit, the reason for the rising breakdown field with increasing pressure in HF discharges is the same as in the DC case. As the pressure increases, the electron mean free path, $\lambda = <v_e>/\nu_m^e$, decreases. Since at these high pressures, most of energy losses result from recoil collisions, the field increases directly proportionally to the pressure as expressed in equation (5.88).

The remarkable feature of the breakdown curves for HF fields at low pressures is the fact that the greater the electrode spacing, the easier it becomes to cause a breakdown. This is a result of the breakdown condition is established from the balance between energy gained from the field and electron loss by diffusion.

As the electrode spacing becomes smaller, the diffusion loss is greater and the field needs to increase to make up for the increased loss. Finally, eliminating the field between equations (5.88) and (5.95) obtained at the high and low pressure limits, respectively, we may calculate the pressure at which breakdown will occur most easily

$$p\lambda = \frac{2\pi c\,\sqrt{Mu_I}\,(k_B T_0)^2}{m\,(\nu_m^e/n_o)^2}\,\frac{1}{p\Lambda}\,, \tag{5.96}$$

where $\lambda = 2\pi c/\omega$ denotes the wavelength of the applied field. Neglecting the small variation of the gas temperature, this equation is of the type

$$p\lambda = \frac{\text{const.}}{p\Lambda}\,, \tag{5.97}$$

which represents the optimum condition for breakdown.

Curves of gas discharge breakdown as a function of pressure are often plotted for DC fields as Paschen curves in which, for a particular gas, the breakdown voltage V is found to be a function of the product of pressure by the electrode spacing, pd. The same type of variables may be introduced in the HF case, where instead of the breakdown voltage we use the product of the *rms* field by the diffusion length, $E_{rms}\Lambda$, and for pd we use $p\Lambda$. In the case of HF phenomena we have one more variable than in DC case, namely the frequency ω, and this may be introduced through the variable $p\lambda$ in equation (5.97). Accordingly, we have for the high pressure limit using equation (5.88)

$$E_{rms}\Lambda \simeq \frac{m}{e\,k_B T_o}\sqrt{\frac{2\,u_I}{3\,M}}\,\frac{\nu_m^e}{n_o}\,p\Lambda, \tag{5.98}$$

while for the low pressure limit we obtain from (5.95)

$$E_{rms}\Lambda \simeq \sqrt{\frac{2}{3}}\,\frac{2\pi c\,u_I\,k_B T_o}{e\,(\nu_m^e/n_o)}\,\frac{1}{p\lambda}. \tag{5.99}$$

However, if we express the electric field in terms of the effective electric field (4.20), we can take care of the field wavelength variations. Since $E_e = E_{rms}$ and $E_e = (\nu_m^e/\omega)\,E_{rms}$ at high and low pressure limits, respectively, we obtain for this latter situation

$$E_e\Lambda \simeq \sqrt{\frac{2}{3}}\,\frac{u_I}{e}\,. \tag{5.100}$$

The concept of effective electric field is strictly correct only when $\nu_m^e(v_e) = \text{const}$, but a similar equation would be valid for different variations of ν_m^e with the electron

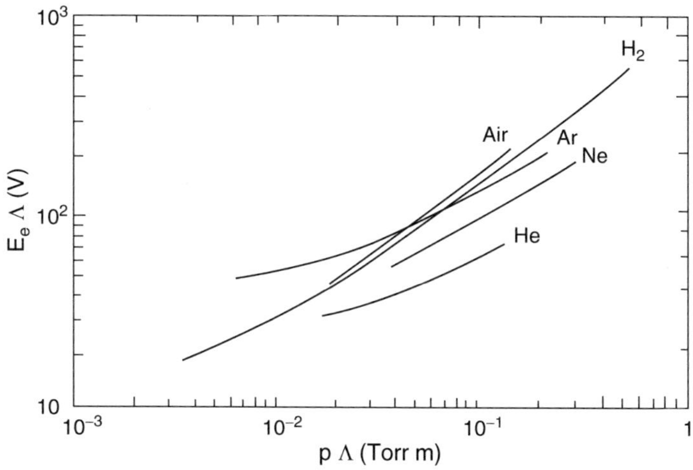

Fig. 5.4 Paschen curves $E_e\Lambda$ vs $p\Lambda$ for high-frequency breakdown in He, Ne, Ar, H_2 and air (1 Torr = 133.3 Pa) (Brown 1956)

velocity. The product $E_e\Lambda$ represents hence the breakdown voltage. Figure 5.4 taken from Brown (1956) shows the Paschen curves for high-frequency breakdown in various gases.

5.2.4 Ionization-Diffusion Plasma Front

Let us analyse now the expansion of a plasma front under the combined effect of diffusion and ionization, for relatively low electron densities in order the space-charge effects may be neglected, and the diffusion may be assumed as free diffusion. In this case, the solution of the continuity equation (5.76) in spherical geometry, for diffusion starting from a source point and constant given energy-averaged ionization frequency, is a Gaussian function with the form

$$n_e(r,t) \;=\; \frac{N_0}{(4\pi D_e t)^{3/2}}\,\exp\left(<\nu_{ion}>\,t - \frac{r^2}{4D_e t}\right)\,. \qquad (5.101)$$

The solution of equation (5.101) verifies the normalization

$$N(t) \;=\; \int_0^{\infty} n_e(r,t)\;4\pi r^2\,dr \;=\; N_0\;\exp\left(<\nu_{ion}>\;t\right) \qquad (5.102)$$

and equalling to zero the argument of the exponential function of (5.101), we observe that this solution represents a front that propagates at a speed

$$v = 2\sqrt{<\nu_{ion}> D_e} \, . \tag{5.103}$$

The characteristic length of the front, defined as $|\nabla n_e/n_e|^{-1}$ in a reference system moving at the speed v is

$$L = \left| \frac{1}{n_e} \frac{\partial n_e}{\partial r} \right|^{-1} = \frac{2D_e}{v} = \sqrt{\frac{D_e}{<\nu_{ion}>}} \, . \tag{5.104}$$

When an initial profile $n_{e0}(r)$ for the electron density exists rather than a source point, we have

$$n_e(r,t) = \frac{1}{(4\pi D_e t)^{3/2}} \int_0^\infty n_{e0}(r') \exp\left(<\nu_{ion}> t - \frac{(r-r')^2}{4D_e t} \right) 4\pi r'^2 \, dr' \, . \tag{5.105}$$

If the initial density decays quicker than exponentially with position, this solution presents a front propagation at the velocity v, in which the sharpness of the front can be defined by the characteristic length L. The expansion of a collisional plasma under the combined effect of ionization and diffusion with the presence of a space-charge field has been treated in Boeuf et al. (2010) and Zhu et al. (2011).

5.3 Electron Attachment

5.3.1 Boltzmann Equation with Ionization and Attachment

As in the case of the electron rate coefficient for excitation of a given state *j* from a state *i* (2.132), the rate coefficients for electron ionization and electron attachment (in $m^3 \, s^{-1}$) are, respectively,

$$C_{ion} = <v_e \, \sigma_{ion}(v_e)> = \frac{1}{n_e} \int_{u_I}^\infty v_e \, \sigma_{ion}(v_e) \, f_e^0(v_e) \, 4\pi v_e^2 \, dv_e \tag{5.106}$$

$$C_{att} = <v_e \, \sigma_{att}(v_e)> = \frac{1}{n_e} \int_0^\infty v_e \, \sigma_{att}(v_e) \, f_e^0(v_e) \, 4\pi v_e^2 \, dv_e \, . \tag{5.107}$$

Both may be expressed under the form of the first Townsend coefficient (expressed in m^{-1}) as follows

$$\alpha = \frac{<\nu_{ion}>}{v_{ed}} = \frac{n_o \, C_{ion}}{v_{ed}} \tag{5.108}$$

$$\eta = \frac{<\nu_{att}>}{v_{ed}} = \frac{n_o \, C_{att}}{v_{ed}} \, , \tag{5.109}$$

with $v_{ed} = |\mathbf{v_{ed}}|$, with $\mathbf{v_{ed}} = <\mathbf{v_e}>$, denoting the electron drift velocity (3.18)

$$\mathbf{v_{ed}} = \frac{1}{n_e}\int_0^{\infty} \frac{v_e}{3}\, \mathbf{f_e^1}\, 4\pi v_e^2\, dv_e. \tag{5.110}$$

The difference $(\alpha - \eta)$ represents the net gain of electrons created by length unity. Since $(\alpha - \eta)$ is the rate of change of the number of electrons per length unity along the direction of the electron drift caused by the field, the diffusion terms in equations (5.7) and (5.8) for the isotropic f_e^0 and anisotropic f_e^1 components of the electron velocity distribution function, respectively, under the two term approximation, can be taken into account as in Yousfi et al. (1985) and Itoh et al. (1988) by making the replacements

$$\frac{v_e}{3}\frac{\partial f_e^1}{\partial z} \Longrightarrow \frac{v_e}{3}\,(\alpha - \eta)\, f_e^1 \tag{5.111}$$

$$v_e \frac{\partial f_e^0}{\partial z} \Longrightarrow v_e\,(\alpha - \eta)\, f_e^0. \tag{5.112}$$

Thus, from equations (3.87) and (5.7), with (5.111) introduced into the latter term, we obtain the following expression for the equation describing the evolution of the isotropic component

$$\frac{\partial f_e^0}{\partial t} + \frac{eE}{m}\frac{1}{3v_e^2}\frac{\partial}{\partial v_e}(v_e^2 f_e^1) + \frac{v_e}{3}\,(\alpha - \eta)\, f_e^1 = I^0 + J^0, \tag{5.113}$$

in which $\mathbf{E} = -E\,\mathbf{e_z}$ and where I^0 and J^0 are given by equations (3.85) and (5.75). This latter equation with the inclusion of the term for secondary electron ionization. In the same manner, from equations (3.88) and (5.8), with (5.112) introduced into the latter term, we obtain the following equation for the anisotropic component

$$\frac{\partial f_e^1}{\partial t} + \frac{eE}{m}\frac{\partial f_e^0}{\partial v_e} + v_e\,(\alpha - \eta)\, f_e^0 = -\,\nu_m^e f_e^1\,, \tag{5.114}$$

in which ν_m^e denotes the effective collision frequency for momentum transfer (3.140), here with the inclusion of ionization and attachment frequencies

$$\nu_m^e = \nu_m + \sum_{i,j}(\nu_{ij}^m + \nu_{ji}^m) + \nu_{ion} + \nu_{att}. \tag{5.115}$$

For steady-state conditions of f_e^1, attained in a time $\sim (\nu_m^e)^{-1}$, we obtain from equation (5.114)

$$f_e^1 = -\frac{eE}{m\nu_m^e}\frac{\partial f_e^0}{\partial v_e} - \frac{v_e}{\nu_m^e}\,(\alpha - \eta)\, f_e^0 \tag{5.116}$$

and introducing this expression into (5.113), we obtain the following close equation for the isotropic component

$$\frac{\partial f_e^0}{\partial t} - \left(\frac{eE}{m}\right)^2 \frac{1}{3v_e^2} \frac{\partial}{\partial v_e} \left(\frac{v_e^2}{\nu_m^e} \frac{\partial f_e^0}{\partial v_e}\right) - \frac{eE}{m} (\alpha - \eta) \frac{1}{3v_e^2} \frac{\partial}{\partial v_e} \left(\frac{v_e^3}{\nu_m^e} f_e^0\right)$$
$$- \frac{eE}{m} \frac{v_e}{3\nu_m^e} (\alpha - \eta) \frac{\partial f_e^0}{\partial v_e} - \frac{v_e^2}{3\nu_m^e} (\alpha - \eta)^2 f_e^0 = I^0 + J^0 . \quad (5.117)$$

Considering now stationary conditions also for f_e^0 and making the substitutions $\nu_m^e = n_o \, v_e \, \sigma_m^e$ and $v_e = \sqrt{2u/m}$, we obtain after division of both members by $-\,2n_o/(mv_e)$

$$\left(\frac{eE}{n_o}\right)^2 \frac{d}{du} \left(\frac{u}{3\sigma_m^e} \frac{df_e^0}{du}\right) + \frac{eE}{n_o} \frac{\alpha - \eta}{n_o} \frac{d}{du} \left(\frac{u}{3\sigma_m^e} f_e^0\right)$$
$$+ \frac{eE}{n_o} \frac{\alpha - \eta}{n_o} \frac{u}{3\sigma_m^e} \frac{df_e^0}{du} + \left(\frac{\alpha - \eta}{n_o}\right)^2 \frac{u}{3\sigma_m^e} f_e^0$$
$$+ \frac{d}{du} \left[\frac{2m}{M} u^2 \sigma_m \left(f_e^0 + k_B T_0 \frac{df_e^0}{du}\right) + 4B_0 \sigma_0 \, u \, f_e^0\right]$$
$$+ \sum_{i,j} \delta_i \left\{(u + u_{ij}) \, \sigma_{ij}(u + u_{ij}) \, f_e^0(u + u_{ij}) - u \, \sigma_{ij}(u) \, f_e^0(u)\right\}$$
$$+ \sum_{j,i} \delta_j \left\{(u - u_{ij}) \, \sigma_{ji}(u - u_{ij}) \, f_e^0(u - u_{ij}) - u \, \sigma_{ji}(u) \, f_e^0(u)\right\}$$
$$+ \frac{1}{\Delta} \left(\frac{u}{\Delta} + u_I\right) \sigma_{ion} \left(\frac{u}{\Delta} + u_I\right) f_e^0 \left(\frac{u}{\Delta} + u_I\right)$$
$$+ \frac{1}{1 - \Delta} \left(\frac{u}{1 - \Delta} + u_I\right) \sigma_{ion} \left(\frac{u}{1 - \Delta} + u_I\right) f_e^0 \left(\frac{u}{1 - \Delta} + u_I\right)$$
$$- u \, \sigma_{ion}(u) \, f_e^0(u) - u \, \sigma_{att}(u) \, f_e^0 = 0 . \quad (5.118)$$

Besides the terms associated with the field and with the electron drift due to the net balance production between ionization and attachment, this equation also includes the contributions of elastic collisions, rotational excitation, inelastic and superelastic collisions, ionization (with terms for the appearance of secondary and scattered electrons and the removal of primary electrons), and the removal of electrons by attachment. With this formulation the drift velocity (in which $\mathbf{v_{ed}} = v_{ed} \, \mathbf{e_z}$) is obtained by substituting (5.116) into (5.110)

$$v_{ed} = -\frac{1}{n_e} \int_0^\infty \frac{v_e}{3} \left(\frac{eE}{m\nu_m^e} \frac{df_e^0}{dv_e} + \frac{v_e}{\nu_m^e} (\alpha - \eta) \, f_e^0\right) 4\pi v_e^2 \, dv_e . \quad (5.119)$$

Identifying here the electron mobility (3.50) and the free diffusion coefficient (5.12), we really obtain

$$v_{ed} \;=\; \mu_e\, E \;-\; (\alpha - \eta)\, D_e\,. \tag{5.120}$$

This expression is in conformity with the electron particle current density $\mathbf{\Gamma_e} = \Gamma_e\, \mathbf{e_z}$ given by

$$\Gamma_e \;=\; n_e\, \mu_e\, E \;-\; D_e\, \frac{dn_e}{dz}\,, \tag{5.121}$$

in which the electron number density presents the exponential growth

$$n_e(z) \;=\; n_e(0)\, \exp((\alpha - \eta) z)\,. \tag{5.122}$$

We note that when $(\alpha - \eta) > 0$ a density gradient in the z direction is formed, from which a drift velocity with opposite direction arises given by the second term of equation (5.120).

Integrating now equation (5.113) over all velocity space and using again the expression (5.110) for the drift velocity, one obtains the continuity equation

$$\frac{\partial n_e}{\partial t} \;+\; n_e\, (\alpha - \eta)\, v_{ed} \;=\; n_e\, (<\!v_{ion}\!> \;-\; <\!v_{att}\!>)\,, \tag{5.123}$$

whose equation under stationary conditions expresses the electron particle balance equation under the form

$$\frac{\alpha - \eta}{n_o}\; v_{ed} \;=\; C_{ion} \;-\; C_{att}\,, \tag{5.124}$$

where $(\alpha - \eta)/n_o$ is usually termed effective reduced Townsend ionization coefficient.

On the other hand, the equation for energy conservation can be obtained multiplying both sides of equation (5.113) by the electron energy $u = \frac{1}{2}\, m v_e^2$ and integrating over the energy space

$$\frac{\partial}{\partial t}(n_e < u >) \;=\; P_E \;-\; P_{el} \;-\; P_{rot} \;-\; P_{inel} \;-\; P_{ion} \;-\; P_{att} \;-\; P_{flow}\,. \tag{5.125}$$

The first term on the right-hand side member represents the power gain from the field per volume unity, $P_E = e n_e v_{ed} E$, while the other terms represent the power lost in recoil collisions, rotational excitation, inelastic collisions of first and second kinds, ionization, attachment, and the divergence of the energy flux due to the non-null ionization-attachment balance along the discharge axis. This last term has the expression

$$P_{flow} = \int_0^\infty u \, \frac{v_e}{3} \, (\alpha - \eta) \, f_e^1 \, 4\pi v_e^2 \, dv_e$$
$$= - (\alpha - \eta) \int_0^\infty u \, \frac{v_e}{3} \left(\frac{eE}{m\nu_m^e} \frac{\partial f_e^0}{\partial v_e} + \frac{v_e}{\nu_m^e} (\alpha - \eta) \, f_e^0 \right) 4\pi v_e^2 \, dv_e \, . \tag{5.126}$$

We note that as long as the rates of ionization and attachment do not exactly balance each other, it exists an energy current along the discharge axis associated with the local non-conservation of the electron number density. When $(\alpha - \eta) > 0$ there exists a local electron energy loss within a given axis slab $[z, z + dz]$, since the outward flux at $z + dz$ is larger than the flux entered at z position. In the opposite situation, $(\alpha - \eta) < 0$, we have $P_{flow} < 0$, so that this term represents now a local gain of energy for the electrons in the interior of the axis slab $[z, z + dz]$.

The equation (5.118) is solved by finite differencing the electron energy axis and matrix inversion. A guess for $(\alpha - \eta)$ is first introduced in equation (5.118) and then an iterative procedure is used until a final convergence for this parameter is achieved through equation (5.124).

5.3.2 *Boltzmann Calculations for* SF_6 *with a Buffer Gas*

Because of its outstanding electrical and physical properties, sulfur hexafluoride (SF_6) is the most studied electronegative (electron attaching) gas in the literature with a huge number of papers published dealing with experimental determinations of dielectric properties and Boltzmann's calculations. SF_6 is widely used in the electric power industry as an insulation medium for high-voltage gas circuit breakers. It presents excellent arc-quenching properties and a breakdown strength (minimum applied electric field that results in breakdown) three times higher than air at atmospheric pressure. It is a very stable electronegative, non-toxic, and non-flammable gas, and in its normal state, SF_6 is chemically inert. Moreover, due to the high etching rate of silicon in SF_6 plasma, the fabrication of integrated circuits is another use of SF_6. However, apprehension has been arising about the fact that SF_6 has been identified as a potent global warming greenhouse gas, so that nowadays alternative solutions have been investigated. Considering the relatively poor dielectric strength of environment-friendly pure gases and gas mixtures such as air, an alternative solution is to mix SF_6 with another gas as possible substitute gases for pure SF_6 (Cliteur et al. 1999a,b; Pinheiro and Loureiro 2002; Li et al. 2012; Wang et al. 2013).

As indicated above, a possible solution to avoid or to reduce the greenhouse problem is the replacement of SF_6 with several mixtures of SF_6 with different buffer gases, such as an inert gas, nitrogen, and carbon dioxide. The extension of the stationary Boltzmann equation (5.118) to a gas mixture is straightforward

$$
\left(\frac{eE}{n_o}\right)^2 \frac{d}{du}\left(\frac{u}{3\sum_s \delta_s\, \sigma^e_{s(m)}}\, \frac{df^0_e}{du}\right) + \frac{eE}{n_o}\, \frac{\alpha-\eta}{n_o}\, \frac{d}{du}\left(\frac{u}{3\sum_s \delta_s\, \sigma^e_{s(m)}}\, f^0_e\right)
$$

$$
+ \frac{eE}{n_o}\, \frac{\alpha-\eta}{n_o}\, \frac{u}{3\sum_s \delta_s\, \sigma^e_{s(m)}}\, \frac{df^0_e}{du} + \left(\frac{\alpha-\eta}{n_o}\right)^2 \frac{u}{3\sum_s \delta_s\, \sigma^e_{s(m)}}\, f^0_e
$$

$$
+ \sum_s \delta_s\, \frac{d}{du}\left[\frac{2m}{M_s}\, u^2\, \sigma_{s(m)}\left(f^0_e + k_B T_0\, \frac{df^0_e}{du}\right)\right] + \delta_2\, \frac{d}{du}\left[4 B_0 \sigma_0\, u\, f^0_e\right]
$$

$$
+ \sum_s \delta_s \Bigg\{ \sum_{i,j} \delta_i \left\{(u+u_{s(ij)})\, \sigma_{s(ij)}(u+u_{s(ij)})\, f^0_e(u+u_{s(ij)}) - u\, \sigma_{s(ij)}(u)\, f^0_e(u)\right\}
$$

$$
+ \sum_{j,i} \delta_j \left\{(u-u_{s(ij)})\, \sigma_{s(ji)}(u-u_{s(ij)})\, f^0_e(u-u_{s(ij)}) - u\, \sigma_{s(ji)}(u)\, f^0_e(u)\right\}
$$

$$
+ \frac{1}{\Delta}\left(\frac{u}{\Delta} + u_{s(I)}\right) \sigma_{s(ion)}\left(\frac{u}{\Delta} + u_{s(I)}\right) f^0_e\left(\frac{u}{\Delta} + u_{s(I)}\right)
$$

$$
+ \frac{1}{1-\Delta}\left(\frac{u}{1-\Delta} + u_{s(I)}\right) \sigma_{s(ion)}\left(\frac{u}{1-\Delta} + u_{s(I)}\right) f^0_e\left(\frac{u}{1-\Delta} + u_{s(I)}\right)
$$

$$
- u\, \sigma_{s(ion)}(u)\, f^0_e(u) \Bigg\} - \delta_1\, u\, \sigma_{att}(u)\, f^0_e = 0\,. \tag{5.127}
$$

Equation (5.127) is written for a mixture of two gases $s = (1, 2)$, in which $s = 1$ is SF_6 (the only with attachment considered here) and $s = 2$ is for the other gas (a rare gas, N_2 or CO_2, see Pinheiro and Loureiro 2002). δ_s represents the fractional mixture composition (i.e. $\delta_1 + \delta_2 = 1$). The term for rotational excitation exists only for N_2, while the term for the superelastic collisions $j \to i$ exists with some significance only for the case of vibrational levels of N_2 or CO_2. On the other hand, in the case of a gas mixture, the equation (5.124) writes as follows

$$
\frac{\alpha-\eta}{n_o}\, v_{ed} = \sum_s \delta_s\, C_{s(ion)} - \delta_1\, C_{att}\,. \tag{5.128}
$$

The electron cross sections for SF_6 shown in Fig. 5.5 have been obtained from Itoh et al. (1988, 1993). They include cross sections for momentum transfer, total vibrational excitation, total electronic excitation, ionization, and five individual cross sections for electron attachment with formation of the negative ions SF_6^-, SF_5^-, F^-, SF_4^- and F_2^-. See also Hayashi and Nimura (1984) about the importance of the attachment cross section for F^- formation in the effective Townsend ionization coefficient.

Figure 5.6 shows the effective reduced Townsend ionization coefficient, $(\alpha - \eta)/n_o$, obtained in Pinheiro and Loureiro (2002), as a function of the reduced electric field E/n_o, in pure SF_6, using an energy partition $\Delta = 0.5$ between the secondary and scattered electrons. These results have been obtained using the formulation presented above, in which a two-term expansion is used

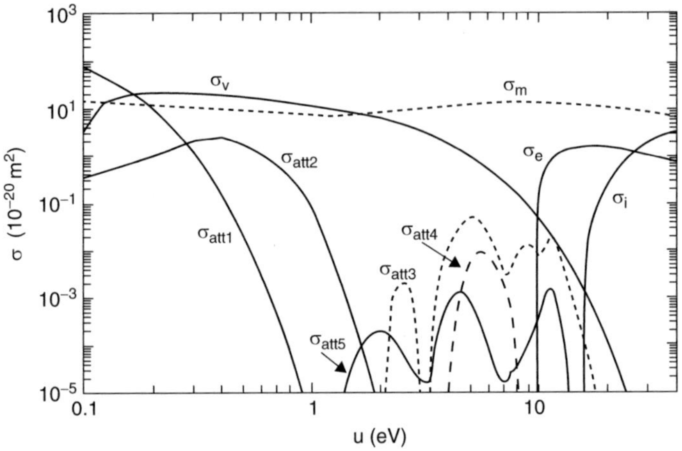

Fig. 5.5 Electron cross sections for SF_6 as a function of the electron energy as follows: σ_m, momentum transfer (*dotted line*); σ_v, total vibrational excitation; σ_e, total electronic excitation; σ_i, ionization; σ_{att1} to σ_{att5}, electron attachment with formation of SF_6^- (*full*), SF_5^- (*full*), F^- (*dotted*), SF_4^- (*broken*), and F_2^- (*full*) ions, respectively

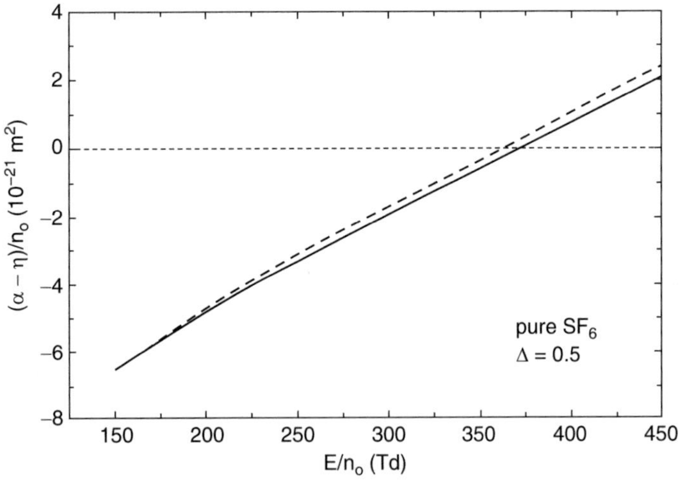

Fig. 5.6 Effective ionization Townsend coefficient, $(\alpha - \eta)/n_o$, in pure SF_6 obtained in Loureiro and Pinheiro (2002), *full line*, and in Itoh et al. (1993), *broken line*, this latter using a six-term Boltzmann equation method

for the electron energy distribution function, in the range $E/n_o = 150 - 450\,\text{Td}$ ($1\,\text{Td} = 1 \times 10^{-21}\,\text{V m}^2$). For comparison it is also plotted in Fig. 5.6 the data obtained in Itoh et al. (1993) using a six-term Boltzmann method. As pointed out in Pinheiro and Loureiro (2002), the two-term approximation is valid at low and intermediate E/n_o values despite the fact that SF_6 is a strongly electronegative gas.

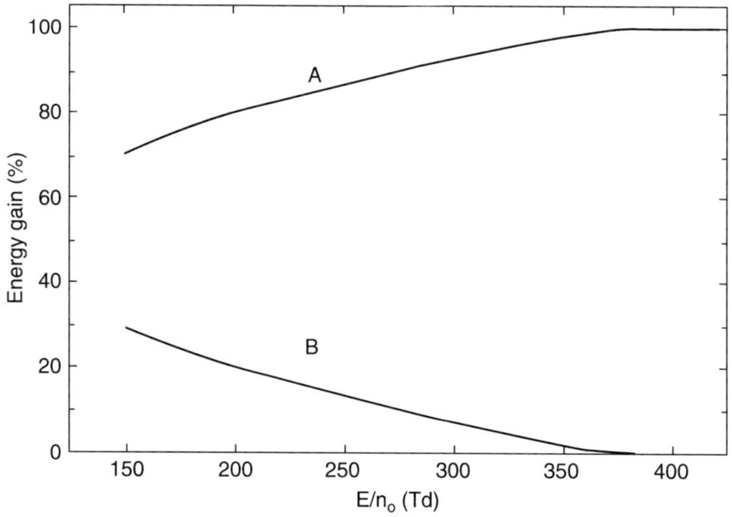

Fig. 5.7 Fractional power gained by the electrons in pure SF_6 from the field (*A*) and from the axial drift current as $(\alpha - \eta) < 0$ (*B*) (Pinheiro and Loureiro 2002)

Figure 5.6 shows that $(\alpha - \eta)/n_o < 0$ for $E/n_o < 370\,\text{Td}$, which means that the divergence of the electron particle flow along the drift direction given by equation (5.126) is negative. In this case, since $P_{flow} < 0$ in equation (5.125), the drift produces a local gain of energy for the electrons in the interior of a given axis slab $[z, z + dz]$. Figure 5.7 shows the fractional power gained by the electrons from the field and the power gained locally by the electrons from the axial drift current, this latter due to the existence of a negative ionization-attachment balance. The contribution P_{flow} represents 41.3 % of the power gained from the field at $E/n_o = 150\,\text{Td}$ (i.e. 29.2 % of the total power) and it rapidly increases as E/n_o decreases below 150 Td (not shown in the figure). On the other hand, the fractional power transferred by the electrons to SF_6 molecules is represented in Fig. 5.8. The various curves are for the power lost in recoil (elastic) collisions, inelastic collisions (vibrational plus electronic excitation), ionization, attachment, and the power lost from the electron particle flow when $(\alpha - \eta) > 0$, this last term for $E/n_o > 370\,\text{Td}$ only. Figure 5.8 shows that the power transferred into vibrational and electronic excitation is always larger than $\sim 60\,\%$ ($\sim 80\,\%$ and $\sim 62\,\%$ at 400 Td and 600 Td, respectively).

Figure 5.9 shows the effective reduced Townsend ionization coefficient $(\alpha - \eta)/n_o$, calculated in Pinheiro and Loureiro (2002), for different mixture compositions in SF_6–He, as a function of the reduced electric field E/n_o. As in Figs. 5.6, 5.7, and 5.8, the electron energy share between the primary scattered and secondary electrons after an ionizing collision is assumed $\Delta = 0.5$. Swarm data from pulsed Townsend experiments determined in Urquijo et al. (2001) and Xiao et al. (1999) are also plotted for comparison. The data show a fairly uniform trend

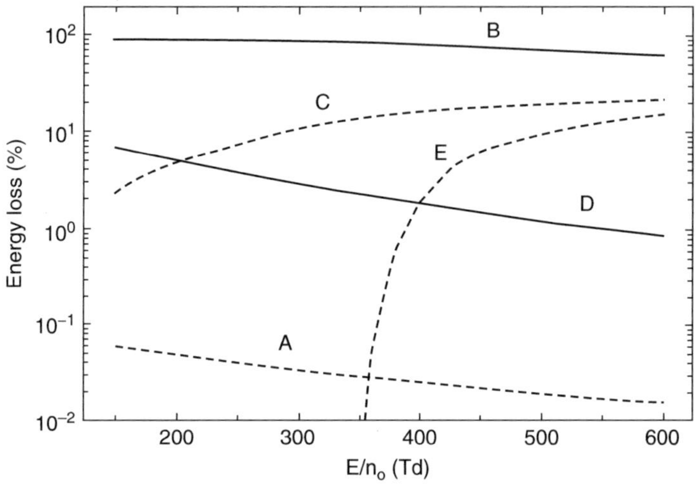

Fig. 5.8 Fractional power lost by the electrons in pure SF_6 due to elastic collisions (*A*), vibrational plus electronic excitation (*B*), ionization (*C*), attachment (*D*), and from the drift current as $(\alpha - \eta) > 0$ (*E*) (Pinheiro and Loureiro 2002)

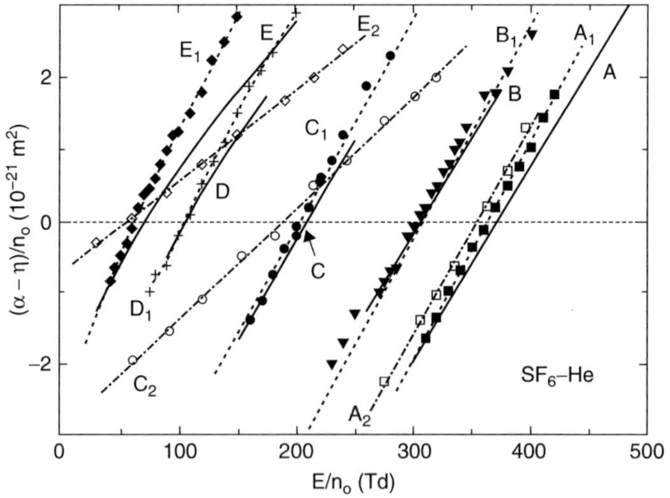

Fig. 5.9 Effective reduced Townsend ionization coefficient in SF_6—He: (*A*) pure SF_6; (*B*) 20 %He; (*C*) 50 %He; (*D*) 80 %He; (*E*) 90 %He. Data points: Urquijo et al. (2001) — (■, A_1) pure SF_6, (▼, B_1) 20 %He, (•, C_1) 50 %He, (+, D_1) 80 %He, and (♦, E_1) 90 %He; Xiao et al. (1999) — (□, A_2) pure SF_6, (○, C_2) 50 %He, and (◇, E_2) 90 %He (Pinheiro and Loureiro 2002)

to increase with He addition for fixed E/n_o values, which is consistent with the decreasingly electronegative character of the mixture. In pure SF_6, the calculated data for $(\alpha - \eta)/n_o$ are in better agreement with the measured data of Urquijo et al. (2001) than with those of Xiao et al. (1999). The agreement is always rather good

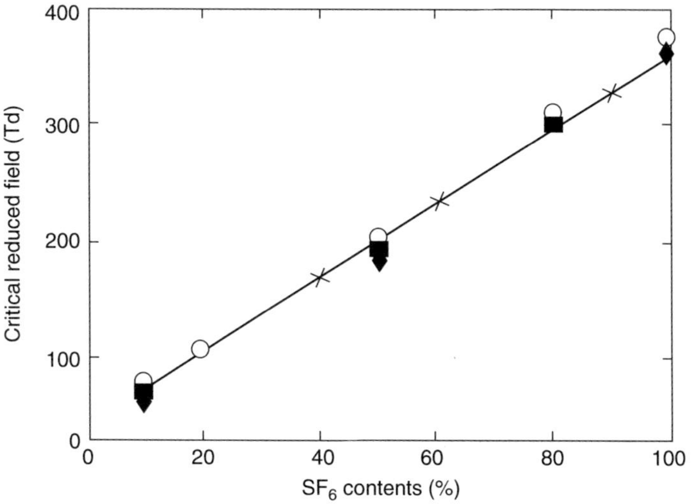

Fig. 5.10 Critical reduced electric field strengths, $(E/n_o)_0$, as a function of SF_6 percentage in a SF_6–He mixture calculated in Wang et al. (2013) (*full line with x*). For comparison they are also plotted the calculated data of Pinheiro and Loureiro (2002) (○) and the experimental data of Urquijo et al. (2001) (■) and Xiao et al. (1999) (◆) (Wang et al. 2013)

with Urquijo et al. (2001) as the He percentage increases, in particular at low E/n_o values (i.e. for $(\alpha - \eta)/n_o < 0$), except in the case of a SF_6–90 %He mixture where the calculated data have a less pronounced slope for the nearly linear variation of $(\alpha - \eta)/n_o$ with E/n_o. In a SF_6–90 %He mixture the predicted data are generally lower than the measured data of Urquijo et al. (2001) (by a factor of 2 at 100 Td). On the other hand, significant deviations emerge at SF_6–50 %He for the data reported in Xiao et al. (1999) when compared with Pinheiro and Loureiro (2002). The disagreement becomes progressively larger with increasing He percentage and in this case the slope of the straight-line obtained by fitting $(\alpha - \eta)/n_o$ data is significantly less pronounced in Xiao et al. (1999) than that predicted in Pinheiro and Loureiro (2002). A comparison between calculated and measured data in SF_6 mixtures with Xe, CO and N_2 is also presented in Pinheiro and Loureiro (2002).

Recent calculated critical reduced electric field strengths, $(E/n_o)_0$, for which ionization exactly balances attachment, $(\alpha - \eta) = 0$, taken from Wang et al. (2013), are plotted in Fig. 5.10 for the case of a SF_6–He mixture, as a function of SF_6 percentage. For comparison they are also plotted the calculated data of Pinheiro and Loureiro (2002) as well as experimental results from Urquijo et al. (2001) and Xiao et al. (1999). The calculated data of Wang et al. (2013) show excellent agreement with Pinheiro and Loureiro's calculations and fairly good agreement with the measurements of Urquijo et al. (2001) and Xiao et al. (1999).

As pointed out in Yousfi et al. (1985) the attachment coefficient may be very sensitive to the value assumed for the energy partition after an ionizing collision Δ between the secondary and the scattered electrons, in particular as this share of

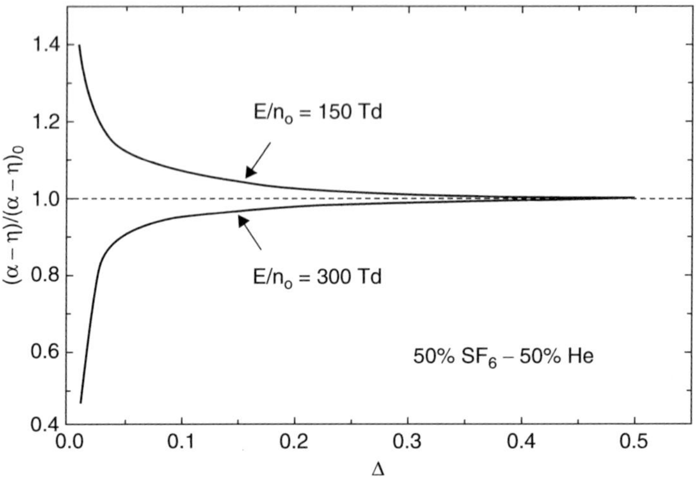

Fig. 5.11 $(\alpha - \eta)/(\alpha - \eta)_0$ as a function of the electron energy partition Δ in a SF_6–50 %He mixture, for $E/n_o = 150$ Td and 300 Td, with $(\alpha - \eta)_0$ denoting the reference results at $\Delta = 0.5$ (Pinheiro and Loureiro 2002)

energy is large: $\Delta \to 0$; $(1-\Delta) \to 1$ (Itoh et al. 1988). Figure 5.11 shows the effect of Δ on the effective Townsend ionization coefficient $(\alpha - \eta)/n_o$, calculated in a SF_6–50 %He mixture with $E/n_o = 150$ and 300 Td. The plotted results correspond to the ratio $(\alpha - \eta)/(\alpha - \eta)_0$, with $(\alpha - \eta)_0$ denoting the reference value at $\Delta = 0.5$ used in Figs. 5.6 to 5.10. In a SF_6–50 %He mixture the critical value at which ionization exactly balances attachment occurs at $E/n_o \sim 200$ Td, so that we have $(\alpha - \eta) < 0$ and $(\alpha - \eta) > 0$ at 150 Td and 300 Td, respectively.

Appendices

A.5.1 Expansion of the Boltzmann Equation in Spherical Harmonics

In this appendix we present the derivation of the term with the gradient in the space of positions of the Boltzmann equation (5.1) using the expansion in spherical harmonics (Cherrington 1980; Delcroix 1963, 1966). In this case using the expansion (5.2) we write the Boltzmann equation as

$$\sum_l \frac{\partial f_e^l}{\partial t} P_l + \sum_l v_{ez} \frac{\partial f_e^l}{\partial z} P_l = \sum_l I^l(f_e^l) P_l, \tag{5.129}$$

having assumed the anisotropy directed along the z axis. Since $v_{ez} = v_e \cos\theta$ and making use of the recursion relation (3.209), we may write

$$\sum_l \frac{\partial f_e^l}{\partial t} P_l + \sum_l v_e \frac{\partial f_e^l}{\partial z} \left(\frac{l+1}{2l+1} P_{l+1} + \frac{l}{2l+1} P_{l-1} \right) = \sum_l I^l(f_e^l)\, P_l. \quad (5.130)$$

The second left-hand side term can be rewritten in terms of the polynomial P_l as follows

$$\sum_l \frac{\partial f_e^l}{\partial t} P_l + \sum_l v_e \left(\frac{l}{2l-1} \frac{\partial f_e^{l-1}}{\partial z} + \frac{l+1}{2l+3} \frac{\partial f_e^{l+1}}{\partial z} \right) P_l = \sum_l I^l(f_e^l)\, P_l, \quad (5.131)$$

so that the equation of order l is

$$\frac{\partial f_e^l}{\partial t} + v_e \left(\frac{l}{2l-1} \frac{\partial f_e^{l-1}}{\partial z} + \frac{l+1}{2l+3} \frac{\partial f_e^{l+1}}{\partial z} \right) = I^l(f_e^l), \quad (5.132)$$

from which the equations (5.7) and (5.8) are immediately obtained.

A.5.2 Longitudinal Diffusion

In the time of flight experiments for determining the electron mobility μ_e, the presence of an axial electric field makes the diffusion to be different in directions perpendicular and parallel to the field (Parker and Lowke 1969; Wedding et al. 1985; Nakamura 1987). The continuity equation, in the assumption of null electron source or sink, is

$$\frac{\partial n_e}{\partial t} + (\nabla . \mathbf{\Gamma_e}) = 0, \quad (5.133)$$

in which the electron particle current density is

$$\mathbf{\Gamma_e} = -D_{eT} \nabla_T n_e - D_{eL} \nabla_z n_e - n_e \mu_e \mathbf{E}, \quad (5.134)$$

assuming the applied electric field directed along the z axis, and being D_{eT} and D_{eL} the transverse and longitudinal electron diffusion coefficients. Inserting (5.134) into (5.133) and considering $\mathbf{E} = -E\, \mathbf{e_z}$, we obtain

$$\frac{\partial n_e}{\partial t} - D_{eT} \left(\frac{\partial^2 n_e}{\partial x^2} + \frac{\partial^2 n_e}{\partial y^2} \right) - D_{eL} \frac{\partial^2 n_e}{\partial z^2} + \mu_e E \frac{\partial n_e}{\partial z} = 0, \quad (5.135)$$

whose solution for motion and spread of a pulse, starting from the origin at $t = 0$, is

$$n_e(x, y, z, t) = \frac{N_0}{4\pi D_{eT} t \sqrt{4\pi D_{eL} t}} \exp\left(-\frac{x^2+y^2}{4D_{eT}t}\right) \exp\left(-\frac{(z-\mu_e E t)^2}{4D_{eL}t}\right). \quad (5.136)$$

This solution is normalized as follows

$$\int_{-\infty}^{\infty}\int_{-\infty}^{\infty}\int_{-\infty}^{\infty} n_e(x, y, z, t)\, dx\, dy\, dz = N_0. \quad (5.137)$$

If a short pulse of N_0 electrons starts from $z = 0$ at $t = 0$, in a transverse infinite medium, the density $n_e(z, t)$ is

$$n_e(z, t) = \frac{N_0}{\sqrt{4\pi D_{eL} t}} \exp\left(-\frac{(z - v_{ed}\, t)^2}{4D_{eL}t}\right), \quad (5.138)$$

being $v_{ed}{=}\mu_e E$ the electron drift velocity. This density passes through a maximum at

$$t_M = \frac{z}{v_{ed}} \sqrt{1 + \left(\frac{D_{eL}}{v_{ed}\, z}\right)^2} - \frac{D_{eL}}{v_{ed}^2}. \quad (5.139)$$

In Pitchford and Phelps (1982) and Phelps and Pitchford (1985) we may find the calculation of D_{eL} from an expansion in powers of the spatial gradients of the electron distribution function $f_e(\mathbf{r}, \mathbf{v_e}, t)$ in cylindrical geometry

$$f_e(\mathbf{r}, \mathbf{v_e}, t) = f(\mathbf{v_e})\, n_e(\mathbf{r}, t) - \frac{g_r(\mathbf{v_e})}{n_o}\frac{\partial n_e}{\partial r} - \frac{g_z(\mathbf{v_e})}{n_o}\frac{\partial n_e}{\partial z}, \quad (5.140)$$

where $f(\mathbf{v_e})$, $g_r(\mathbf{v_e})$, and $g_z(\mathbf{v_e})$ are velocity distributions normalized such that

$$\int_{\mathbf{v_e}} f(\mathbf{v_e})\, \mathbf{dv_e} = 1\,; \quad \int_{\mathbf{v_e}} g_r(\mathbf{v_e})\, \mathbf{dv_e} = \int_{\mathbf{v_e}} g_z(\mathbf{v_e})\, \mathbf{dv_e} = 0, \quad (5.141)$$

and n_o is the gas number density. The continuity equation is then

$$\frac{\partial n_e}{\partial t} - D_{eT}\frac{1}{r}\frac{\partial}{\partial r}\left(r\frac{\partial n_e}{\partial r}\right) - D_{eL}\frac{\partial^2 n_e}{\partial z^2} + v_{ed}\frac{\partial n_e}{\partial z} = 0. \quad (5.142)$$

Using the two-term expansion in Legendre polynomials for the velocity dependent functions (5.141), we obtain the following expression for the electron drift velocity in accordance with (3.171)

$$v_{ed} = \int_{\mathbf{v_e}} v_e \cos\theta\ f(\mathbf{v_e})\ \mathbf{dv_e} = -\frac{E}{n_o}\frac{e}{3}\sqrt{\frac{2}{m}}\int_0^{\infty}\frac{u}{\sigma_m^e(u)}\frac{df^0}{du}\,du, \quad (5.143)$$

whereas the transverse and longitudinal electron diffusion coefficients, in two-term approximation, are given by (see Pitchford and Phelps 1982; Phelps and Pitchford 1985)

$$n_o\, D_{eT} = \int_{\mathbf{v_e}} v_e\ \sin\theta\ g_r(\mathbf{v_e})\ \mathbf{dv_e} = \frac{1}{3}\sqrt{\frac{2}{m}}\int_0^\infty \frac{u}{\sigma_m^e(u)} f^0(u)\, du \tag{5.144}$$

$$\begin{aligned} n_o\, D_{eL} &= \int_{\mathbf{v_e}} v_e\ \cos\theta\ g_z(\mathbf{v_e})\ \mathbf{dv_e} = \frac{1}{3}\sqrt{\frac{2}{m}}\int_0^\infty u\, g_z^1(u)\, du \\ &= n_o\, D_{eT} - \frac{E}{n_o}\frac{e}{3}\sqrt{\frac{2}{m}}\int_0^\infty \frac{u}{\sigma_m^e(u)}\frac{dg_z^0}{du}\, du - v_{ed}\int_0^\infty g_z^0(u)\,\sqrt{u}\, du\,. \end{aligned} \tag{5.145}$$

The functions $f^0(u)$ and $g_z^0(u)$ are normalized such that

$$\int_0^\infty f^0(u)\,\sqrt{u}\,du = 1\,;\quad \int_0^\infty g_z^0(u)\,\sqrt{u}\,du = 0\,. \tag{5.146}$$

The transverse diffusion coefficient agrees with the free diffusion coefficient previously reported through equation (3.193) and derived later on in (5.12), whereas $D_{eL} \equiv D_{eT}$ as $g_z^0(u) \to 0$ throughout (i.e. for null axial density gradients).

Exercises

Exercise 5.1. Determine the power lost in ionization from equation (5.75).

Resolution: The equation for electron energy conservation (5.125) is obtained multiplying both sides of the equation for evolution of the isotropic component of the velocity distribution function (5.113) by the electron energy $u = \frac{1}{2}\, mv_e^2$ and integrating over the whole velocity space

$$\frac{\partial}{\partial t}(n_e < u >) + \ldots\ldots\ldots\ldots\ldots = \int_0^\infty u\,(I^0 + J^0)\, 4\pi v_e^2\, dv_e\,.$$

Then, the term for P_{ion} in equation (5.125) is given by

$$P_{ion} = -\int_0^\infty u\, J_{ion}^0\, 4\pi v_e^2\, dv_e\,,$$

with J_{ion}^0 expressed by equation (5.75). Thus, we may write

$$P_{ion} = -\int_0^\infty u \, \frac{2}{m v_e} \, n_o \Bigg\{ \left(\frac{u}{\Delta} + u_I\right) \sigma_{ion}\left(\frac{u}{\Delta} + u_I\right) f_e^0\left(\frac{u}{\Delta} + u_I\right) \frac{1}{\Delta}$$
$$+ \left(\frac{u}{1-\Delta} + u_I\right) \sigma_{ion}\left(\frac{u}{1-\Delta} + u_I\right) f_e^0\left(\frac{u}{1-\Delta} + u_I\right) \frac{1}{1-\Delta}$$
$$- u \, \sigma_{ion}(u) \, f_e^0(u) \Bigg\} \; 4\pi v_e^2 \, dv_e \, .$$

Replacing v_e with the electron energy u and using for simplification the electron energy distribution function (EEDF) normalized such that

$$n_e = \int_0^\infty F(u) \, \sqrt{u} \, du \, ,$$

in which

$$F(u) = \frac{4\pi}{m} \sqrt{\frac{2}{m}} \, f_e^0(v_e) \, ,$$

we find

$$P_{ion} = -\sqrt{\frac{2}{m}} \, n_o \Bigg\{ \int_0^\infty u' \left(\frac{u'}{\Delta} + u_I\right) \sigma_{ion}\left(\frac{u'}{\Delta} + u_I\right) F\left(\frac{u'}{\Delta} + u_I\right) \frac{1}{\Delta} \, du'$$
$$+ \int_0^\infty u'' \left(\frac{u''}{1-\Delta} + u_I\right) \sigma_{ion}\left(\frac{u''}{1-\Delta} + u_I\right) F\left(\frac{u''}{1-\Delta} + u_I\right) \frac{1}{1-\Delta} \, du''$$
$$- \int_{u_I}^\infty u^2 \, \sigma_{ion}(u) \, F(u) \, du \Bigg\} \, .$$

Making the replacements $u = u'/\Delta + u_I$, $u' = \Delta \, (u - u_I)$, and $du' = \Delta \, du$ in the first term, and $u = u''/(1-\Delta) + u_I$, $u'' = (1-\Delta) \, (u - u_I)$, and $du'' = (1-\Delta) \, du$ in the second term, we obtain

$$P_{ion} = -\sqrt{\frac{2}{m}} \, n_o \Bigg\{ \int_{u_I}^\infty (u - u_I) \, u \, \sigma_{ion}(u) \, F(u) \, \Delta \, du$$
$$+ \int_{u_I}^\infty (u - u_I) \, u \, \sigma_{ion}(u) \, F(u) \, (1-\Delta) \, du$$
$$- \int_{u_I}^\infty u^2 \, \sigma_{ion}(u) \, F(u) \, du \Bigg\}$$
$$= \sqrt{\frac{2}{m}} \, n_o \int_{u_I}^\infty u_I \, u \, \sigma_{ion}(u) \, F(u) \, du$$
$$= n_o \, n_e \, u_I \, C_{ion} \, ,$$

in which

$$C_{ion} \;=\; \frac{1}{n_e}\;\sqrt{\frac{2}{m}}\int_{u_I}^{\infty} u\,\sigma_{ion}(u)\,F(u)\,du$$

is the electron ionization rate coefficient.

Exercise 5.2. Determine the power lost in electron attachment.

Resolution: The power lost in electron attachment is obtained from

$$P_{att} \;=\; -\int_0^{\infty} u\,J^0_{att}\;4\pi v_e^{\,2}\;dv_e\;,$$

where J^0_{att} has the same form as the third term of equation (5.75) but for attachment

$$J^0_{att} \;=\; -\,\frac{2}{mv_e}\;n_o\;u\;\sigma_{att}\,f_e^0\;.$$

Substituting and replacing v_e with the electron energy u, we obtain

$$P_{att} \;=\; \frac{8\pi}{m^2}\;n_o\int_0^{\infty} u^2\;\sigma_{att}(u)\,f_e^0(u)\;du\;.$$

Using now the EEDF $F(u)$ as defined in Exercise 5.1, we may write

$$P_{att} \;=\; \int_0^{\infty} u^{3/2}\;\nu_{att}(u)\;F(u)\;du\;,$$

with $\nu_{att}(u)$ denoting the electron attachment frequency given by

$$\nu_{att}(u) \;=\; n_o\;v_e\;\sigma_{att}(v_e) \;=\; n_o\;\sqrt{\frac{2u}{m}}\;\;\sigma_{att}(u)\;.$$

Replacing $\nu_{att}(u)$ in P_{att}, we obtain at the end

$$P_{att} \;=\; n_e < u\;\nu_{att} > \;.$$

Exercise 5.3. Write the expressions for the electron drift velocity (5.119) and for the power brought to the electron drift by the ionization-attachment balance (5.126) in terms of the EEDF $F(u)$ used in the previous exercises.

Resolution: From equation (5.119) we obtain

$$v_{ed} \;=\; -\,\frac{1}{n_e n_o}\;\sqrt{\frac{2}{m}}\int_0^{\infty}\frac{u}{3\sigma_m^e}\left(eE\,\frac{dF}{du}\;+\;(\alpha-\eta)\;F\right)du\;,$$

with σ_m^e denoting the effective electron cross-section for momentum transfer, while from (5.126) we find

$$P_{flow} = -\sqrt{\frac{2}{m}}\,\frac{\alpha-\eta}{n_o}\int_0^\infty \frac{u^2}{3\sigma_m^e}\left(eE\,\frac{dF}{du} + (\alpha-\eta)\,F\right)du\,.$$

In the case of pure SF_6, P_{flow} is < 0 at $E/n_o < 370\,\text{Td}$ and $P_{flow} > 0$ at $E/n_o > 370\,\text{Td}$, so that in the first situation P_{flow} represents a gain energy term.

References

W.P. Allis, Motions of ions and electrons, in *Handbuch der Physik*, vol. 21, ed. by S. Flügge (Springer, Berlin, 1956), pp. 383–444

J.-P. Boeuf, B. Chaudhury, G.Q. Zhu, Theory and modeling of self-organization and propagation of filamentary plasma arrays in microwave breakdown at atmospheric pressure. Phys. Rev. Lett. **104**(1), 015002 (2010)

S.C. Brown, Breakdown in gases: alternating and high-frequency fields, in *Handbuch der Physik*, vol. 22, ed. by S. Flügge (Springer, Berlin, 1956), pp. 531–575

B.E. Cherrington, *Gaseous Electronics and Gas Lasers* (Pergamon Press, Oxford, 1980)

G.J. Cliteur, K. Suzuki, K.C. Paul, T. Sakuta, SF_6-N_2 circuit-breaker arc modelling around current zero: I. Free-recovery simulation using a collisional-radiative plasma model. J. Phys. D: Appl. Phys. **32**, 478–493 (1999a)

G.J. Cliteur, K. Suzuki, K.C. Paul, T. Sakuta, SF_6-N_2 circuit-breaker arc modelling around current zero: II. Arc-circuit-interaction simulation using Boltzmann analysis. J. Phys. D: Appl. Phys. **32**, 494–503 (1999b)

J.-L. Delcroix, *Physique des Plasmas: Volumes 1 and 2*. Monographies Dunod (Dunod, Paris, 1963, 1966) (in French)

J.V. DiCarlo, M.J. Kushner, Solving the spatially dependent Boltzmann's equation for the electron-velocity distribution using flux corrected transport. J. Appl. Phys. **66**(12), 5763–5774 (1989)

M. Hayashi, T. Nimura, Importance of attachment cross-sections of F^- formation for the effective ionisation coefficients in SF_6. J. Phys. D: Appl. Phys. **17**, 2215–2223 (1984)

T. Holstein, Energy distribution of electrons in high frequency gas discharges. Phys. Rev. **70**(5–6), 367–384 (1946)

H. Itoh, Y. Miura, N. Ikuta, Y. Nakao, H. Tagashira, Electron swarm development in SF_6: I. Boltzmann equation analysis. J. Phys. D: Appl. Phys. **21**, 922–930 (1988)

H. Itoh, T. Matsumura, K. Satoh, H. Date, Y. Nakao, H. Tagashira, Electron transport coefficients in SF_6. J. Phys. D: Appl. Phys. **26**, 1975–1979 (1993)

X. Li, H. Zhao, S. Jia, Dielectric breakdown properties of SF_6-N_2 mixtures in the temperature range 300−3000 K. J. Phys. D: Appl. Phys. **45**, 445202 (2012)

A.D. MacDonald, D.U. Gaskell, H.N. Gitterman, Microwave breakdown in air, oxygen, and nitrogen. Phys. Rev. **130**(5), 1841–1850 (1963)

Y. Nakamura, Drift velocity and longitudinal diffusion coefficient of electrons in nitrogen and carbon monoxide. J. Phys. D: Appl. Phys. **20**, 933–938 (1987)

J.H. Parker Jr., J.J. Lowke, Theory of electron diffusion parallel to electric fields. I. Theory. Phys. Rev. **181**(1), 290–301 (1969)

A.V. Phelps, L.C. Pitchford, Anisotropic scattering of electrons by N_2 and its effect on electron transport. Phys. Rev. A **31**(5), 2932–2949 (1985)

M.J. Pinheiro, J. Loureiro, Effective ionization coefficients and electron drift velocities in gas mixtures of SF_6 with He, Xe, CO_2 and N_2 from Boltzmann analysis. J. Phys. D: Appl. Phys. **35**, 3077–3084 (2002)

L.C. Pitchford, A.V. Phelps, Comparative calculation of electron-swarm properties in N_2 at moderate E/N values. Phys. Rev. A **25**(1), 540–554 (1982)

W.R.L. Thomas, The determination of the total excitation cross section in neon by comparison of theoretical and experimental values of Townsend's primary ionization coefficient. J. Phys. B: Atomic Mol. Phys. Ser. 2 **2**, 551–561 (1969)

J. Urquijo, E. Basurto, J.L. Hernández-Ávila, Effective ionization electron and ion transport in SF_6—He mixtures. J. Phys. D: Appl. Phys. **34**, 2151–2159 (2001)

W. Wang, X. Tu, D. Mei, M. Rong, Dielectric breakdown properties of hot SF_6/He mixtures predicted from basic data. Phys. Plasmas **20**(11), 113503 (2013)

A.B. Wedding, H.A. Blevin, J. Fletcher, The transport of electrons through nitrogen gas. J. Phys. D: Appl. Phys. **18**, 2361–2373 (1985)

D.M. Xiao, L.L. Zhu, Y.Z. Chen, Electron swarm parameters in SF_6 and helium gas mixtures. J. Phys. D: Appl. Phys. **32**, L18–L19 (1999)

S. Yoshida, A.V. Phelps, L.C. Pitchford, Effect of electrons produced by ionization on calculated electron-energy distributions. Phys. Rev. A **27**(6), 2858–2867 (1983)

M. Yousfi, P. Ségur, T. Vassiliadis, Solution of the Boltzmann equation with ionisation and attachment: application to SF_6. J. Phys. D: Appl. Phys. **18**, 359–375 (1985)

G.Q. Zhu, J.-P. Boeuf, B. Chaudhury, Ionization-diffusion plasma front propagation in a microwave field. Plasma Sources Sci. Technol. **20**(3), 035007 (2011)

Chapter 6
Presence of Space Charge Fields

This chapter deals with different aspects of electron kinetics when, besides the applied external field, a space-charge electric field also exists. Thus, this chapter includes: *(i)* The diffusion of electrons under the presence of a space-charge field, with one or several positive ions, and including the presence an external magnetic field; *(ii)* The Boltzmann equation for a discharge when a space-charge field created by a negatively charged wall with respect to the plasma is considered; *(iii)* The determination of maintenance field of a microwave discharge; *(iv)* The afterglow of a microwave discharge; *(v)* The analysis of non-local effects created by radial space-charge fields in a discharge.

6.1 Electron Transport with Space Charge Fields

6.1.1 Ambipolar Diffusion

In Sect. 5.1 we have assumed that the electrons were free to diffuse through a uniform neutral background gas and that the ions have no substantial effect on this process. However, this free diffusion can only occur in very low density plasmas where the Coulomb interactions can be neglected. At moderate and high density plasmas, such as it commonly exists in a discharge, this is definitely not the case and we need to consider the diffusion under the effects of space-charge electric fields.

In a plasma the Coulomb interactions will tend to maintain overall space-charge neutrality $|n_e - n_i| \ll n_e$. However, in the presence of gradients in the plasma densities, both the electrons and ions will tend to diffuse into the regions of lower densities. Due to the large differences between the masses of electrons and ions, the diffusion coefficients are very different, and the electrons will tend to diffuse

J.M.A.H. Loureiro, J. de Amorim Filho, *Kinetics and Spectroscopy of Low Temperature Plasmas*, Graduate Texts in Physics,
DOI 10.1007/978-3-319-09253-9_6

more rapidly than the ions, leaving an excess of positive charge behind and creating an electric field due to the space-charge separation. This space-charge field will tend to retard the electron diffusion and accelerate the ion diffusion, so that space-charge neutrality is maintained in most of the regions. Under these conditions, the electrons and ions will diffuse at the same rate as determined by a common diffusion coefficient termed ambipolar diffusion coefficient.

As we have seen in Sect. 1.3.1, the field of a positive ion in a plasma is reduced by a factor $\exp(-r/\lambda_D)$ due to the electron shielding, being λ_D the Debye length given by

$$\lambda_D = \sqrt{\frac{\epsilon_0 \, u_k}{e^2 n_e}}, \tag{6.1}$$

with $u_k = eD_e/\mu_e$ denoting the characteristic energy (5.14), and where $u_k = k_B T_e$ in the case of electrons with a Maxwellian velocity distribution at temperature T_e. If the plasma is contained in a vessel of dimensions greater than λ_D, the ions are able to hold their shielding electrons and the diffusion of electrons and ions will occur coupled each other with an ambipolar diffusion coefficient. On the contrary, if the dimensions are smaller than λ_D, the electrons will diffuse independently of the ions, with no modifications relatively to free diffusion. These limits correspond to high and low electron densities, respectively.

Another distinction takes place with the two limits of high and low gas concentrations depending if the vessel is large or small compared to the mean free path. The first situation has been treated in Allis and Rose (1954) and Allis (1956) and this is more closely related to the diffusion process. The second one has been considered in Bernstein and Holstein (1954) and it corresponds to a collisionless regime.

Let us consider now the case where the gas density is large enough to a collisional regime can exist. According to equation (3.62) in Sect. 3.1.5, the particle current densities of electrons and ions are given by

$$\mathbf{\Gamma_e} = -\, D_e \, \nabla n_e \; - \; n_e \, \mu_e \, \mathbf{E_s} \tag{6.2}$$

$$\mathbf{\Gamma_i} = -\, D_i \, \nabla n_i \; + \; n_i \, \mu_i \, \mathbf{E_s}, \tag{6.3}$$

in which we assume that D_e and D_i are space constants and $\mathbf{E_s}$ denotes the space-charge electric field created in the plasma. This field is determined by Poisson's equation

$$(\nabla . \, \mathbf{E_s}) = \frac{e \, (n_i - n_e)}{\epsilon_0}. \tag{6.4}$$

To these equations we must still add the stationary continuity equation (2.152) for electrons and ions

$$(\nabla . \mathbf{\Gamma_e}) = I_0 \tag{6.5}$$

$$(\nabla . \mathbf{\Gamma_i}) = I_0, \tag{6.6}$$

being $I_0 = n_e <\nu_{ion}>$ the ionization source term (2.161), equal for electrons and ions, resulting from electron-ion pair production by electron ionization collisions.

From equations (6.5) and (6.6), we have

$$(\nabla . \mathbf{\Gamma_e}) = (\nabla . \mathbf{\Gamma_i}) \tag{6.7}$$

and a simple solution may be obtained making two simple, though not rigorously exact, assumptions. The first is the assumption of congruence

$$\mathbf{\Gamma_e} = \mathbf{\Gamma_i} , \tag{6.8}$$

which is suggested, but not required by (6.7), and the second one is the assumption of proportionality

$$\frac{\nabla n_e}{n_e} = \frac{\nabla n_i}{n_i} . \tag{6.9}$$

This latter is rigorously valid in the limit of low densities, in which $\mathbf{E_s}$ is vanishingly small and the electrons and ions satisfy independently the same diffusion equation, although obviously with different diffusion coefficients; and in the opposite limit of high densities, when the difference between the electron and ion densities is small compared with either one of the densities, and the diffusion becomes ambipolar.

Let us consider first the limit of perfect ambipolar diffusion. Here, the electrons and ions diffuse exactly at the same rate, so that $\mathbf{v_{ed}} = \mathbf{v_{id}} = \mathbf{v_d}$, and from equation (6.8) we still have $n_e = n_i = n$. Eliminating $\mathbf{E_s}$ in equations (6.2) and (6.3) we obtain

$$n \, \mathbf{v_d} = - D_a \nabla n , \tag{6.10}$$

being D_a the ambipolar diffusion coefficient defined by

$$D_a = \frac{D_i \, \mu_e + D_e \, \mu_i}{\mu_e + \mu_i} . \tag{6.11}$$

On the other hand, the space-charge field is obtained by subtracting equations (6.2) and (6.3) one from another

$$\mathbf{E_s} = - \frac{D_e - D_i}{\mu_e + \mu_i} \frac{\nabla n}{n} . \tag{6.12}$$

Since $eD_e/\mu_e = u_k \equiv k_B T_e$ and $eD_i/\mu_i = k_B T_i$, with $\mu_i \ll \mu_e$, we still obtain from equation (6.11)

$$D_a = D_i \frac{1 + T_e/T_i}{1 + \mu_i/\mu_e} \simeq D_i \left(1 + \frac{T_e}{T_i}\right) . \tag{6.13}$$

When $T_i \ll T_e$, we further have $D_a \simeq D_i\ (T_e/T_i)$, whereas when $T_i \simeq T_e$ we obtain $D_a \simeq 2\ D_i$ but this latter case does not correspond to low temperature plasmas. Equation (6.11) can still be rewritten as

$$\frac{D_a}{D_e} = \frac{\mu_i}{\mu_e} \frac{1 + T_i/T_e}{1 + \mu_i/\mu_e} \tag{6.14}$$

showing that in the limit $T_i \ll T_e$, we have $D_a/D_e \simeq \mu_i/\mu_e \ll 1$. Finally, from (6.12) we may also write the space-charge electric field as follows

$$\mathbf{E_s} \simeq -\frac{k_B T_e}{e} \frac{\nabla n}{n} , \tag{6.15}$$

showing that $k_B T_e/e$ is a measure of the potential drop associated with the space-charge field. This field sharply increases in the vicinity of a discharge wall due to the rapid decrease of density gradient in consequence of the recombination of charged species on the wall.

6.1.2 Transition from Free to Ambipolar Diffusion

Let us consider now the transition from low to high charged species densities, that is the transition from free to ambipolar electron diffusion. This analysis has been presented in Allis (1956). Here, the assumption of congruence (6.8), $\mathbf{\Gamma_e} = \mathbf{\Gamma_i} = \mathbf{\Gamma}$, is still assumed, but we no longer assume $\mathbf{v_{ed}} = \mathbf{v_{id}}$ and $n_e = n_i$. Considering also the assumption of proportionality (6.9), we obtain by eliminating $\mathbf{E_s}$ in equations (6.2) and (6.3)

$$\mathbf{\Gamma} = -D_{se} \nabla n_e , \tag{6.16}$$

with

$$D_{se} = \frac{D_i\ \mu_e + D_e\ \mu_i}{n_e\ \mu_e + n_i\ \mu_i} n_i \tag{6.17}$$

denoting an effective diffusion coefficient for electrons, assuring the transition from free to ambipolar electron diffusion. Using now the ambipolar diffusion coefficient (6.11), we may still write D_{se} as follows

$$D_{se} = D_a \left(1 + \frac{\mu_e\,(n_i - n_e)}{n_e\,\mu_e + n_i\,\mu_i}\right)$$
$$= D_a \left(1 + \mu_e\,\frac{\rho}{\sigma_c}\right), \tag{6.18}$$

in which we have introduced the space-charge density and the plasma conductivity

$$\rho = e\,(n_i - n_e) \tag{6.19}$$
$$\sigma_c = e\,(n_e\,\mu_e + n_i\,\mu_i)\,. \tag{6.20}$$

Equation (6.18) shows that D_a is the minimum of D_{se}. The space-charge field is then obtained substituting (6.16) in (6.2)

$$\mathbf{E_s} = -\,\frac{D_e - D_{se}}{\mu_e}\,\frac{\nabla n_e}{n_e}\,. \tag{6.21}$$

Using Poisson's equation we obtain the following expression for the space-charge density

$$\rho = \epsilon_0\,(\nabla.\,\mathbf{E_s}) = -\,\epsilon_0\,\nabla.\left(\frac{D_e - D_{se}}{\mu_e}\,\frac{\nabla n_e}{n_e}\right)$$
$$= \epsilon_0\,\frac{D_e - D_{se}}{\mu_e}\left[\left(\frac{|\nabla n_e|}{n_e}\right)^2 - \frac{\nabla^2 n_e}{n_e}\right] \tag{6.22}$$

and defining the diffusion length Λ by

$$\nabla^2 n_e = -\,\frac{n_e}{\Lambda^2}\,, \tag{6.23}$$

we find for the space-charge density at the centre where $\nabla n_e = 0$

$$\rho_0 = \frac{\epsilon_0}{\Lambda^2}\,\frac{D_e - D_{se}}{\mu_e}\,. \tag{6.24}$$

Obviously $\rho_0 \to 0$ when $D_{se} \to D_e$ and ρ_0 has its maximum value when $D_{se} \to D_a$.

The relation between the space-charge density $\rho_0 = e\,(n_{i0} - n_{e0})$ and the Debye length $\lambda_D = \sqrt{\epsilon_0\,D_e/(e\,n_{e0}\,\mu_e)}$ at the centre is

$$\frac{n_{i0} - n_{e0}}{n_{e0}} = \frac{D_e - D_{se}}{D_e}\left(\frac{\lambda_D}{\Lambda}\right)^2. \tag{6.25}$$

When the Debye length equals the diffusion length and simultaneously $D_{se} \ll D_e$, the central ion density is approximately twice the electron density. On the other hand, as n_{e0} increases, the Debye length decreases, and $n_{i0} \sim n_{e0}$. The effective diffusion coefficient (6.18) can be expressed in terms of the central conductivity σ_{c0} substituting equation (6.24)

$$D_{se} = D_a \left(1 + \mu_e \frac{\rho_0}{\sigma_{c0}}\right)$$

$$= D_a \frac{D_e + \Lambda^2\sigma_{c0}/\epsilon_0}{D_a + \Lambda^2\sigma_{c0}/\epsilon_0}, \tag{6.26}$$

which shows clearly the two limits D_e and D_a for the effective diffusion coefficient, as the plasma conductivity increases from 0 to ∞.

Since $\mu_e \gg \mu_i$, we have $\sigma_{c0} \simeq en_{e0}\mu_e$ and $\lambda_D = \sqrt{\epsilon_0 D_e/\sigma_{c0}}$, so that (6.26) can still be expressed under the form

$$D_{se} = D_a \frac{1 + (\Lambda/\lambda_D)^2}{D_a/D_e + (\Lambda/\lambda_D)^2}, \tag{6.27}$$

showing that D_{se} tends to D_e or D_a, when $\Lambda/\lambda_D \to 0$ or $\Lambda/\lambda_D \to \infty$, respectively.

The equations (6.26) and (6.27) are not always valid. If the proportionality assumption held rigorously the space-charge density would have a maximum at the centre. However equation (6.22) shows that it exists a minimum at the centre given by equation (6.24). At low conductivities the error results from the fact of $\mathbf{E_s}$ has been obtained in equation (6.2) from a small difference of two large quantities $\mathbf{\Gamma_e}$ and $-D_e\ \nabla n_e$, so that (6.22) is false. At high conductivities one cannot conclude that ρ is proportional to n_e in taking the small difference $n_i - n_e$. Nevertheless, in neither limit is the assumption of proportionality (6.9) itself very wrong. Only in the transition between the two limits it does not work. Figure 6.1 taken from Allis (1956) shows the comparison between the solutions obtained for D_{se}/D_e, as a function of $\Lambda^2\sigma_{c0}/(\epsilon_0 D_e) \equiv (\Lambda/\lambda_D)^2$, by solving numerically equations (6.2), (6.3), and (6.4) and considering the approximate solution (6.26) and (6.27). Figure 6.1 shows that D_{se}/D_e decreases from ~ 1 to $\sim D_a/D_e$, as the normalized central conductivity $\Lambda^2\sigma_{c0}/(\epsilon_0 D_e)$ increases in the range $0.02 - 100$.

Finally, in a similar manner to the case of a magnetic field treated in equation (5.52), one can define an effective diffusion length to take into account the space-charge effects. Assuming $\Lambda_s^2 \sim D_{se}^{-1}$ and $\Lambda^2 \sim D_e^{-1}$, we obtain the following relation from equation (6.26)

$$\Lambda_s^2 = \Lambda^2 \frac{1 + \Lambda^2\sigma_{c0}/(\epsilon_0 D_a)}{1 + \Lambda^2\sigma_{c0}/(\epsilon_0 D_e)}. \tag{6.28}$$

Here, $\Lambda_s = \Lambda$, when $\sigma_{c0} \to 0$, and $\Lambda_s = \sqrt{D_e/D_a}\ \Lambda \gg \Lambda$, when $\sigma_{c0} \to \infty$. The effect of space-charge on the electron diffusion is to increase the effective size of the volume occupied by the plasma.

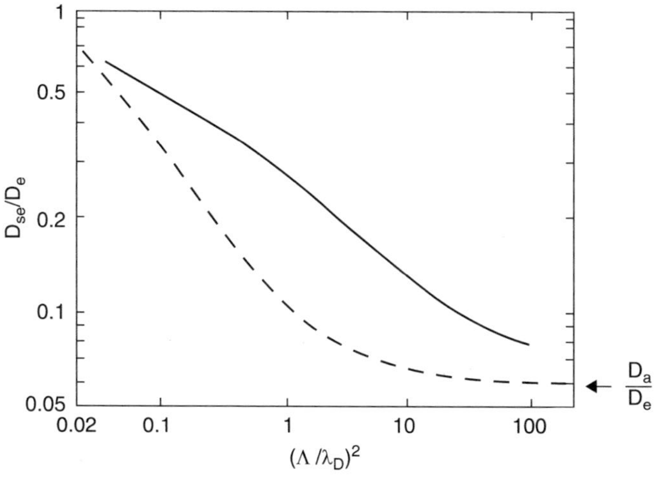

Fig. 6.1 Transition from free to ambipolar electron diffusion as the Debye length decreases. The ratio D_{se}/D_e is plotted against $(\Lambda/\lambda_D)^2$ for the case of the numerical solution of diffusion equations (*full curve*) and for the approximate solution (6.26) and (6.27) (*broken*) (Allis 1956)

6.1.3 Ambipolar Diffusion with Several Positive Ions

In many cases it may occur the formation of several positive and negative ions in a plasma, such as N^+, N_2^+, N_3^+ and N_4^+, or O^+, O_2^+ and O^-, in the case of a nitrogen, or an oxygen discharge, respectively. As only positive ions exist the diffusion is ambipolar in the limit of high charged species concentrations. Thus, let us consider the case of a plasma containing electrons and two types of positive ions (Cherrington 1980). As in (6.2) and (6.3), we have now

$$\mathbf{\Gamma_e} = -D_e \nabla n_e - n_e \mu_e \mathbf{E_s} \tag{6.29}$$

$$\mathbf{\Gamma_1} = -D_1 \nabla n_1 + n_1 \mu_1 \mathbf{E_s} \tag{6.30}$$

$$\mathbf{\Gamma_2} = -D_2 \nabla n_2 + n_2 \mu_2 \mathbf{E_s}\,. \tag{6.31}$$

As before, we assume space-charge neutrality $n_e = n_1 + n_2$ and equal fluxes of electrons and positive ions $\mathbf{\Gamma_e} = \mathbf{\Gamma_1} + \mathbf{\Gamma_2}$, i.e. the assumption of congruence. If we make the further assumption of proportionality for the densities

$$\frac{\nabla n_e}{n_e} = \frac{\nabla n_1}{n_1} = \frac{\nabla n_2}{n_2}\,, \tag{6.32}$$

we obtain the following diffusion coefficients under perfect ambipolar diffusion conditions ($\mathbf{v_e} = \mathbf{v_1} = \mathbf{v_2}$)

$$D_{a,e} = \frac{n_1\,(\mu_e\,D_1 + \mu_1\,D_e) \;+\; n_2\,(\mu_e\,D_2 + \mu_2\,D_e)}{n_e\,\mu_e + n_1\,\mu_1 + n_2\,\mu_2} \tag{6.33}$$

$$D_{a,1} = \frac{n_e\,(\mu_e\,D_1 + \mu_1\,D_e) \;+\; n_2\,(\mu_2\,D_1 - \mu_1\,D_2)}{n_e\,\mu_e + n_1\,\mu_1 + n_2\,\mu_2} \tag{6.34}$$

$$D_{a,2} = \frac{n_e\,(\mu_e\,D_2 + \mu_2\,D_e) \;+\; n_1\,(\mu_1\,D_2 - \mu_2\,D_1)}{n_e\,\mu_e + n_1\,\mu_1 + n_2\,\mu_2}\,, \tag{6.35}$$

satisfying equations of the type (6.16), whereas for the space-charge field we obtain

$$\mathbf{E_s} = -\,\frac{n_e\,D_e - \;(n_1\,D_1 + n_2\,D_2)}{n_e\,\mu_e + n_1\,\mu_1 + n_2\,\mu_2}\;\frac{\nabla n_e}{n_e}\,. \tag{6.36}$$

These expressions can be extended to more than two positive ions as follows:

$$D_{a,e} = \frac{\mu_e\,\sum_i n_i\,D_i + D_e\,\sum_i n_i\,\mu_i}{n_e\,\mu_e + \sum_i n_i\,\mu_i} \tag{6.37}$$

$$D_{a,i} = \frac{n_e\,(\mu_e\,D_i + \mu_i\,D_e) \;+\; \sum_{j\neq i} n_j\,(\mu_j\,D_i - \mu_i\,D_j)}{n_e\,\mu_e + \sum_i n_i\,\mu_i} \tag{6.38}$$

$$\mathbf{E_s} = -\,\frac{n_e\,D_e - \;\sum_i n_i\,D_i}{n_e\,\mu_e + \sum_i n_i\,\mu_i}\;\frac{\nabla n_e}{n_e}\,. \tag{6.39}$$

Returning to the case of two ions only, we note that some simplifications can be made by assuming that the ions are in thermal equilibrium with the neutral gas at temperature T_o. This assumption implies that $\mu_2 D_1 = \mu_1 D_2$ and if we further assume $D_e \gg (D_1, D_2)$ and $\mu_e \gg (\mu_1, \mu_2)$, we obtain from equations (6.33), (6.34), and (6.35)

$$D_{a,e} = \frac{n_1\,D_{a,1} + n_2\,D_{a,2}}{n_e} \tag{6.40}$$

$$D_{a,1} = D_1\left(1 + \frac{T_e}{T_o}\right) \tag{6.41}$$

$$D_{a,2} = D_2\left(1 + \frac{T_e}{T_o}\right)\,. \tag{6.42}$$

These equations show that the ions do not influence directly each other but both influence the electrons and vice-versa. Both $D_{a,1}$ and $D_{a,2}$ are independent of the concentrations but $D_{a,e}$ is not.

The derivation of the diffusion coefficients involves a restrictive approximation, the assumption of proportionality (6.32). Thus, if such assumption is not valid it is not possible to derive diffusion coefficients independent of the charged species concentrations and of their gradients. Under such circumstances the use of a diffusion coefficient to represent the ratio between a given particle flux and a density gradient is of questionable validity. It is preferable to retain the space-charge field explicitly in equations (6.29), (6.30), and (6.31). The consideration of negative ions is still more complicated and the restricted concept of ambipolar diffusion cannot be applied (Ferreira et al. 1988).

6.1.4 Ambipolar Diffusion with a Magnetic Field

The assumption of congruence (6.8) follows from (6.7) when the flow is irrotational as it is true in the absence of a magnetic field. However, when a magnetic field is present such assumption is invalid and the whole derivation, including the expression for the ambipolar diffusion coefficient, cannot be conducted (Allis 1956).

When the magnetic field is large enough for flow anisotropies to be significant, at least for electrons, and if the only gradients are perpendicular to the magnetic field, one may expect to have an identical situation in the plane perpendicular to the magnetic field direction as that obtained in all directions in the absence of magnetic field (Monroe 1973; Zhilinskii and Tsendin 1980; Golant et al. 1980; Vidal et al. 1999), in which the free diffusion coefficients and mobilities should be replaced with coefficients and mobilities under the action of a magnetic field. When a magnetic field exists the equivalent equation to (6.11) is now

$$D_{a\perp} \;=\; \frac{D_{i\perp}\,\mu_{e\perp} + D_{e\perp}\,\mu_{i\perp}}{\mu_{e\perp} + \mu_{i\perp}}\,, \tag{6.43}$$

where the transverse free diffusion coefficients are the perpendicular diagonal elements D_1 in equation (5.67)

$$D_{e\perp} \;=\; \frac{1}{1+(\mu_e B)^2}\,D_e \tag{6.44}$$

$$D_{i\perp} \;=\; \frac{1}{1+(\mu_i B)^2}\,D_i\,, \tag{6.45}$$

and the transverse mobilities are

$$\mu_{e\perp} \;=\; \frac{eD_{e\perp}}{k_B T_e} \;=\; \frac{1}{1+(\mu_e B)^2}\,\mu_e \tag{6.46}$$

$$\mu_{i\perp} \;=\; \frac{eD_{i\perp}}{k_B T_i} \;=\; \frac{1}{1+(\mu_i B)^2}\,\mu_i\,. \tag{6.47}$$

The derivation of a final expression for $D_{a\perp}$ is now straightforward. Replacing equations (6.44) to (6.47) in (6.43), we obtain

$$D_{a\perp} = \frac{D_i\,\mu_e + D_e\,\mu_i}{\mu_e\,(1+(\mu_i B)^2) + \mu_i\,(1+(\mu_e B)^2)} = \frac{D_{a\|}}{1+\mu_e\mu_i B^2}\,, \quad (6.48)$$

with $D_{a\|} = D_a$ denoting the parallel ambipolar diffusion coefficient (or the ambipolar diffusion coefficient in the absence of a magnetic field) given by equation (6.11). In case of high magnetic fields $\mu_e B \gg 1$, and because of $D_a \simeq (\mu_i/\mu_e)D_e$, we obtain from (6.44)

$$D_{e\perp} \simeq \frac{D_e}{(\mu_e B)^2} \simeq \frac{D_a}{\mu_e\mu_i B^2}\,, \quad (6.49)$$

whereas from (6.48) we have

$$D_{a\perp} \simeq \frac{D_a}{\mu_e\mu_i B^2} \simeq D_{e\perp}\,. \quad (6.50)$$

In this case, the magnetic confinement is predominant as compared with the confinement due to the space-charge electric field.

It follows from equation (6.48) that it is possible to define a perpendicular and a parallel ambipolar diffusion coefficient and that, in principle, it would be possible to write the following equation for electron diffusion under the conjoint action of a space-charge electric field and a magnetic field, with the latter oriented along the z axis

$$\frac{\partial n_e}{\partial t} - D_{a\perp}\nabla_\perp^2 n_e - D_a\,\frac{\partial^2 n_e}{\partial z^2} = n_e <\nu_{ion}>\,. \quad (6.51)$$

However, if the ratio between the two diffusion coefficients significantly differs from unity, the equation (6.7), $(\nabla . \mathbf{\Gamma_e}) = (\nabla . \mathbf{\Gamma_i})$, imposed by the equally of the source terms for electrons and ions, does not allow to assume the hypothesis of congruence: $\mathbf{\Gamma_e} = \mathbf{\Gamma_i}$. As a matter of fact, various experimental results indicate that the actual effective perpendicular diffusion coefficient is greater than the coefficient given by equation (6.48) by several orders of magnitude (Lieberman and Lichtenberg 1994; Vidal et al. 1999).

Because of the ion mass is so much larger than the electron mass, the increasing of the magnetic field dramatically decreases the electron transport perpendicular to the magnetic field well before any significant decrease occurs in the ion transport, while along the parallel direction electrons and ions diffuse at the same rate, under ambipolar diffusion conditions. As first pointed out by Simon (1955), some specific simplifications for some special cases can explain how one can have perpendicular transport to the magnetic field much greater than that given by equation (6.48).

As presented by Lieberman and Lichtenberg (1994) to obtain the diffusion coefficient obtained by Simon (1955), let us consider a finite plasma in which the length L along the B field in z direction is much larger than the length l across the field. Even for $L \gg l$ the more rapid diffusion along the B field is usually important and cannot be neglected. Using for simplicity rectangular coordinates, with the y direction taken of infinite extent in order the charged species densities may be assumed constant, we may write the continuity equations for electrons and ions

$$\frac{\partial n_e}{\partial t} - D_{e\perp} \frac{\partial^2 n_e}{\partial x^2} - \mu_{e\perp} \frac{\partial (n_e E_{sx})}{\partial x} - D_e \frac{\partial^2 n_e}{\partial z^2} - \mu_e \frac{\partial (n_e E_{sz})}{\partial z} = n_e <\nu_{ion}> \tag{6.52}$$

$$\frac{\partial n_i}{\partial t} - D_{i\perp} \frac{\partial^2 n_i}{\partial x^2} + \mu_{i\perp} \frac{\partial (n_i E_{sx})}{\partial x} - D_i \frac{\partial^2 n_i}{\partial z^2} + \mu_i \frac{\partial (n_i E_{sz})}{\partial z} = n_e <\nu_{ion}> , \tag{6.53}$$

being E_{sx} and E_{sz} the components of the space-charge electric field. A limiting case can be considered depending of the amplitude of E_{sx}. For $E_{sx}\ l \ll k_B T_e/e$, and $< k_B T_i/e$, the perpendicular mobility terms in (6.52) and (6.53) are smaller compared to the perpendicular diffusion terms. Dropping these mobility terms, multiplying (6.52) by μ_i and (6.53) by μ_e and adding the two equations, we obtain when $n_e = n_i = n$

$$\frac{\partial n}{\partial t} - D_{S\perp} \frac{\partial^2 n}{\partial x^2} - D_a \frac{\partial^2 n}{\partial z^2} = n <\nu_{ion}> , \tag{6.54}$$

where $D_{S\perp}$ is the Simon diffusion coefficient perpendicular to the **B** field given by

$$D_{S\perp} = \frac{D_{i\perp}\ \mu_e + D_{e\perp}\ \mu_i}{\mu_e + \mu_i} , \tag{6.55}$$

different hence from (6.43), whereas D_a is the usual parallel ambipolar diffusion coefficient given by equation (6.11).

Since $\mu_e \gg \mu_i$ and normally $D_{i\perp} > D_{e\perp}$, equation (6.55) simplifies to $D_{S\perp} \simeq D_{i\perp}$. With this approximation the diffusion equation (6.54) becomes

$$\frac{\partial n}{\partial t} - D_{i\perp} \frac{\partial^2 n}{\partial x^2} - D_a \frac{\partial^2 n}{\partial z^2} = n <\nu_{ion}> . \tag{6.56}$$

The perpendicular loss of ions is hence by free diffusion. Physically this corresponds to a situation in which the electrons, flowing along the **B** field lines, almost completely remove the negative charge that produces the space-charge field component E_{sx}. Then, due to the presence of the **B** field, the electrons preferentially flow out along the field and the ions flow out across the field, being $\boldsymbol{\Gamma}_\mathbf{e} \neq \boldsymbol{\Gamma}_\mathbf{i}$ and existing a net current flow in the wall. Here, Lieberman and Lichtenberg (1994) call attention that if the electron flow along the **B** field lines is impeded by inertial or collisional effects, or if it exists a substantial perpendicular ion acceleration such that $E_{sx}\, l > k_B T_i/e$, the perpendicular ion diffusion term in (6.53) is smaller than the mobility term and the preceding derivation of $D_{S\perp}$ is invalid. However, a complete analysis of this subject is beyond the purposes of this textbook.

6.2 Boltzmann Equation with Space Charge Fields

6.2.1 Boltzmann Equation for a Discharge

Let us assume now that the total electric field acting on the electrons is composed by the axial DC electric field of the positive column of a glow discharge, $\mathbf{E} = -E\,\mathbf{e_z}$, and by the radial space-charge field created by diffusion of charged species to the cylindrical discharge tube wall, $\mathbf{E_s} = E_s\,\mathbf{e_r}$. Considering, as in Sect. 5.1.2, the first term approximation for the anisotropies of the electron distribution function and that $\mathbf{f_e^1}$ reaches steady-state conditions, equation (5.20) takes the form

$$\mathbf{f_e^1} = -\frac{eE}{m\nu_m^e}\frac{\partial f_e^0}{\partial v_e}\,\mathbf{e_z} + \frac{eE_s}{m\nu_m^e}\frac{\partial f_e^0}{\partial v_e}\,\mathbf{e_r} - \frac{v_e}{\nu_m^e}\,\nabla f_e^0\,, \tag{6.57}$$

being the gradient of the isotropic component of the electron distribution function $\nabla f_e^0 = \partial f_e^0/\partial r\ \mathbf{e_r}$, with $\partial f_e^0/\partial r < 0$, radially oriented towards the discharge axis. The electron current density of electrons obtained from equation (6.57) is hence

$$\begin{aligned}\mathbf{\Gamma_e} &= \int_0^\infty \frac{v_e}{3}\,\mathbf{f_1^e}\,4\pi v_e^2\,dv_e \\ &= n_e\,\mu_e\,E\,\mathbf{e_z} - \left(n_e\,\mu_e\,E_s + D_e\,\frac{\partial n_e}{\partial r}\right)\mathbf{e_r}\,,\end{aligned} \tag{6.58}$$

having assumed here the local field approximation in which the ratio of the isotropic component of the electron distribution function to the electron number density, f_e^0/n_e, is independent of position and therefore the electron free diffusion coefficient, $D_e = < v_e^2/(3\nu_m^e) >$, is independent of the radial coordinate. In a quasi-neutral discharge ($\lambda_D \ll \Lambda$, being λ_D the electron Debye length and Λ the diffusion length of the discharge vessel), the space-charge field $\mathbf{E_s}$ is relatively large and it nearly cancels the radial diffusion term, being small the net radial electron current to the discharge tube wall due to the very nature of the ambipolar diffusion regime.

Substituting equation (6.57) into equation (5.21), in which $\mathbf{E}$ is replaced with $\mathbf{E} + \mathbf{E_s}$

$$\frac{\partial f_e^0}{\partial t} - \frac{e}{m}\frac{1}{3v_e^2}\frac{\partial}{\partial v_e}\left(v_e^2\,((\mathbf{E}+\mathbf{E_s})\,.\,\mathbf{f_e^1})\right) + \frac{v_e}{3}\left(\nabla.\,\mathbf{f_e^1}\right) = I^0 + J^0\,, \tag{6.59}$$

we obtain

$$\frac{\partial f_e^0}{\partial t} + \frac{1}{v_e^2} \frac{\partial}{\partial v_e} \left(v_e^2 \left(g_E^0 + g_{E_s}^0 + g_r^0\right)\right) + \frac{e v_e}{3 m \nu_m^e} \nabla . \left(\frac{\partial f_e^0}{\partial v_e} \mathbf{E_s} \right)$$
$$- \frac{v_e^2}{3 \nu_m^e} \nabla^2 f_e^0 = I^0 + J^0 , \tag{6.60}$$

in which the continuous electron flow components in velocity space are given by

$$g_E^0 = - \frac{1}{3 \nu_m^e} \left(\frac{eE}{m} \right)^2 \frac{\partial f_e^0}{\partial v_e} > 0 \tag{6.61}$$

$$g_{E_s}^0 = - \frac{1}{3 \nu_m^e} \left(\frac{eE_s}{m} \right)^2 \frac{\partial f_e^0}{\partial v_e} > 0 \tag{6.62}$$

$$g_r^0 = \frac{v_e}{3 \nu_m^e} \frac{eE_s}{m} \frac{\partial f_e^0}{\partial r} < 0 . \tag{6.63}$$

The first component has already been obtained in equation (3.92) and it takes into account the acceleration of electrons due to the axial electric field, whereas the other two represent, respectively, the radial acceleration due to the space-charge field and the slowing down of electrons due to radial movement by diffusion against the space-charge field. These latter two components also nearly cancel each other due to the ambipolar diffusion, so that they may be always dropped from equation (6.60). The third and fourth terms on the left-hand side member of equation (6.60) exhibit a different behaviour, since the space-charge field only exists in the third term, being the fourth one a term describing pure diffusion transport.

The stationary equation obtained from (6.60) to describe a steady-state plasma is hence

$$\frac{1}{v_e^2} \frac{\partial}{\partial v_e} \left(v_e^2 \, g_E^0\right) + \frac{v_e}{3 \nu_m^e} \nabla . \left(\frac{\partial f_e^0}{\partial v_e} \frac{e\mathbf{E_s}}{m} - v_e \nabla f_e^0 \right) = I^0 + J^0 . \tag{6.64}$$

In a quasi-neutral discharge with $\lambda_D \ll \Lambda$, the space-charge field is strong enough in order the ambipolar diffusion exists and, consequently, the combined effect of the terms on the space of positions in equation (6.64) can also be neglected. This is the reason why both terms are often neglected in discharge modelling. This Boltzmann equation is usually termed "homogeneous Boltzmann equation'', as it only accounts for the motion of electrons in velocity (or energy) space resulting from the applied field and the collisions (elastic and inelastic) with the neutral gas. Since the collisional terms $I^0 + J^0$ depend on the pressure linearly, and the diffusion term together with the space-charge field term depend on it inversely, the neglecting of this latter term is so much justifiable as the pressure increases.

When the combined effect of diffusion with the space-charge field is neglected in equation (6.64), one must also neglect, for consistency, the new electrons produced by ionization, since the term for diffusion under the effects of the space-charge field exactly compensates, under steady-state conditions, for the rate of appearance of new electrons. In this case, ionization must be treated like any other excitation process. As discussed in Sect. 3.4.1 the electron energy distribution function (EEDF) normalized through equation (3.166) can be then obtained as a function of the reduced electric field E/n_o, with n_o denoting the gas number density, the gas temperature T_o to take into account the small heating of electrons in recoil collisions with non-frozen molecules (usually a very weak dependence), and of the fractional population concentrations $\delta_i = n_i/n_o$ and $\delta_j = n_j/n_o$, with n_i and n_j denoting the concentrations of individual states. In the case of inelastic and superelastic collisions for excitation and de-excitation of vibrational levels, it is useful to introduce a new variable characterising the degree of vibrational excitation with the form of a vibrational temperature T_υ (see Loureiro and Ferreira 1986). The rotational collisions may be taken into account through a continuous approximation, which does not introduce any new variable but only the characteristic parameters B_0 and σ_0 for a given molecular gas (see Sect. 3.3.3). Finally, due to the relative weak population concentrations of the excited electronic states, only the inelastic collisions from the electronic ground-state are usually necessary to be taken into account, whereas the superelastic collisions are neglected for the electronic states.

When the term for diffusion under the effects of the space-charge field is neglected, the continuity equation is not embodied in equation (6.64), since we obtain a null identity as both members of the Boltzmann equation are multiplied by $4\pi v_e^2$ and integrated over the whole velocity space. Thus, the determination of the sustaining electric field for the discharge needs to be obtained externally to equation (6.64). In this case we should introduce the space-charge field into equation (6.2) to obtain the electron flow current density (6.16)

$$\mathbf{\Gamma_e} = -D_{se}\, \nabla n_e \qquad (6.65)$$

and then to introduce this expression into the continuity equation (5.77) to get, under steady-state conditions, the electron rate balance equation

$$-D_{se}\, \nabla^2 n_e = n_e <\nu_{ion}> . \qquad (6.66)$$

Here, D_{se} is, as we have seen in Sect. 6.1.2, an effective electron diffusion coefficient, assuring the transition from free to ambipolar diffusion, and which tends to the ambipolar diffusion coefficient (6.11) when $\lambda_D \to 0$.

Assuming that n_e vanishes at the wall due to electron recombination, we may replace $\nabla^2 n_e = -n_e/\Lambda^2$, being Λ the diffusion length, as we have seen in Sect. 5.1.3, with $\Lambda = R/2.405$ in the case of a cylindrical discharge tube of radius R. We obtain hence the following relation from equation (6.66)

$$\frac{D_{se}}{\Lambda^2} = <\nu_{ion}> , \qquad (6.67)$$

which expresses the fact that, under steady-state conditions, the rate of diffusion losses, represented here in the form D_{se}/Λ^2, must be exactly balanced by the averaged ionization rate. Noting that $D_{se} \propto 1/n_o$ through $D_e = < v_e^2/(3\nu_m^e) > \propto 1/n_o$, $n_o D_{se}$ is independent of the gas density, so that (6.67) can be still rewritten under the form

$$\frac{n_o D_{se}}{(n_o \Lambda)^2} = C_{ion} , \qquad (6.68)$$

where $C_{ion} = < v_e \ \sigma_{ion} > = < \nu_{ion} > /n_o$ is the electron ionization rate coefficient. Since $n_o D_{se}$ and C_{ion} are obtained by integrating the EEDF, both become functions of the independent variables E/n_o and T_v, and in a smaller extent also of T_o, so that equation (6.68) embodies a discharge characteristic for the maintenance field, e.g. under the form of curves of E/n_o against $n_o \Lambda$ for constant values of T_v. In the case of electrons given by a Maxwellian energy distribution, $n_o D_{se}$ and C_{ion} are functions of T_e only, so that equation (6.68) permits to obtain a relationship of T_e against $n_o \Lambda$. In this latter case T_e and E/n_o can be linked each other through the electron energy balance equation (see Exercises 3.1 and 3.2 at the end of Chap. 3).

Let us consider now the case where the electron density is not high enough in order a space-charge electric field may be created capable to compensate nearly the diffusion term in (6.64). If further, the gas pressure is not sufficiently large for the collisions dominate, the term due to the space-charge field can be neglected in (6.64) but the diffusion term cannot. We obtain in this case an equivalent equation as that for breakdown (5.78), in which the appearance of new electrons needs to be considered in the inelastic collision term J^0

$$\frac{1}{v_e^2} \frac{\partial}{\partial v_e} \left(v_e^2 \ g_E^0 \right) - \frac{v_e^2}{3 \nu_m^e} \nabla^2 f_e^0 = I^0 + J^0 . \qquad (6.69)$$

From equation (6.69) one easily observes that the condition for the diffusion may be neglected is the pressure and the diffusion length to be large enough to have

$$\frac{v_e^2}{3 \nu_m^e \Lambda^2} f_e^0 \ll \nu_e f_e^0 , \qquad (6.70)$$

with ν_e denoting here the characteristic frequency for energy relaxation (4.1). If the collisions are predominantly inelastic, $\nu_e \simeq \nu_m^e$, the above inequality results in $\Lambda \gg v_e/\nu_m^e$, which shows that the condition is the electron diffusion length to be much larger than the mean free path.

The continuity equation obtained by integrating both members of (6.69) in velocity space, considering further $\nabla^2 n_e = - n_e/\Lambda^2$, may be written now as

$$\frac{n_o D_e}{(n_o \Lambda)^2} = C_{ion} , \qquad (6.71)$$

where D_e is the free electron diffusion coefficient, which expresses the relationship for determining the discharge characteristics $E/n_o(n_o\Lambda, T_v)$. Since $D_e > D_{se}$, and in particular $D_e \gg D_a$, the electron ionization rate coefficient capable to compensate the electron loss rate by diffusion is now significantly larger than in the case when the space-charge field exists and, consequently, the reduced sustaining electric field E/n_o needs to be significantly larger as well.

Furthermore, when the space-charge electric field does not significantly exist the diffusion term cannot be neglected in equation (6.69), so that the Boltzmann equation needs to be solved by considering a guess for the input parameters E/n_o and $n_o\Lambda$ (or E/n_o, $n_o\Lambda$ and T_v in case of a molecular gas). The validity of our choice is checked after through equation (6.71) (or through equation (6.71) together with a system of rate balance equations for the populations of the vibrational levels). In the case of an atomic gas, where the inelastic collisions may be considered from the ground-state only and the superelastic collisions may be neglected, and assuming the gas temperature sufficiently low to be neglected in the recoil collision term (3.85), the Boltzmann equation can be written in a form similar to equation (3.165), with the inclusion of a new term taking into account the diffusion

$$-\left(\frac{eE}{n_o}\right)^2 \frac{\partial}{\partial u}\left(\frac{u}{3\sigma_m^e}\,\frac{\partial f_e^0}{\partial u}\right) + \frac{u}{3\sigma_m^e}\,\frac{f_e^0}{(n_o\Lambda)^2} = \frac{2m}{M}\,\frac{\partial}{\partial u}\left(u^2\sigma_m f_e^0\right)$$
$$+\sum_{j>0}\left\{(u+u_j)\,\sigma_j(u+u_j)\,f_e^0(u+u_j) - u\,\sigma_j(u)\,f_e^0(u)\right\}. \qquad (6.72)$$

The use of partial derivates in equations (6.69) and (6.72) is due to the fact that the electron distribution function also includes the dependence on space coordinates: $f_e^0(\mathbf{r}, v_e)$ or $f_e^0(\mathbf{r}, u)$. However, the writing of (6.72) implies the assumption of the local field approximation (LFA), in which f_e^0/n_e becomes independent of space (see Sect. 5.1.2).

Finally, it is worth noting here that the Paschen curves for breakdown are embodied in equation (6.71). In fact, assuming the breakdown voltage as $V_b = E\,d$, being d the run away distance for electrons, we may write $V_b = (E/n_o)\,(n_od)$ and since E/n_o is a function of n_od only, we obtain a characteristic of the type $V_b(n_od)$.

6.2.2 Boltzmann Equation for a Microwave Discharge

Let us consider now the situation of a discharge produced and sustained by an high-frequency electric field of amplitude E_0 and frequency ω, much larger than the characteristic frequency for energy relaxation ν_e given by equation (4.1), in order the isotropic component of the electron velocity distribution function may be assumed time-independent (Rose and Brown 1955; Ferreira and Loureiro 2000). When a time constant space-charge electric field $\mathbf{E}_s$ is also present, in result of a negatively charged discharge tube wall, the electron Boltzmann equation takes the following form using equations (4.22) and (6.64)

$$-\frac{1}{v_e^2}\frac{\partial}{\partial v_e}\left(\frac{v_e^2}{3\nu_m^e}\,\frac{1}{2}\left(\frac{eE_0}{m}\right)^2\frac{1}{1+(\omega/\nu_m^e)^2}\,\frac{\partial f_e^0}{\partial v_e}\right)$$
$$+\frac{v_e}{3\nu_m^e}\,\nabla.\left(\frac{\partial f_e^0}{\partial v_e}\,\frac{e\mathbf{E_s}}{m}\;-\;v_e\,\nabla f_e^0\right)=I^0+J^0\,. \qquad (6.73)$$

As analysed before for a DC discharge, when the electron density is sufficiently high the space-charge field is large and the term with $\mathbf{E_s}$ nearly cancels the diffusion term, being their difference small at sufficiently high pressures as compared with the other terms in equation (6.73). In this case equation (6.73) becomes independent of the coordinate $\mathbf{r}$, so that the isotropic component of the electron velocity distribution function $f_e^0(\mathbf{r}, v_e)$ exhibits the same dependence on the space coordinate as the electron density $n_e(\mathbf{r})$, that is the LFA is verified. The equation so obtained is the electron homogeneous Boltzmann equation.

Writing equation (6.73) in terms of the electron energy $u = \frac{1}{2}\,mv_e^{\,2}$, we obtain as in equation (4.24), after multiplication of both members by $\sqrt{u}$, and in case of an atomic gas with inelastic collisions from the ground-state only and neglecting superelastic collisions

$$-\frac{d}{du}\left[\frac{2}{3}\,u^{3/2}\,\nu_m^e\,u_c\,\frac{df_e^0}{du}\;+\;\frac{2m}{M}\,u^{3/2}\,\nu_m\left(f_e^0+k_BT_o\,\frac{df_e^0}{du}\right)\right]$$
$$=\sum_{j>0}\left\{\sqrt{u+u_j}\;\nu_j(u+u_j)\,f_e^0(u+u_j)\;-\;\sqrt{u}\;\nu_j(u)\,f_e^0(u)\right\}\,, \qquad (6.74)$$

where u_c represents the time-averaged energy gain per collision defined in (4.25)

$$u_c(u)\;=\;\frac{e^2E_0^2}{2m\,(\nu_m^e(u)^2+\omega^2)}\,. \qquad (6.75)$$

As we have seen in equation (4.20), when the effective collision frequency for momentum transfer is independent of electron energy, we may use the concept of effective electric field given by

$$E_e\;=\;\frac{1}{\sqrt{1+(\omega/\nu_m^e)^2}}\,\frac{E_0}{\sqrt{2}}\,, \qquad (6.76)$$

and the time-averaged energy gain per collision is given by

$$u_c\;=\;\frac{\overline{P_E(t)}}{n_e\,\nu_m^e}\;=\;\frac{e^2E_e^{\,2}}{m\,\nu_m^{e\,2}}\,, \qquad (6.77)$$

being $\overline{P_E(t)}$ the time-averaged power absorbed per volume unit. However, when ν_m^e is a function of the electron energy, the concept of effective electric field, which embodies all the dependences on the applied field (amplitude and frequency), is no longer applicable. In such a case, an explicit dependence on ω arises from equation (6.74), through the energy $u_c(u)$, which can be rewritten now as follows

$$u_c(u) = \frac{e^2}{2m}\left(\frac{E_0}{n_o}\right)^2 \frac{1}{(\nu_m^e(u)/n_o)^2 + (\omega/n_o)^2}, \tag{6.78}$$

being n_o the gas number density. We can then conclude that $u_c(u)$ is a function of two independent parameters E_0/n_o and ω/n_o. Substituting the expressions for the collision frequencies ν_m^e, ν_m, and ν_j in equation (6.74), and dividing both members by $n_o\sqrt{2/m}$, we obtain

$$-\frac{d}{du}\left[\frac{2}{3}\,u^2\sigma_m^e\,u_c\,\frac{df_e^0}{du} + \frac{2m}{M}\,u^2\sigma_m\left(f_e^0 + k_B T_o\,\frac{df_e^0}{du}\right)\right]$$
$$= \sum_{j>0}\left\{(u+u_j)\,\sigma_j(u+u_j)\,f_e^0(u+u_j) - u\,\sigma_j(u)\,f_e^0(u)\right\}. \tag{6.79}$$

The isotropic component of the electron velocity distribution function is hence function of the above two independent parameters and also of T_o, when we include the small electron heating of the order of T_o/T_e due to elastic collisions with non-frozen atoms.

When we consider the homogeneous electron Boltzmann equation one must also neglect the secondary electrons produced by ionization, since the term for diffusion under the effects of the space-charge field exactly compensates the appearance of new electrons. The ionization must be then treated like an ordinary excitation process by considering only the primary scattered electrons. The neglect of the diffusion and of the source of new electrons altogether is permissible as long as the energy losses associated with the ionization are much smaller than the losses resulting from excitation and elastic recoil collisions, which is satisfied in most of the discharges at not too low pressures. However, for the purposes of discharge modelling one must take into account the electron balance equation (6.68) independently of (6.79), since the condition for plasma maintaining is no longer contained in the Boltzmann equation.

When using equation (6.79), the modelling of a microwave discharge can proceed through two successive steps. First, the homogeneous Boltzmann equation (6.79) is solved as a function of the independent variables E_0/n_o and ω/n_o, and the electron transport and the collisional rate coefficients are calculated from f_e^0 (in particular, the electron ionization rate coefficient, C_{ion}, and the reduced electron free and ambipolar diffusion coefficients, n_oD_e and n_oD_a). Second these data are inserted into the equation for the discharge maintenance, in which (6.68) corresponds to its simplest form, allowing to obtain a relationship between the independent variables E_0/n_o, ω/n_o and $n_o\Lambda$, which constitutes the discharge characteristic for the maintenance field. Notice, however, that equation (6.68) involves the resolution of the continuity equation coupled to the momentum transfer equations for electrons and ions, taking into account appropriate boundary conditions for the plasma, so that this second step involves the solution of a boundary-value problem from which the relationship $E_0/n_o(\omega/n_o, n_o\Lambda)$ is obtained as an eigenvalue solution.

As discussed above, the concept of effective electric field (6.76) is not applicable as the collision frequency ν_m^e depends on the electron energy. However, in practice, it is useful to replace the pair of the reduced variables E_0/n_o and ω/n_o with another pair E_e/n_o and ω/ν_{ce}, where E_e represents an effective electric field defined as in equation (6.76), but using instead of $\nu_m^e(u)$ a constant effective collision frequency ν_{ce} for momentum transfer of electrons with a given energy. The choice of ν_{ce} is rather arbitrary, but it seems preferable to choose a frequency $\nu_{ce} = \nu_m^e(u^*)$ for an energy u^* of the order, or not far from, the mean electron energy $<u>$. In this way the limiting situations of $\omega \ll \nu_{ce}$ and $\omega \gg \nu_{ce}$ have a clear physical meaning, as they correspond to many and to few collisions per oscillation, respectively, for the bulk of electrons. In the limit $\omega \ll \nu_{ce}$, we have $E_e/n_o \rightarrow E_0/(\sqrt{2}\ n_o)$ and the electron distribution f_e^0 is exactly the same as that produced in a DC field with the same amplitude as the *r.m.s.* HF field, $E_0/\sqrt{2}$. This occurs as long as the frequency ω is sufficiently high in order the electron distribution is not time-modulated, i.e. as long as $\omega > \nu_e$, being ν_e the characteristic frequency for energy relaxation. In the opposite limit of $\omega \gg \nu_{ce}$, we have $E_e/n_o \rightarrow (\nu_{ce}/\omega)E_0/(\sqrt{2}\ n_o)$ and the electron distribution f_e^0 depends on the reduced variable E_0/ω alone.

Although the electron rate coefficients and the transport parameters are function of E_e/n_o and ω/ν_{ce}, it is possible to choose individual effective collision frequencies ν_{ce} for each gas, such as the ionization rate coefficient can be expressed (in the range 10–100 Td) as a function of E_e/n_o alone for all values of ω/ν_{ce}. In argon, we find $\nu_{ce}/n_o = 2.0 \times 10^{-13}\,\text{m}^3\,\text{s}^{-1}$ (Ferreira and Loureiro 1984), which corresponds to electrons with an energy $u^* = 8.5\,\text{eV}$, and for nitrogen this value is not very distant $\nu_{ce}/n_o = 2.4 \times 10^{-13}\,\text{m}^3\,\text{s}^{-1}$ (Ferreira and Loureiro 1989), which in this case corresponds to electrons of 2.1 eV or 2.9 eV, or even 10.4 eV, due to the particular shape of $\nu_m^e(u)$ in N_2 (a sharp maximum at ~2 eV and a monotonic growth at higher energies; see Fig. 4.2).

Figure 6.2 shows the electron ionization rate coefficient in argon as a function of the parameter E_e/n_o for various values of ω/ν_{ce} from $\ll 1$ to $\gg 1$. With the value assumed here for ν_{ce}, the ionization rate coefficient at low electron densities is practically a function of the sole parameter E_e/n_o. However, at higher n_e/n_o values this behaviour is no more true because in this case the total ionization rate must also include stepwise ionization from the metastable state $Ar(^3P_2)$ (Ferreira et al. 1985).

The behaviour exhibited in Fig. 6.2 just results from the shape of the EEDF and it cannot be explained by any fundamental reason. Choosing E_e/n_o in this form, all of the electron energy averaged quantities must depend also on an additional parameter proportional to the ratio ω/n_o, such as ω/ν_{ce}, except in the limiting cases $\omega/\nu_{ce} \ll 1$ and $\gg 1$. Only the dependence of the ionization rate coefficient on ω/ν_{ce} turns out to be quite small using the present definition, at least in the range

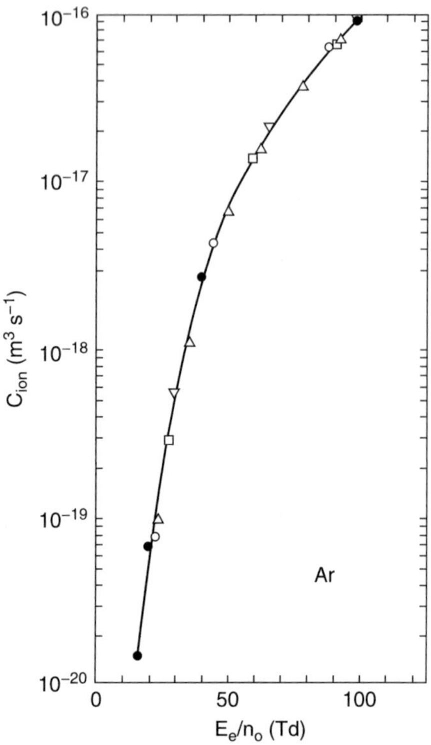

Fig. 6.2 Ionization rate coefficient in Ar, as a function of the effective electric field to the gas number density, for the following values of ω/ν_{ce}: (○) $\ll 1$, (▽) 0.053; (□) 0.425; (△) 0.8; (●) $\gg 1$. The *full curve* represents the best fit to these data (Ferreira and Loureiro 1984)

of E_e/n_o mentioned here. Thus E_e is not a true effective electric field as defined for gases in which $\nu_m^e(u)$ is constant but constitutes a useful parameter for the analysis of the discharge characteristics.

In the case of a molecular gas, such as nitrogen, the situation is more complicated since we need to include as well the effect of electron superelastic collisions for de-excitation of vibrational levels $N_2(X\,^1\Sigma_g^+, v>0)$, which is taken into account through an additional independent parameter for the population distribution in the manifold of vibrational levels. This parameter is the vibrational temperature of the so-called vibrational distribution function (VDF), see e.g. Loureiro and Ferreira (1986), so that the ionization rate coefficient function of the independent parameters E_0/n_o, ω/n_o and T_υ, may be expressed in terms of E_e/n_o and T_υ only if we use the concept of effective electric field. Figure 6.3 shows the ionization rate coefficient in N_2 as a function of the ratio E_e/n_o, assuming $\nu_{ce}/n_o = 2.4 \times 10^{-13}\,m^3\,s^{-1}$ (Ferreira and Loureiro 1989), for $T_\upsilon = 4000\,K$ and $T_\upsilon = T_o = 400\,K$.

The practical interest of these features arises primarily in the calculations of theoretical characteristics for the steady-state maintenance field of HF discharges from the balance between the electron production and loss rates (6.68). Since this latter equation establishes a relationship between the variables E_0/n_o, ω/n_o and $n_o\Lambda$, and because C_{ion} is function of E_e/n_o only, the characteristic for the

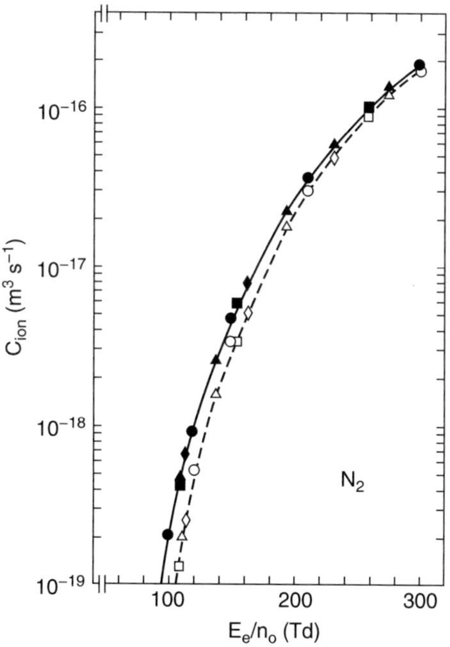

Fig. 6.3 Ionization rate coefficient in N_2, as a function of the effective electric field to the gas number density, for $T_v = 4000$ K (*full curves*) and $T_v = 400$ K (*broken curves*), and the following values of ω/ν_{ce}: (●,○) $\ll 1$, (▲,△) 0.42; (◆,◇) 0.83; (■,□) 1.67 (Ferreira and Loureiro 1989)

maintenance field of an HF discharge may be expressed as $E_e/n_o(n_o\Lambda, \omega/\nu_{ce})$, or as $E_e/n_o(n_o\Lambda, \omega/\nu_{ce}, T_v)$ in case of a molecular gas. Figure 6.4 shows the characteristics E_e/n_o against n_oR, for HF argon discharges in cylindrical geometry, for various values of ω/ν_{ce} from $\ll 1$ to $\gg 1$, assuming predominant direct ionization from the electronic ground-state Ar(^{1}S$_0$). The data points on this figure were derived from experiments with surface wave produced discharges (Ferreira et al. 1987). Due to the neglecting of step-wise ionization from the metastable states, the calculated data overestimate the reduced maintenance field.

It is worth noting at this point that equation (6.68) as it stands is independent of the nature of the applied field (HF or DC) and of the particular set-up used to apply it. It constitutes a quite general steady-state condition applicable to all discharges provided only that direct ionization and loss by diffusion under space-charge field effects are the dominant processes and the field can be assumed approximately uniform throughout the plasma. However, the nature of the field and its frequency ω affect the electron and ion transport (the free diffusion coefficient and the mobility of electrons still continue to be function of both E_e/n_o and ω/ν_{ce}, contrary to C_{ion} which is function of E_e/n_o only as a result of our choice for ν_{ce}), so that the effective diffusion coefficient D_{se} depends on ω/ν_{ce}.

The analysis of the discharge ionization-loss balance shows that E_e/n_o is a function of the external parameters $n_o\Lambda$ and ω/ν_{ce}, provided the pressure is sufficiently low to direct ionization and loss by diffusion are the principal mechanisms. Thus the mean power absorbed per electron at unit gas density, $\overline{P_E(t)}/(n_en_o)$, is also function

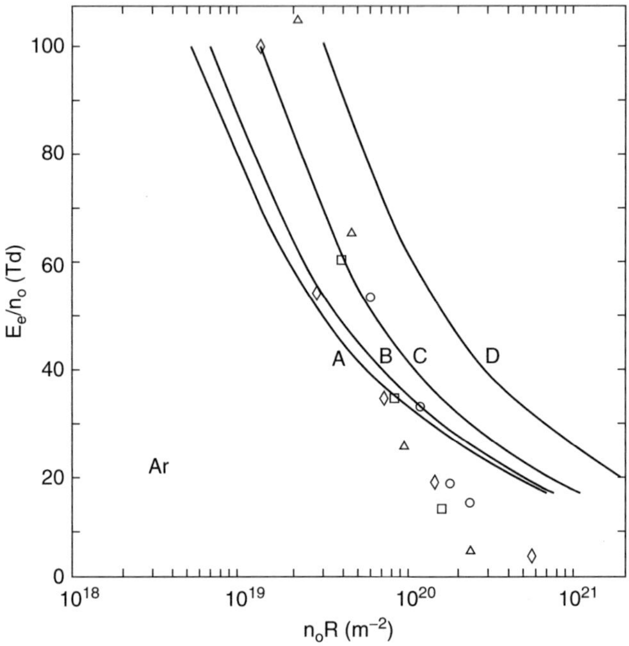

Fig. 6.4 Characteristics of the maintenance field for cylindrical argon discharges, assuming predominant direct ionization, for the following values of ω/ν_{ce}: (*A*) $\gg$ 1, (*B*) 0.8; (*C*) 0.15; (*D*) $\ll$ 1. Data points from experiments on surface wave discharges (Ferreira et al. 1987)

of the same parameters. We remember here that according to equations (4.15)

$$\overline{P_E(t)} \;=\; \frac{1}{2}\,\mathrm{Re}\{\overline{\sigma}_{ce}\}\,E_0^2\;, \tag{6.80}$$

being $\mathrm{Re}\{\overline{\sigma}_{ce}\}$ the real part of the electron conductivity (4.16) given by

$$\mathrm{Re}\{\overline{\sigma}_{ce}\} \;=\; -\,\frac{2}{3}\,\frac{e^2 n_e}{m}\int_0^\infty u^{3/2}\,\frac{\nu_m^e}{{\nu_m^e}^2+\omega^2}\,\frac{df}{du}\,du\;, \tag{6.81}$$

in which we have used the EEDF normalized such as

$$\int_0^\infty f(u)\,\sqrt{u}\,du \;=\; 1\;. \tag{6.82}$$

The quantity $\overline{P_E(t)}/(n_e n_o)$ depends hence on the variables E_0/n_o and ω/n_o only

$$\frac{\overline{P_E(t)}}{n_e n_o} \;=\; -\,\frac{1}{3}\,\frac{e^2}{m}\left(\frac{E_0}{n_o}\right)^2\int_0^\infty u^{3/2}\,\frac{(\nu_m^e/n_o)}{(\nu_m^e/n_o)^2+(\omega/n_o)^2}\,\frac{df}{du}\,du \tag{6.83}$$

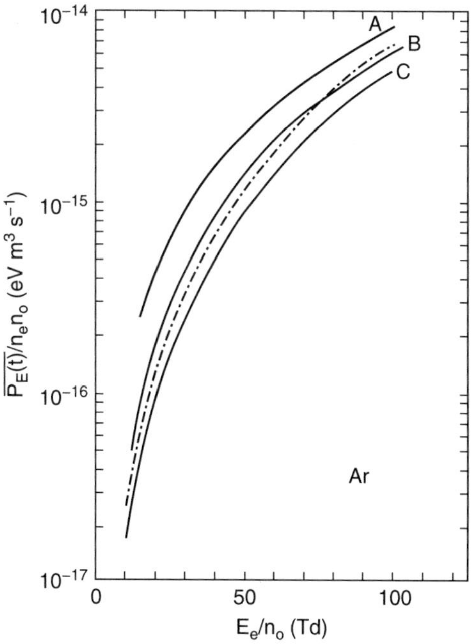

Fig. 6.5 Mean absorbed power per electron at unit gas density in Ar, as function of E_e/n_o, at low electron densities, for the following values of ω/ν_{ce}: (*A*) $\ll 1$; (*B*) 0.8; (*C*) $\gg 1$. The *chain curve* is for $\omega/\nu_{ce} \gg 1$ with a Maxwellian distribution (Ferreira and Loureiro 1984)

and as before this set can be transformed to $(E_e/n_o, \omega/\nu_{ce})$, or to $(n_o\Lambda, \omega/\nu_{ce})$ because of $E_e/n_o(n_o\Lambda, \omega/\nu_{ce})$.

Figure 6.5 shows the mean absorbed power per electron at unit gas density calculated for argon from the solutions to the Boltzmann equation, as a function of the ratio of the effective electric field to gas number density, at low electron densities, from $\omega/\nu_{ce} \ll 1$ to $\omega/\nu_{ce} \gg 1$.

Figure 6.6 shows the characteristics $\overline{P_E(t)}/(n_e n_o)$ against $n_o R$ calculated for argon from the solutions to the homogeneous electron Boltzmann equation and the ionization-loss balance assuming predominant direct ionization. The data points shown on the figure were obtained from experiments with surface wave produced discharges at frequencies ranging from 210 MHz up to 2.45 GHz, using different discharge tube diameters from a few millimeters to a few centimeters (Dervisevic et al. 1983). For a given $n_o R$ product, the values of $\overline{P_E(t)}/(n_e n_o)$ are much lower at $\omega \gg \nu_{ce}$, or at $\omega = 0.8\nu_{ce}$, which corresponds to the conditions of the experiments, than at $\omega \ll \nu_{ce}$. There is reasonable agreement between theory and experiment, the discrepancies being most likely caused by the neglect of stepwise ionization and by the non-uniformity of the field in the surface wave discharges. The best agreement is obtained assuming a Maxwellian distribution (curve M), with temperatures obtained from the electron energy balance equation, which can be justified having in view the high electron densities achieved in these experiments.

Finally, Fig. 6.7 shows the time-averaged absorbed power per electron at unit gas density against E_e/n_o in nitrogen, for different values of ω/ν_{ce} and T_v.

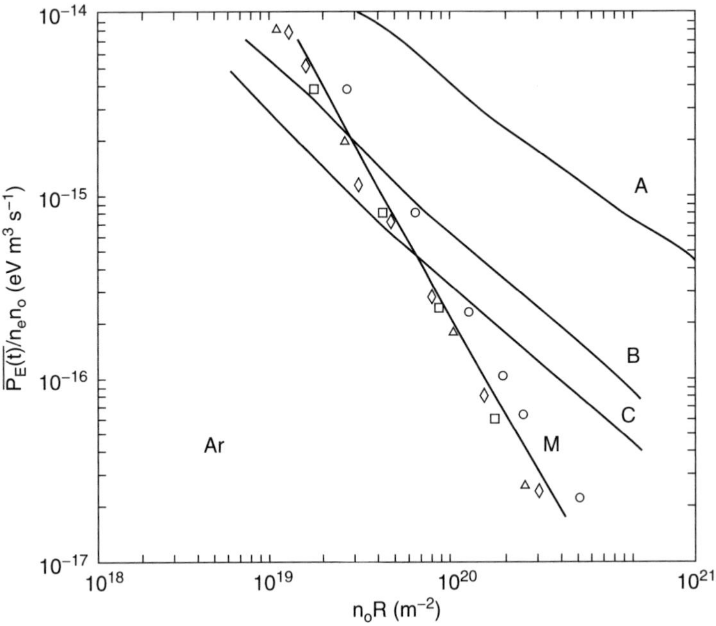

Fig. 6.6 Characteristics of $\overline{P_E(t)}/(n_e n_o)$ against $n_o R$ for cylindrical argon discharges, assuming predominant direct ionization, for the following values of ω/ν_{ce}: *(A)* $\ll 1$; *(B)* 0.8; *(C)* $\gg 1$. *Curve M* is for a Maxwellian distribution. Data points are from experiments on surface wave discharges (Ferreira et al. 1987)

Fig. 6.7 Mean absorbed power per electron at unit gas density in N_2, as function of E_e/n_o, for $T_v = 4000$ K (*full curves*) and $T_v = T_o = 400$ K (*broken curves*), and for the following values of ω/ν_{ce}: (A) $\ll 1$; (B) 0.83; (C) $\gg 1$ (Ferreira and Loureiro 1989)

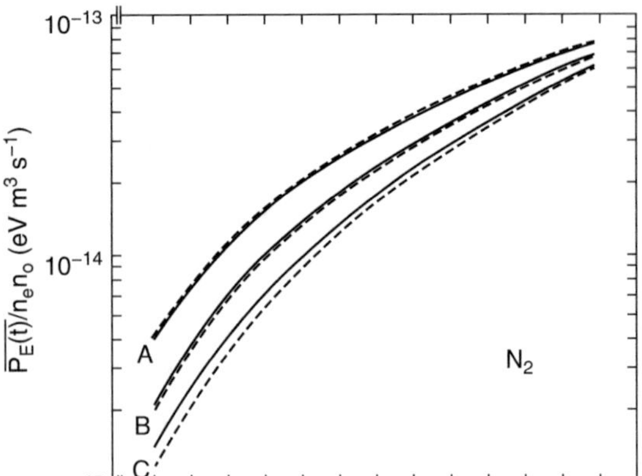

One notes that $\overline{P_E(t)}/(n_e n_o)$ decreases as ω/ν_{ce} increases, as it occurs in argon, whereas only minor modifications result from varying T_v. This can be understood since the changes produced by T_v have a strong effect on the tail of the EEDF but induce only minor changes in the bulk distribution.

6.2.3 *Afterglow of a Microwave Discharge*

When the electric field is cut-off in a discharge a complex kinetic relaxation process immediately initiates for the electrons, ions and neutral excited species, which ends when the final equilibrium is achieved. In the case of a flowing discharge in a tube, the post-discharge can be observed, even when the field is still switched on, after the end of the discharge in a well separated region, which depends of the gas flow rate. The emission of particular lines or bands may occur in the post-discharge as a result of complex interplay mechanisms between the different charged and neutral species along the relaxation process, which obviously depend of the nature of the gas and the post-discharge time or position. The occurrence of the emissions leads to the appearance of an afterglow, in which the initial plasma generated species de-excite and participate in secondary chemical reactions. In consequence some species have shorter and others longer lifetimes. Depending on the gas composition, superelastic collisions may continue to sustain the plasma in the afterglow for a while by releasing the energy stored in rovibronic degrees of freedom of the molecules. Especially in molecular gases, the plasma chemistry in the afterglow significantly differs from the plasma glow.

For the electron distribution function two different mechanisms take place in the decaying plasma, besides the very rapid depletion of the high-energy tail just after the switch off of the field due to electron inelastic collisions. The first is the reduction of the space-charge electric field created in the discharge producing thus a transition from ambipolar to free electron diffusion along the post-discharge. The second is the prevalence of dissociative electron-ion recombination until the late or remote afterglow where the final extinction of the active medium takes place.

The electron distribution function in the earlier instants of the afterglow of a stationary HF discharge can be determined by assuming a zero electric field at $t = 0$ in the time-dependent electron Boltzmann equation (Guerra et al. 2001)

$$\frac{\partial f_e^0}{\partial t} - \frac{1}{v_e^2}\frac{\partial}{\partial v_e}\left(\frac{v_e^2}{3\nu_m^e}\frac{1}{2}\left(\frac{eE_0}{m}\right)^2\frac{1}{1+(\omega/\nu_m^e)^2}\frac{\partial f_e^0}{\partial v_e}\right)$$

$$+\frac{v_e}{3\nu_m^e}\,\nabla.\left(\frac{\partial f_e^0}{\partial v_e}\frac{e\mathbf{E_s}}{m} - v_e\,\nabla f_e^0\right) = I^0 + J^0\,. \tag{6.84}$$

This equation is identical to (6.73) written for the case of a time-varying electric field of frequency ω sufficiently high to the electron distribution function remains time-independent in the discharge, but here we also introduce the term for the time variation of the electron distribution in the post-discharge as the electric field is cut-off. By multiplying both members of the Boltzmann equation by $4\pi v_e^2$ and integrating over the whole velocity space, we obtain the continuity equation as in (6.66)

$$\frac{dn_e}{dt} - D_{se}\,\nabla^2 n_e = -\,n_e\,<\nu_{rec}>, \tag{6.85}$$

but here instead of the term for secondary electron production by electron impact ionization, which can be neglected in the afterglow, we include a term for electron-ion recombination with the frequency ν_{rec}. This term plays an important role in the final stages of the afterglow (Gritsinin et al. 1996) when the electron and ion densities decay altogether with the same rate. This decay occurs according to the electron-ion volume recombination law $1/n_e(t) \simeq t$ (see Exercise 6.6).

The fall of the electric field to zero at $t = 0$ is obviously an approximation (Borysow and Phelps 1994; Dhali and Low 1988), but the time interval at which this occurs is very short as compared with the characteristic times of the phenomena under analysis. Here, we are mainly concerned with the time variation of the electron energy distribution function due to elastic and inelastic collisions, and with the variation of the electron density due to diffusion under the effects of the space-charge field $\mathbf{E_s}$, and due to electron-ion recombination. In particular, the effects produced by the field $\mathbf{E_s}$ are progressively decreased as the time evolves and the electron density decreases.

The time-dependent Boltzmann equation in the form expressed by equation (6.84) is difficult to solve, even numerically, due to the presence of the space-charge field term, and of the dependence of f_e^0 on the configuration and velocity spaces. In order to determine $\mathbf{E_s}$ self-consistently, Poisson's equation must be used, coupled to the continuity and momentum transport equations for electrons and ions. This task involves a great deal of computational work, so that we will consider here a much simpler approach which consists in considering in the Boltzmann equation directly the effective diffusion coefficient (6.26) describing the smooth transition from ambipolar to free diffusion as the electron density decreases (Guerra et al. 2001)

$$D_{se} = D_a\,\frac{D_e + \Lambda^2 e n_e \mu_e/\epsilon_0}{D_a + \Lambda^2 e n_e \mu_e/\epsilon_0}. \tag{6.86}$$

Here, $D_a \sim (\mu_i/\mu_e)D_e$ is the ambipolar diffusion coefficient, μ_i and μ_e are the ion and electron mobilities, and $\Lambda = R/2.405$ is the characteristic diffusion length for a cylindrical plasma container of radius R, obtained by replacing $\nabla^2 n_e$ with $\sim -\,n_e/\Lambda^2$, which is consistent with the assumption that n_e vanishes at the wall. In the limit $\lambda_D \ll \Lambda$, with $\lambda_D = \sqrt{\epsilon_0 u_k/e^2 n_e}$ denoting the electron Debye length and $u_k = eD_e/\mu_e$ the characteristic energy, we obtain $D_{se} \to D_a$, whereas at the opposite limit $\lambda_D \gg \Lambda$, we have $D_{se} \to D_e$. Equation (6.86) hence allows one to cover the whole bridge between nearly ambipolar diffusion (in the earlier and intermediate instants of the afterglow) and free diffusion (in the late or remote afterglow) by considering the decreasing of n_e and the corresponding variations of D_e and μ_e.

With the concept of effective diffusion coefficient and keeping in mind that the continuity equation (6.85) results from the integration of the Boltzmann equation in velocity space, we will return to the Boltzmann equation (6.84) and assume that the third term on the left-hand side member may be replaced with a composed term accounting for diffusion under the presence of the space-charge field as follows

$$\frac{\partial f_e^0}{\partial t} - \frac{1}{v_e^2}\frac{\partial}{\partial v_e}\left(\frac{v_e^2}{3\nu_m^e}\frac{1}{2}\left(\frac{eE_0}{m}\right)^2 \frac{1}{1+(\omega/\nu_m^e)^2}\frac{\partial f_e^0}{\partial v_e}\right) + \frac{v_e^2}{3\nu_m^e \Lambda^2}\frac{D_{se}}{D_e} f_e^0 = I^0 + J^0 . \tag{6.87}$$

When this equation is integrated in velocity space, we obtain in the third term of the left-hand side member using $D_e = <v_e^2/3\nu_m^e>$

$$\frac{D_{se}}{\Lambda^2 D_e}\int_0^\infty \frac{v_e^2}{3\nu_m^e} f_e^0 \, 4\pi v_e^2 \, dv_e = D_{se}\frac{n_e}{\Lambda^2} ,$$

obtaining thus equation (6.85). When $D_{se} \sim D_a$ at the earlier instants of the afterglow, the term for diffusion is small and it may usually be neglected as in the discharge. On the contrary, when $D_{se} \rightarrow D_e$ in the later stages of the afterglow, the diffusion term cannot be neglected and we obtain the same equation (6.69) as in the breakdown.

Writing equation (6.87) in terms of the electron energy u, we obtain the following equation for the case of a molecular gas

$$\begin{aligned}\frac{\partial f_e^0}{\partial t} &- \frac{1}{\sqrt{u}}\frac{\partial}{\partial u}\left[\frac{2}{3}u^{3/2}\nu_m^e u_c \frac{\partial f_e^0}{\partial u}\right] + \frac{2}{3m}\frac{u}{\nu_m^e \Lambda^2}\frac{D_{se}}{D_e} f_e^0 \\ &= \frac{1}{\sqrt{u}}\frac{\partial}{\partial u}\left[\frac{2m}{M}u^{3/2}\nu_m\left(f_e^0 + k_B T_o \frac{\partial f_e^0}{\partial u}\right) + 4B_0\nu_0\sqrt{u} f_e^0\right] \\ &+ \frac{1}{\sqrt{u}}\sum_{i,j}\left[\sqrt{u+u_{ij}}\,\nu_{ij}(u+u_{ij})\, f_e^0(u+u_{ij}) - \sqrt{u}\,\nu_{ij} f_e^0\right] \\ &+ \frac{1}{\sqrt{u}}\sum_{j,i}\left[\sqrt{u-u_{ij}}\,\nu_{ji}(u-u_{ij})\, f_e^0(u-u_{ij}) - \sqrt{u}\,\nu_{ji} f_e^0\right] \\ &- \nu_{rec} f_e^0 ,\end{aligned} \tag{6.88}$$

where u_c is the time-averaged energy gain per collision (6.75), $\nu_0 = n_o\sqrt{2u/m}\,\sigma_0$ is the frequency for rotational exchanges induced by electron impact including both inelastic and superelastic collisions (3.160), and $\nu_{rec} = n_i\sqrt{2u/m}\,\sigma_{rec}$ is the frequency for electron-ion recombination, with n_i denoting the ion number density, and σ_{rec} the electron-ion recombination cross section. As we have stated before the production of secondary electrons by electron impact ionization may be neglected in the afterglow.

The equation (6.88) may be transformed to a set of coupled ordinary differential equations by a finite differencing of the electron energy axis into n cells of width Δu, using a scheme close to that indicated in Sect. 3.4.2, and employing the following algorithm for time-evolution

$$\sqrt{u}\,\mathbf{I}\,.\,\mathbf{F}(t+\Delta t) \;=\; (\Delta t\,\mathbf{A} \;+\; \sqrt{u}\,\mathbf{I})\,.\,\mathbf{F}(t). \tag{6.89}$$

Here, $\mathbf{F}(t)$ is the column vector of the temporal electron energy distribution function $f_e^0(u,t)$ at the instant t, $\mathbf{A}$ is the time-independent matrix of coefficients obtained from finite differencing all terms of equation (6.88), $\mathbf{I}$ is the identity matrix, with $\sqrt{u}\,\mathbf{I} = \sqrt{u_i}\,\delta_{ij}$, $u_i = (i-1/2)\,\Delta u$ denotes the energy in the middle of the cell i–th of the energy grid, and Δt is the time step.

The analysis of the time relaxation of $f_e^0(u,t)$ in the post-discharge is analysed by setting $f_e^0(u,0) = n_e\,f(u)$ as the initial condition, with n_e and $f(u)$ denoting the electron density and the EEDF in the discharge. The latter normalized as follows

$$\int_0^\infty f(u)\,\sqrt{u}\,du \;=\; 1, \tag{6.90}$$

verifying therefore

$$f(u) \;=\; \frac{4\pi}{m}\,\sqrt{\frac{2}{m}}\,\frac{f_e^0(u,0)}{n_e}. \tag{6.91}$$

The electron density is updated during the time-evolution process in D_{se}/D_e given by equation (6.86), by updating $D_e(t)$ and $\mu_e(t)$ at each instant, and in $\nu_{rec}(t)$, where it is assumed $n_i(t) = n_e(t)$. Equation (6.89) evolves with an initial time-step $\Delta t = 10^{-10}$ s for the first interactions and then the predictor-corrector method of Adams-Bashforth-Moulton is used for adjusting the stepsize (see e.g. Fox and Mayers 1968).

Figure 6.8 shows the EEDF $f_e^0(u,t)/n_e(t)$, multiplied by the normalization factor (6.91) in order to have $\int_0^\infty f_e^0(u,t)\sqrt{u}\;du = n_e(t)$, calculated at different instants in the time-interval $t = 1\times10^{-7} - 1\times10^{-4}$ s, in the afterglow of a flowing microwave discharge at $\omega/(2\pi) = 2.45$ GHz in nitrogen, for typical operating conditions corresponding to a Pyrex tube of inner radius $R = 0.8$ cm, at the pressure $p = 266$ Pa (i.e. 2 Torr) and electron density $n_e(0) = 5\times10^{17}\,\mathrm{m}^{-3}$. The electron density is usually modified in the discharge by varying the injected microwave power. The typical values of the gas temperature at present conditions are of the order of $T_o = 1000$ K, so that this value has been assumed constant here. Figure 6.8 shows the depletion of the high-energy tail of the EEDF at the earlier instants of the afterglow as a result of electron inelastic collisions.

Although the EEDF is largely modified at short times Fig. 6.9 shows that the electron density remains practically unchanged up to $t \sim 10^{-5}$ s due to the large

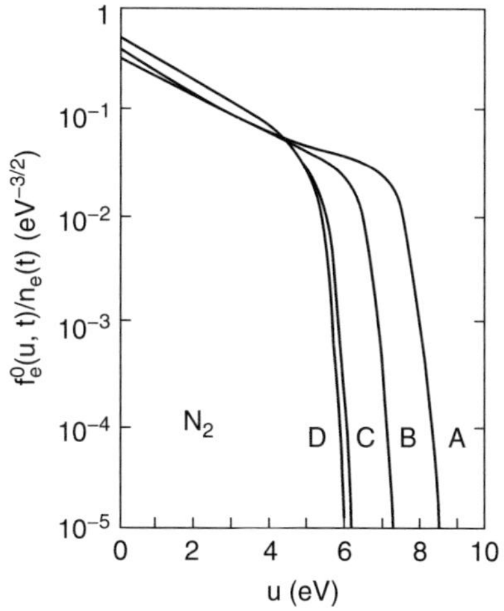

Fig. 6.8 Electron energy distribution function in the post-discharge of an $\omega/(2\pi) = 2.45$ GHz microwave discharge in N_2, with $p = 266$ Pa and $n_e(0) = 5 \times 10^{17}$ m^{-3}, for the following instants in the afterglow: (*A*) 1×10^{-7} s; (*B*) 1×10^{-6} s; (*C*) 1×10^{-5} s; (*D*) 1×10^{-4} s (Guerra et al. 2001)

characteristic times for electron losses by ambipolar diffusion. Figure 6.9 reports the values of $n_e(t)$ calculated from equation (6.88), in which a transition from ambipolar to free diffusion regimes is considered (curve A), as well as the values obtained when free diffusion of electrons is assumed during the whole relaxation process (i.e., assuming $D_{se}/D_e = 1$ in equation (6.88), curve B). The electron density falls sharply at $t \sim 10^{-7} - 10^{-6}$ as the electron diffusion occurs without the presence of the space-charge electric field.

The behaviour exhibited by $n_e(t)$ essentially results from the fact of the effective diffusion coefficient D_{se} remains practically unchanged up to $t \sim 3 \times 10^{-4}$ s, with $D_{se} \sim D_a$, instant at which it sharply increases to D_e, due to the diminution of n_e. Figure 6.10 shows the ratio D_{se}/D_e, given by equation (6.86), calculated at the same conditions as before. The free diffusion is completely achieved at $t \sim 4 \times 10^{-4}$ s. The present results are qualitatively in line with the measurements realized in Marković et al. (1997) using the breakdown time delay technique, in which it was observed a rapid transition from ambipolar to free diffusion at t $\sim$ 20 ms.

The creation of secondary electrons by electron impact ionization in the post-discharge may be neglected due to the rapid depletion of the EEDF. However, additional ionization may be produced by Penning ionization reactions. In the case of nitrogen, these reactions occur through collisions of the metastables states $N_2(A\ ^3\Sigma_u^+)$ and $N_2(a'\ ^1\Sigma_u^-)$ as follows (Brunet et al. 1983; Brunet and Rocca-Serra 1985; Berdyshev et al. 1988)

$$\begin{aligned} N_2(A\ ^3\Sigma_u^+) + N_2(a'\ ^1\Sigma_u^-) &\rightarrow N_2^+ + N_2 + e \\ &\rightarrow N_4^+ + e \end{aligned} \qquad (6.92)$$

$$\begin{aligned} N_2(a'\ ^1\Sigma_u^-) + N_2(a'\ ^1\Sigma_u^-) &\rightarrow N_2^+ + N_2 + e \\ &\rightarrow N_4^+ + e\,. \end{aligned} \qquad (6.93)$$

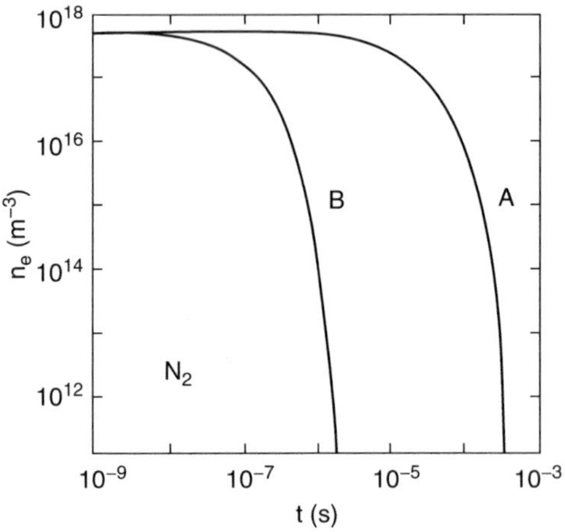

Fig. 6.9 Temporal evolution of the electron number density $n_e(t)$ in a N_2 post-discharge, calculated for the same conditions as in Fig. 6.8, using the reference model indicated here (*A*) and assuming free diffusion of electrons during the whole relaxation process (*B*) (Guerra et al. 2001)

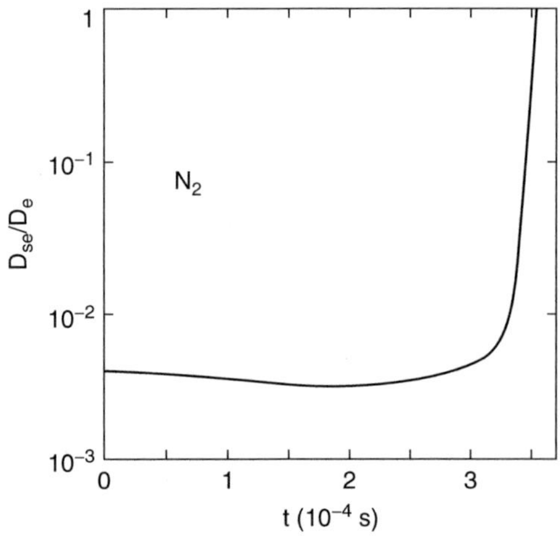

Fig. 6.10 Temporal evolution of the ratio of the effective electron diffusion coefficient to the free diffusion coefficient in a N_2 post-discharge, for the same conditions as in Figs. 6.8 and 6.9. The nearly flat region for D_{se}/D_e corresponds to time-interval where the ambipolar diffusion prevails (Guerra et al. 2001)

These reactions lead to the introduction of an additional source term on the right-hand side member of equation (6.88)

$$I_{ion} = \frac{1}{\sqrt{u}} \left(R^1_{ion}\, \delta(u_1) + R^2_{ion}\, \delta(u_2) \right) , \tag{6.94}$$

with R^1_{ion} and R^2_{ion} representing the rates (in m^{-3} s^{-1}) of the Penning or associative ionization reactions (6.92) and (6.93), respectively,

$$R^1_{ion} = [N_2(A)]\, [N_2(a')]\, k^1_{ion} \tag{6.95}$$

$$R^2_{ion} = [N_2(a')]\, [N_2(a')]\, k^2_{ion} , \tag{6.96}$$

and where the δ function obeys to the usual normalization condition $\int_0^\infty \delta(u)\, du = 1$. Estimations for the rate coefficients of the two reactions are $k_{ion}^1 = 1 \times 10^{-17}\,\mathrm{m}^3\,\mathrm{s}^{-1}$ and $k_{ion}^2 = 5 \times 10^{-17}\,\mathrm{m}^3\,\mathrm{s}^{-1}$ (Guerra and Loureiro 1997). As a result of the threshold energies of the two states, the electrons are created with the energies $u_1 \simeq 0$ and $u_2 = 1.3\,\mathrm{eV}$. The creation of new electrons permits the description of a very interesting phenomenon, the increase of n_e in the afterglow after an initial stage of decay (Guerra et al. 2004), as experimentally observed in Bogdan et al. (1993), Sadeghi et al. (2001) and Amorim (2005).

6.2.4 Beyond the Local Field Approximation

The analysis carried out in Chap. 5 and previous sections of Chap. 6 assumes the local-field approximation (LFA); see Sect. 5.1.2. According to this approximation the variation in space of positions of the isotropic component of the electron velocity distribution function $f_e^0(\mathbf{r}, v_e)$ takes place through the electron density $n_e(\mathbf{r})$, so that the ratio $f_e^0(\mathbf{r}, v_e)/n_e(\mathbf{r})$ is independent of $\mathbf{r}$. However, in some situations this approximation is not valid and an accurate description of the nonequilibrium electron kinetics needs to be taken into account in both the velocity and position spaces. The electron kinetics in inhomogeneous plasma regions is then governed by relaxation processes in the space of positions and a distinctly non-local behaviour in the electron distribution function arises.

The basic equation to study the electron kinetics in nonuniform regions of a stationary unmagnetized plasma is the inhomogeneous, time-independent electron Boltzmann equation (3.6)

$$\left(\mathbf{v_e} \cdot \frac{\partial f_e}{\partial \mathbf{r}}\right) - \frac{e}{m}\left(\mathbf{E} \cdot \frac{\partial f_e}{\partial \mathbf{v_e}}\right) = \left(\frac{\partial f_e}{\partial t}\right)_{e-o}, \tag{6.97}$$

which determines the electron distribution function $f_e(\mathbf{r}, \mathbf{v_e})$ and where the collision term in the right-hand side member includes elastic, inelastic, and ionization processes. The electric field is assumed nonuniform $\mathbf{E}(\mathbf{r})$ and e and m denote the absolute value of the charge and the mass of the electrons.

If the electric field and the inhomogeneity in the plasma are parallel to a fixed space direction, for instance the z axis, with $\mathbf{E}(z) = -\,E(z)\;\mathbf{e_z}$, the electron distribution function becomes symmetrical around the field and can be given through an expansion in Legendre polynomials $L_n(\cos\theta)$, with θ representing the direction of $\mathbf{v_e}$ with respect to the z axis (3.9). In the lowest approximation order, the so-called two-term approximation (3.11), the electron distribution function $f_e(z, \mathbf{v_e})$ becomes a function $f_e(z, v_e, \cos\theta)$ as follows

$$f_e(z, \mathbf{v_e}) \simeq f_e^0(z, v_e) + f_e^1(z, v_e)\,\cos\theta\,. \tag{6.98}$$

The substitution of this expansion into the Boltzmann equation (6.97) leads to the equation system (3.87), (5.7) and (3.88), (5.8) for the stationary isotropic and first anisotropic components

$$\frac{v_e}{3}\frac{\partial f_e^1}{\partial z} + \frac{eE}{m}\frac{1}{3v_e^2}\frac{\partial}{\partial v_e}\left(v_e^2 f_e^1\right) = I^0(f_e^0) + J^0(f_e^0) \tag{6.99}$$

$$v_e \frac{\partial f_e^0}{\partial z} + \frac{eE}{m}\frac{\partial f_e^0}{\partial v_e} = -\nu_m^e f_e^1 . \tag{6.100}$$

The collision operators for elastic and inelastic collisions in (6.99) are given by equations (3.85) and (3.134), while the effective collision frequency for momentum transfer ν_m^e is given by (3.139).

Substituting now the velocity by the electron energy $u = \frac{1}{2}\, m v_e^2$ and dividing both members of equation (6.99) by $\sqrt{2/mu}$ and both members of equation (6.100) by $\sqrt{2u/m}$, we obtain the following set of equations

$$\frac{\partial}{\partial z}\left(\frac{u}{3} f_e^1\right) + eE\,\frac{\partial}{\partial u}\left(\frac{u}{3} f_e^1\right) = \frac{2m}{M}\frac{\partial}{\partial u}\left(u^2\, n_o\, \sigma_m\, f_e^0\right)$$
$$+ \sum_{j>0} n_o\{(u+u_j)\,\sigma_j(u+u_j)\, f_e^0(z, u+u_j) - u\,\sigma_j(u)\, f_e^0(z,u)\} \tag{6.101}$$

$$\frac{\partial f_e^0}{\partial z} + eE\,\frac{\partial f_e^0}{\partial u} = -\, n_o\, \sigma_m^e\, f_e^1 . \tag{6.102}$$

For brevity only inelastic electron collisions of the first kind with ground-state atoms are considered and the small heating for electrons due to elastic collisions with non-frozen atoms is neglected. $\sigma_m^e = \nu_m^e/(n_o v_e)$ represents the effective electron collision cross-section for momentum transfer.

The isotropic component of the electron distribution function normalized such as

$$n_e(z) = \int_0^\infty f_e^0(z, v_e)\, 4\pi v_e^2\, dv_e = \frac{4\pi}{m}\sqrt{\frac{2}{m}}\int_0^\infty f_e^0(z,u)\,\sqrt{u}\, du \tag{6.103}$$

may be now renormalized as follows

$$n_e(z) = \int_0^\infty F_e^0(z,u)\,\sqrt{u}\, du . \tag{6.104}$$

For consistency the anisotropic component $f_e^1(z,u)$ should be renormalized to $F_e^1(z,u)$ in the same way. With this normalization, the electron particle current density $\mathbf{\Gamma_e}(z) = \Gamma_e(z)\,\mathbf{e_z}$ (3.18) is expressed as

$$\Gamma_e(z) = \frac{1}{3}\sqrt{\frac{2}{m}}\int_0^\infty u\, F_e^1(z,u)\, du , \tag{6.105}$$

with $F_e^1(z,u)$ obtained from (6.102), and being the stationary electron particle balance equation (5.77) as follows

$$\frac{d}{dz}\Gamma_e(z) = 0, \tag{6.106}$$

because the creation of secondary electrons is neglected in (6.101).

On the other hand, multiplying both members of equation (6.101) by the electron energy u (before to have making the division by $\sqrt{2/mu}$) and integrating over all velocity space, we obtain the stationary equation for energy conservation

$$\frac{d}{dz}J_E(z) = P_E(z) - P_{el}(z) - P_{inel}(z)\,, \tag{6.107}$$

The left-hand side member represents the divergence of the energy current density $\mathbf{J_E}(z) = J_E(z)\,\mathbf{e_z}$ defined as follows

$$J_E(z) = \frac{1}{3}\sqrt{\frac{2}{m}}\int_0^\infty u^2\,F_e^1(z,u)\,du\,, \tag{6.108}$$

while the terms on the right-hand side member of (6.107) account for the power gain from the field $P_E(z)$ (3.116)

$$P_E(z) = (\mathbf{J_e}\,.\,\mathbf{E}) = J_e(z)\,E(z)\,. \tag{6.109}$$

with $\mathbf{J_e} = -\,e\,\mathbf{\Gamma_e}$ denoting the electron current density, the power lost in elastic collisions $P_{el}(z)$ (3.115), and the power lost in inelastic collisions $P_{inel}(z)$ (3.151). In a nearly homogeneous plasma a nearly complete compensation should occur, at each space position z, between the input power from the electric field and the power losses by elastic and inelastic collisions, being small the divergence of the energy current density in the power balance equation (6.107). When such compensation does not occur non-local effects need to be taken into account in the determination of the EEDF and the LFA is no more valid.

As firstly noticed by Bernstein and Holstein (1954), a suitable form of equations (6.101) and (6.102) for determining the isotropic and the anisotropic components, in position and energy spaces simultaneously, is obtained by replacing the electron kinetic energy u by the total energy $\epsilon = u - e\,V(z)$ (Tsendin 1974; Winkler et al. 1997), with $V(z) = \int_0^z E(z')\,dz'$ denoting the electric potential, in which $\mathbf{E} = -\,E(z)\,\mathbf{e_z}$. With such transformation ϵ is independent of the coordinate z and the kinetic energy becomes a function of z and ϵ, $u(z,\epsilon) = \epsilon + e\,V(z)$. Then, the left-hand side members of equations (6.101) and (6.102) become, using the normalization condition (6.104),

$$\frac{\partial}{\partial z}\left(\frac{u}{3}\,F_e^1\right) + \frac{\partial}{\partial u}\left(\frac{u}{3}\,F_e^1\right)\frac{\partial u}{\partial z} \tag{6.110}$$

$$\frac{\partial F_e^0}{\partial z} + \frac{\partial F_e^0}{\partial u}\frac{\partial u}{\partial z}. \tag{6.111}$$

Replacing now $F_e^0(z,u)$ and $F_e^1(z,u)$ with the new functions dependent on (z,ϵ), $\tilde{F}_e^0(z,\epsilon)$ and $\tilde{F}_e^1(z,\epsilon)$, the equations (6.101) and (6.102) can be written under the form

$$\frac{\partial}{\partial z}\left(\frac{u}{3}\,\tilde{F}_e^1\right) = \frac{2m}{M}\frac{\partial}{\partial \epsilon}\left(u^2\, n_o\, \sigma_m\, \tilde{F}_e^0\right)$$
$$+\sum_{j>0} n_o\{(u+u_j)\,\sigma_j(u+u_j)\,\tilde{F}_e^0(z,\epsilon+u_j) \;-\; u\,\sigma_j(u)\,\tilde{F}_e^0(z,\epsilon)\} \tag{6.112}$$

$$\frac{\partial \tilde{F}_e^0}{\partial z} = -\,n_o\,\sigma_m^e\,\tilde{F}_e^1. \tag{6.113}$$

The anisotropic component $\tilde{F}_e^1(z,\epsilon)$ obtained from (6.113) may be now inserted into equation (6.112) obtaining thus the parabolic equation

$$-\frac{\partial}{\partial z}\left(\frac{u}{3n_o\sigma_m^e}\frac{\partial \tilde{F}_e^0}{\partial z}\right) = \frac{2m}{M}\frac{\partial}{\partial \epsilon}\left(u^2\, n_o\, \sigma_m\, \tilde{F}_e^0\right)$$
$$+\sum_{j>0} n_o\{(u+u_j)\,\sigma_j(u+u_j)\,\tilde{F}_e^0(z,\epsilon+u_j) \;-\; u\,\sigma_j(u)\,\tilde{F}_e^0(z,\epsilon)\}. \tag{6.114}$$

The equation (6.114) describes the evolution of the isotropic distribution function $\tilde{F}_e^0(z,\epsilon)$ and has to be solved as an initial-boundary-value problem on a region whose boundaries are determined by the spatial course of the electric field (Winkler et al. 1997). From the mathematical point of view the parabolic problem has to be completed by appropriate boundary conditions. In the case of cylindrical geometry the equivalent equation to (6.114) takes the form of an elliptic equation (Golubovskii et al. 2015).

A large simplification of the kinetic study of inhomogeneous plasmas is reached when the spatial evolution of the electron kinetic quantities consists of a sequence of nearly homogeneous states. In this case the kinetic quantities can be determined by solving the much simpler homogeneous kinetic Boltzmann equation for the sequence of the electric field strengths of each nearly homogeneous state. Such a treatment is the so-called LFA. However in this case a rapid spatial establishment of the electron distribution function into homogeneous states would be required. This means that the corresponding relaxation length of the electrons should be very short in comparison with the spatial change of the electric field (see Sect. 5.1.2).

A situation where the LFA is quite often questionable is the radial electron diffusion under the effect of a radial space-charge electric field, $\mathbf{E_s} = E_s\,\mathbf{e_r}$, in a microwave discharge at low pressures. The non-local behaviour results from the space-charge field and the spatial diffusion terms are simultaneously taken into account in the inhomogeneous Boltzmann equation. Considering the complete form of the Boltzmann equation as referred in equation (6.60) for the case of a microwave discharge, we may write

$$-\frac{1}{v_e^2}\frac{\partial}{\partial v_e}\left[\frac{v_e^2}{3\nu_m^e}\left(\frac{e}{m}\right)^2\left(E_e^2+E_s^2\right)\frac{\partial f_e^0}{\partial v_e}-\frac{v_e^3}{3\nu_m^e}\frac{e}{m}\left(\mathbf{E_s}\,.\nabla f_e^0\right)\right]$$
$$+\frac{v_e}{3\nu_m^e}\frac{e}{m}\nabla.\left(\frac{\partial f_e^0}{\partial v_e}\mathbf{E_s}\right)-\frac{v_e^2}{3\nu_m^e}\nabla^2 f_e^0 = I^0(f_e^0)+J^0(f_e^0)\,, \qquad (6.115)$$

where the local effective microwave electric field is

$$E_e(\mathbf{r},v_e) = \frac{1}{\sqrt{1+(\omega/\nu_m^e(v_e))^2}}\frac{E_0(\mathbf{r})}{\sqrt{2}}\,. \qquad (6.116)$$

The different terms in the left-hand side member of (6.115) account for heating by the HF electric field and by the space-charge field, transport under the effects of the space-charge field, and spatial diffusion, while in the right-hand side member we have the terms for energy losses by elastic and inelastic collisions.

Rewritten (6.115) in terms of the electron energy and assuming the dependence of E_s on the radial coordinate, we obtain after multiplication of both members by $\sqrt{u}$, using $F_e^0(\mathbf{r},u)$ normalized according to equation (6.104), and cylindrical geometry

$$-\frac{2}{3m}\frac{\partial}{\partial u}\left[\frac{u^{3/2}}{\nu_m^e}\left\{e^2E_e^2\frac{\partial F_e^0}{\partial u}-eE_s\left(\frac{\partial F_e^0}{\partial r}-eE_s\frac{\partial F_e^0}{\partial u}\right)\right\}\right]$$
$$-\frac{2}{3m}\frac{u^{3/2}}{\nu_m^e}\frac{1}{r}\frac{\partial}{\partial r}\left[r\left(\frac{\partial F_e^0}{\partial r}-eE_s\frac{\partial F_e^0}{\partial u}\right)\right] = \frac{2m}{M}\frac{\partial}{\partial u}\left(u^{3/2}\,\nu_m\,F_e^0\right)$$
$$+\sum_{j>0}\left[\sqrt{u+u_j}\,\nu_j(u+u_j)\,F_e^0(r,u+u_j)-\sqrt{u}\,\nu_j(u)\,F_e^0(r,u)\right]\,. \qquad (6.117)$$

If the pressure is sufficiently high or the tube radius R sufficiently large for the characteristic energy relaxation length $\lambda_e = v_e/\nu_e$ satisfies $\lambda_e \ll R$, the LFA is valid. The effects of diffusion under the presence of space-charge field are small, and the electron distribution function is determined by the same expressions as in a homogeneous plasma. In this case the isotropic component $F_e^0(r,u)$ can be factorized as $n_e(r)\times(F_e^0(r,u)/n_e(r))$, with the ratio $F_e^0(r,u)/n_e(r)$ depending on the local reduced effective electric field strength, E_e/n_o. On the contrary, in the opposite limit $\lambda_e \gg R$, the isotropic component $F_e^0(r,u)$ is totally non-local.

Bernstein and Holstein (1954) have noticed that in a positive column of radius R the kinetic equation for $F_e^0(r,u)$ can be reduced to a two-dimensional diffusion equation along the radial coordinate r and along the total energy $\epsilon = u - eV(r)$, where $V(r) = -\int_0^r E_s(r')\,dr'$ denotes the radial space-charge electric potential. If the plasma scale R is small, the total energy ϵ remains nearly constant during the electron radial displacement and the electron distribution function $F_e^0(r,u)$ becomes totally non-local. The fundamental paper of Bernstein and Holstein (1954) was forgotten for 20 years, until a similar solution was rediscovered by Tsendin (1974). For electrons trapped by a potential well $V(r)$, the solution depends only on the total energy ϵ and the kinetic Boltzmann equation can be averaged over the coordinate r. So the radial plasma inhomogeneity and the radial electric field can be excluded from the kinetic equation and it can be formally reduced to the same form as in the familiar case of a homogeneous plasma. The only difference between the local case and this totally non-local case lies in the fact that the totally non-local isotropic component of the electron distribution function depends on the total energy ϵ, instead of the kinetic energy u, such as it has already been considered in equations (6.112) and (6.113). The development of such idea was one the most brilliant legacies of Lev Tsendin (Tsendin 1974, 1982a,b, 2011; Kolobov 2013). Later on the same idea has also been treated and improved in Kortshagen (1993, 1994) and more recently in Golubovskii et al. (2015).

When the timescale of spatial diffusion is much shorter than the timescale of diffusion in energy space, the electrons, confined by the radial space-charge electric potential $V(r) < 0$ (with $V(r=0) = 0$), are moving with an almost constant total energy $\epsilon = u - eV(r)$. Thus, the accessible region for the electrons in the phase plane (r, ϵ) is bounded by $\epsilon \geq -eV(r)$. Figure 6.11 shows that electrons with a total energy ϵ_1 can access the discharge cross section up to their turning point radius r_1 such as $\epsilon_1 = -eV(r_1)$. The curve $u_j - eV(r)$ determines the lower limit of (r, ϵ) plane above which the electrons are able to perform an inelastic collision with the threshold energy u_j. In the case of electrons of total energy ϵ_2, the region where these exciting collisions take place is $r \leq r_2$, with r_2 determined by $(\epsilon_2 - u_j) = -eV(r_2)$. At $r = 0$ the total energy coincides with the kinetic energy.

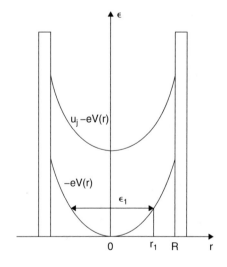

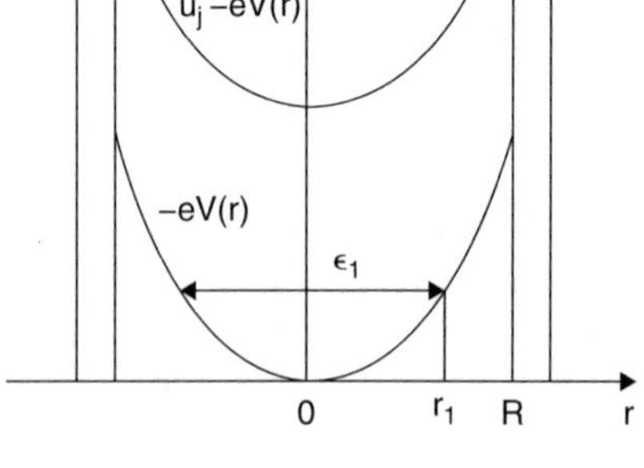

Fig. 6.11 Phase plane (r, ϵ) for the trapped electrons. The accessible region is bounded by $\epsilon \geq -eV(r)$. The turning point r_1 for electrons of total energy ϵ_1 is defined as $\epsilon_1 = -eV(r_1)$. The *curve* $u_j - eV(r)$ determines the region on (r, ϵ) plane above which the electrons are able to perform an inelastic collision with threshold energy u_j (Kortshagen 1993)

The situation addressed by the totally non-local approach is that of a discharge at sufficiently low pressure, so that the energy relaxation length of the electrons exceeds the discharge dimensions. Since electron displacement occurs faster than energy relaxation, the spatial fluxes at different ϵ are practically independent. In the fully collisionless case the total energy ϵ is a constant of motion. Thus, it is reasonable to express equation (6.117) in the total energy rather than in the kinetic energy (Bernstein and Holstein 1954; Tsendin 1974). Being $\epsilon = u - eV(r)$ the total energy, one obtains $u(r,\epsilon) = \epsilon + eV(r)$, and therefore $\partial u/\partial r = -eE_s$, so that substituting $F_e^0(r,u)$ by the distribution function $\tilde{F}_e^0(r,\epsilon)$, one obtains

$$\frac{\partial F_e^0}{\partial r} - eE_s \frac{\partial F_e^0}{\partial u} = \frac{\partial F_e^0}{\partial r} + \frac{\partial F_e^0}{\partial u}\frac{\partial u}{\partial r} \equiv \frac{\partial \tilde{F}_e^0}{\partial r}, \tag{6.118}$$

whereas $\partial F_e^0/\partial u \equiv \partial \tilde{F}_e^0/\partial \epsilon$. Since the left-hand side member of equation (6.117) can be written as follows

$$-\frac{2}{3m}\frac{\partial}{\partial u}\left(\frac{u^{3/2}}{\nu_m^e}\, e^2 E_e^{\,2}\, \frac{\partial F_e^0}{\partial u}\right)$$
$$-\frac{2}{3m}\frac{1}{r}\left(\frac{\partial}{\partial r} - eE_s \frac{\partial}{\partial u}\right)\left[r\, \frac{u^{3/2}}{\nu_m^e}\left(\frac{\partial F_e^0}{\partial r} - eE_s \frac{\partial F_e^0}{\partial u}\right)\right], \tag{6.119}$$

equation (6.117) takes the following final form using the distribution function $\tilde{F}_e^0(r,\epsilon)$ (Kortshagen 1993, 1994)

$$-\frac{2}{3m}\left[\frac{\partial}{\partial \epsilon}\left(\frac{u^{3/2}}{\nu_m^e}\, e^2 E_e^{\,2}\, \frac{\partial \tilde{F}_e^0}{\partial \epsilon}\right) + \frac{1}{r}\frac{\partial}{\partial r}\left(r\, \frac{u^{3/2}}{\nu_m^e}\, \frac{\partial \tilde{F}_e^0}{\partial r}\right)\right] = \frac{2m}{M}\frac{\partial}{\partial \epsilon}\left(u^{3/2}\, \nu_m\, \tilde{F}_e^0\right)$$
$$+ \sum_{j>0}\left[\sqrt{u+u_j}\; \nu_j(u+u_j)\, \tilde{F}_e^0(r, \epsilon + u_j) - \sqrt{u}\; \nu_j(u)\, \tilde{F}_e^0(r,\epsilon)\right]. \tag{6.120}$$

The kinetic energy is then a function of the radial coordinate, $u(r)$, as well as the effective electric field is $E_e(r, u(r))$ in the case of $E_0(r)$ in equation (6.116).

For sufficiently low pressures the second term on the left-hand side of equation (6.120) is the dominant term. For the idealized case of collisionless motion of electrons in a confining space-charge potential without accelerating electric field, the distribution function is independent of the spatial coordinate, i.e. $\tilde{F}_e^0(r,\epsilon) = \tilde{F}_e^0(\epsilon)$, because the total energy is a constant of motion. Thus, one may assume that $\tilde{F}_e^0(r,\epsilon)$ is radially dependent only to a first order correction

$$\tilde{F}_e^0(r,\epsilon) = \tilde{F}_{e,0}^0(\epsilon) + \tilde{F}_{e,1}^0(r,\epsilon). \tag{6.121}$$

The non-local approach is achieved by considering different times scales for spatial diffusion and energy diffusion. In cases where the energy relaxation length is sufficiently large and the electric field is not too high, the spatial diffusion is much faster than the energy diffusion. This fact justifies an averaging of equation (6.120) over the radial cross section, accessible for the electrons with a certain total energy ϵ. Using this assumption it is possible to obtain a simple ordinary differential equation for $\tilde{F}^0_{e,0}(\epsilon)$ by integrating equation (6.120) over the discharge cross section. However, for electrons with a small total energy not the whole discharge cross section is accessible but only the part up to a maximum radius $r^*(\epsilon)$, which is the turning point of the electron motion defined by $u(r^*) = 0$ or $\epsilon = -eV(r^*)$ (see Fig. 6.11).

The resulting equation by this integration procedure is an ordinary differential equation for the electron distribution function $\tilde{F}^0_{e,0}(\epsilon)$

$$-\frac{\partial}{\partial \epsilon}\left(\overline{D}(\epsilon)\,\frac{\partial \tilde{F}^0_{e,0}(\epsilon)}{\partial \epsilon} + \overline{G}(\epsilon)\,\tilde{F}^0_{e,0}(\epsilon)\right)$$
$$= \sum_{j>0}\left[\overline{H}_j(\epsilon + u_j)\,\tilde{F}^0_{e,0}(\epsilon + u_j) - \overline{H}_j(\epsilon)\,\tilde{F}^0_{e,0}(\epsilon)\right] \tag{6.122}$$

with the radially averaged quantities

$$\overline{D}(\epsilon) = \frac{2}{3m}\int_0^{r^*(\epsilon)} \frac{u(r)^{3/2}}{\nu^e_m(u(r))}\, e^2\, E_e^2(r, u(r))\, r\, dr \tag{6.123}$$

$$\overline{G}(\epsilon) = \frac{2m}{M}\int_0^{r^*(\epsilon)} u(r)^{3/2}\, \nu_m(u(r))\, r\, dr \tag{6.124}$$

$$\overline{H}_j(\epsilon) = \int_0^{r_j(\epsilon)} \sqrt{u(r)}\, \nu_j(u(r))\, r\, dr\,. \tag{6.125}$$

Here, $r^*(\epsilon)$ is the turning point radius and $r_j(\epsilon)$ is the maximum radius for which the j-th inelastic process is possible, i.e. $u(r_j(\epsilon)) = u_j$. The physical meaning of this integration is obvious (Kortshagen 1993, 1994). Since the electrons are moving everywhere across their accessible cross section before they considerably change their energy, the values obtained at one position are distributed all over the accessible cross section. This physical picture is hence in opposition to the assumption of locality. The radial average equation (6.122) is formally equivalent to the homogeneous Boltzmann equation and thus may be easily solved with the same numerical techniques. The spatial dependence on r coordinate is not lost but only hidden in the definition of the total energy $\epsilon = u - eV(r)$.

If one is interested in radially resolved parameters like the electron density, electron rate coefficients, etc., it is necessary to know the dependence on the electron kinetic energy. The electron energy distribution function at a certain radial position r_0 may be easily obtained from the distribution function of total energy as follows

$$F^0_e(r_0, u) = \tilde{F}^0_{e,0}(\epsilon = u - eV(r_0))\,. \tag{6.126}$$

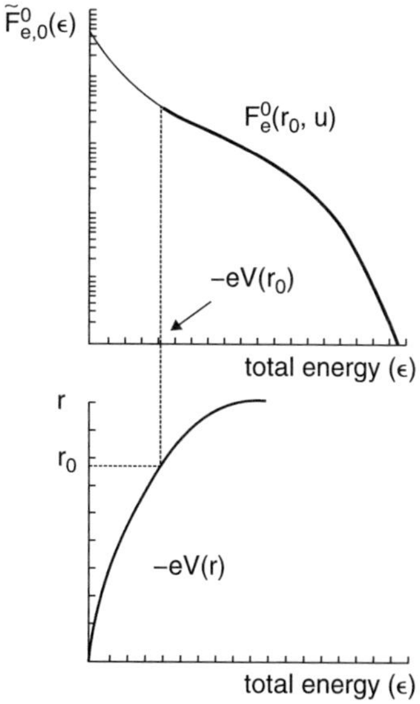

Fig. 6.12 Scheme for finding the electron distribution function of kinetic energy $F^0_e(r, u)$ (*bold* part of the *curve* in *upper diagram*) at a position r_0 from the electron distribution function of total energy $\tilde{F}^0_{e,0}(\epsilon)$ (*whole curve* in *upper diagram*) and the space-charge potential (*lower diagram*) (Kortshagen 1994)

At the radial position r_0 the potential energy $-\ eV(r_0)$ constitutes a threshold for the electrons. Electrons with a total energy less than this value are reflected by the space potential before reaching r_0. Electrons with a higher total energy can reach this position with the kinetic energy $u(r_0) = \epsilon + eV(r_0)$. Thus, $F^0_e(r, u)$ is obtained from $\tilde{F}^0_{e,0}(\epsilon)$ by cutting the low energy part with $\epsilon < -\ eV(r_0)$ and using the value $-\ eV(r_0)$ as the new zero of the kinetic energy scale.

Although obtained from the spatially averaged kinetic equation (6.122), the spatial information is fully included in the electron distribution function of total energy $\tilde{F}^0_{e,0}(\epsilon)$ in combination with the space-charge potential (see Fig. 6.12). The distribution of the electron density in the space-charge potential is then easily obtained

$$n_e(r) = \int_0^\infty F^0_e(r, u)\ \sqrt{u}\ du$$
$$= \int_{-eV(r)}^\infty \tilde{F}^0_{e,0}(\epsilon)\ \sqrt{\epsilon + eV(r)}\ d\epsilon\ , \tag{6.127}$$

while the ionization rate (in m^{-3} s^{-1}) is

$$R_{ion}(r) = n_e(r)\ <\nu_{ion}>(r)\ = \int_{u_I}^\infty \nu_{ion}(u)\ F^0_e(r, u)\ \sqrt{u}\ du$$
$$= \int_{u_I - eV(r)}^\infty \nu_{ion}(\epsilon + eV(r))\ \tilde{F}^0_{e,0}(\epsilon)\ \sqrt{\epsilon + eV(r)}\ d\epsilon\ , \tag{6.128}$$

with u_I denoting the ionization threshold energy.

The validity of this fully non-local approach is based on the assumption that the spatial diffusion term in the inhomogeneous Boltzmann equation (6.117) is the dominant term. As pointed out by Bernstein and Holstein (1954) this condition is fulfilled when the space-charge electric field is larger than the effective maintaining electric field strength.

Exercises

Exercise 6.1. In a glow discharge of tube radius R the electron density varies radially with the form

$$n_e(r) = n_{e0}\, J_0\left(\frac{r}{\Lambda}\right) ,$$

where r is the radial coordinate of the cylindrical system, J_0 is the Bessel function of first kind and order zero, and $\Lambda = R/2.405$ is the cylindrical diffusion length, with 2.405 representing the first root of $J_0(x)$. Determine the space-charge electric field under perfect ambipolar diffusion conditions.

Resolution: The space charge electric field is given by equation (6.15)

$$\mathbf{E_s} = -\frac{k_B T_e}{e}\frac{\nabla n_e}{n_e} ,$$

so that using the selected identity of Bessel functions

$$J_n'(x) = \frac{1}{2}\,[J_{n-1}(x) - J_{n+1}(x)] ,$$

with

$$J_{-n}(x) = (-1)^n\, J_n(x) ,$$

we obtain

$$J_0'(x) = \frac{1}{2}\,[-J_1(x) - J_1(x)] = -\,J_1(x) .$$

We may write hence

$$\nabla n_e = n_{e0}\,\frac{d}{dr}\left(J_0\left(\frac{r}{\Lambda}\right)\right)\mathbf{e_r} = n_{e0}\,\left(-J_1\left(\frac{r}{\Lambda}\right)\right)\frac{1}{\Lambda}\,\mathbf{e_r}$$

and therefore

$$\frac{\nabla n_e}{n_e} = -\frac{J_1(r/\Lambda)}{J_0(r/\Lambda)}\frac{2.405}{R}\,\mathbf{e_r}\ .$$

The space-charge electric field is then given by

$$\mathbf{E_s}(r) = \frac{k_B T_e}{e}\frac{2.405}{R}\frac{J_1(2.405\,r/R)}{J_0(2.405\,r/R)}\,\mathbf{e_r}\ .$$

At $r = R, J_0(2.405) = 0$ corresponds to the first root of the zero order Bessel function, while for the first order function we have $J_1(2.405) = 0.5191$. Then, the space-charge field sharply increases near the wall.

Exercise 6.2. Estimate the deviation from charge neutrality implied by the space charge field obtained in Exercise 6.1

Resolution: The presence of a space charge field implies that the charge neutrality is only approximately valid. Using the local Gauss law

$$(\nabla.\,\mathbf{E_s}) = \frac{\rho}{\epsilon_0}\ ,$$

where $\rho = e\ (n_i - n_e)$ is the space-charge density, we may write using cylindrical coordinates

$$n_i - n_e = \frac{\epsilon_0}{e}\frac{1}{r}\frac{d}{dr}(r\,E_s)\ .$$

Substituting the space-charge field obtained in Exercise 6.1 and making the replacement $x = 2.405\ r/R$, we obtain

$$n_i - n_e = \frac{\epsilon_0 k_B T_e}{e^2}\left(\frac{2.405}{R}\right)^2\frac{1}{x}\frac{d}{dx}\left(x\,\frac{J_1(x)}{J_0(x)}\right)\ .$$

The derivative of the term inside brackets is done using the recurrence formulae for Bessel functions

$$(x\,J_1(x))' = x\,J_0(x)$$
$$J_0'(x) = -\,J_1(x)\ ,$$

allowing to obtain at the end

$$n_i - n_e = \frac{\epsilon_0 k_B T_e}{e^2}\left(\frac{2.405}{R}\right)^2\left(1 + \left(\frac{J_1(x)}{J_0(x)}\right)^2\right)\ .$$

Dividing now by $n_e(x) = n_{e0}\, J_0(x)$, we obtain the relative deviation from neutrality

$$\frac{n_i - n_e}{n_e} = 5.784 \left(\frac{\lambda_D}{R}\right)^2 \frac{1 + (J_1(x)/J_0(x))^2}{J_0(x)} ,$$

with

$$\lambda_D = \sqrt{\frac{\epsilon_0 k_B T_e}{e^2 n_{e0}}}$$

denoting the Debye length at the centre.

Exercise 6.3. Find the electron density at the centre of a cylindrical discharge of radius R, current intensity I, and electron drift velocity $\mathbf{v_{ed}}$, assuming a Bessel radial density profile.

Resolution: The current intensity is expressed as

$$I = \int_0^R J_e(r)\, 2\pi r\, dr ,$$

where $J_e(r) = e\, n_e(r)\, |\mathbf{v_{de}}|$ is the electron current density, $n_e(r) = n_e(0)\, J_0(r/\Lambda)$ the electron density profile, $n_e(0)$ the density at the centre, $\Lambda = R/2.405$ the diffusion length, and $|\mathbf{v_{ed}}|$ is assumed radially constant. We obtain hence the following expression

$$I = 2\pi\, e\, n_e(0)\, |\mathbf{v_{ed}}| \left(\frac{R}{2.405}\right)^2 \int_0^{2.405} J_0(x)\, x\, dx ,$$

in which the integral is

$$\int_0^{2.405} J_0(x)\, x\, dx = 2.405\, J_1(2.405) = 2.405 \times 0.5191 ,$$

allowing to obtain at the end

$$n_e(0) = \frac{0.737\, I}{e\, |\mathbf{v_{ed}}|\, R^2} .$$

Exercise 6.4. Determine the ionization rate coefficient necessary to maintain a discharge over the cross section of a cylindrical discharge of radius R, for the case of a gas with reduced mobility of ions $n_o \mu_i$, under ambipolar diffusion conditions, and electrons at temperature T_e much larger than the temperature of ions.

Resolution: The steady-state rate balance equation for the electron number density (6.66) is as follows

$$-\,D_a\,\nabla^2 n_e \;=\; n_e\,n_o\,C_{ion}\;,$$

being C_{ion} the ionization rate coefficient ($= <\nu_{ion}>/n_o$), D_a the ambipolar diffusion coefficient, and n_o the gas number density. Owing to the cylindrical geometry of the problem, we may write

$$-\,D_a\,\frac{1}{r}\,\frac{d}{dr}\left(r\,\frac{dn_e}{dr}\right) \;=\; n_e\,n_o\,C_{ion}\;,$$

where r is the distance to the axis of the discharge cylinder. The solution of this equation for the boundary condition $n_e(R) = 0$ is

$$n_e(r) \;=\; n_{e0}\,J_0\left(\frac{2.405\,r}{R}\right)\,,$$

where n_{e0} is the electron number density at the axis, and J_0 the Bessel function of first kind and order zero. Since $\nabla^2 n_e \;=\; -\;n_e/\Lambda^2$, with $\Lambda \;=\; R/2.405$ denoting the diffusion length, we obtain the following relationship between the plasma parameters and the size of the system

$$5.784\,\frac{n_o D_a}{(n_o R)^2} \;=\; C_{ion}\,.$$

On the other hand, when $T_e \gg T_i$ the ambipolar diffusion coefficient (6.13) takes the form

$$D_a \;\simeq\; \mu_i\,\frac{k_B T_e}{e}\,,$$

so that the equation for the plasma maintenance is

$$5.784\,\frac{n_o \mu_i}{(n_o R)^2}\,\frac{k_B T_e}{e} \;=\; C_{ion}\,.$$

Since the ionization rate coefficient is determined by the reduced electric field E/n_o, this equation can be used to find the longitudinal electric field in the positive column. The ionization rate coefficient should be then equal to the average frequency, at the gas number density, for electron escape to the tube walls. Since T_e depends on E/n_o as well, we obtain a characteristic for the plasma maintenance with the form $E/n_o(n_o R)$.

Exercise 6.5. Estimate the mean lifetime of an electron in a cylindrical discharge tube under ambipolar diffusion conditions.

Resolution: The distribution of the electron density over the tube cross section is

$$n_e(r) \;=\; n_{e0}\, J_0\left(\frac{2.405\, r}{R}\right) ,$$

so that the flux of electrons at the discharge tube wall is

$$\begin{aligned}\Gamma_e &= -\,D_a \left(\frac{dn_e}{dt}\right)_R \;=\; -\,D_a\, n_{e0}\,\frac{2.405}{R}\left(\frac{d}{dx}J_0(x)\right)_{2.405}\\ &= D_a\, n_{e0}\,\frac{2.405}{R}\,J_1(2.405)\,.\end{aligned}$$

On the other hand, the total number of electrons in the tube element with the length L is

$$\begin{aligned}N_e &= L\int_0^R n_e(r)\, 2\pi\, r\, dr \;=\; 2\pi L\, n_{e0}\left(\frac{R}{2.405}\right)^2\int_0^{2.405} J_0(x)\, x\, dx\\ &= 2\pi L\, n_{e0}\,\frac{R^2}{2.405}\,J_1(2.405)\,.\end{aligned}$$

The number of electrons per time unit from this tube element is $2\pi RL\,\Gamma_e$, so that the mean time spent by an electron in the discharge tube is

$$\tau \;=\; \frac{N_e}{2\pi RL\,\Gamma_e} \;=\; \frac{R^2}{5.784\, D_a}\,.$$

This lifetime is obviously the inverse of the frequency for escape of electrons by diffusion found in Exercise 6.4. In fact, from

$$\frac{\partial n_e}{\partial t} \;-\; D_a\,\nabla^2 n_e \;=\; 0\,,$$

we may write

$$\frac{\partial n_e}{\partial t} \;+\; \nu_{dif}\, n_e \;=\; 0\,,$$

from which we may conclude

$$\nu_{dif} \;=\; 5.784\,\frac{D_a}{R^2} \;=\; \tau^{-1}\,.$$

Exercise 6.6. Determine the electron distribution over the cross section of a cylindrical tube of radius R, for the case of electrons produced in a narrow region $r_0 \ll R$ around the axis of the tube.

Resolution: This situation can occur due to the variation of the gas temperature with the distance from the tube axis (Smirnov 1981). If the discharge current is not too small, the gas in the discharge tube is heated up owing to electron collisions. The heat flux is then directed towards the tube walls and the gas temperature is larger at the axis. Since the gas pressure is radially uniform the number density of the gas molecules at the axis is lower than near the walls. Then the collisions between the electrons and the gas molecules occur less frequently at the axis and the average energy of electrons at the axis is higher. Since ionization is determined by collisions of fast electrons of the energy tail of the electron distribution function, even a small difference in the gas temperature can result in a appreciable difference between the ionization frequencies at the axis and the tube walls.

Solving the stationary rate balance equation for the electron density in cylindrical geometry

$$-D_a \frac{1}{r}\frac{d}{dr}\left(r\frac{dn_e}{dr}\right) = n_e\, n_o\, C_{ion} ,$$

in the region $r \leq r_0$, assuming $n_e(r) \simeq n_{e0}$ in the right-hand side member due to the smallness of r_0, we obtain

$$n_e(r) = n_{e0}\left(1 - \frac{n_o\, C_{ion}}{4D_a}\, r^2\right) ,$$

whereas in the region $r_0 \leq r \leq R$, where $C_{ion} = 0$, we obtain considering the boundary condition $n_e(R) = 0$

$$n_e(r) = n_e(r_0)\, \frac{\ln(R/r)}{\ln(R/r_0)} .$$

Matching the above solutions at $r = r_0$, where the electron density and its derivate (that is, the flux) are continuous, we obtain

$$n_e(r_0) = n_{e0}\left(1 - \frac{n_o\, C_{ion}}{4D_a}\, r_0^2\right)$$

$$n_{e0}\, \frac{n_o\, C_{ion}}{2D_a}\, r_0 = \frac{n_e(r_0)}{\ln(R/r_0)}\, \frac{1}{r_0}$$

and relating the two expressions to each other, we obtain the following condition for the plasma maintenance

$$\frac{2D_a}{n_o\, C_{ion}\, r_0^2} = \frac{1}{2} + \ln\left(\frac{R}{r_0}\right) .$$

This condition is equivalent to that of Exercise 6.4. It relates the equality between the rates of electron ionization and electron escape from the volume where the electrons are generated.

Exercise 6.7. A discharge is maintained between two infinite plates. Determine the distribution of electrons in the discharge gap and the condition for the plasma maintenance, in the case of the electrons are not totally recombined at the wall (that is, the probability ξ for electron recombination at the wall differs from unity). Assume further that the electron ionization rate coefficient does not vary over the discharge cross section.

Resolution: This problem with a slab geometry satisfies a rate balance equation equivalent to that of last exercise

$$- D_a \frac{d^2 n_e}{dx^2} = n_e \, n_o \, C_{ion},$$

with x denoting the distance from the central plane of symmetry parallel to the walls of the discharge gap. Since $n_e(x) = n_e(-x)$ due to the symmetry of the problem, the above equation has the solution

$$n_e(x) = n_{e0} \cos\left(\sqrt{\frac{n_o \, C_{ion}}{D_a}}\, x\right),$$

where n_{e0} is the electron density at the symmetry plane. In the limiting case, where the probability of recombination at the walls is $\xi = 1$, we have $n_e(x) = 0$ at $x = \pm\, L/2$, being L the distance between the walls, and we obtain

$$\sqrt{\frac{n_o \, C_{ion}}{D_a}} \frac{L}{2} = \frac{\pi}{2}.$$

On the contrary, when $\xi < 1$, the flux of charged particles recombining at the walls is (see Appendix A.1.1 and Sect. 6.1.1)

$$\xi \, \frac{n_e(L/2) <v_i>}{4} = \xi \, \frac{n_{e0} <v_i>}{4} \cos\left(\sqrt{\frac{n_o \, C_{ion}}{D_a}} \frac{L}{2}\right),$$

being $<v_i>$ the average velocity of ions near the walls in a layer with thickness of the order of the mean free path of ions. Since ionization in the vicinity of the walls is insignificant, the flux of electrons recombining at the walls is

$$(\Gamma_e)_{L/2} = - D_a \left(\frac{dn_e}{dx}\right)_{L/2} = n_{e0} \sqrt{n_o \, C_{ion} \, D_a} \, \sin\left(\sqrt{\frac{n_o \, C_{ion}}{D_a}} \frac{L}{2}\right).$$

Equalling both expressions, we obtain

$$\cot\left(\sqrt{\frac{n_o\, C_{ion}}{D_a}}\,\frac{L}{2}\right) = \frac{4}{\xi\, <v_i>}\,\sqrt{n_o\, C_{ion}\, D_a}\,.$$

At the wall the cotangent function is $\ll 1$ being its argument close to $\pi/2$

$$\cot\eta \simeq \sin\left(\frac{\pi}{2} - \eta\right) \simeq \frac{\pi}{2} - \eta\,,$$

so that we obtain

$$\frac{\pi}{2} - \sqrt{\frac{n_o\, C_{ion}}{D_a}}\,\frac{L}{2} = \frac{4}{\xi\, <v_i>}\,\sqrt{n_o\, C_{ion}\, D_a}$$

and therefore

$$\sqrt{\frac{n_o\, C_{ion}}{D_a}}\,\frac{L}{2} = \frac{\pi/2}{1 + 8D_a/(\xi\, <v_i>\, L)}\,.$$

The second term in the denominator of the right-hand side member is a corrective term relatively to the case where the electron number density is zero at the wall. Let us estimate now this term. Since $D_a \sim <v_i>^2/\nu_i \sim <v_i>\, \lambda_i$, being λ_i the mean free path of ions, the corrective term is of the order of $\lambda_i/(\xi L)$, so that when $\lambda_i \ll \xi L$, we obtain the same expression for the discharge maintenance law as the electron number density is null at the wall.

Exercise 6.8. Determine the expressions of electron density decay in a plasma afterglow controlled by: *(i)* electron-ion recombination and electron diffusion, with a constant diffusion coefficient; *(ii)* electron-ion recombination only.

Resolution: *(i)* When the loss terms for electron density in a plasma afterglow are electron-ion recombination, with $n_e(t) = n_i(t)$ at each instant, and electron diffusion, the electron density obeys to the relation (6.85), written here as

$$\frac{dn_e}{dt} + \frac{D_{se}}{\Lambda^2}\, n_e = -\,\alpha\, n_e^2\,,$$

where D_{se} is the effective electron diffusion coefficient, Λ is the diffusion length of fundamental-mode, and α the electron-ion recombination coefficient. Neglecting the dependence of D_{se} on n_e given by equation (6.86), we may write

$$\frac{dn_e}{dt} = -\,\beta\, n_e - \alpha\, n_e^2\,,$$

being $\beta = D_{se}/\Lambda^2 = \text{const}$.

Making the substitution $y = n_e^{-1}$, we obtain the linear differential equation

$$\frac{dy}{dt} = \beta\, y + \alpha\,,$$

and multiplying both members by $e^{-\beta t}$, we can still write

$$\frac{d}{dt}\left(e^{-\beta t} y\right) = \alpha\, e^{-\beta t}\,.$$

This equation can be integrated giving (Borysow and Phelps 1994)

$$\frac{n_e(0)}{n_e(t)} = e^{\beta t} - \frac{\alpha}{\beta}\, n_e(0)\left(1 - e^{\beta t}\right),$$

or still under the form

$$\frac{n_e(t)}{1 + (\alpha/\beta)\, n_e(t)} = \frac{n_e(0)}{1 + (\alpha/\beta)\, n_e(0)}\; e^{-\beta t}\,.$$

(ii) On the other hand, when only electron-recombination exists, we easily obtain (Gritsinin et al. 1996)

$$\frac{n_e(0)}{n_e(t)} = 1 + \alpha\, n_e(0)\, t\,.$$

References

W.P. Allis, Motions of ions and electrons, in *Handbuch der Physik*, vol. 21, ed. by S. Flügge (Springer, Berlin, 1956), pp. 383–444

W.P. Allis, D.J. Rose, The transition from free to ambipolar diffusion. Phys. Rev. **93**(1), 84–93 (1954)

J. Amorim, Levis-Rayleigh and pink afterglow. IEEE Trans. Plasma Sci. **23**(2), 368–369 (2005)

A.V. Berdyshev, I.V. Kochetov, A.P. Napartovich, Ionization mechanism in a quasisteady glow discharge in pure nitrogen. Sov. J. Plasma Phys. **14**, 438–440 (1988)

I.B. Bernstein, T. Holstein, Electron energy distributions in stationary discharges. Phys. Rev. **94**(6), 1475–1482 (1954)

L.S. Bogdan, S.M. Levitskii, E.V. Martysh, Nonmonotonic variation of the electron density in the decaying plasma of a pulsed discharge in nitrogen. Tech. Phys. **38**(6), 532–534 (1993)

J. Borysow, A.V. Phelps, Electric field strengths, ion energy distributions, and ion density decay for low-pressure, moderate-current nitrogen discharges. Phys. Rev. E **50**(2), 1399–1415 (1994)

H. Brunet, J. Rocca-Serra, Model for a glow discharge in flowing nitrogen. J. Appl. Phys. **57**(5), 1574–1581 (1985)

H. Brunet, P. Vincent, J. Rocca-Serra, Ionization mechanism in a nitrogen glow discharge. J. Appl. Phys. **54**(9), 4951–4957 (1983)

B.E. Cherrington, *Gaseous Electronics and Gas Lasers* (Pergamon Press, Oxford, 1980)

E. Dervisevic, E. Bloyet, C. Laporte, P. Leprince, J. Marec, M. Pouey, S. Saada, Energy balance in microwave discharges, in *Proceedings 16th International Conference on Phenomena in Ionized Gases (ICPIG)*, Düsseldorf, vol. 4, 1983, pp. 468–469
S.K. Dhali, L.H. Low, Transient analysis of bulk nitrogen glow discharge. J. Appl. Phys. **64**(6), 2917–2926 (1988)
C.M. Ferreira, J. Loureiro, Characteristics of high-frequency and direct-current argon discharges at low pressures: a comparative analysis. J. Phys. D: Appl. Phys. **17**, 1175–1188 (1984)
C.M. Ferreira, J. Loureiro, Electron excitation rates and transport parameters in high-frequency N_2 discharges. J. Phys. D: Appl. Phys. **22**, 76–82 (1989)
C.M. Ferreira, J. Loureiro, Electron kinetics in atomic and molecular plasmas. Plasma Sources Sci. Technol. **9**, 528–540 (2000)
C.M. Ferreira, J. Loureiro, A. Ricard, Populations in the metastable and the resonance levels of argon and stepwise ionization effects in a low-pressure argon positive column. J. Appl. Phys. **57**(1), 82–90 (1985)
C.M. Ferreira, J. Loureiro, A.B. Sá, The modelling of high-frequency discharges at low and intermediate pressure, in *Proceedings 18th International Conference on Phenomena in Ionized Gases (ICPIG)*, Swansea, vol. Invited papers, 1987, pp. 220–230
C.M. Ferreira, G. Gousset, M. Touzeau, Quasi-neutral theory of positive columns in electronegative discharges. J. Phys. D: Appl. Phys. **21**, 1403–1413 (1988)
L. Fox, D.F. Mayers, *Computing Methods for Scientists and Engineers* (Clarendon Press, Oxford, 1968)
V.E. Golant, A.P. Zhilinsky, I.E. Sakharov, *Fundamentals of Plasma Physics* (Wiley, New York, 1980)
Yu. Golubovskii, D. Kalanov, S. Gorchakov, D. Uhrlandt, Nonlocal electron kinetics and spectral line emission in the positive column of an argon glow discharge. Plasma Sources Sci. Technol. **24**, 025028 (2015)
S.I. Gritsinin, I.A. Kossyi, V.P. Silakov, N.M. Tarasova, The decay of the plasma produced by a freely localized microwave discharge. J. Phys. D: Appl. Phys. **29**, 1032–1034 (1996)
V. Guerra, J. Loureiro, Electron and heavy particle kinetics in a low-pressure nitrogen glow discharge. Plasma Sources Sci. Technol. **6**, 361–372 (1997)
V. Guerra, P.A. Sá, J. Loureiro, Relaxation of the electron energy distribution function in the afterglow of a N_2 microwave discharge including space-charge field effects. Phys. Rev. E **63**, 046404 (2001)
V. Guerra, P.A. Sá, J. Loureiro, Kinetic modeling of low-pressure nitrogen discharges and post-discharges. Eur. Phys. J. Appl. Phys. **28**, 125–152 (2004)
V.I. Kolobov, Advances in electron kinetics and theory of gas discharges. Phys. Plasmas **20**, 101610 (2013)
U. Kortshagen, A non-local kinetic model applied to microwave produced plasmas in cylindrical geometry. J. Phys. D: Appl. Phys. **26**, 1691–1699 (1993)
U. Kortshagen, Experimental evidence on the nonlocality of the electron distribution function. Phys. Rev. E **49**(5), 4369–4380 (1994)
M.A. Lieberman, A.J. Lichtenberg, *Principles of Plasma Discharges and Materials Processing* (John Wiley, New York, 1994)
J. Loureiro, C.M. Ferreira, Coupled electron energy and vibrational distribution functions in stationary N_2 discharges. J. Phys. D: Appl. Phys. **19**, 17–35 (1986)
V. Lj. Marković, Z. Lj. Petrović, M.M. Pejović, Modelling of charged particle decay in nitrogen afterglow. Plasma Sources Sci. Technol. **6**, 240–246 (1997)
R.L. Monroe, Ambipolar drift, deformation and diffusion of a plasma in a magnetic field. Can. J. Phys. **51**(5), 564–573 (1973)
D.J. Rose, S.C. Brown, High-frequency gas discharge plasma in hydrogen. Phys. Rev. **98**(2), 310–316 (1955)
N. Sadeghi, C. Foissac, P. Supiot, Kinetics of $N_2(A\ ^3\Sigma_u^+)$ molecules and ionization mechanisms in the afterglow of a flowing N_2 microwave discharge. J. Phys. D: Appl. Phys. **34**, 1779–1788 (2001)

A. Simon, Ambipolar diffusion in a magnetic field. Phys. Rev. **98**(2), 317–318 (1955)
B.M. Smirnov, *Physics of Weakly Ionized Gases (Problems and Solutions)* (Mir Publishers, Moscow, 1981) (from 1978 Russian edition)
L.D. Tsendin, Energy distribution of electrons in a weakly ionized current-carrying plasma with a transverse inhomogeneity. Sov. Phys. JETP **39**(5), 805–810 (1974)
L.D. Tsendin, Electron distribution function of weakly ionized plasmas in nonuniform electric fields. I. Weak fields (energy balance determined by nearly elastic collisions). Sov. J. Plasma Phys. **8**(1), 96–100 (1982a)
L.D. Tsendin, Electron distribution function in a weakly ionized plasma in an inhomogeneous electric field. II. Strong fields (energy balance determined by inelastic collisions). Sov. J. Plasma Phys. **8**(2), 228–233 (1982b)
L.D. Tsendin, Analytical approaches to glow discharge problems. Plasma Sources Sci. Technol. **20**, 055011 (2011)
F. Vidal, T.W. Johnston, J. Margot, M. Chaker, O. Pauna, Diffusion modeling of an HF argon plasma discharge in a magnetic field. IEEE Trans. Plasma Sci. **27**(3), 727–745 (1999)
R. Winkler, G. Petrov, F. Sigeneger, D. Uhrlandt, Strict calculation of electron energy distribution functions in inhomogeneous plasmas. Plasma Sources Sci. Technol. **6**, 118–132 (1997)
A.P. Zhilinskii, L.D. Tsendin, Collisional diffusion of a partially-ionized plasma in a magnetic field. Sov. Phys. Uspekhi **23**(7), 331–355 (1980)

Part II
Plasma Spectroscopy

Chapter 7
Classification of Equilibria in Plasmas

Plasmas in complete thermodynamic equilibrium only exist in very special conditions like, for example, in the interior of stars. Plasmas are rarely found in equilibrium in laboratory because very often the walls enclosing such plasmas are transparent to the radiation fields over a wide range of frequencies. The radiation escapes and the fulfilling of the Planck radiation law is not achieved. At this point it is not possible to describe the radiation field by a unique temperature. Another source of disequilibrium is that walls have lower temperatures than the inner regions of plasmas and only one thermodynamic temperature does not exist. Other factor that contributes to the instability is the diffusion that causes the out flux of particles, changing the local density of the excited states influencing the production of charged particles for example.

Although most of plasmas are not found in thermal equilibrium (TE) by the reasons presented above, the conditions to establish the TE may exist in a small volume element with the exception of the Planck's radiation equilibrium. In this last case the plasmas are said in local thermodynamic equilibrium (LTE) described by the laws presented in Table 7.2. The approach of TE is very powerful to describe the local plasma even without Planck's radiation equilibrium and the approximation for local equilibrium, i.e. Maxwell, Boltzmann and Saha, may be applied. The radiation plays a key role in the establishment of LTE in low electron density plasmas. Deviations from LTE may be observed when the radiation field is weak.

In this chapter we will present and discuss basic concepts of most relevant collisional-radiative models usually found in low-temperature plasmas. It will be presented the simplest equilibrium, i.e. the corona model, followed by a description of the excitation saturation balance (ESB) and the partial local Saha equilibrium (pLSE). The deviation from thermodynamic equilibrium leads to different plasmas with different classifications and distinct properties. LTE has been treated in many books and research articles and for further details the reader should referred to related literature, for example, van der Mullen's work (van der Mullen 1990).

J.M.A.H. Loureiro, J. de Amorim Filho, *Kinetics and Spectroscopy of Low Temperature Plasmas*, Graduate Texts in Physics,
DOI 10.1007/978-3-319-09253-9_7

7.1 Introduction

Atoms and ions immersed in a plasma bring a lot of information about the medium. This information can be assessed by the atomic state distribution function (ASDF) which describes how atoms and ions are distributed over their internal energy states. The ASDF gives a good description of how the energy is deposited in the plasma as a result of the microscopic processes that regulate the ionization, excitation, particle transport, and so on.

Passive optical diagnostics like optical emission and absorption spectroscopy may be employed to determine the ASDF and from this distribution important parameters of the plasma, such as electronic density and temperature, may be deduced. From the ASDF the equilibrium state of the plasma may be characterized.

Collisional radiative models may help, together with experimental results, in the establishment of the relationship between the ASDF form and the elementary processes underlying the kinetics of the plasma. Although numerical collisional radiative (CR) models exist and are well employed today, in the present chapter emphasis will be given to analytical approximations in order to present to the students a general view of the influence of deviations from thermodynamic equilibrium. We are interested in the classification of the atomic discharge plasmas with respect to the excitation criteria.

In this chapter the key role played by the electrons in the excitation kinetics is presented with emphasis in the dominant electron excitation kinetics (EEK) plasmas with a high degree of ionization, typically $>10^{-4}$. However, departures from equilibrium leading to the so called local partial equilibrium may be useful to describe plasmas with a lower degree of ionization.

Electrons have reduced mass and high velocity and as charged particles they receive energy from the electric field and through collisions with heavy particles deposit part of their energy in the medium. This process influences the form to the ASDF as the result of the interaction and the competition between bound and free electrons.

Proper balances are defined as the ones where the forward and the corresponding backward processes, in the framework of microreversibility, are in equilibrium. This chapter begins with the treatment of thermodynamic equilibrium (TE) which is presented as a collection of proper balances. Four types of equilibrium are fundamental to the establishment of the ASDF. The first one is called the Maxwell's balance which is the result of kinetic energy transfer between particles in collisions. The second one is due to excitation and deexcitation of the atomic levels and is known as Boltzmann's balance. The Saha's balance is the type of equilibrium that regulates the ionization and the recombination of charges in the plasma, and the last one is the Planck's balance for interaction between matter and radiation. Generally, the Planck's balance is the first to be lost due to the escape of radiation from the medium.

Improper balances occur when forward and backward processes are not in equilibrium. In this case there is no inverse process for a given reaction or if it exists it is less effective. A good example is the corona equilibrium where a given excited level is produced by electron collisions with ground state atoms and the radiation is lost from the medium leading to a disequilibrium.

In plasmas, the electron excitation kinetics modulates the shape of the atomic distribution function characterizing each improper balance. They are associated with the parameters governing the macroscopic behavior of the plasmas such as in ionizing plasmas where the improper balance creates a flow of ionization along the state levels of the system. The transport of the charged particles out of the plasma is enhanced by the production and transport of charged particles. Otherwise, when the recombination is more significant than ionization, the plasma is called recombining plasma and the improper balance is governed by the overpopulation of certain state levels due to recombination reactions.

In this chapter the deviation from the thermodynamic equilibrium is discussed in Sect. 7.2 while proper and improper balances are presented in Sect. 7.3. The features and domains of micro reversibility are shown in Sect. 7.4. Maxwell, Boltzmann, Saha and Planck equilibria are presented and discussed in Sects. 7.5, 7.6, 7.7 and 7.8 respectively.

7.2 Deviation from Thermodynamic Equilibrium

The plasma in macroscopic state may be fully characterized by temperature, number density and chemical composition. In the thermodynamic equilibrium (TE) the knowledge of the temperature is an important parameter to access many information of the system with the help of statistical mechanics.

At macroscopic level, the laws of statistical mechanics govern the behavior of particles while at microscopy level the nature of collisions and the energies involved change the scenario picture. In this case the velocity distribution of particles is given by a Maxwell distribution while the distribution of population among the various excited states is described by a Boltzmann distribution. The densities of ionic states are described by a Saha distribution and Planck's radiation law governs distribution of photons according to their energy. In TE the temperature is the key parameter that determines all these distributions.

It can be stated that in TE condition each microscopic process has its inverse in a sense that the direct and reverse processes are in equilibrium. This is the definition of Detailed Balancing (DB).

The four types of balances related to the establishment of the Maxwell, Saha, Boltzmann, and Planck distributions are presented schematically in Table 7.1. These are proper balances because, as stated before, the forward processes are compensated by backward ones establishing a microscopic equilibrium in TE condition where the distribution functions are known.

The departure from the TE leads to a modification of ASDF changing the equilibrium in Saha and Boltzmann balances. Here we are supposing that Planck's balance is responsible for this equilibrium departure. In fact, in most discharge

Table 7.1 Illustration of the proper balances

Balance	Reaction
Maxwell	$X + Y \overset{M}{\leftrightarrow} X + Y$
	$E_X + E_Y = (E_X + \Delta E) + (E_Y - \Delta E)$
Boltzmann	$X + A_l + E_{lu} \overset{B}{\leftrightarrow} X + A_u$
	deexcitation $\leftarrow - \rightarrow$ *excitation*
Saha	$X + A_p + \lvert E_p \rvert \overset{S}{\leftrightarrow} X + A_l^+ + e$
	recombination $\leftarrow - \rightarrow$ *ionization*
Planck	$A_u \overset{P}{\leftrightarrow} A_l + h\nu$
	absorption $\leftarrow - \rightarrow$ *spontaneous emission*
	$A_u + h\nu \rightarrow A_l + 2h\nu$
	$\rightarrow$ *stimullated emission*

Maxwell balance (M), Boltzmann balance (B), Saha balance (S) and Planck balance (P). A_u is an atom with energy u, E is the kinetic energy of the particles and E_{lu} and $|E_p|$ are the increase of internal energy and ionization potential respectively

plasmas this balance is the first to be lost because of radiation escaping from the medium. It was shown in reference (van der Mullen 1990) that the distribution laws valid in TE can be derived using the principle of DB. Non-thermodynamic equilibrium systems, where some proper balances still take place, can be described by the DB principle.

Once the production and loss of radiation and particle number take place and if the inverse processes no more occur, improper balances begin to be important, leading to the departure of TE. The replacement of proper balances by improper ones modifies the shape of the ASDF. It can be seen in Table 7.2 that the escape of radiation from the plasma affects the Planck's equilibrium and may influence other balances as will be shown latter in this chapter. However if the leak of radiation is small, when compared with the energy involved in collisions of heavy particles, the Maxwell, Boltzmann and Saha balances remain in equilibrium and electrons, ions and neutrals are thermalized in the same temperature, i.e. $T_e = T_H$, where T_e is the electron temperature and T_H the heavy particles temperature.

The escape of radiation from the plasma may be accompanied of species gradients and of their variation with time. In this context, it can be found that locally the system remains in equilibrium. In this Local Thermal Equilibrium (LTE) slow variations in plasma conditions may be permitted but the balances should be reestablished instantaneously. In this last case, the characteristic times of reactions in balance need to be short if compared with plasma time constant. In LTE the radiation temperature T_ν and matter temperature T_m are different, *i.e.* $T_\nu \neq T_m = T_e = T_H$, and the kinetics of both, radiation and matter, are decoupled. The radiance temperature T_ν is now different from that of the heavy particles. The *radiance temperature* is the temperature of a black-body radiator that has the same spectral radiance, according to Planck's law.

Table 7.2 Scheme of the departures from equilibrium

Equilibrium condition	Balances	Temperature	Remarks
TE	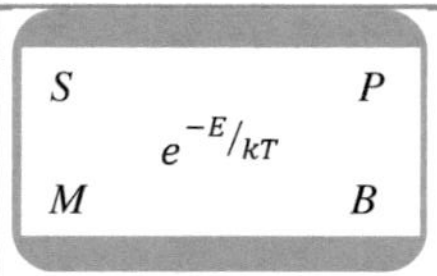	$T_\nu = T_H = T^{exc} = T_e$	*Temporal and spatial distributions are steady-state* $\left(\partial/\partial t = 0 = \nabla\right)$
LTE	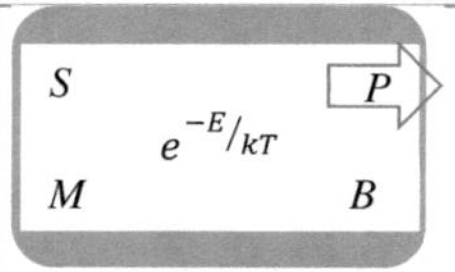	$T_\nu \neq T_H = T^{exc} = T_e$	*Temporal and spatial distributions are decoupled* $\left(\partial/\partial t \neq 0 \neq \nabla\right)$ *matter ↔ radiation*
2-T plasma	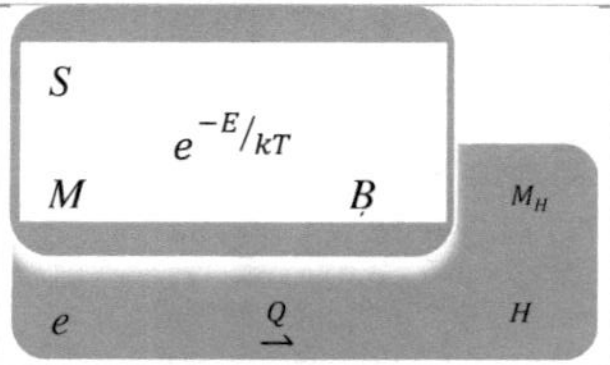	$T^{exc} = T_e$ and $T_e \neq T_H$	*Different forces acting in different particles heavy particles ↔ electrons*
QSS	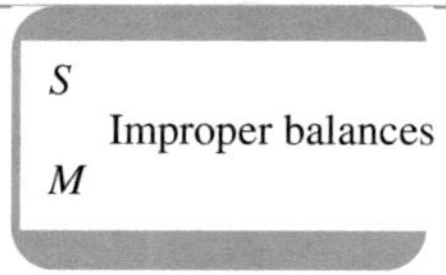	$T^{exc} = T_e$ *for top levels*	*Escape of particles top levels ↔ bottom leves*
pLS_eE (top levels)		$T^{exc} \neq T_e$ *for bottom levels*	
QSSS (bottom levels)			

The departure from the equilibrium becomes more severe as we move downward. Thermodynamic equilibrium (TE), local thermodynamic equilibrium (LTE), two-temperature plasma (2-T plasma), quasi-steady state (QSS), quasi-steady state solution (QSSS) and electron-ruled partial local Saha equilibrium (pLS_eE)

The density of spectral energy $u\ (\nu, T_{rad})$ is given by the Planck's law when the radiation field is in equilibrium, i.e. the absorption and emission are in equilibrium for a given frequency ν. In this case the system is in thermodynamic equilibrium.

$$u\ (\nu, T_{rad}) = \frac{8\pi h\nu^3}{c^3} \frac{1}{e^{h\nu \big/ k_B T_{rad}}} \tag{7.1}$$

Another important issue in LTE condition is the spatial and temporal decoupling $\left({}^{\partial}\big/_{\partial t} \neq 0 \neq \nabla\right)$ of radiation and matter. The electrons receive energy from the electromagnetic field and transfer it through collisions to heavy particles. Since the rate of energy transfer is not very effective due to the small mass ratio m_e/M, the tendency is that each group, electrons and heavy particles, achieves in equilibrium separately much more faster than the whole system.

To establish LTE a high density of particles is needed in order that collisional transitions dominate over radiative transitions between all states, otherwise the leak of thermal radiation will cause deviations from equilibrium distribution. For radiative decay rates to induce less than 10 % of departure from LTE, the collisional rates must be at least about ten times the radiative rates. A rule-of-thumb for collisional transitions to be dominant may be written as (McWhirter 1965):

$$n_e \gg 1.6 \times 10^{12} \left(T_e(eV)^{\frac{1}{2}} (\Delta E\ (p, q)\ (eV))\right)^3\ \text{cm}^{-3} \tag{7.2}$$

T_e is the electron temperature and ΔE is the excitation potential of the level p from level q.

Plasmas that do not comply with LTE laws are said to be in non-equilibrium and are also named non-LTE plasmas. Their properties can only be described by evaluation of individual reaction processes between particles. In general, the Boltzmann equation is solved to access the electron energy distribution function (EEDF) coupled to rate balance equations for heavy particles to describe the kinetics of the system.

Electron temperature is the temperature that describes, through Maxwell's law, the kinetic energy distribution of the free electrons. In this situation where electrons and heavy particle kinetics are decoupled, we have a two-temperature (2-T) plasma, and the ASDF is the result of the competition between Boltzmann and Saha equilibrium balances.

If the electron density is high, i.e. the ionization degree is high, the electrons determine the shape of the ASDF and the important parameter is the electron temperature, which imposes the internal energy states of heavy particles.

If the gradient of particles in the plasma becomes significant, the transport of charged particles along the distances may disturb a lot the local equilibrium of charges perturbing the Saha balance. Electrons and ions leaving the volume cannot contribute to the establishment of local balance between ionization and recombination. However, this picture is not the same for all levels of the ASDF.

Some of them may be in Saha equilibrium while others are not. Actually the upper levels close to the continuum present rates of ionization and recombination that can really justify a Saha balance.

7.3 Proper and Improper Balances

The thermodynamic equilibrium is not attained in real plasmas produced in laboratory due to loss of radiation and/or particles from the medium. In each plasma system the problem consists in evaluate the balances regulating the equilibrium or the departure from it. In order to analyze the deviations from the thermodynamic equilibrium we will present the principle of DB in the light of microscopic reversibility (MR). This permits to characterize the nonequilibrium state due to the rupture of balances listed above. If one or more of these balances are improper, a nonequilibruim state is established. At microscopic level it characterizes the particle changes from one quantum state to another. This elementary proper balance is in equilibrium and the product of elementary concentrations is equal to that of reactants, i.e. respecting the mass conservation. This is the basis for any statistical law of TE and it is helpful to explain when only a part of the system is in equilibrium. This partial equilibrium may be found in Maxwell, Boltzmann, Saha and Planck balances and is denoted in the literature by pLM_eE, pLB_eE, pLS_eE and pLPE respectively. The lower index *e* is an indication that the balance is ruled by the influence of electrons.

In Fig. 7.1 the interdependence between balances is shown. When the medium losses radiation the Planck's balance is affected. This escape of radiation perturbs the balance of transition $1 \leftrightarrow 2$, i.e. the Boltzmann balance, involving the ground and the first excited state. The lack of re-absorption of resonant radiation in this transition favors an overpopulation of the level n_1. The Maxwell's equilibrium is also perturbed once the loss of energy for electrons with energy $\geq E_{12}$ is not balanced by the inverse reaction, i.e. superelastic collisions, and the EEDF is also affected mainly for high-energy electrons found in the tail of the distribution.

On the other hand, in the high-lying levels the energy gap is small and the energy exchanges between electrons and the atoms in these states are small. These electrons belong to the bulk of the EEDF and are in equilibrium among them, due to the very effective collisions, assuring the Maxwell's equilibrium for these electrons. In this situation, it can be said that the equilibrium is partial local Maxwell pLM_eE. As the energy gap of top levels and energy of the bulk electrons are close one have also pLB_eE and pLS_eE. In the example presented above where the levels at the bottom of the energy term are excited by electron impact and the radiation produced by de-excitation is lost from the medium we have the corona balance (CB).

Properties of each improper balance were presented by van der Mullen (1990) according to Fig. 7.1. The Corona balance (CB), which is characterized by excitation from the ground state level and destruction of excited levels by emission of radiation, is found in plasmas with low electron density.

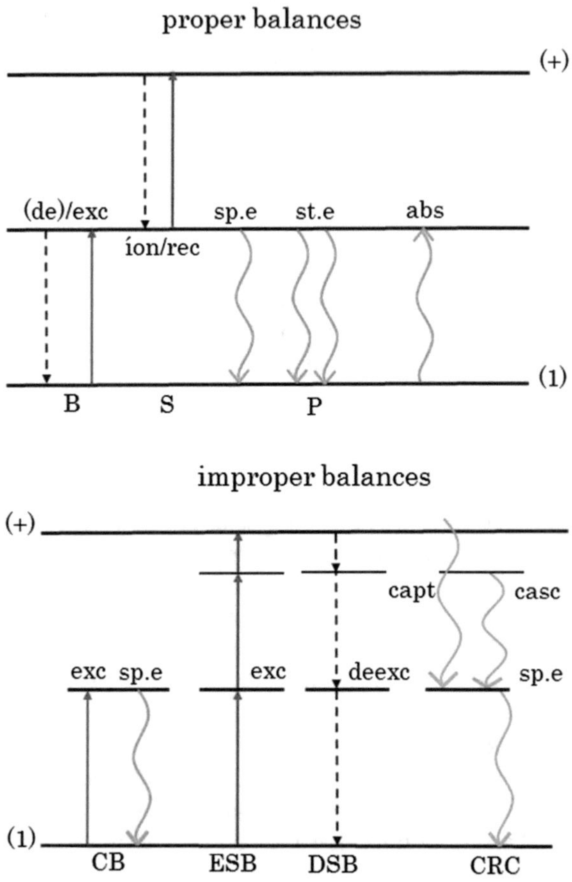

Fig. 7.1 Comparison between proper and improper balances; Boltzmann (B), Saha (S), and Planck (P). Improper balances are: Corona (CB), excitation saturation balance (ESB), de-excitation saturation balance (DSB) and capture radiative cascade (CRC). de-excitation (Dexc), ionization/recombination (ion/re), spontaneous emission (Spe), stimulated emission (Ste), absorption (Abs), capture (Cap), cascade (Casc) (van der Mullen 1990)

The excitation from one excited level to an adjacent higher-energy level is found in excitation saturation balance (ESB). In the class of recombining plasmas an improper balance is achieved by de-excitation from higher adjacent levels (DSB) where one has the inward transport of charged particles. Another example of recombining plasmas is the one regulated by the capture and cascade (CRC) producing radiation due decay from the continuum of higher excited levels also presenting the inward transport of charged particles or irradiated plasmas.

The ASDF is modified by the forces acting in the particles, by the transport of radiation and particles, and by the time evolution of the medium. The plasmas with a high-degree of ionization has high electron density which is the main player that governs the kinetics of the system; this is why these plasmas are called electron excitation kinetic plasma (EEK). As the transport of radiation and particles out of the plasma increase, the system becomes more and more governed by improper balances and the upper part of the energy terms are regulated by pLS_eE balance and the electrons from the bulk of the EEDF have a key role in regulating the kinetics. Otherwise, the energy levels of the bottom of the energy scheme present a population distribution that is out of equilibrium and the time evolution of these densities is much smaller than the characteristic time for plasma growth and decay. In this case the collisional and radiative processes determine the density of each excited level characterizing the ASDF. To obtain the ASDF a collisional-radiative model needs to be used to obtain a quasi-steady state solution (QSSS) for this system. The ASDF form is the net result of improper balances which are associated with processes in microscopic and macroscopic levels.

Figure 7.2 shows a typical case of relationship between the energy levels and the EEDF. The top and bottom levels are associated with parts of the EEDF. The bulk electrons with lower energy are in resonance with energies involved to maintain the pLB_eE and pLS_eE balances at top. The escape of radiation and the excitation of lower levels modify the tail of the EEDF. In low electron density discharges, the EEDF tail is depleted perturbing the pLM_eE. The departure from Planck's balance perturbs the Boltzmann, Saha and Maxwell balances. In the case of tail electrons, the perturbation in Maxwell balance of bulk electrons is weak. In this case only a part of the EEDF assures the partial LM_eE which maintains the Saha equilibrium for levels with $p > p^*$ of the atomic energy diagram.

7.4 Microscopic Reversibility

The rate constants and cross sections for forward and reverse reactions are related through the principle of detailed balancing expressing the time reversibility of the equations of motion in a collision. Consider a collision process according to Fig. 7.3.

A and B particles collide, having internal energy states E_i, reduced mass μ and relative velocity v, leading to particles C and D having internal energy states E_i', reduced mass μ' and relative velocity v'. The fundamental physical principle of invariance under time reversal gives us the relationship between the forward and backward processes;

$$A + B \underset{k_{CD}}{\overset{k_{AB}}{\rightleftarrows}} C + D \tag{7.3}$$

This relationship is called microscopic reversibility, see Chap. 2, Sect. 2.3. The rate constants for forward and backward reactions are related by the detailed balance principle, which presents the time reversibility of the collision processes. Supposing the distribution function of velocities of particles under collision Maxwellian and equating the forward and reverse rates, it can be found that (McWhirter 1965).

$$m_R^2 \, g_A \, g_B \, v_R^2 \, \sigma\left(v_R\right) = m'^2_R \, g_C \, g_D \, v'^2_R \, \sigma'\left(v'_R\right) \tag{7.4}$$

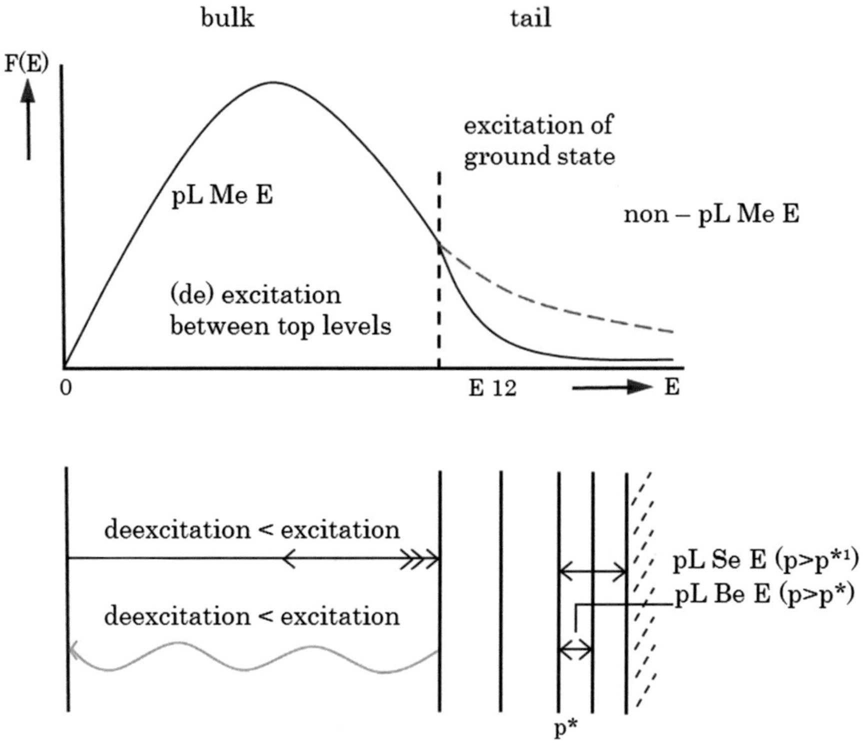

Fig. 7.2 Relationship between the EEDF (bulk and tail electrons) and the ASDF (*bottom* and *top* levels) with respect to the improper balances

Fig. 7.3 Collision process between particles A and B resulting particles C and D

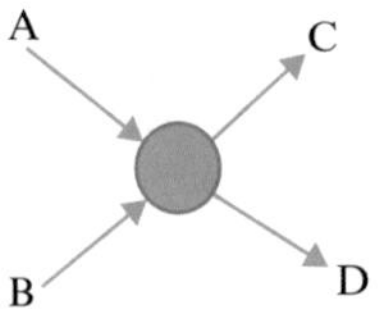

m_R and $m_R^{'}$ are the reduced masses of particles A and B, and C and D respectively. The cross section for inelastic collision A + B is $\sigma(v_R)$ of A and B and $\sigma'(v_R^{'})$ for the reverse reaction C + D; g_A, g_B, g_C and g_D are the degeneracies, or statistical weights, of the energy levels of particles A, B, C and D respectively.

By the conservation of energy, we have (Smirnov 1981):

$$\frac{1}{2} m_R v_R^2 = \frac{1}{2}\, m'^2_R v'^2_R + \mathrm{E}_a \tag{7.5}$$

E_a is the threshold energy. Integrating the expression (7.4) in v_R, assuming a Maxwellian velocity distribution, the ratio between f K_{AB} to K_{CD} can be obtained:

$$\frac{K_{AB}}{K_{CD}} = \left(\frac{m_R'}{m_R}\right)^{3/2} \frac{g_C\, g_D}{g_A\, g_B}\, e^{-E_a/kT} \tag{7.6}$$

relating the forward and backward reaction rates as a function of reduced masses, degeneracies of the levels involved and the Boltzmann factor.

If the particle A is found in internal state α and B in state β, before collision, the equation for macroscopic balance (7.3) may be replaced with an equation where the number of particles per quantum state is:

$$n_A(\alpha)\, n_B(\beta) = n_A(\alpha')\, n_B(\beta') \tag{7.7}$$

$n_A(\alpha)$, $n_B(\beta)$ are the number of particles A and B in α and β states respectively. The Eq. (7.7) shows the elementary reactions between particles A and B. In this case the particle that leaves a quantum state through a certain pathway equals the one that arrives at the same quantum state via the reverse process. This is the principle of microscopic reversibility.

7.5 Maxwell Balance

The equilibrium established due to elastic collisions between particles is called Maxwell's equilibrium. There is no change in the internal energy of particles and only momentum is exchanged. The elastic collisions in Maxwell's balance may be represented by:

$$A\,(E_\alpha) + B\,(E_\beta) \overset{M_H}{\leftrightarrow} A\,(E_\alpha') + B\,(E_\beta') \tag{7.8}$$

As it was presented in Chap. 3, the kinetic theory shows that when the equilibrium is achieved the energy distribution of particles is a Maxwell distribution given by:

$$f(u) = \frac{2}{\sqrt{\pi}(kT)^{3/2}} \exp\left(-\frac{u}{kT}\right) \tag{7.9}$$

where $f(u)$ is normalized so that:

$$\int_0^\infty f(u)\,\sqrt{u}\,du = 1 \tag{7.10}$$

In thermal equilibrium the system is characterized by a sole temperature T with the different constituents of the gas mixture having the same common temperature while the translational states are populated according to $\exp\left(-\frac{u}{kT}\right)$.

As the mass difference between electrons and heavy particles is big, the energy transfer in these collisions is very low. Electrons receive energy from the electric field while heavy particles receive energy through collisions with electrons and redistribute it in collisions among them. Thus, electrons and heavy particles are subject to different processes of energy balance and so that two populations with different energies will arise characterizing a two-temperature plasma. Electrons exchange more efficiently energy in collisions between electrons:

$$e\ (\alpha) + e\,(\beta) \overset{M_e}{\leftrightarrow} e\,(\alpha') + e\left(\beta'\right) \tag{7.11}$$

while heavy particles distribute energy more efficiently through reactions like (7.8). In this case the medium will be characterized by distributions with two-temperatures, one for electrons T_e and another for heavy particles T_H. Collisions between electrons and heavy particles present a perturbation to the medium influencing the energy equipartition.

As plasmas have limited dimensions as in the laboratory, energy may escape from the medium inducing the appearance of disequilibrium. T_H and T_e may be locally in equilibrium but not in the whole system and the electrons may be characterized by a local temperature in a local Maxwell equilibrium $\text{LM}_\text{e}\text{E}$ while heavy particles are in $\text{LM}_\text{H}\text{E}$.

The tail of the EEDF as shown in Fig. 7.2 is modified due to improper balances while the bulk of the distribution is negligibly perturbed and the equilibrium is assured due electron-electron collisions. In this case T_e is kept for the bulk electrons charactering a partial local Maxwell equilibrium for electrons $\text{pLM}_\text{e}\text{E}$.

7.6 Boltzmann Balance

Consider the inelastic and superelastic collisions given by:

$$X\,(\alpha) + A_i\,(\beta) + E_{lu} \overset{B}{\leftrightarrow} X\left(\alpha'\right) + A_f\left(\beta'\right) \tag{7.12}$$

$$deexcitation \leftarrow\ \rightarrow excitation$$

In the excitation process, the particle X with an internal energy state α collides with the particle A_i in an initial state i with internal energy state β, and as a result it is excited to a final state f with energy β'. In plasmas, the particle X may be an electron, ion, atom or molecule. The backward reaction is the de-excitation process. This type of inelastic reaction is a characteristic of the Boltzmann balance. The translational energy of particle X is converted into internal energy of particle A resulting in a population distribution of energy levels given by:

$$\frac{n_A(f)}{n_A(i)} = e^{-\frac{E_{if}}{kT}} \tag{7.13}$$

where E_{if} is the energy difference between the initial state i and the final state f. In this case the translational energy distributions of particles X and A are Maxwellian with the same temperature. There is no change of masses during the reaction and ionization does not occur.

The electrons receive energy from the electromagnetic field and due to their small mass are accelerated attaining high energy. The reaction rates of electron induced transitions are higher than those induced by the ions, atoms or molecules. Even at low ionization degrees the electron collisions are dominant. It was shown (van der Mullen 1990) that for ionization degrees (n_e/n_a) higher than 10^{-4} the collisions with electrons are more important that those with neutrals. Thus the most important processes for distribution or redistribution of atoms over their excited states are those induced by collisions with electrons.

Otherwise for very low ratio $\frac{E_{if}}{kT} \ll 1$ the role of ions and atoms may appear. In this case there are two types of Boltzmanm balances, i.e. one regulated by electrons with temperature T_e and other ruled by heavy particles with temperature T_H, characterizing a two-temperature Boltzmann balance.

As the ratio $m_e/M_A \ll 1$, the Boltzmann relation given by (7.13) is regulated by the electron temperature T_e which imposes the electron temperature on the ASDF. The temperature that results in a Boltzmann distribution for the population of the energy level distribution is called excitation temperature. The excitation temperature is the temperature that describes, through Boltzmann's law, the relative population distribution of atoms or molecules over their energy levels.

In a real case condition only part of the atomic energy levels is described by a Boltzmann balance in equilibrium. De-exciting electron collisions with low-lying levels, radiative escape and other disturbing processes may result that B_e is applicable only for part of the ASDF, generally for the high-levels part characterizing a partial Local Boltzmann Equilibirum pLB$_e$E.

7.7 Saha Balance

In a gas at high temperature, the collisions between atoms may ionize some of them resulting in production of charges, i.e. electrons.

The Saha equation describes the degree of ionization of a plasma as a function of the temperature, density, and ionization energies of the constituent atoms. The Saha equation only holds for weakly ionized plasmas for which the Debye length is large. The screening of the coulomb charge of ions and electrons by other charged particles causes the reduction of the ionization potentials but in general, this effect is negligible.

In the Saha balance the process of ionization in the plasma is a forward process and the backward one is the recombination,

$$X\ (\alpha) + A_p\,(\beta) + \left|E_p\right| \overset{S}{\leftrightarrow} X\left(\alpha'\right) + A_l^+\left(\beta'\right) + e \tag{7.14}$$

recombination ← → *ionization*

In the Saha equation an electron is created by the ionization process, which is different of the Boltzmann balance where the number of particles is conserved. To derive the Saha equation one can start from the Boltzmann distribution, considering initially two-bound states l and u of a given atom or ion:

$$\frac{N_u}{N_l} = \frac{g_u}{g_l}\,exp^{-\frac{E_u - E_l}{kT}} \tag{7.15}$$

g_u and g_l are the statistical weights of the upper and lower levels respectively.

If the upper state is taken into the continuum in Eq. (7.15) we are dealing with the creation of a pair electron-ion. The generalization of (7.15) can be done for states in the continuum with energy E_k and wavenumber k. Note that here k is the wavenumber and not the Boltzmann constant!

Now the number of upper states N_u needs to be replaced by dn_k, which is the number of free electron-ion pairs with electrons in the wavenumber interval between k and $k + dk$ and g_u replaced by dg_u of the free electrons in this interval.

Supposing a normalized cube volume with side length L, the number of wavenumber modes in the interval k and $k + dk$ in a solid angle $d\Omega$ is given by:

$$dN_{k\Omega} = \left(\frac{L}{2\pi}\right)^3 k^2 dk d\Omega \tag{7.16}$$

for each of the two directions of polarization of the electromagnetic waves

Substituting (7.16) in (7.15) and taking into account the considerations discussed above to generalize (7.15) we find:

$$\frac{dN_k}{N_l} = \frac{Vg_i}{\pi^2 g_l} exp^{-\frac{E_k - E_l}{kT}} k^2 dk \tag{7.17}$$

Considering that $E_k = \hbar^2 k^2/2m_e$, $N_e = \int dN_k$ and integrating over k we get:

$$\frac{N_e}{N_l} = \frac{Vg_i}{\pi^2 g_l} \exp\left(\frac{E_l}{kT}\right) \int_0^\infty k^2 \exp\left(-\frac{\hbar^2 k^2}{2m_e kT}\right) dk \tag{7.18}$$

which results in,

$$\frac{N_e}{N_l} = \frac{Vg_i}{\pi^2 g_l}\left(\frac{m_e kT}{2\pi\hbar^2}\right)^{3/2} \exp\left(\frac{E_l}{kT}\right) \tag{7.19}$$

An ion-free electron pair is formed after the ionization reaction. The average volume occupied by the resulting ion-free electron pair is roughly $V = 1/N_i$. Taking into account that the Bohr radius a_0 and the ionization energy of the hydrogen atom E_H are given by (Demtroeder 2010):

$$a_0 = \frac{\varepsilon_0 h^2}{\pi \mu e^2}; \quad E_H = \frac{\mu e^4}{8\varepsilon_0^2 h^2} \tag{7.20}$$

ε_0 is the free space permittivity, h is the Planck's constant, and μ is the reduced ion-free electron mass. Substituting (7.20) into (7.19) we obtain the Saha equation:

$$\frac{N_e N_i}{N_l} = \frac{2g_i}{g_l a_0^3}\left(\frac{kT}{4\pi E_H}\right)^{3/2} \exp\left(-\frac{E_{li}}{kT}\right) \tag{7.21}$$

The Eq. (7.21) was proposed by Saha in 1921 (Griem 2005) to explain the spectra measured from stars and so to determine their temperature. In LTE the partition function of an atom or ion may be written as:

$$Z(T) = \sum\nolimits_n g_n \exp\left(-\frac{E_n}{kT}\right) \tag{7.22}$$

Substituting (7.22) into (7.21) the Saha equation may also be written as:

$$\frac{N_e N_i}{N_l} = \frac{2Z_i(T)}{Z(T) a_0^3}\left(\frac{kT}{4\pi E_H}\right)^{3/2} \exp\left(-\frac{E_\infty}{kT}\right) \tag{7.23}$$

where $Z_i(T)$ is the ion partition function and E_∞ is the ionization potential of the atom. This Saha equation is useful to relate the densities of electrons, ions and neutrals in LTE plasmas. Departures from LTE should be taken with care in low-density plasmas as it will be discussed later on.

The Saha balance is valid when the electron density is high enough to assure a collisional ionization rate larger than the radiative rates. When an atom is immersed in a high density plasma a reduction of the ionization potential occurs due to the presence of the surrounding charges. This effect limits the number of terms occurring in a partition. The reduction in the ionization density may be then calculated supposing an isolated atom or ion with a charge Z−1 (Griem 2005).

The reduction of the ionization potential ΔE is given by:

$$\Delta E_{\infty} = \frac{ze^2}{4\pi\varepsilon_0\lambda_D} \tag{7.24}$$

λ_D is the Debye length presented in Chap. 1 and is equal to $\lambda_D = \left(\varepsilon_0 kT_e \big/ e^2 n_0\right)^{1/2}$ or $\lambda_D\,(\mathrm{cm}) = 743\sqrt{T_e(eV) \big/ n_e\,(\mathrm{cm}^{-3})}$. The Saha equation with the ionization correction becomes:

$$\frac{N_e N_i}{N_l} = \frac{2Z_i(T)}{Z(T)a_0^3}\left(\frac{kT}{4\pi E_H}\right)^{3/2} \exp\left(-\frac{E_{\infty} - \Delta E_{\infty}}{kT}\right) \tag{7.25}$$

The competition between electrons and heavy particles regulate the establishment of Saha equilibrium, where ionization and recombination compensate each other, leading to the appearance of a two-temperature plasma, i.e. with one temperature for electrons and other for heavy particles. As the electrons are dominant in the Saha balance the temperature to be considered in the Saha Eq. (7.25) is the electron temperature.

In plasmas actually the difference between temperatures of atoms and ions is small because the masses are practically the same which makes the energy exchange very effective. Moreover, in the presence of Saha equilibrium the fast exchange of atoms into ions and vice-versa tends to equalize the ion and atom temperatures. The rate of ionization collisions increases strongly for levels near the continuum preserving the Saha balance.

The same cannot be said to low-lying levels where recombination is impaired. This situation characterizes a local Saha equilibrium mediated by electrons in a part of the ASDF which is named partial local electron ruled Saha equilibrium $\mathrm{pLS_eE}$.

A parameter normally used to characterize the ASDF is b which relates the density of a given p level to its value as if the system was in Saha equilibrium;

$$b(p) = \frac{n(p)}{n^S(p)} \tag{7.26}$$

This expression is important to quantify the departure of the level density from the Saha equilibrium.

7.8 Planck's Balance

While Maxwell, Boltzmann and Saha balances deal with interactions between particles, the Planck's balance regulates matter-radiation interaction which may be represented by:

$$A_u \overset{P}{\leftrightarrow} A_l + h\nu \qquad (7.27)$$

absorption $\leftarrow \rightarrow$ *spontaneous emission*

$$A_u + h\nu \rightarrow A_l + 2h\nu \qquad (7.28)$$

$\rightarrow$ *stimulated emission*

The first approach to describe this equilibrium is the use of Boltzmann relation (7.13); $n_\nu = n(u)/n(l) = \exp(-h\nu/kT)$ which is only correct for high values of $h\nu/kT$. n_ν is the number of photons at frequency ν. The Wien's approximation, also called Wien's law, is used to describe the spectrum of thermal radiation or blackbody radiation. This law was first derived by Wilhelm Wien in 1896. The equation does accurately describe the intensity of the emitted radiation in the short wavelength range, i.e. at the high frequency limit, of the spectrum of thermal emission from objects, but it fails to accurately fit the experimental data for long wavelengths or low frequency emission.

Wien derived his law from thermodynamic concepts. The Wien's law may be written as:

$$I(\nu,\ T) = \frac{2h\nu^3}{c^2} \exp - \frac{h\nu}{kT} \qquad (7.29)$$

where $I(\nu, T)$ is the amount of energy emitted per unit surface area, time, solid angle, and frequency unities, at frequency ν, being T the temperature of the blackbody. The peak value of this curve, as determined by taking the derivative and solving for zero, occurs at frequency $\nu_{max} = 5.88 \times 10^{10} T(K)$.

As the Wien approximation failed to describe the whole spectrum of thermal radiation at long wavelengths, Planck proposed a law based on quantum mechanics to describe completely the radiation spectrum as a function of frequency and temperature of the emitting body:

$$I(\nu,\ T) = \frac{2h\nu^3}{c^2} \frac{1}{e^{\frac{h\nu}{kT}} - 1} \qquad (7.30)$$

The Wien approximation may be derived from Planck's law by assuming $h\nu \gg kT$. In this case;

$$\frac{1}{e^{\frac{h\nu}{kT}} - 1} \cong e^{-\frac{h\nu}{kT}} \qquad (7.31)$$

and Planck's law approximately equals the Wien approximation at high frequencies.

In plasmas created in laboratory the dimensions of the vessel are limited and consequently the radiation may escape from the medium causing the departure from equilibrium. This radiation leaks without giving rise to absorption or stimulated emission so that it is an important source of disequilibrium. Another important aspect is that the radiation created in one part of the plasma may be absorbed in another region so that the radiation transfer cannot be treated as a local event. To do a proper treatment of this non-local phenomenon we need to introduce the concept of escape factor in order to handle it locally.

A spontaneously emitted photon may be reabsorbed; an atom initially in the lower state is put into an upper state, reducing the apparent decay rate of this level. The escape factor is defined as the ratio of the effective radiative decay rate to the radiative decay rate of a spontaneously emitted photon by an isolated atom (Drawin and Emard 1973):

$$A_{eff} = \Lambda\, A_{ul} \tag{7.32}$$

A_{eff} is the effective emission rate, A_{ul} the spontaneous emission rate from the upper to the lower level and Λ is the escape factor. If $\Lambda = 1$ the plasma is optically thin for the transition $u \rightarrow l$ and all radiation produced in this transition escapes from the medium. On the contrary if $\Lambda = 0$ all radiation produced is re-absorbed inside the medium and the Planck balance is established. If $\Lambda < 0$ absorption exceeds the total emission and if $\Lambda > 1$ stimulated emission is dominant.

If the plasma is homogeneous, in any real situation, the equivalent linewidth needs to be evaluated. Another important issue are the lineshape and the geometry of the reactor.

In a simple approach the escape factor Λ_{ij} for a transition $(i \rightarrow j)$ may be written as:

$$\Lambda_{ij} = e^{-\sigma_{ij} n l} \tag{7.33}$$

where σ_{ij} is the photon absorption cross section for the transition $(i \rightarrow j)$, n is the density of the absorber and l is the characteristic length of the medium.

The absorption cross section σ_{ij} may be calculated by (Drawin and Emard 1973):

$$\sigma_{ij} = \frac{1}{2}\frac{\sqrt{\pi}e^2}{m_e}\frac{1}{\nu_{ij}}\sqrt{\frac{M}{T}}\sqrt{\frac{N}{2R}} f_{ij} \tag{7.34}$$

N is the Avogrado number and R is the ideal gas constant. Introducing the constants values, we have:

$$\tau = \sigma_{ij} n l = 5.13 \times 10^{-15} \times f \times \lambda \left(\mathring{A}\right) \times \left(\frac{M}{T(K)}\right)^{1/2} \times n \left(\mathrm{cm}^{-3}\right) \times l\ (\mathrm{cm}) \tag{7.35}$$

When the medium is optically thin the photons escape from the plasma inducing a non-thermal equilibrium condition. As explained before the escape of radiation emitted by transitions involving the lower levels modifies the equilibrium once the excitation is not compensated by the re-absorption of radiation inside the plasma, which in turn modifies the tail of the EEDF. The population density of upper levels is then governed by the radiation field. This imposes a partial local Planck equilibrium pLPE since it only applies to a part of the ASDF.

It was discussed in this chapter the conditions that takes the thermodynamically equilibrium away. It was presented the Maxwell, Boltzmann, Saha and Plank balances and showed the different steps for departure from TE. The balances are close coupled and the departure from equilibrium of one balance affects the others.

Particles are responsible for the establishment of Maxwell, Boltzmann and Saha equilibrium and in discharge plasmas the most important particle to describe these equilibria are the electrons. Planck's balance is ruled by radiation escape from the medium, which of course affects the other balances regulated by material particles.

Exercises

Exercise 7.1. Injection of neutral beam of atoms is a technique employed to heat a plasma and initiate a sustainable nuclear reaction. A technique usually employed consists in accelerating a beam of charge particles and after neutralize them. Suppose that a monoenergetic beam of protons with 10 keV passes through a gas cell of 50 cm long containing hydrogen at 0.1 Pa at 300 K. What fraction of protons are neutralized after passing through the cell? It is given the reaction coefficient for collisions between the protons and the hydrogen atoms $\sigma v = 1 \times 10^{-14} \mathrm{m^3 s^{-1}}$.
Resolution: The number of protons at a distance x in a slab inside the cell may be written as:

$dn_1 = -n_1(x)\, n_2\, \sigma\, dx$, where n_1 is the proton density and n_2 is the density of neutral hydrogen atoms. After integration we have:

$n_1(x) = n_1(0) \exp(-n_2 \sigma x)$. Discarding charge exchange reactions, the fraction that becomes neutralized is:

$$\frac{n_1(0) - n_1(x)}{n_1(0)} = 1 - \exp(-n_2 \sigma x)$$

The proton velocity is $v = \left(\frac{3kT}{m}\right)^{1/2} = 1.7 \times 10^6$ m/s. The cross section for proton-atom collision is $\frac{\sigma v}{v} = \frac{1 \times 10^{-14}\ \mathrm{m^3 s^{-1}}}{1.7 \times 10^6\ \mathrm{m/s}} = 5.88 \times 10^{-20} \mathrm{m^2}$. The density of hydrogen atoms in the cell is $n_2 = \frac{p}{kT} = \frac{0.1\ \mathrm{Pa}}{1.38 \times 10^{-23} \times 300\ \mathrm{K}} = 2.42 \times 10^{19} \mathrm{m^{-3}}$. The fraction of neutralized protons is $1 - \exp\left(-2.42 \times 10^{19} \times 5.88 \times 10^{-20} \times 0.5\right) = 0.51$.
Exercise 7.2. Calculate the ionization degree of a hydrogen plasma in Local Thermodynamic equilibrium at a temperature of 1 eV and pressure of 100 Pa.
Resolution: The total density of particles is $n_t = \frac{p}{kT} = \frac{100\ \mathrm{Pa}}{1.38 \times 10^{-23} \times 11{,}600\ \mathrm{K}} = 6.25 \times 10^{20} \mathrm{m^{-3}}$

The Saha Eq. (7.21) may be used considering $g_i = g_n$

$$\frac{n_e n_i}{n_n} = \frac{2}{a_0^3}\left(\frac{kT}{4\pi E_H}\right)^{3/2} \exp\left(-\frac{E_H}{kT}\right);$$

$$\frac{n_i^2}{n_n} = \frac{2}{\left(0.56 \times 10^{-10}\right)^3}\left(\frac{1}{4\pi \times 13.6}\right)^{3/2} \exp\left(-\frac{13.6}{1}\right)$$

Considering that the plasma is macroscopically neutral, i.e. $n_e = n_i$, $\frac{n_i^2}{n_n} = 6.32 \times 10^{21}\ \mathrm{m}^{-3}$; but $n_t = n_n + n_i$ so $n_i^2 = 6.32 \times 10^{21}\ \left(6.25 \times 10^{20} - n_i\right)$, $n_i^2 + 6.32 \times 10^{21} n_i - 3.95 \times 10^{42} = 0$; thus $n_i = 5.75 \times 10^{20}\mathrm{m}^{-3}$

The ionization degree is $\frac{n_i}{n_n} = \frac{5.75\times10^{20}}{6.25\times10^{20}} = 0.92$, so that 92 % of particles are ionized.

Exercise 7.3. A system is in thermodynamic equilibrium with a temperature T. Find the average number of photons n in a given state with energy $h\nu$.

Resolution: The number of photons in a given state is given by $\sum_n n \exp\left(-{h\nu n}/{kT}\right)$. Thus the average number of photons in one state with energy $h\nu$ is:

$$\bar{n} = \frac{\sum_n n \exp\left(-{h\nu n}/{kT}\right)}{\sum_n \exp\left(-{h\nu n}/{kT}\right)} = \frac{1}{\exp\left(-{h\nu n}/{kT}\right) - 1}$$ which is the Planck's distribution.

Exercise 7.4. Consider a hydrogen plasma in LTE with a temperature of 12,000 K and electron density of $10^{14}\mathrm{cm}^{-3}$. Calculate the principal quantum number from which the LTE may be assumed. If the electron density is $10^{10}\mathrm{cm}^{-3}$ which is the level with the same temperature? If the temperature is lowered to 6,000 K with $n_e = 10^{12}\mathrm{cm}^{-3}$ which is the new minimum quantum number to assure LTE?

Resolution: Using Eq. (7.2);

$$n_e \gg 1.6 \times 10^{12}\left(T_e(\mathrm{eV})^{\frac{1}{2}}(\Delta E\ (p,q)\ (\mathrm{eV}))\right)^3\ \mathrm{cm}^{-3}$$

we can obtain:

$$\Delta E \ll \left(\frac{n_e}{1.6 \times 10^{12}\left(T_e(\mathrm{eV})^{\frac{1}{2}}\right.}\right)^{1/3}$$

For $n_e = 10^{14}\mathrm{cm}^{-3}$ and $T_e = 12{,}000\ \mathrm{K} = 1.03\ \mathrm{eV}$, $\Delta E \ll 3.95\ \mathrm{eV}$.

The energy of a given level of the hydrogen atom is:

$$E\,(\mathrm{eV}) = \frac{-13.6}{n^2}$$

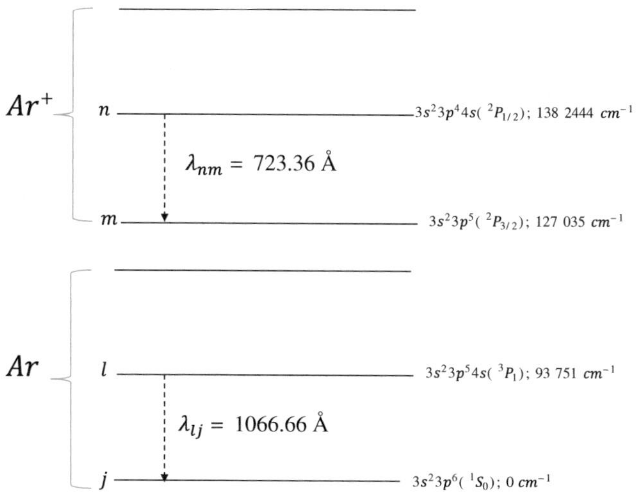

Fig. 7.4 Energy levels of Ar and Ar^+

So, the energy differences between adjacent levels are:

n	E	$\lvert\Delta E\rvert$
1	−13.6	0
2	−3.4	10.2
3	−1.51	1.89
4	−0.85	0.66
5	−0.54	0.31
6	−0.38	0.16

So the minimum quantum number from which the level for the establishment of LTE may be assumed as $n = 3$.

For $n_e = 10^{10}\text{cm}^{-3}$ and $T_e = 12{,}000$ K $= 1.03$ eV, $\Delta E \ll 0.395$ eV. In this case the minimum quantum number to assure LTE is $n = 5$.

and for $n_e = 10^{12}\text{cm}^{-3}$ and $T_e = 6{,}000$ K $= 0.58$ eV, $\Delta E \ll 0.954$ eV. In this case the minimum quantum number to assure LTE is $n = 4$.

Exercises 7.5–7.9. We are interested in the study of an argon plasma at atmospheric pressure. This plasma is created by an electrical discharge in a tube with *i.d.* 4 mm with an electronic density $n_e = 2.1 \times 10^{17}\text{cm}^{-3}$. The goal is to determine the electron temperature using the ratio between two emission lines; the first is the line from the neutral excited argon atom, transition $3s^23p^54s\ (^3P_1) \rightarrow 3s^23p^6\ (^1S_0)$, $\lambda = 1{,}066.66$ Å and the second line from the single ionized argon ion, transition $3s^23p^44s\ (^2P_{1/2}) \rightarrow 3s^23p^5\ (^2P_{3/2})$, $\lambda = 723.36$ Å. The energy levels of Ar and Ar^+ are shown in the Fig. 7.4.

The oscillator strengths are 0.061 and 0.18 for lines 1,066.66 Å and 723.36 Å, respectively. In this case, we suppose that the plasma is homogeneous, stationary, in LTE and optically thin for the two lines considered.

Exercise 7.5. Considering the intensities of emitted lines for two systems with different charges and the expressions for Boltzmann and Saha distributions, find the ratio between the intensities for lines with wavelengths 1,066.66 Å and 723.36 Å. Consider that the reduction of the ionization potential of Ar is negligible.

Resolution: The intensity of an emitted line by the Ar^+ ion may be written as:

$$I_{nm}\left(Ar^+\right) = C\, N_n A_{nm} h\nu_{nm}$$

C is a constant that relies on the calibration of the optical system, N_n is the density of the upper level of the transition, A_{nm} is the transition probability for spontaneous emission ($n \to m$), h is the Planck's constant and ν_{nm} the frequency of the photon emitted during transition $n \to m$.

The transition probability is given by:

$$A_{nm} = \frac{8\pi e^2 \nu_{nm}^2}{mc^3} f_{nm}^{em}$$

Substituting in the expression for $I_{nm}(Ar^+)$ we obtain, taking into account that $f_{nm}^{em} = \frac{g_n}{g_m} f_{nm}^{abs}$ and the Boltzmann distribution:

$$I_{nm}\left(Ar^+\right) = K\,\frac{1}{\lambda_{nm}^3}\,\frac{g_n^{z+1}}{g_m^{z+1}} f_{nm}^{abs} N_m^{z+1} = K\,\frac{1}{\lambda_{nm}^3}\,\frac{g_n^{z+1}}{g_m^{z+1}} f_{nm}^{abs} N_1^{z+1}\,\frac{g_m^{z+1}}{g_1^{z+1}} \exp -\frac{E_m^{z+1} - E_1^{z+1}}{kT_e}$$

$$I_{nm}\left(Ar^+\right) = K\,\frac{1}{\lambda_{nm}^3}\,\frac{g_n^{z+1}}{g_1^{z+1}} f_{nm}^{abs} N_1^{z+1} \exp -\frac{E_m^{z+1} - E_1^{z+1}}{kT_e}$$

For the emitted line of Ar ($l \to j$) we have:

$$I_{lj}(Ar) = K\,\frac{1}{\lambda_{lj}^3}\,\frac{g_l^z}{g_j^z} f_{lj}^{abs} N_j^z$$

Substituting the Saha expression ($\frac{n_j^{z+1} n_e}{n_j^z} = 2\,\frac{g_j^{z+1}}{g_j^z}\left(\frac{2\pi\hbar^2}{mkT_e}\right)^{-3/2} \exp -\left(\frac{E_j^{z+1}-E_j^z}{kT_e}\right)$):

In this case the lower state of transition $j = 1$

$$I_{lj}(Ar) = K\,\frac{1}{\lambda_{lj}^3}\,\frac{g_l^z}{g_j^z} f_{lj}^{abs} N_j^z = K\,\frac{1}{\lambda_{lj}^3}\,\frac{g_l^z}{g_j^z} f_{lj}^{abs} N_1^{z+1}\,\frac{g_j^z}{2g_1^{z+1}}\left(\frac{2\pi\hbar^2}{mkT_e}\right)^{3/2} n_e \exp \frac{E_j^{z+1} - E_1^z}{kT_e}$$

$E_j^{z+1} - E_1^z$ is the ionization energy of argon because $E_1^z = 0$.

Now we can calculate the line ratio:

$$\frac{I_{nm}\left(Ar^{+}\right)}{I_{lj}(Ar)}=2\,\frac{\lambda_{lj}^{3}}{\lambda_{nm}^{3}}\frac{g_{n}^{z+1}}{g_{l}^{z}}\frac{f_{nm}^{abs}}{f_{lj}^{abs}}\left(\frac{2\pi\hbar^{2}}{mkT_{e}}\right)^{-3/2}\frac{1}{n_{e}}\exp-\left(\frac{E_{m}^{z+1}+E_{1}^{z+1}-E_{j}^{z+1}}{kT_{e}}\right)$$

Exercise 7.6. Determine, taking into account the data given in the figure for the energy levels of Ar and Ar^{+}, the electron temperature knowing that the measured ratio between the lines is ${I_{Ar^+}}/{I_{Ar}} = 4.72\times 10^{3}$.

Resolution: The statistical weights $(2J+1)$ of levels g_n, g_m, g_l and g_j are 2, 4, 3 and 1 respectively. Knowing that $1\ \text{eV} = 8065.73\ \text{cm}^{-1}$, the electron temperature may be found by the expression determined in the exercise 7.5:

$$4.72\times 10^{3}=2\left(\frac{1,066.66}{723.36}\right)^{3}\frac{2}{3}\,\frac{0.18}{0.061}\left(\frac{6.28\times\left(1.054\times 10^{-27}\right)^{2}}{9.11\ 10^{-28}\times 1.38\ 10^{-12}T_{e}}\right)^{-3/2}$$

$$\frac{1}{2.1\ 10^{17}}\exp-\left(\frac{17.14+15.75-11.62}{kT_{e}}\right)$$

To satisfy this relation the electronic temperature should be $T_e = 4.6$ eV.

Exercise 7.7. Using the electron temperature found in the preceding exercise and knowing that the electron density is $n_e = 2\times 10^{17}\text{cm}^{-3}$, calculate the number densities of Ar and Ar^{+}. Assume the plasma composed by only these two species.

Resolution: Since $n_e = n^{+} = 2\times 10^{17}\text{cm}^{-3}$ from Saha equation we have:

$$\frac{n_e n_i}{n_0}=\frac{2g_i}{g_l a_0^3}\left(\frac{kT}{4\pi E_H}\right)^{3/2}\exp\left(-\frac{E_{li}}{kT}\right);$$

$$\frac{n_i^2}{n_0}=\frac{2}{\left(0.56\times 10^{-10}\right)^{3}}\frac{4}{1}\left(\frac{4.6}{4\pi\times 15.75}\right)^{3/2}\exp\left(-\frac{15.75}{4.6}\right)$$

$$\frac{n_i^2}{n_0}=6.01\ 10^{21}\text{cm}^{-3}\quad n_0=7.34\ 10^{12}\text{cm}^{-3}$$

Exercise 7.8. The initial hypothesis of a plasma in LTE is justified for the above Ar transition?

Resolution: Using Eq. (7.2);

$$n_e \gg 1.6\,x\,10^{12}\left(T_e(eV)^{\frac{1}{2}}(\Delta E\ (p,q)\ (eV))^{3}\ cm^{-3}\right.$$

As $T_e = 4.6\ eV$,

$$n_e \gg 3.43\times 10^{12}(\Delta E\ (p,q)\ (\text{eV}))^{3}\ \text{cm}^{-3}$$

For the Ar transition $\Delta E\ (p,q) = 11.62$ eV

$$n_e\left(2 \times 10^{17}\text{cm}^{-3}\right) \gg 5.38 \times 10^{15}\ \text{cm}^3$$

Exercise 7.9. Is this plasma optically thin for the above Ar transition?
Resolution: Using the escape factor formula given by Eq. (7.35);

$$\tau = 5.13 \times 10^{-15} \times f \times \lambda\left(\mathring{A}\right) \times \left(\frac{M}{T(K)}\right)^{1/2} \times n\ \left(\text{cm}^{-3}\right)\ x\ l\ (\text{cm})$$

$$= 5.13\ 10^{-15} \times 0.061 \times \left(\frac{40}{4.6 \times 11,600}\right)^{1/2} \times 7.34 \times 10^{12} \times 0.4 = 2.51\ 10^{-5}$$

$$\Lambda = e^{-\tau} \sim 1$$

The escape factor is $\sim$100 % so the medium is optically thin for this transition.

References

W. Demtroeder, *Atoms, Molecules and Photons: An Introduction to Atomic-, Molecular- and Quantum Physics*, 2nd edn. (Springer, Berlin, 2010)

H.W. Drawin, F. Emard, Optical escape factors for bound-bound and free-bound radiation. Beitr. Plasmaphysik **13**, 143 (1973)

H.R. Griem, *Principles of Plasma Spectroscopy* (Cambridge University Press, Cambridge, 2005)

R.P. McWhirter, in *Plasma Diagnostic Techniques*, ed. by R.H. Huddlestone, S.L. Leonard (Academic, New York, 1965)

B.M. Smirnov, *Physics of Weakly Ionized Gases* (MIR Publ., Moscow, 1981). Revised from 1978 edition

J.A.M. van der Mullen, Excitation equilibria in plasmas: a classification. Phys. Rep. **191**, 109–220 (1990)

Chapter 8
Emission Spectroscopy

Plasmas, hot or cold, are sources of radiation. The plasmas being created from low-pressure to atmospheric pressure, in thermodynamic equilibrium or not, emit in a broad region of the electromagnetic spectra, from vacuum ultraviolet to infrared. The emission spectroscopy employed in the analysis of the emitted radiation is a non-intrusive technique with good spatial and temporal resolutions. Emission spectroscopy may be also helpful to determine some plasma parameters as gas temperature, vibrational temperature and fractional composition of the medium for example. With a good optical arrangement, local measurements can be done for these parameters with fast photo-detectors like photomultipliers tubes. Spectroscopic measurements of plasma parameters and species concentrations can be done in real-time, with resolution of some nanoseconds, which allows to follow their evolutions. These properties make the emission spectroscopy a powerful tool to probe, for example, surface treatments by plasma. Another good example of using emission spectroscopy in industry is the end-point detection of some etching reactors, used in microelectronic, helping us in the stop of the etching process.

The aim of this chapter is to show the possibilities of optical emission spectroscopy, their limitations and interpretation of spectra. We begin with notions of line radiation and some reminders of atomic and molecular physics without being exhaustive about the subject. Readers interested in a deeper treatment should search more specialized textbooks (Demtröder 2010; Herzberg 1950). In the subsequent sections, we present notions of spontaneous, absorption and stimulated emissions and a brief discussion of molecular bands. Applications of optical emission spectroscopy, to infer some plasma parameters, such as actinometry, titration, gas and vibrational temperature measurements and electronic density measurement by line broadening are presented.

The original version of this chapter was revised. An erratum to this chapter can be found at DOI 10.1007/ 978-3-319-09253-9_12

J.M.A.H. Loureiro, J. de Amorim Filho, *Kinetics and Spectroscopy of Low Temperature Plasmas*, Graduate Texts in Physics,
DOI 10.1007/978-3-319-09253-9_8

8.1 Line Radiation

As said before in the introduction of this chapter, the radiation plays a key role in plasma diagnostics and its classification is of fundamental importance in order to may be used as a plasma diagnostic tool. From quantum mechanics the radiation due to transitions between different energy levels of a quantum system is given by $E_i - E_f = \hbar\omega$. If radiation is being emitted after transition between two discrete levels $E_f < E_i$ it is classified as emission. Otherwise, if absorption of a quantum is taking place $E_f > E_i$, see Fig. 8.1, we are talking about absorption. In this case a transition is characterized by a change of dipole momentum.

As a phenomenon that occurs between two discrete energy levels, neither emission nor absorption of photons can take place between two free electrons; one of them needs to be bound to an atomic or molecular structure.

It is important to classify the radiation involved in electron transitions according to the nature of the free-electron and the energy of bounded system. Electrons in the electric field of an ions, as well as the transitions between energy levels of a neutral atom, are shown in Fig. 8.2. When both states E_f and E_i belong to the continuum, the transition is said to be free-free. Then, part of the electron kinetic energy is lost to the field created by the positive ion or during the interaction with neutrals. In this case the emitted radiation is in the infrared region and it is called of a continuum bremsstrahlung.

Transitions that occur when the initial state of the electron is in the continuum of energy and the final state is a bound-state, of an atom, for example, is called bound-free. This kind of process occurs during an electron recombination or a dielectronic transition and the radiation so emitted is in the continuum range. Electron-neutral collisions belong also to this kind of transitions.

Bound-bound transitions are related to discrete energy levels and they result in emission or absorption of radiation from ultraviolet to infrared region of the electromagnetic spectrum.

From the optical emission spectrum it is possible to analyze the line profiles and to have access to plasma parameters that contribute to the line broadening.

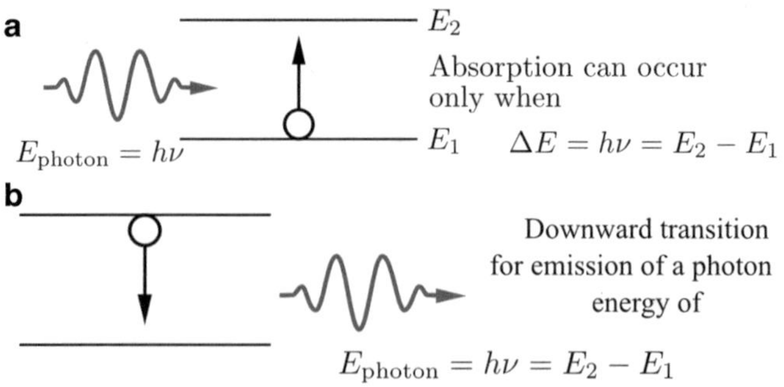

Fig. 8.1 Radiative transition between two energy levels

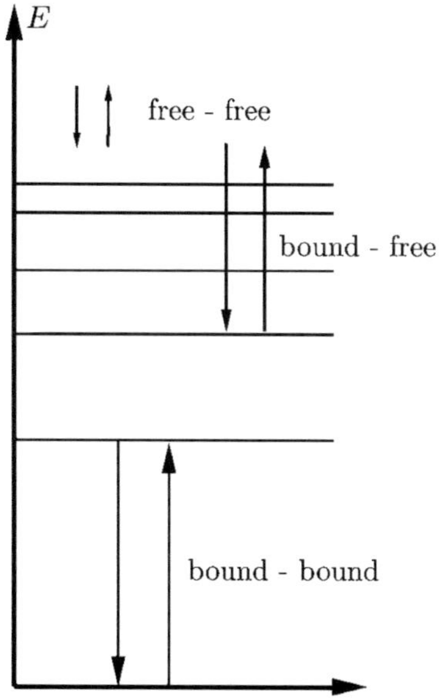

Fig. 8.2 Energy levels and energy of free-free, free-bound and bound-bound transitions

Many works have been published since the beginning of the twentieth century to understand the mechanisms that broaden, for example, the lines of Balmer series of hydrogen.

A spectral line extends over a range of frequencies, not with a single frequency, having nonzero linewidth. In this chapter we will study the main reasons that affect the broadening of a given line, making the central frequency may be shifted from its nominal wavelength. There are several reasons for such broadening and shift. The reasons may be divided into two broad categories; broadening due to local conditions and broadening due to extended conditions (non-local effects). Here we will discuss only broadening due to local effects once they are the most important in low-temperature plasmas.

Broadening due to local conditions results from effects that hold in a small region around the emitting element, usually small enough to assure local thermodynamic equilibrium. Broadening due to extended conditions may result from changes in the spectral distribution of the radiation as it traverses its path to the observer. The broadening due to local effects may be natural, Doppler and by pressure effects (Stark, van der Waals and Resonance).

8.1.1 Natural Broadening

Natural broadening has its origin on the uncertainty principle due to the finite lifetime of an excited state linked with the uncertainty of its energy. A short lifetime will have a large energy uncertainty and a broad emission. This broadening effect results in a non-shifted Lorentzian profile. In a first approach, a damped harmonic oscillator can classically describe it. This damping is weak and may be represented by a constant γ, in the equation bellow;

$$x(t) = x_0 e^{\left(-\frac{\gamma}{2}t\right)} \cos \omega t \tag{8.1}$$

where $\omega \cong \omega_0$; here ω_0 is the angular frequency at line center. Then according to the classical electromagnetic theory a vibrating electric charge is continually damped by energy radiation. The energy of the oscillator expressed by Eq. (8.1) is given by (White 1934);

$$E = E_0 e^{(-\gamma t)} \tag{8.2}$$

where,

$$\gamma = \frac{8\pi^2 e^2 \nu_0^2}{3mc^3} \tag{8.3}$$

Equation (8.3) e is the electron charge and m its mass. The intensity of the radiated energy may be represented by the following Lorentzian profile (Fig. 8.3):

$$I(\nu) = \frac{\gamma}{2\pi} \frac{1}{4\pi^2(\nu_0 - \nu)^2 + (\gamma/2)^2} \tag{8.4}$$

From Eq. (8.4) it can be seen that the half of the maximum intensity $I_{max}(\nu)/2$, is achieved when $I(\nu) = I_{max}(\nu)/2 = 1/\pi\gamma$, then:

$$4\pi^2(\nu_0 - \nu)^2 + (\gamma/2)^2 = \gamma^2/2 \rightarrow \ \Delta\nu_n = 2(\nu_0 - \nu) = \gamma/2\pi \tag{8.5}$$

The line profile of natural broadening, which is a Lorentzian, has full-width-at-half-maximum (FWHM) given by:

$$\Delta\lambda_{n=} \frac{2\pi c}{\omega_0^2}\gamma = \frac{4\pi e^2}{3mc^2} = 1.16\ 10^{-4}\text{Å} \tag{8.6}$$

which is negligible.

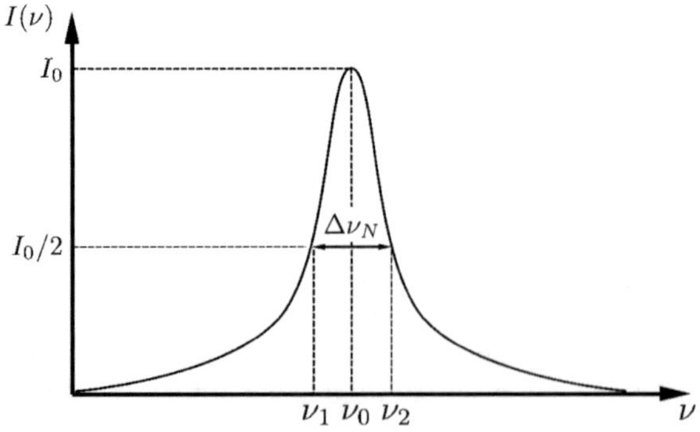

Fig. 8.3 Line profile of natural broadening

In a general way the natural broadening may be calculated by:

$$\Delta E \Delta t = \frac{h}{4\pi} \tag{8.7}$$

or

$$\Delta \nu_s = \frac{1}{4\pi \tau_s} \tag{8.8}$$

where τ_s is the mean lifetime of the transition s to i, and is inversely proportional to Einstein's coefficient A_s. The FWHM of the natural broadening may be written as:

$$\Delta\lambda_{natural} = \frac{\lambda_{si}^2}{4\pi c}\left(\sum\nolimits_{n<s} A_{sn} + \sum\nolimits_{m<i} A_{im}\right) \tag{8.9}$$

λ_{si} is the wavelength of transition s to i, and n and m are lower energy levels in transitions from s and i, respectively.

8.1.2 Doppler Broadening

The thermal motion of atoms (or molecules) causes Doppler broadening of the line emitted (or absorbed) by an atom (or molecule). As they have a distribution of velocities, each photon emitted will be "red"- or "blue"-shifted by this effect depending on the velocity of the atom relatively to the observer. Supposing that u is the atom velocity and θ the angle between u and the observer direction, the frequency of the emitted radiation is shifted by (White 1934):

$$\frac{\Delta\nu}{\nu_0} = \frac{u \cos\theta}{c} \tag{8.10}$$

where ν_0 is the line center frequency when $u = 0$, $\nu = \nu_0 \pm \Delta\nu$ the observed frequency and c the velocity of light. Since the spectral line is a combination of all the emitted radiation effects, as higher the temperature of the gas is, broader the spectral line emitted from that gas will be. This Doppler broadening effect is described by a Gaussian profile and there is no associated shift. The intensity $I(\nu)$ of the emitted radiation as a function of frequency, considering a Maxwellian distribution function of velocities, is given by:

$$I(\nu) = I_0 \exp\left[-\frac{Mc^2}{2RT\nu^2}(\nu - \nu_0)^2\right] \tag{8.11}$$

where M is the atomic mass of the emitter, R is the universal constant of gases and T is the temperature.

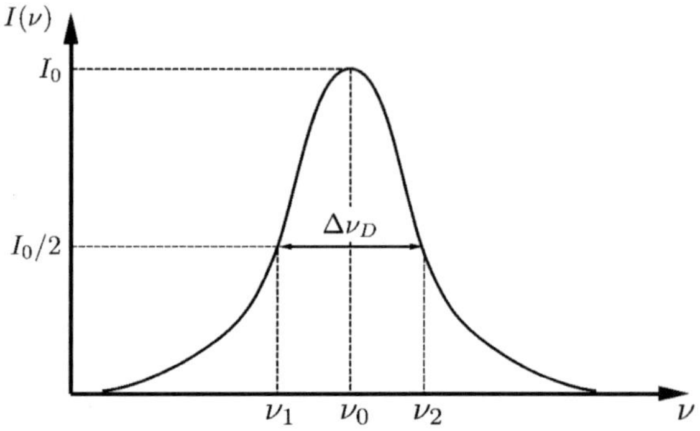

Fig. 8.4 Gaussian line profile due to Doppler broadening

Figure 8.4 shows a typical lineshape of a Doppler broadened line. In this figure, ν_1 and ν_2 the frequencies for $I_0/2$ and $\Delta\nu = \upsilon_2 - \upsilon_1$ is the FWHM which, from Eq. (8.11) is given by:

$$\Delta\nu_D = \nu_2 - \nu_1 = 2\,\frac{\nu_0}{c}\sqrt{\frac{2RT}{M}\ln 2} = 7.17\ 10^{-7}\nu_0\sqrt{\frac{T}{M}} \tag{8.12}$$

which can be written as;

$$\Delta\lambda_D = 7.17\ 10^{-7}\lambda_0\sqrt{\frac{T(K)}{M\ \left(\frac{g}{mol}\right)}} \tag{8.13}$$

8.1.3 Pressure Broadening

The nearby particles may affect the radiation emitted by an atom. Two limiting cases regulate how the interactions occur:

1. *Impact pressure broadening* or *collisional broadening*: The collision of neighbor particles with the emitting atom interrupt the emission process, and by shortening the characteristic times for the process, they increase the uncertainty in the energy emitted. The duration of a collision in the impact pressure broadening theory needs to be much shorter than the lifetime of the emitting process. This effect depends on both the density and the temperature of the gas. The broadening effect is described by a Lorentzian profile and there may be an associated shift.
2. *Quasi-static pressure broadening*: The presence of other particles, like quasi-static ions around the emitter, shifts the energy levels of the radiating particle, resulting in the change of frequency of the emitted radiation. The duration of the collision is much longer than the lifetime of the emission process. This effect depends on the density of the gas, but is rather insensitive to temperature. The form of the line profile is determined by the functional form of the perturbing

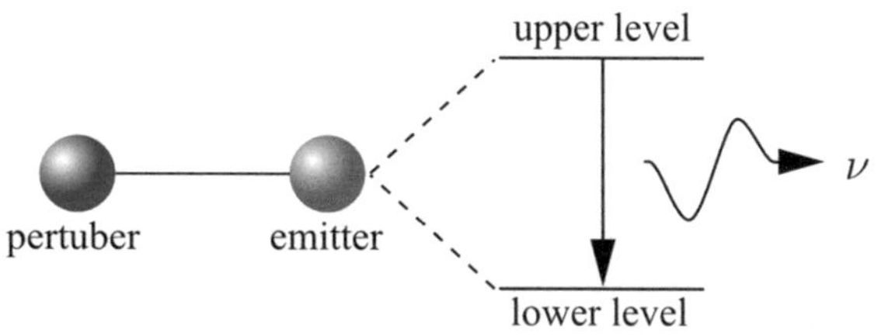

Fig. 8.5 Long-range interaction between a perturber and an emitter atom

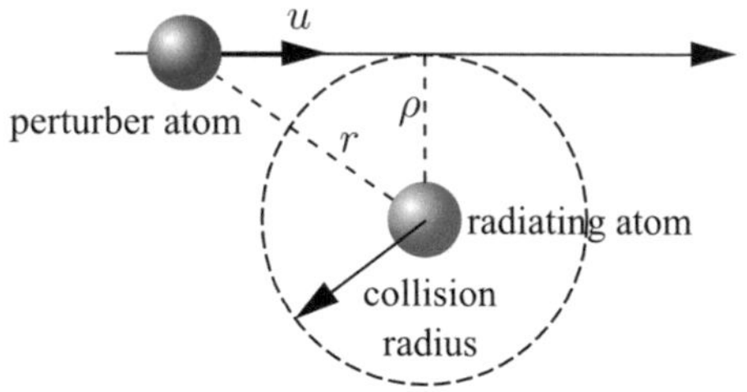

Fig. 8.6 Collision between a perturber and an emitter atom

force with respect to distance from the perturbing particle. There may also be a shift in the line center. A stable distribution is a general expression for the lineshape resulting from quasi-static pressure broadening.

In the present introductory text, we will describe the main broadening mechanisms using impact pressure broadening or collisional broadening theory because it is simple and help us to depict the physics involved in the line broadening by collisions.

8.1.3.1 Impact Broadening

Consider a system formed by a perturber atom and another excited atom as it can be seen in Fig. 8.5, where r is the interatomic distance.

Considering that a perturbation of a given energy level (ΔE) of an emitting particle is caused by another pertuber particle in its neighborhood at a given distance r, then we may write (Kuhn 1969):

$$\Delta E = h\Delta \nu \sim r^{-n} \tag{8.14}$$

The perturbation induced by this collision changes the frequency of the emitted photon according to:

$$\Delta \nu = C_n r^{-n} \tag{8.15}$$

here C_n is a constant due to the integrations of Eq. (8.14).

These kind of collisions are classified as long-range due to the Coulombian interaction. The shortest distance between the collider and the perturber is referred as impact parameter ρ, see Fig. 8.6.

Figure 8.6 depicts the collision between the perturber and the emitter atom. The change in the phase of the photon frequency during collision is a function of the parameter ρ, and the velocity u.

For small variation in the potential energy of the system, during the collision, it can be considered that the perturber atom is in a linear and uniform movement. This approximation is known as classical approximation path. A collision occurs when a phase change ε is verified. This phase shift may be calculated by:

$$\int_{-\infty}^{+\infty} \Delta\omega_0(t)dt = \varepsilon \tag{8.16}$$

substituting Eqs. (8.14) and (8.15) into (8.16) gives:

$$\varepsilon = 2\pi \int_{-\infty}^{+\infty} C_n r^{-n}(t)dt \tag{8.17}$$

From Fig. 8.6 it can be seen that $u^2t^2 + \rho^2 = r^2$ and the Eq. (8.17) may be written as:

$$\varepsilon = 2\pi \int_{-\infty}^{+\infty} C_n \frac{1}{[u^2t^2 + \rho^2 = r^2]^{n/2}} \tag{8.18}$$

Knowing that: $2\pi \int_{-\infty}^{+\infty} C_n \frac{1}{[u^2t^2 + \rho^2 = r^2]^{n/2}} = \frac{2\pi C_n a_n}{u\rho^{n-1}}$, then;

$$\varepsilon = \frac{2\pi C_n a_n}{u\rho^{n-1}} \tag{8.19}$$

The impact parameter may be expressed by:

$$\rho = \left(\frac{2\pi C_n a_n}{u\varepsilon}\right)^{1/(n-1)} \tag{8.20}$$

where (Kuhn 1969):

$$a_n = \sqrt{\pi}\frac{\Gamma\left[(n-1)/2\right]}{\Gamma\left[(n)/2\right]} = \pi, 2, \frac{\pi}{2}, \frac{4}{3}, \frac{3\pi}{8}, \ldots \quad \text{n} = 2, 3, 4, 6, \ldots \tag{8.21}$$

From Eq. (8.20) it can be seen that when the phase shift decreases, the impact parameter increases. The condition for the highest value of the impact parameter, i.e. lowest phase change, defines the Weisskopf's radius ρ_0, or optical collision radius (Griem 1997);

$$\rho_0 = \left(\frac{2\pi C_n a_n}{u}\right)^{1/(n-1)} \tag{8.22}$$

The mean free path λ of a particle before the collision takes place is defined by :

$$\lambda = \frac{1}{\sigma N_p} \tag{8.23}$$

N_p is the density of perturbing particles and σ is the collision cross section. Taking into account that the Weisskopf's radius defines, when a collision takes place, and knowing that a classical transversal cross section is defined by $\sigma = \pi\rho_0^2$, the mean frequency between collisions may be given by:

$$\nu_{col} = \frac{1}{\tau_{col}} = \pi\rho_0^2 N_P u \tag{8.24}$$

The impact broadening theory supposes that the collisions are adiabatic, i.e. the energy perturbations are only due to collisional processes. Applying the uncertainty principle:

$$\Delta E \tau_{col} = \frac{h}{4\pi} \tag{8.25}$$

$$\Delta\nu = \frac{\nu_{col}}{4\pi} = \frac{1}{4}\rho_0^2 N_P u \tag{8.26}$$

Substituting Eq. (8.22) into (8.26) it results:

$$\Delta\nu = \frac{1}{4} N_P u \left(\frac{2\pi C_n a_n}{u}\right)^{2/(n-1)} \tag{8.27}$$

as the mean particle velocity is given by $\left(\frac{8k_BT}{\pi\,\mu_{red}}\right)^{1/2}$, where k_B is the Boltzmann constant and μ_{red} is the reduced mass the Eq. (8.27) may be written as:

$$\Delta\lambda = \frac{1}{4c} N_P \lambda^2 (2\pi C_n a_n)^{2/(n-1)} \left(\frac{8k_BT}{\pi\,\mu_{red}}\right)^{(n-3)/2(n-1)} \tag{8.28}$$

From the Eq. (8.28) the FWHM due to collisions may be calculated by the impact theory. Pressure broadening may also be classified by the nature of the perturbing force as follows:

- *Van der Waals broadening* occurs when the emitting particle is being perturbed by van der Waals forces. For the quasi-static case, the line profile is given by a van der Waals profile when colliders are of different types. The energy shift as a function of distance may be estimated taking into account the Lennard-Jones potential as it will be shown.

- *Resonance broadening* occurs when the perturbing particle is of the same type as the emitting particle, which introduces the possibility of an energy exchange process.
- *Linear Stark broadening* occurs via the linear Stark effect which results from the interaction of an emitter with an electric field, which causes a shift in energy linear with the field strength.

A quadratic Stark broadening occurs via the quadratic Stark effect which results from the interaction of an emitter with an electric field causing a shift in energy quadratic with the field strength. This case will not be discussed in the present book.

8.1.3.2 Van der Waals Broadening

The Van der Waals forces are well represented by Lennard-Jones potential which is proportional to r^{-6}, $n = 6$ in Eq. (8.28), resulting in:

$$\Delta\lambda_{waals} = \frac{1}{4c} N_P \lambda^2 (2\pi C_6 a_6)^{2/5} \left(\frac{8k_B T}{\pi\, \mu_{red}} \right)^{3/10} \tag{8.29}$$

In this equation C_6, is given by (Griem 1997):

$$C_6 = e^2 \alpha R_j^2 \tag{8.30}$$

α is the mean polarizability of the atom and R_j^2 the difference between the squares of the coordinates:

$$\alpha = \frac{9}{2} \left(\frac{h^2 \varepsilon_0}{\pi m_e e^2} \right)^3 \tag{8.31}$$

$$R_j^2 = \frac{1}{2} \frac{E_H}{E_\infty - E_j} \left[5 \frac{z^2 E_H}{E_\infty - E_j} + 1 - 3l_j \left(l_j + 1 \right) \right] \ ; \ \begin{cases} j = s, \ upper\ level \\ \\ j = i, \ lower\ level \end{cases} \tag{8.32}$$

h is the Planck constant, e the electron charge, m_e the electron mass, E_H is the ionization potential of hydrogen atom, E_∞ ionization energy of the emitter, l_j the orbital quantum number, and z the effective charge.

Combining Eqs. (8.30), (8.31) and (8.32) and considering that $a_6 = 3\pi/8$, after Eq. (8.21):

$$\Delta\lambda_{waals} = \frac{1}{4c} N_P \lambda^2 \left(2\pi \frac{27\pi e^2}{16} \left(\frac{h^2 \varepsilon_0}{\pi m_e e^2} \right)^3 R_j^2 \right)^{2/5} \left(\frac{8k_B T}{\pi\, \mu_{red}} \right)^{3/10} \tag{8.33}$$

Using the perfect gas law the Eq. (8.33) may be written as:

$$\Delta\lambda_{waals} = \frac{1}{4c}\lambda_{si}^{2}\left(\frac{27h^{5}{\varepsilon_0}^{3}R_j^{2}}{8\pi {m_e}^{3}e^{4}}\right)^{2/5}\left(\frac{8k_BT_g}{\pi}\right)^{3/10}\frac{P}{k_BT}\sum\nolimits_{p}\frac{\chi_p}{{\mu_{red}}^{3/10}} \tag{8.34}$$

χ_p is the percentage of perturbing particles in the medium, λ_{si} is the wavelength of the transition from s (upper level) to i (lower level).

8.1.3.3 Resonance Broadening

In a resonance transition the potential where the emitting particle is immersed is proportional do r^{-3} (Kuhn 1969), $n = 3$. So, the broadening of an emitted line is according to Eq. (8.28) given by:

$$\Delta\lambda_{res} = \frac{1}{4c}N_P\lambda^{2}\left(2\pi C_3 a_3\right) \tag{8.35}$$

In this case C_3 is (Griem 1997):

$$C_3 = \frac{e^{2}f_{fi}}{32\pi^{3}\varepsilon_0 m_e\lambda_{fi}}\sqrt{\frac{g_f}{g_i}} \tag{8.36}$$

λ_{fi} is the transition wavelength between the initial i and final state f with statistical weights g_i and g_f respectively and f_{fi} is the oscillator strength of the transition. Substituting the Eq. (8.36) into (8.35) we obtain the FWHM width for a line broadening due to resonance collisions, which is Lorentzian:

$$\Delta\lambda_{res} = \frac{e^{2}f_{fi}}{16\pi^{2}\varepsilon_0 m_e c}n_f\lambda_{fi}\sqrt{\frac{g_f}{g_i}} \tag{8.37}$$

n_f is the perturber particles density in the fundamental state.

8.1.3.4 Stark Broadening

The first and more complete study about Stark broadening was realized by Griem (1964) in a systematic way, initially for hydrogen atoms. It was employed a classical theory for Stark broadening, by considering the microfields generated by charges around the emitter. In these first study the contributions of many collisions of a statistical ensemble of quasi-independent charged particles resulting in the broadening of the Balmer lines were taken into account. A general expression used in the literature is given by Griem (1964):

$$\lambda\ (nm) = 2.5\ 10^{-10}\ \alpha_{if}\left(n_e,\ T_e\right)n_e^{2/3} \tag{8.38}$$

the parameter $\alpha_{if}(n_e, T_e)$ is found in the Griem's book for each chemical element (not only hydrogen) for each transition i to f as a function of electron temperature T_e (K) and density n_e $\left(\text{cm}^{-3}\right)$.

A more complete study was recently realized by Gigosos and Cardenoso (Gig-Card model) (Gigosos and Cardenoso 1996) in which the Schrödinger equation was solved for the system atom-emitter, electrons and ions, with the ion dynamic taken into account. The authors give tables as a function of electron temperature (until 10^7 K), density (from 10^{14} cm^{-3} to 10^{18}cm^{-3}) and the ratio between the masses of emitter and perturber.

8.2 Spontaneous, Induced Emission and Absorption

An atom in an initial state $|i\rangle$ with an eigenvalue E_i may absorb photons of energy $h\nu$ from the electromagnetic field of intensity $I(\nu)$, as a function of frequency ν, and brought to an upper state $|u\rangle$ with a rate $W_{i,u}\left(s^{-1}\right)$:

$$W_{i,u}^{abs}\left(s^{-1}\right) = B_{i,u} I(\nu) \tag{8.39}$$

here $B_{i,u}$ is the Einstein coefficient for absorption. $I(\nu)$ is proportional to number of the photons and $h\nu$ is the photon energy, being $I(\nu) = n(\nu) h\nu$. The absorption of n photons depletes the radiation field by $nh\nu$.

Another process occurs when the radiation field induce transitions, from the upper state $|u\rangle$ to a lower state $|i\rangle$ leading to the emission of a photon with energy $\Delta\text{E} = h\nu = E_u - E_i$. This mechanism increases the photon number and both the initial and final photon emitted have the same direction. The rate of the induced emission of photons is:

$$W_{u,i}^{stim}\left(s^{-1}\right) = B_{u,i} I(\nu) \tag{8.40}$$

$B_{u,i}$ is the Einstein coefficient for induced emission.

Another kind of emission, called fluorescence, occurs spontaneously without the need of an external electric field. The emitted photon can take any arbitrary direction in any mode of the radiation field. The spontaneous probability of emission per second is given by:

$$W_{u,i}^{sp.\ em}\left(s^{-1}\right) = A_{u,i} \tag{8.41}$$

$A_{u,i}$ is the Einstein coefficient for spontaneous emission. It depends only of states $|u\rangle$ and $|i\rangle$ and is independent of the radiation field. Figure 8.7 depicts the processes of absorption, induced emission and spontaneous emission of radiation.

Supposing a system of two-levels in stationary conditions, the absorption rate will equal the induced and spontaneous emissions. If the density of particles populating the upper state $|u\rangle$ is N_u and of the lower state $|i\rangle$ is N_i:

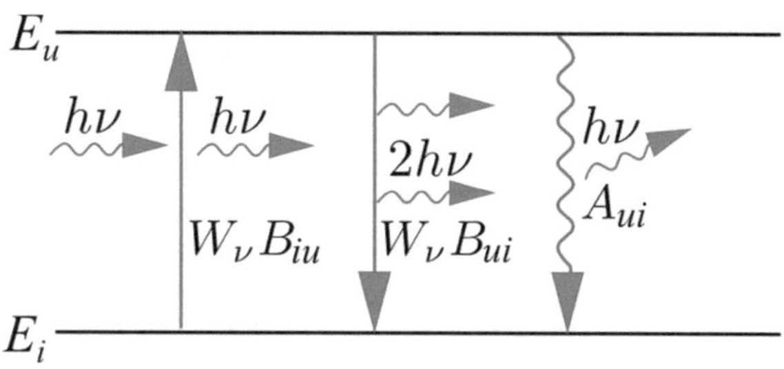

Fig. 8.7 Absorption, induced emission and spontaneous emission of radiation in a two-level system

$$B_{i,u} I(\nu) N_i = (B_{u,i} I(\nu) + A_{u,i})\ N_u \tag{8.42}$$

The relationship between N_i and N_u at thermal equilibrium is:

$$\frac{N_u}{N_i} = \frac{g_i}{g_u} e^{-\frac{E_u - E_i}{KT}} \tag{8.43}$$

Substituting (8.43) in (8.42) it results for $I(\nu)$:

$$I(\nu) = \frac{A_{u,i}/B_{u,i}}{(g_i/g_u)(B_{i,u}/B_{u,i})\, e^{h\nu/kT} - 1} \tag{8.44}$$

From Planck's radiation law the energy density of a given thermal radiation field is:

$$I(\nu) = \frac{8\pi h\nu^3}{c^3} \frac{1}{e^{h\nu/kT} - 1} \tag{8.45}$$

Inserting (8.45) into (8.44) and using microreversibility:

$$B_{u,i} = \frac{g_i}{g_u} B_{i,u} \tag{8.46}$$

$$A_{u,i} = \frac{8\pi h\nu^3}{c^3} B_{u,i} \tag{8.47}$$

The selection rules regulate the possible transitions in atomic and molecular systems. In these transitions energy and angular momentum need to be conserved according to the symmetry of the system. A transition occurs between an excited state $|u\rangle$ to a lower state $|i\rangle$ if the dipole moment of the transition is not zero:

$$D_{u,i} = e \int \psi_u^* \vec{r}\ \psi_i \neq 0 \tag{8.48}$$

$D_{u,i}$ is transition dipole moment, ψ_u^* is the complex conjugate of the wave function of the state $|u\rangle$, ψ_i is the wave function of the lower state $|i\rangle$, and $\vec{r}$ is the characteristic distance of the transition, between the states involved in the transition.

A deeper discussion about the transition dipole moment is out of the scope of the present book. Readers more interested should search in a more specialized book (Herzberg 1950). For our purpose, i.e. to use in emission spectroscopy to diagnose plasmas, the notations of the principal quantum number, angular moment and the selection rules will be presented.

The excited atomic levels are determined by the solution of the Schrödinger equation where the eigenfunctions are the wavefunctions and the eigenvalues are the energies of the levels. These wavefunctions are characterized by three quantum numbers; the principal quantum number n, the quantum number related to the orbital angular moment l, the magnetic quantum number m and the spin of electrons s. The range of possible l numbers are $l = 0,\ 1,\ 2, \ldots (n-1)$, for m $= -l \leq m \leq +l$ and $s = +1/2, -1/2$. Labeling of levels with angular quantum number l, the maximum number of electrons and degeneracy are shown in Table 8.1.

Wavefunctions for one electron are represented by one miniscule letter according to convection. When talking about light atoms, spins and angular quantum numbers combine according to Russell-Saunders in a LS coupling. In a paper published in 1925, Henry Norris Russell, a Princeton astronomer, and Frederick Albert Saunders, a Harvard physicist, the spin-orbit coupling was proposed for atoms with small number of electrons. For example, two valence electrons couple together to give a resultant L and the spins of the same electrons couple together to form S. Both L and S will in turn be coupled to form J, a vector representative of the total angular momentum of the atom. In atoms with higher number of electrons, the spin and angular momentum of each electron interact resulting in a j vector for this electron. Summing over all electrons it results the total angular momentum of the atom. This case is known as jj-coupling.

A level is represented by the electrons in the valence shell which is characterized by three quantum number L, S and J as:

$$^{2S+1}L_J \tag{8.49}$$

2S + 1 is the multiplicity of the level which is basically defined by the spin.

Table 8.1 Labeling of levels with angular quantum number l, magnetic quantum number m, name, maximum number of electrons and degeneracy

l	Letter	Name	m	Maximum number of electrons	Degeneracy
0	s	sharp	0	2	1
1	p	principal	$-1 \leq m \leq +1$	6	3
2	d	diffuse	$-2 \leq m \leq +2$	10	5
3	f	fundamental	$-3 \leq m \leq +3$	14	7
4	g		$-4 \leq m \leq +4$	18	9
5	h		$-5 \leq m \leq +5$	22	11

Table 8.2 Selection rules for electric dipole transitions

Selection rule	Comment
$\Delta l = \pm 1$	Atoms with one electron
$\Delta L = 0, \pm 1$	Atoms with many electrons L.S coupling; $0 \nrightarrow 0$
$\Delta S = 0$	
$\Delta M = 0, \pm 1$	
$\Delta J = 0, \pm 1$	$\Delta J = 0 \nrightarrow 0$

When a given level is excited it will be de-excited by "allowed transitions", i.e. the ones that obey the selection rules. The following rules, should be verified if transitions due to the changing of electric dipole occur (Table 8.2):

Besides electric dipole transitions there are also allowed transitions by electric quadrupole and magnetic dipole. They are much weaker than the electric dipole transitions by orders of magnitude and they become important when the electric dipole transitions are forbidden. Calculations show that probabilities for magnetic dipole and quadrupole electric transitions are 10^{-5} and 10^{-8} of dipole transition probabilities, respectively (Herzberg 1950). Electric quadrupole and magnetic dipole transition selection rules will not be discussed in the present book but readers interested may look in specialized literature (White 1934).

To calculate the total angular momentum of an atom J, we need to obtain the total orbital angular momentum $L = \sum_{i=1}^{Z} l_i$, the total spin $S = \sum_{i=1}^{Z} s_i$ and then to add L and S, i.e. $J = L + S$; here l_i and s_i are the orbital and spin angular momenta of individual electrons. This spin-orbit coupling is in the picture of Russell-Saunders (L.S) coupling valid for atoms with $1 \leq Z \leq 30$ electrons. For elements with atomic number greater than 30, spin-orbit coupling becomes more significant and the j-j coupling scheme is more appropriate to be used.

Four quantum numbers characterize completely a state of an atom. They are L, S, J and M, where $|L - S| \leq J \leq L + S$ and $-J \leq M \leq J$. Thus, there are $2S + 1$ values of J when $J > S$ and $2L + 1$ values when $L < S$. As the energy of the levels relies on J, the corresponding levels split into $(2J + 1)$-multiplet. This is important to determine the lowest energy level of a given atom configuration.

In a multiplet state it is useful to invoke the following Hund's rules to evaluate the order of energy levels:

1. The lowest energy level corresponds to the state with largest spin S;
2. Among the states with a given value of S, the lowest level is the one with the largest L;
3. For a subshell in which less than half is full, the lowest state is the one corresponding to $J = |L - S|$, and for a subshell more than half full the lowest state corresponds to $J = L + S$.

In the next section it will be discussed the molecular bands and structures. The knowing of the molecular structures and transitions is useful as support of diagnostics tools for studing experimentally low-temperature plasmas.

8.3 Molecular Bands and Structure

In this part of the chapter, we will present the main properties of molecular states. The description is not exhaustive and more specialized literature should be consulted for further details about the molecular structure (Steinfeld 1974). The interaction potential between atoms in diatomic molecules will be discussed to introduce the rotation and vibration of molecules. Electronic states and transitions will be presented briefly to familiarize the reader with spin-orbit interaction according to Hund's coupling schemes.

8.3.1 Interaction Potential in Diatomic Molecules

If two atoms combine to form a molecule, electrons of individual atoms rearrange in a way to equilibrate the charge distribution of the valence level. To accommodate the charge density, there is an attraction between the two nuclei and an interaction between negative charges of two atoms. In this charge distribution rearrangement, a stable molecule is formed. This spatial configuration of charges is accommodated in such way that the total energy is minimum.

When the atoms of a diatomic molecule are separated to intranuclear distances R larger than $r_A + r_B$, the interaction potential is regulated by van der Waals forces that scale as R^{-6} and the attraction between the nuclei is weak. On the other hand, for $R < r_A + r_B$ there is an overlap of the electron shells of A and B atoms and the electrostatic interaction due to the increased electron density between the two nuclei results in attraction forces that scale as R^{-12}.

The empirical potential of Lenard-Jones describes well the total range of intranuclear distances, as it is illustrated in Fig. 8.8:

$$E_{pot}^{LJ}(R) = \frac{a}{R^{12}} - \frac{b}{R^6} \tag{8.50}$$

where a and b are two constants that depend on the nature of atoms A and B determined by fitting the experimentally obtained potential curve or by very precise calculations using the Schrödinger equation.

It follows from Eq. (8.50) that $V_{pot}^{LJ}(R) = 0, R_0 = (a/b)^{1/6}$. The minimum of the potential curve of Fig. 8.8 is obtained doing $dE/dR = 0$. This minimum point of the potential curve gives the equilibrium distance $R_e = (2a/b)^{1/6} = 2^{1/6}R_0$. The binding energy at R_e is $E_B = -E_{pot}(R_e) = b^2/4a$.

Another good approximation for the diatomic molecular potential is the Morse potential introduced by Philip M. Morse in 1929. It is a convenient model for the potential energy of a diatomic molecule. It constitutes an approximation for the vibrational structure of the molecule because it explicitly includes the effects of bond breaking, such as the existence of unbound states. It also accounts for the anharmonicity of the oscillator. The Morse potential can also be used to model other

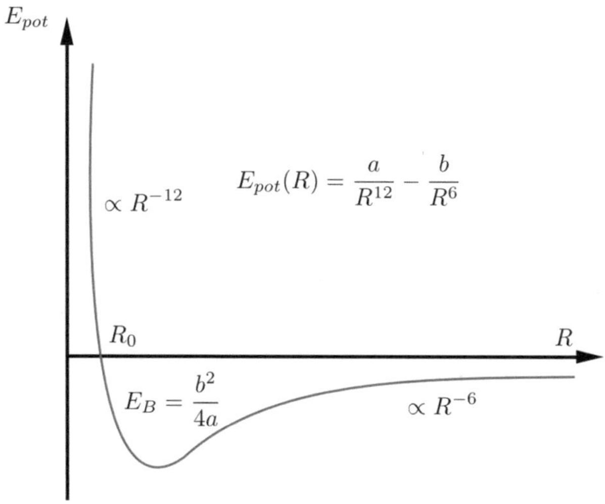

Fig. 8.8 Lenard-Jones potential

interactions such as the interaction between an atom and a surface. The proposed empirical potential has the form:

$$E_{pot}^{LJ}(R) = E_B\left[1 - e^{-a(R-R_e)}\right]^2 \tag{8.51}$$

8.3.2 Adiabatic Approximation

The nuclei, due to their large masses, move slower than electrons in the shell. The electrons can change their position in very fast manner when the nuclei present small deviations of their position. In other words, the vibration and rotation of the nuclei do not change during an electronic transition. The wave function of electrons is $\psi(R, r)$, where R is the nuclei coordinates and r the electrons one. The r coordinates of electrons are little affected by the changes in the nuclei positions once the kinetic energy of nuclei are much smaller than that of electrons. This kinetic energy of the nuclei can be regarded as a small perturbation in the Hamiltonian of the rigid molecule H_0. So the total wave function of the nuclei and electrons may be written as:

$$\psi\left(r_i, R_n\right) = \chi\left(R_n\right) . \varphi\left(r_i\right) \tag{8.52}$$

The electron position is r_i while the nuclear coordinate is R_n. $\chi(R_n)$, the nuclear wave function, may be considered independent of $\varphi(r_i)$ in a first approximation, where the electronic wave function of a rigid molecule has a fixed nuclear position R_n. The nuclear motion is independent of the electron motion and in this approximation the coupling between them may be neglected. This approximation, which is useful to solve problems for the electronic ground states of many molecules, is known as adiabatic approximation. It was introduced in 1927 by Max Born and Robert Oppenheimer and is named in the literature as Born-Oppenheimer approximation.

8.3.3 *Rigid Rotor*

Suppose a diatomic molecule with atomic masses M_a and M_b attached by an hypothetical rigid bar massless. This molecule may rotate around an axis that pass through the center of mass of the system, perpendicular to the internuclear axis with angular velocity ω, see Fig. 8.9.

The rotational energy of the rigid rotator is given by:

$$E_{rot} = \frac{1}{2} I\,\omega^2 = \frac{J^2}{2I} \tag{8.53}$$

the moment of inertia is $I = m_a r_a^2 + m_b r_b^2$, and the reduced mass $M = M_a.M_b/M_a + M_b$. The rotational angular momentum is $J = I\omega$. The contribution of electrons to this momentum is small due to their reduced mass as well as the rotation around the inter-nuclear axis. From quantum mechanics we know that $|J|^2 = J\,(J+1)\,\hbar^2$, where only discrete values of energy are permitted. The rotational energy at equilibrium distance R_e is represented by:

$$E_{rot} = \frac{J\,(J+1)\,\hbar^2}{2MR_e^2} \tag{8.54}$$

In spectroscopy is usually to express the rotational term values as $F(J) = E(J)/hc$;

$$F(J) = \frac{J\,(J+1)\,\hbar^2}{2hcMR_e^2} = B_e J\,(J+1) \tag{8.55}$$

where B_e is the rotational constant;

$$B_e\,\left(\mathrm{cm}^{-1}\right) = \frac{\hbar}{4\pi cMR_e^2} \tag{8.56}$$

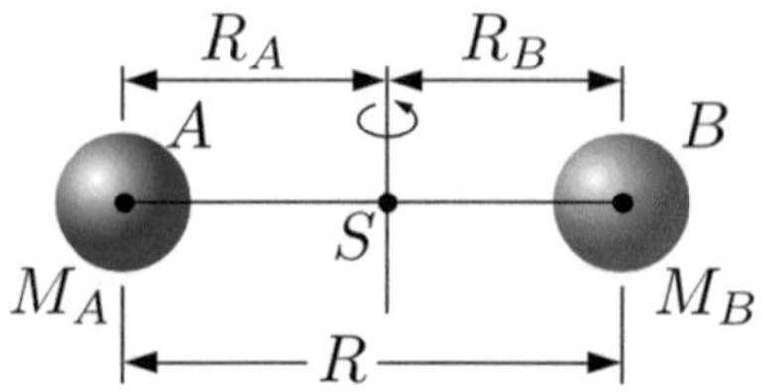

Fig. 8.9 Sketch of a rigid-rotor diatomic molecule

The absorbed radiation (falling in the spectral frequencies of $10^9 - 10^{13}$ Hz) between two adjacent levels may be written as:

$$\overline{\upsilon_{rot}}(J) = (E\,(J+1) - E(J))\,/hc = 2B_e\,(J+1) \tag{8.57}$$

The wavelengths associated with rotational transitions are observed in the range $\lambda = 10^{-5} - 10^{-1}$ m, i.e. microwave range.

8.3.4 Non-Rigid Rotor

In a real molecule the rotation induced centrifugal distortion increases the inter-nuclear distance and modifies the moment of inertia. The centrifugal force $M\omega^2 R$ must be compensated by a restoring force $-k\,(R - R_e)$ in order to keep the atoms together. As $J = I\omega = \sqrt{J\,(J+1)\,\hbar^2}$, it can be easily shown that:

$$R = R_e + \frac{J\,(J+1)\,\hbar^2}{MkR^3} \tag{8.58}$$

by expression (8.58) it can be seen that the centrifugal force, due to molecular rotation, has an effect of widening the equilibrium distance. The term-values are then corrected according to:

$$F(J) = B_e J\,(J+1) - D_e[J\,(J+1)]^2 + H_e[J\,(J+1)]^3 + \cdots \tag{8.59}$$

where the rotational constants are given by:

$$B_e\ \left(\mathrm{cm}^{-1}\right) = \frac{\hbar}{4\pi cMR_e^2},\quad D_e\ \left(\mathrm{cm}^{-1}\right) = \frac{\hbar^3}{4\pi ckM^2R_e^6},\quad H_e\ \left(\mathrm{cm}^{-1}\right) = \frac{3\hbar^5}{4\pi ck^2M^3R_e^{10}} \tag{8.60}$$

Usually the correction to second order is enough for most part of applications.

8.3.5 Vibration of Diatomic Molecules

Besides the rotation, the molecules vibrate. Around the minimum energy of the potential, in a first approximation we may think in a parabola with $E_{pot} \cong 1/2k_r(R - R_e)^2$, where k_r is a constant of a hypothetical spring that connects the two atoms. The eigenvalues of the vibrational energy are determined by the solution of the Schrödinger equation:

$$\frac{d^2\psi_{vib}}{dr^2} + \frac{2M}{\hbar^2}\left(E - \frac{1}{2}\omega^2 M\right)\psi_{vib} = 0 \tag{8.61}$$

where $\omega = \sqrt{k_r/M}$. The solution of this equation results in the energy levels of the harmonic oscillator:

$$E(v) = \left(v + \frac{1}{2}\right)\hbar\omega \tag{8.62}$$

the energy is quantized and depends on the vibrational quantum number $v = 0, 1, 2, \ldots$ Note that at $v = 0$ the energy is $E\,(v = 0) = \frac{1}{2}\hbar\omega$ and not zero. This called the "zero-point" energy in contrast with the zero energy of the rotator. If the vibrator was at zero energy at $v = 0$ of the ground electronic state a violation of the Heisenberg uncertainty relation would appear.

The exact solution of the Schrödinger equation gives the vibrational energy eigenvalues for vibrations that are larger and the elongation distance is far from R_e:

$$E(v) = \hbar\omega_0\left(v + \frac{1}{2}\right) - \frac{\hbar^2{\omega_0}^2}{4E_B}\left(v + \frac{1}{2}\right)^2 \tag{8.63}$$

E_B is the binding energy of the rigid rotator. The energy separation between adjacent vibrational levels decreases with the increase of the vibrational quantum numbers and is given by:

$$\Delta E(v) = E\,(v + 1) - E(v) = \hbar\omega\left[1 - \frac{\hbar\omega}{2E_B}\,(v + 1)\right] \tag{8.64}$$

In spectroscopy it is usually to express the vibrational term values as $G(J) = E(v)/hc$;

$$G(v) = \omega_e\left(v + \frac{1}{2}\right) - \omega_e x_e\left(v + \frac{1}{2}\right)^2 \tag{8.65}$$

where the vibrational constants are:

$$\omega_e = \frac{\omega_0}{2\pi c} \quad \text{and} \quad \omega_e x_e = \frac{\hbar\omega_0^2}{8\pi c E_B} \tag{8.66}$$

A characteristic frequency of vibrations is 10^{14} Hz and for rotations is 10^{11} Hz. During one rotation around 1000 vibrations take place resulting in a change of nuclear distance during one period of rotation.

Due to the centrifugal distortions, the rotational constants need to be corrected:

$$B_v(v) = B_e - \alpha_e\left(v + \frac{1}{2}\right) + \gamma_e\left(v + \frac{1}{2}\right)^2 + \cdots \tag{8.67}$$

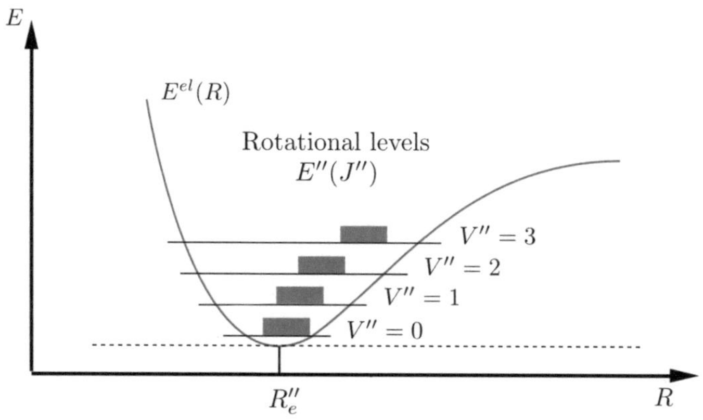

Fig. 8.10 Rotational and vibrational levels in a given electronic state of a diatomic molecule

$$D_v(v) = D_e - \beta_e\left(v + \frac{1}{2}\right) + \delta_e\left(v + \frac{1}{2}\right)^2 + \cdots \tag{8.68}$$

where $\alpha_e \ll B_e$ and $\beta_e \ll D_e$. In most cases the use of the first expansion term is enough.

8.3.6 Rotational-Vibrational Spectra

Transition between ro-vibrational levels belonging to the same electronic state are responsible for emissions in the infrared region of the spectra from 1 to 20 μm while pure rotational transitions, i.e. within the same vibrational and electronic states, give origin to emissions in the microwave region of the spectra that falls from 0.1 to 10 cm (Fig. 8.10).

For hetero-nuclear diatomic molecules there is a permanent dipole in the molecule and the radiation can be absorbed or emitted by the molecules. For homonuclear diatomic molecules, this is not the case and pure ro-vibrational spectra cannot occur. Only ro-vibratinal transitions between different electronic states are possible.

The term value of a molecule modelled as a rotor-vibrating, in a given electronic state, can be written as:

$$T(v,J) = G(v) + F_v(J) = \left[\omega_e\left(v+\tfrac{1}{2}\right) - \omega_e x_e\left(v+\tfrac{1}{2}\right)^2 + \omega_e y_e\left(v+\tfrac{1}{2}\right)^3 + \cdots\right] + \left[B_v J(J+1) - D_v[J(J+1)]^2 + H_v[J(J+1)]^3 - \cdots\right] \tag{8.69}$$

For the transitions is important to know the selection rules. The vibrational quantum level v can change by any integral amount, although $\Delta v = \pm 1$ gives by far the most intense lines of transitions. The sign + for convention stands for absorption of photons while the minus represents the emission of radiation. Of course $\Delta v = 0$ is related to pure rotational transitions. Higher order transitions such as $\Delta v = \pm 2, \pm 3, \ldots$, called overtones may be observed in some experiments. The rotational quantum number J can change only by unity (Herzberg 1950).

In the spectroscopy the upper energy state of a transition is labeled with prime as v', J' while the lower state with double prime v'', J''. Transitions with $\Delta J = J' - J'' = +1$ are called R-transitions whereas those with $\Delta J = J' - J'' = -1$, P-transitions. The ensemble of rotational transitions of a given vibrational transition form in the spectrum what is called a vibrational band. The frequencies of a ro-vibrational transition, in cm^{-1}, are given by:

$$\upsilon(v', J') - \nu(v'', J'') = \nu_0 + \left[B_v' J'(J'+1) - D_v'[J'(J'+1)]^2\right] - \left[B_v'' J''(J''+1) - D_v''[J''(J''+1)]^2\right] \quad (8.70)$$

ν_0 is the band origin of the transition. The transition $\Delta J = 0$ is the origin of Q-transitions or branch and it does not exist in pure ro-vibrational transitions of diatomic molecules. It appears in ro-vibrational transitions between electronic states with different total angular momentum as will be discussed the Sect. 8.3.7.

8.3.7 Electronic States and Transitions in Diatomic Molecules

The total energy of a molecule in a given electronic, vibrational and rotational state is given by:

$$E = E_e + E_v + E_r \quad (8.71)$$

or in terms of wave-numbers units, i.e. term values;

$$T = T_e + G + F \quad (8.72)$$

here the model of the vibrating rotator is used to represent the diatomic molecule in different electronic states.

The transitions between two electronic states are given by:

$$\nu = T' - T'' = (T_e' - T_e'') + (G' - G'') + (F' - F'') \quad (8.73)$$

Figure 8.11 depicts the structure of molecular transitions between two electronic states. It can be seen the rotational and vibrational structures formed during the electronic transitions as well as ro-vibrational spectra inside an electronic state. In the same manner pure rotational transitions in a given electronic and vibrational state are illustrated.

As the F terms are small if compared with G ones and can be neglected in order to obtain a coarse structure of the electronic transition. This rotationless structure is also called vibrational structure since only υ_ν is significant. The following formula represents these all-possible transitions between the different vibrational levels of the two participating electronic states as depicted in Fig. 8.11:

$$\nu = \nu_e + \left[\omega_e'\left(v' + \frac{1}{2}\right) - \omega_e{}'x_e{}'\left(v' + \frac{1}{2}\right)^2 + \omega_e{}'y_e{}'\left(v' + \frac{1}{2}\right)^3 + \cdots\right]$$
$$- \left[\omega_e{}''\left(v'' + \frac{1}{2}\right) - \omega_e{}''x_e{}''\left(v'' + \frac{1}{2}\right)^2 + \omega_e{}''y_e{}''\left(v'' + \frac{1}{2}\right)^3 + \cdots\right] \quad (8.74)$$

When a transition occurs between the same upper vibrational state ν' with different lower states v'' a v''-progression appears. The v''-progressions extend from the $v' = 0$ (first band) toward longer wavelengths. On the other hand another series of progressions may be observed when different transitions from v' levels fall in the same lower state v''. This is called a v'-progression and extend from the first band $v' = 0$ towards shorter wavelengths.

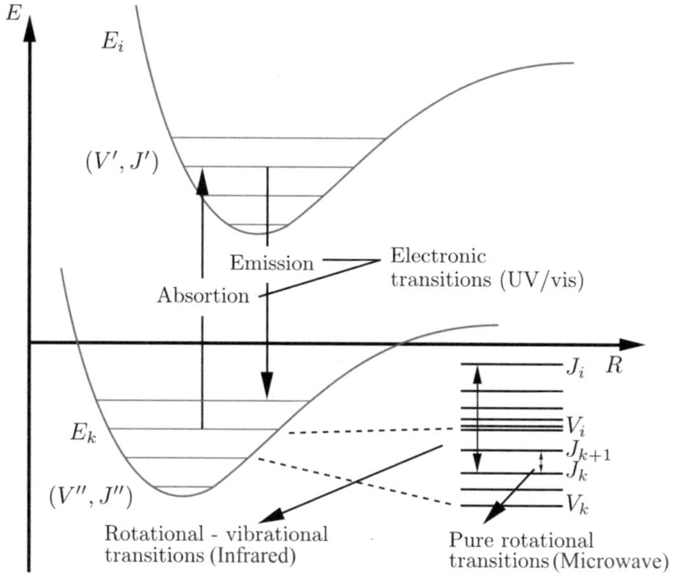

Fig. 8.11 Structure of molecular transitions between two electronic states

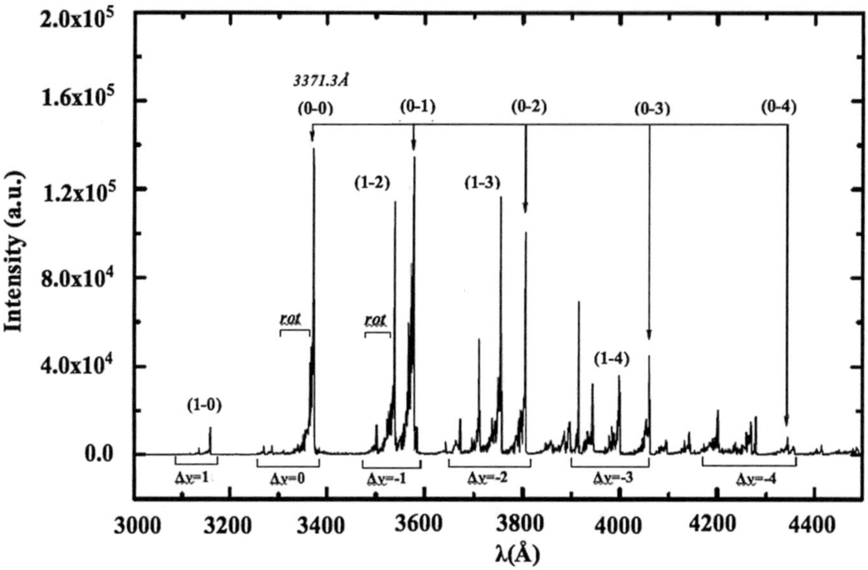

Fig. 8.12 Emission spectrum of the second positive system of nitrogen, transition $N_2\left(C^3\Pi_u \rightarrow B^3\Pi_g\right)$, recorded in a positive column of a direct current discharge. Discharge current of 20 mA and pressure of 2.0 Torr (Nagai 2004)

Sequences are group of transitions that have the same $\Delta v = v' - v''$. The bands of a sequence draw close together and are a characteristic of C_2, CN and N_2 spectra for example.

Figure 8.12 depicts a transition $N_2\left(C^3\Pi_u \rightarrow B^3\Pi_g\right)$, of the nitrogen molecule where it can be seen clearly the $\Delta v = 1, 0, -1, -2, -3, -4$ sequences that are grouped in a short region of the blue spectrum. It can be remarked many rotational bands belonging to different bands of sequences.

It is not the aim of this book to go deep inside in the structure of progressions and sequences but more details on progressions and bands may be found in classical textbooks (Herzberg 1950; Steinfeld 1974). The knowledge of these structures may be useful in optical diagnostic of molecular plasmas.

Until now the contribution of rotation to the energy term was neglected and only electronic and vibrational energies were taken into account. Now we are interested in the exploration of rotational structure belonging to electronic-vibrational transitions. The terms of wave-numbers units, i.e. term values may be written as:

$$\nu = \nu_0 + F'\left(J'\right) - F''\left(J''\right) \tag{8.75}$$

the quantity ν_0, called the band origin or the zero line, is equal to $\nu_e + \nu_v$ and is constant for a given electronic-vibrational transition while $F'(J')$ and $F''(J'')$ are the rotational terms of the upper and the lower state respectively. For a non-rigid vibrating rotator the term values may be written:

$$F_v(J) = B_v J\,(J+1) + (A - B_v)\,\Lambda^2 - D_v[J\,(J+1)]^2 \tag{8.76}$$

here A is a constant and Λ is the electronic angular moment component projection in the inter-nuclear axis. The term $(A - B_v)\,\Lambda^2$ is constant for a given vibrational level in an electronic state and may be neglected in calculations (Herzberg 1950). All-possible transitions between the different rotational levels of the two participating vibrational-electronic states are given by:

$$\begin{aligned} \nu = \nu_0 &+ \left[B_v{}'J'\,(J'+1) - D_v{}'[J'\,(J'+1)]^2 + \ldots\right] \\ &- \left[B_v{}''J''\,(J''+1) - D_v{}'[J''\,(J''+1)]^2 + \ldots\right] \end{aligned}$$

If at least one the two states has $\Lambda \neq 0$ the selection rule for J is:

$$\Delta J = J' - J'' = 0, \pm 1 \tag{8.78}$$

If the transition occurs between two $^1\Sigma$ states the $\Delta J = 0$ is forbidden.

As $\Delta J = 0, \pm 1$, three series of lines may be expected during a rotational transition between levels belonging to different electronic states. These series or branches are given by, neglecting higher orders terms D_v:

$$R\ branch;\ \nu = \nu_0 + 2B_v' + \left(3B_v' - B_v''\right) J + \left(B_v' - B_v''\right) J^2 = R(J) \tag{8.79}$$

$$Q\ branch;\ \nu = \nu_0 \qquad + \left(B_v' - B_v''\right) J + \left(B_v' - B_v''\right) J^2 = Q(J) \tag{8.80}$$

$$P\ branch;\ \nu = \nu_0 \qquad - \left(B_v' + B_v''\right) J + \left(B_v' - B_v''\right) J^2 = P(J) \tag{8.81}$$

The Q branch appears only when $\Lambda \neq 0$, i.e. it is not present in $^1\Sigma \rightarrow {}^1\Sigma$ transitions. When the Q branch is present, it overlaps the P and R ones resulting in more complicated spectrum.

Due to quadratics terms in Eqs. (8.79), and (8.81), one of the two branches turns back given origin to a band head, see Fig. 8.12. It occurs usually in most bands falling in the visible and ultraviolet regions of the spectrum. If $B_v' - B_v''$ is negative, the band head occurs in the R branch. The band is degraded or shaded toward red. On the other hand if $B_v' - B_v''$ is positive the band head falls on the P branch and is degraded toward the violet.

The intensity of the lines/bands emitted is proportional to the population in a given excited state N_i:

$$I_{em} = N_i\,A_{ij} h\nu \tag{8.82}$$

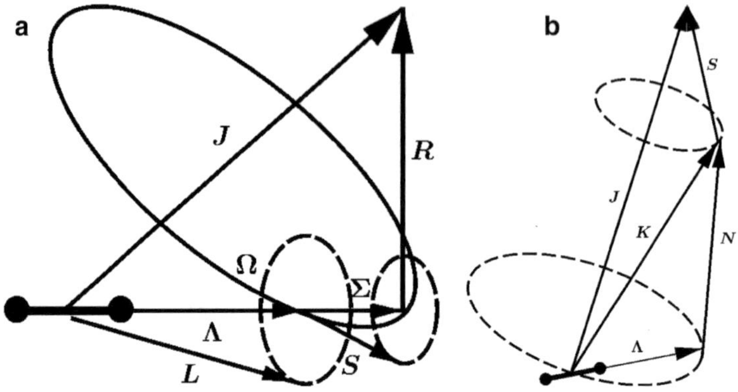

Fig. 8.13 (**a**) Hund's coupling case a and (**b**) Hund's coupling case b. Σ and Λ are the components of spin vector *S* and electronic orbital angular L around the inter-nuclear axis, respectively. R is the angular momentum of nuclear rotation

A_{ij} is the Einstein coefficient for emission between levels i and j. At thermal population the distribution of population of atoms/molecules in the excited level i is $N_i = N_j \frac{g_i}{g_j} e^{-E_{ij}/kT}$. E_{ij} is the energy difference between the i and j levels and g_i and g_j their statistical weight.

The selection rules for electronic dipole transitions are the result dipole changing during a given transition. For the angular momentum conservation, the following rule applies:

$$\Delta\Lambda = 0, \pm 1 \tag{8.83}$$

Due to symmetry properties of the wavefunction others rules become important:

$$+ \leftrightarrow -\text{and}- \leftrightarrow +;\ \text{but not}+ \not\leftrightarrow +\text{and}- \not\leftrightarrow \quad - \tag{8.84}$$

In the case of homonuclear diatomic molecules the following rules hold:

$$u \leftrightarrow g\ \text{and}\ g \leftrightarrow u\ ;\ \text{but not}\ u \not\leftrightarrow u\ \text{and}\ g \not\leftrightarrow g \tag{8.85}$$

u and g stands for ungerade (uneven) and gerade (even) from German.

The composition of different angular momenta in the molecule such as the electron spin, electronic orbital angular and nuclear momenta will result in total angular momentum J of the molecule. When the interaction of nuclear rotation with the electronic motion is very weak the electron spin and electronic orbital angular momenta precess around the line joining the nuclei, see Fig. 8.13a. In this case the mode coupling between moments called Hund's case a.

When $\Lambda = 0$ and $S \neq 0$ the spin vector S is no more coupled to the inter-nuclear axis and the addition of Λ and N form a resultant designed by K. It couples with

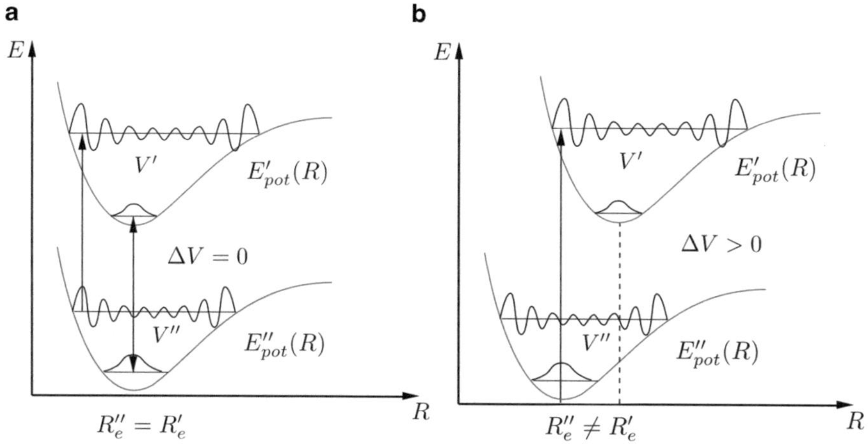

Fig. 8.14 Sketch of the Franck-Condon transition between electronic states with cases with the same R and with R $\neq$ R$'$. (**a**) Transitions with $\Delta v = 0$. (**b**) Transition with $\Delta v \neq 0$

the spin vector *S* to form the total angular momentum *J* of the molecule. This is the Hund's case b.

Hund's case a and b are found in the majority of electronic states of diatomic molecules. The rotational levels of the electronic states are defined according each coupling case. The aim is just to present to the reader the importance of these couplings. There are others Hund's cases of momenta coupling but they are not discussed in this book. More details can be seen in the Herzberg's book (Herzberg 1950).

The vibrational transitions between different electronic states are governed by Franck-Condon factor. It depends on the overlap of vibrational wave functions of the two electronic states. In Fig. 8.14 it can be seen that vertical arrows may represent electronic transitions between vibronic states, belonging to different electronic states, once electron transition is much faster than vibrational and rotational movements.

In the Franck-Condon approximation, the inter-nuclear distance R is the same for the start and final points of the transition. As the photon momentum is very small compared to the one of the vibrating nuclei, the momentum is conserved during the transition. The Franck-Condon transition probability is obtained from quantum mechanical calculation as:

$$P(R)dR = \frac{\psi'_{vib}(R)\psi''_{vib}(R)dR}{\displaystyle\int \psi'_{vib}(R)\psi''_{vib}(R)dR} \tag{8.86}$$

The Eq. (8.86) gives the transition probability in the interval dR around R. To illustrate the applications of atomic and molecular physics to diagnose low-temperature plasmas we will present and discuss, in next sections, some optical emission techniques like; actinometry, titration, rotational and vibrational temperature measurements and determination of electronic density by line Stark broadening.

8.4 Actinometry

Optical diagnostics are simple and easy to implement as tools for noninvasive measurements of nonequilibrium reactive plasmas to study the complex chemistry of the medium. From optical emission spectra it is usually straightforward to identify the emitting species and to follow qualitative changes in plasma properties as a function of external parameters such as gas composition, electric power, etc. Emission spectroscopy provides direct information only on plasma species, like excited atoms or molecules, in electronic ro-vibrational excited states. However, the vast majority of plasma species are in their ground electronic states. Therefore, it is more useful to monitor the concentration of ground state plasma species. For example, relationships between the etch rate of a material and the concentrations of key reactive species are useful for developing kinetic models of etching in reactive plasmas. Laser techniques are available for direct measurements of the concentration of many ground state species, but optical emission spectroscopy is much more widely available due to its simplicity. Thus, there is great interest in methods using optical emission as an indirect source of information about the concentration of ground state reactive species.

Among a wide variety of optical diagnostics techniques, a relatively and simple technique consists in the recording the intensity of lines and bands as a function of wavelength. Others techniques like laser-based ones, i.e. laser induced fluorescence, needs lasers beams as probes to interrogate specific optical transitions and recording of subsequent emissions.

In this section we discuss the simplest and easy technique of optical emission spectroscopy called actinometry in order to illustrate the power of optical diagnostics to study nonequilibrium plasma discharges.

The emission lines intensities of the atoms in the discharge are functions not only of the densities of the emitter excited atoms but they contain contributions from the cross sections for electron impact excitation of the excited levels as well as from the electron energy distribution function.

In order to measure the relative density of a given specie in the ground or metastable level using emission spectroscopy a technique known as actinometry which, under certain conditions, allows one to monitor the concentration using optical emission intensity ratios. This method offers the possibility of using optical emission spectroscopy as a powerful quantitative tool for the analysis of plasma composition. However, the absolute densities determination is difficult when using actinometry due to the uncertainties in the calibration of the experimental set-up. Actinometry is based on the addition of an inert trace gas of known concentration, without disturbing the discharge, to work as a reference. This inert gas added is called actinometer.

Although easy to handle than optical absorption spectroscopy or laser-induced fluorescence, the validity of actinometry is somewhat controversial and the criteria for utilization of the technique and its limits of validity must be verified in each case.

Optical emission actinometry involves the use of emission intensity ratios to monitor the concentration of ground state species. The emission line intensity from the species of interest is divided by the emission intensity of a line from an inert gas, the actinometer, which is added to the plasma in small quantities. This normalization of the emission signal is designed to correct, for changes in emission intensity, which result from variation in plasma electron density, cross section for electron impact excitation and the electron energy distribution function instead of changes in the ground state concentration of the species of interest.

The following three conditions are sufficient to ensure that the emission line intensity ratio of the emitter $X*$ and the actinometer $A*$, i.e. $\frac{I_{X*}}{I_{A*}}$ is proportional to the concentration ratio $\frac{[X]}{[A]}$:

1. $X*$ and $A*$ must be produced by electron impact excitation from the ground state species X and A;
2. $X*$ and $A*$ must decay primarily by photon emission, i.e. a corona equilibrium is assumed; and
3. the electron impact excitation cross sections for X and A must have a similar threshold and shape as a function of electron energy.

When these conditions are satisfied, electron impact excitation is balanced by optical emission, and the proportionality is assumed between the ratio of the emitted lines and the concentrations.

Supposing that these X_i^* states are mainly populated by electron impact from the ground state X:

$$e + X \overset{k_e^{X_i}}{\rightarrow} X_i^* + e \tag{8.87}$$

The de-excitation of excited states X_i^* is either radiative;

$$X_i^* \overset{A_{ij}}{\rightarrow} X_j^* + h\nu_{ij} \tag{8.88}$$

or non-radiative by quenching with species Q;

$$X_i^* + Q \overset{k_Q}{\rightarrow} X + Q \tag{8.89}$$

Empirical observations alone do not provide an understanding of the basis for the success or failure of optical emission actinometry. Such an understanding can be reached through identification of the basic mechanisms responsible for production of the excited state species.

The emission intensity I_{X*} of a transition $X_i^* \rightarrow X_j^*$ is then written as the ratio between production by electron impact and loss by radiative de-excitation and quenching:

$$I_{X*} = C_{cal}\frac{h\nu_{ij}A_{ij}k_e^{X_i}n_e}{\left(\sum A_{ij} + k_Q\,[Q]\right)}[X] \tag{8.90}$$

where n_e is the electronic density, $h\nu_{ij}$ is the emitted photon energy, A_{ij} is the Einstein coefficient of the observed transition $i \rightarrow j$, $\sum A_{ij}$ is the summation of all radiative de-excitation processes starting from the ith level and C_{cal} represents a constant which is dependent of the detection system response. The coefficient $k_e^{X_i}$ can be expressed by the following:

$$k_e^{X_i} = \left(\frac{2e}{m}\right)^{1/2}\int_{u_{th}}^{\infty} v(u)\sigma_i(u)f(u)^{1/2}du \tag{8.91}$$

where $v(u)$ is the electron velocity, $\sigma_i(u)$ is the collision cross section with threshold u_{th} for the excitation of the ith level and $f(u)$ is the electron energy distribution function (EEDF), normalized such that $\int_0^{\infty} f(u)u^{1/2}du = 1$. Substituting the Eq. (8.91) into (8.90) and doing the ratio of the emission intensity I_{X*} of a given line of the atom and the emission intensity I_{A*} of the actinometer, we obtain:

$$\frac{I_{X*}}{I_{A*}} = \frac{C_{X*}\,[X]\displaystyle\int_{u_{th}}^{\infty} v\,(u)\,\sigma_{iX*}\,(u)f\,(u)\,u^{1/2}du}{C_{A*}\,[A]\displaystyle\int_{u_{th}}^{\infty} v\,(u)\,\sigma_{iA*}\,(u)f\,(u)\,u^{1/2}du} \sim c\frac{[X]}{[A]} \tag{8.92}$$

$\sigma_{iX*}\,(u)$ and $\sigma_{iA*}\,(u)$ are excitation cross sections of the probed atom and the actinometer and C a proportionality constant. In practice, I_{X*}/I_{A*} may be proportional to *[X]/[A]* over some limited range of plasma parameters. For this reason, as stated above, empirical observations alone do not provide an understanding of the basis for the success or failure of actinometry. An example is the measurement of oxygen atoms, in the ground state, density in a discharge where argon is used as actinometer. Figure 8.15 illustrates the energy levels involved.

As an illustration of actinometry technique we present the detection of oxygen atoms in a low-pressure positive column oxygen discharge, with argon as actinometer. In low-pressure discharges the Ar($2p_1$) excited state is created by direct electron impact upon the ground state. Stepwise excitation from metastable states may be neglected, once these states of Ar are severely destroyed by collisions with oxygen atoms.

The oxygen excited levels O (3p ^{3}P) and O(3p ^{5}P) may be created from direct excitation of atoms O ($2p^{4}\,^3$P) in the ground state or by dissociative excitation by electron collisions from the ground state molecules. This second channel of excited

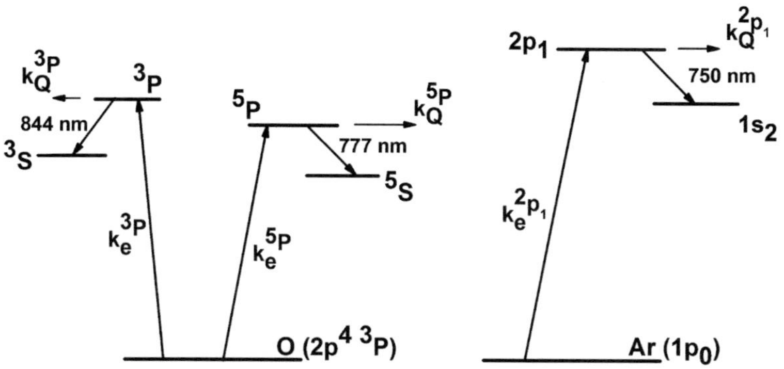

Fig. 8.15 Illustration of the excited levels involved in the determination of oxygen atom concentration in the ground state by actinometry (Pagnon et al. 1995)

atoms creation may cripple the actinometry and the validity conditions need to be determined. Readers interested may look at the work of Pagnon et al. (1995) in order to have an idea of the mechanisms involved.

8.5 Titration

Density of atoms in the ground state may also be determined using a titration technique, which is based on chemiluminescence reactions. It is mainly important to probe atoms in the discharge afterglow or post-discharges (Ricard 1996). The experimental arrangement is depicted in Fig. 8.16 showing plasma region of a molecular gas (A_2) and the molecular reactant gas (B_2) introduced in downstream region, with absence of charges, where a faint glow is produced by the specie AM*.

The absolute density of atom A may be determined by the detection of emitted molecular continuum AM^*:

$$A + M_2 \rightarrow AM^* + M \tag{8.93}$$

Fig. 8.16 Illustration of titration in a flowing afterglow. It can be seen that chemiluminescence reaction gives the origin to the faint glow in the post-discharge

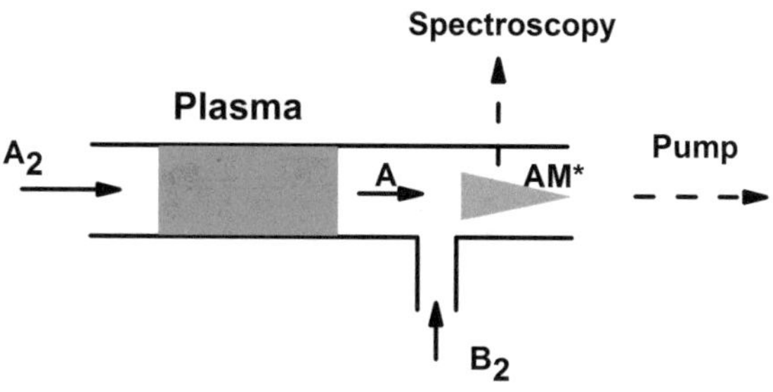

As an example, let's illustrate the detection of N atoms by titration in a flowing post-discharge. The molecular gas used to probe the atoms in this case is NO_β, which reacts with N atoms to form:

$$N + NO \rightarrow N_2 + O \tag{8.94}$$

The O atoms formed react with N atoms and other molecular species M_2 to form $NO\ (B\,^2\Pi)$:

$$O + N + M_2 \rightarrow NO\ \left(B\,^2\Pi\right) + M_2 \tag{8.95}$$

$$NO\ \left(B\,^2\Pi\right) \rightarrow NO\ \left(X\,^2\Pi\right) + h\nu\ \left(NO_\beta\ at\ \lambda = 320\ \text{nm}\right) \tag{8.96}$$

When the flow of NO is weak the reactions (8.95) and (8.96) are dominant and the emission of violet band of NO at 320 nm is strong. However, when the NO flux is greater than N flux, all N atoms are consumed by NO to produce O atoms according to Eq. (8.94) and the dominant reaction is:

$$O + NO \rightarrow NO_2\left(A\,^2B_1\right) \tag{8.97}$$

$$NO_2\ \left(A\,^2B_1\right) \rightarrow NO_2\ \left(X\,^2A_1\right) + h\nu\ \ (NO_2\ at\ \lambda = 575\ \text{nm}) \tag{8.98}$$

NO and O react forming NO_2 in the excited state that radiates a green continuum centered at 575 nm. The equilibrium between N and NO is the extinction point of optical emission, which allows the determination of the N atom density.

8.6 Rotational and Vibrational Temperature Measurements

The intensity of the rotational transitions of a band is proportional to the population of molecules in a given excited state. In this part of the text we will develop a little bit more the expression for emission intensity in order to determine rotational and vibrational temperatures of molecules in gas discharge plasmas.

The intensity of spectral lines in a transition between electronic states, $n'v'J'$ (upper) and $n''v''J''$ (lower) is given by:

$$I_{n''v''J''}^{n'v'J'} = N_{n'v'J'}hc\nu A_{n''v''J''}^{n'v'J'} \tag{8.99}$$

$N_{n'v'J'}$ is the molecular density in the upper state of the transition, $A_{n''v''J''}^{n'v'J'}$ is transition probability of the Einstein's coefficient, J is the rotational quantum number, v the vibrational quantum number and n the quantum number of the electronic state. The Einstein's coefficient $A_{n''v''J''}^{n'v'J'}$ is proportional to the electric dipole matrix of the transition **R** and is given by (Kovacs 1969):

$$A^{n'v'J'}_{n''v''J''} = \frac{64\pi^4 \nu^3}{3h} \frac{\sum \left| R^{n'v'J'}_{n''v''J''} \right|^2}{g_{j'}} \tag{8.100}$$

$g_{j'}$ is the degerencency degree of the upper state of the transition with rotational quantum number J' of multiplicity $2J' + 1$. Assuming that electronic, vibration and rotation motion are independent the summation on Eq. (8.100) may be written as:

$$\sum \left| R^{n'v'J'}_{n''v''J''} \right|^2 = \left| R^{n'n''}_{e} \right|^2 \left| R^{v'v''}_{vib} \right|^2 S_{J'J''} \tag{8.101}$$

$R^{n'n''}_{e}$ is the electronic transition moment, $R^{v'v''}_{vib}$ is the vibration transition moment and $S_{J'J''} = \sum_{-J''}^{j'} \left| R^{J'J''}_{rot} \right|^2$, i.e. the rotational transition moment or the Höln-London factor. The density of molecules in the initial upper state of the transition is given by:

$$N_{n'v'J'} = \frac{N}{Q} \left(2 - \delta_{0,\Lambda'}\right) \left(2J' + 1\right) e^{-\left(T_{n'} + G_{v'} + F_{J'}\right) hc/kT} \tag{8.102}$$

N is the total density of the molecules in the upper state $T_{n'}$ (electronic term), $G_{v'}$ (vibrational term) and $F_{J'}$ (rotational term). T is the temperature and Q the partition function. The factor $(2 - \delta_{0,\Lambda'})$ represents the Λ-splitting, with $\delta_{0,\Lambda'} = 1$ for $\Lambda' = 0$ and $\delta_{0,\Lambda'} = 0$ for $\Lambda' \neq 0$. Then, the intensity $I^{n'v'J'}_{n''v''J''}$ in the Eq. (8.99) may be written now as:

$$I^{n'v'J'}_{n''v''J''} = C \left| R^{n'n''}_{e} \right|^2 \left| R^{v'v''}_{vib} \right|^2 e^{-G_{v'} hc/kT} S_{J'J''} e^{-F_{J'} hc/kT} \tag{8.103}$$

where the constant C is:

$$C = \frac{64\pi^4 c \nu^4}{3} \left(2 - \delta_{0,\Lambda'}\right) e^{-T_{n'} hc/kT} \tag{8.104}$$

The density of molecules in a ro-vibrational level of an electronic transition is:

$$N_{n'v'J'} = C e^{-\left(G_{v'} + F_{J'}\right) hc/kT} \tag{8.105}$$

For a rotational transition in a given vibrational level of an electronic state, the term $C \left| R^{n'n''}_{e} \right|^2 \left| R^{v'v''}_{vib} \right|^2$ is constant;

$$I^{n'v'J'}_{n''v''J''} = C' S_{J'J''} e^{-F_{J'} hc/kT} = C' S_{J'J''} e^{-B'_{v} J'(J'+1) hc/kT} \tag{8.106}$$

here the centrifugal distortion correction was not taken into account.

The rotational temperature in a discharge may be, under certain conditions (Otorbaev et al. 1989), considered as representative of the gas temperature, a parameter of fundamental importance to study the kinetic of the plasma. Using the Eq. (8.106), the rotational temperature may be calculated by:

$$\ln\left(\frac{I^{n'v'J'}_{n''v''J''}}{C'S_{J'J''}}\right) = -\frac{B'_v J'\,(J'+1)\,hc}{kT_g} \tag{8.107}$$

The plot of $\ln\left(\frac{I^{n'v'J'}_{n''v''J''}}{C'S_{J'J''}}\right)$ as a function of $B'_v J'\,(J'+1)\,hc$ gives a straight line whose slope is inversely proportional to the gas temperature. Taking as example the Second Positive System of nitrogen we analyze the transition $v' = 0 \to v'' = 0$ and the branch R to estimate the gas temperature of a positive column direct current discharge. The figure below depicts a typical spectrum of the $N_2\left(C^3\Pi_u \to B^3\Pi_g\right)$ transition;

Figure 8.17 shows a well resolved emission spectrum of the Second Positive System of Nitrogen, transition $N_2\left(C^3\Pi_u,\ v' = 0 \to B^3\Pi_g,\ v'' = 0\right)$ recorded with a very high-resolution monochromator. It can be seen the rotational structure of Q, P and R branches. For $18 \le J' \le 19$ it can be remarked the triplet due to spin-orbit coupling. The R branch is degraded towards the blue part of the spectrum and is the best resolved branch, reason why it is commonly employed in the determination of rotational temperature. The band-head is located in the P branch and the Q branch overlaps the P and R branch for $J' \le 19$.

The assignment of most intense lines in the P, Q and R branches shows that for low J values the three branches overlap, which makes difficult to use these intensities for rotational temperature measurements. Only $J \ge 14$, is a well resolved structure present and can be used in the determination of rotational temperature.

The states $N_2(C^3\Pi_u)$ and $N_2(B^3\Pi_g)$ have electronic orbital angular Π with $\Lambda = 1$ and $\Delta\Lambda = 0$ which give the Höln-London factor (Kovacs 1969):

$$S^R_{J'J''} = \frac{(J''+1+\Lambda'')\,(J''+1-\Lambda'')}{J''+1} \tag{8.108}$$

As the R branch corresponds to the transition $\Delta J = +1, J' = J'' + 1$, so:

$$S^R_{J'} = \frac{(J'+1)\,(J'-1)}{J'} = \frac{J'^2-1}{J'} \tag{8.109}$$

Substituting (8.109) into (8.107) gives:

$$\ln\left(\frac{IJ'}{J'^2-1}\right) = C' - \frac{B'_v J'\,(J'+1)\,hc}{kT_r} \tag{8.110}$$

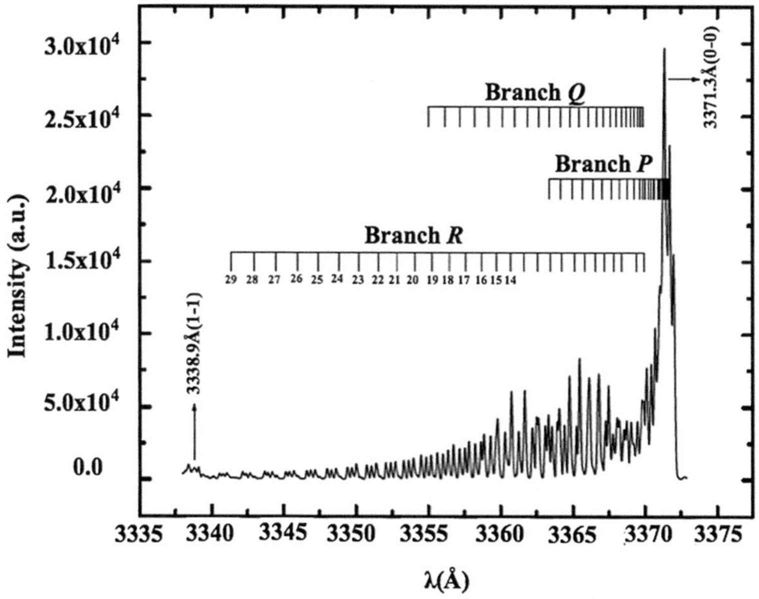

Fig. 8.17 Spectrum of the second positive system of nitrogen, transition $N_2\left(C^3\Pi_u,\ v'=0\rightarrow B^3\Pi_g,\ v''=0\right)$ recorded in a positive column discharge in nitrogen. Discharge current of 20 mA and pressure of 2.0 Torr (Nagai 2004)

C' is a constant. The plot of $\ln\left(\frac{IJ'}{J'^2-1}\right)$ against $J'\,(J'+1)$ enables one to determine the rotational temperature once the slope of a straight line that joins the points is inversely proportional to the temperature. With rotational constants for the state $N_2(C^3\Pi_u)$ (Loftus and Krupenie 1977) the rotational temperature may be determined:

$$T_r = \frac{B'_v J'\,(J'+1)\,hc}{k\ln\left(\frac{IJ'}{J'^2-1}\right)} \tag{8.111}$$

Figure 8.18 shows the experimental points measured in a positive column of a direct current discharge with pressure of 2 Torr and currents of 10, 20, 40 and 50 mA.

As can be seen two regions with different slopes are found. The first one is for the interval $14 \le J' \le 24$ corresponding to the lowest temperature and a second one for $24 \le J' \le 29$ with highest temperature. The lowest one, 370 K for discharge current of 10 mA, is adopted as the gas temperature. The highest, in the case of nitrogen discharges, is attributed to mixing reactions with the $N_2\left(E^3\Sigma_g^+\right)$ state which perturbs the rotational-vibrational distribution of the state $N_2(C^3\Pi_u)$.

From the vibrational spectra the vibrational temperature may be determined. The knowledge of the vibrational temperature is an important parameter to characterize the vibrational distribution function (VDF). The influence of the VDF in the

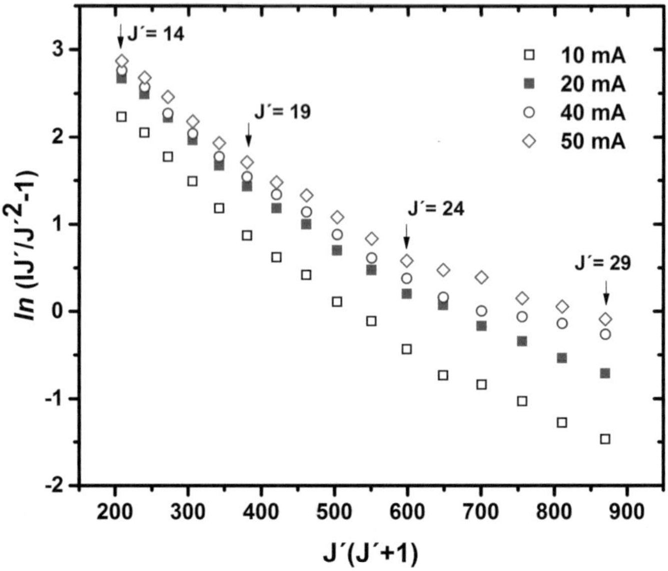

Fig. 8.18 Experimental points obtained from spectra of the second positive system of nitrogen, R branch, pressure of 2 Torr as a function of discharge current. Two regions with different slopes are readily identified

kinetic of the discharge if of fundamental importance to predict the ionization and dissociation rates in discharges. In most molecular gases discharges the VDF is far from equilibrium and is best described by a modified Treanor distribution (Treanor et al. 1968) named Treanor-Gordiets as shown in (Gordiets et al. 1998).

Experimentally the vibrational temperature is determined from the population distribution of the first vibrational levels, which is close to a temperature defined by a Boltzmann distribution of levels population. This temperature is calculated from many emission bands of a given electronic transition in the same sequence as shown in the Fig. 8.19 for the Second Positive System of Nitrogen molecule.

Assuming a Boltzmann distribution for the population of the first vibrational levels and neglecting contributions of terms v^2 and v^3 one can write:

$$ln\frac{N_v}{N_0} = \frac{hc\omega_e\left(v+\frac{1}{2}\right)}{kT_v} \tag{8.112}$$

The slope of the straight line that fits the points in a plot of $ln\frac{N_v}{N_0}$ as a function of $\left(v+\frac{1}{2}\right)$ is inversely proportional to the vibrational temperature. The ratio $\frac{N_v}{N_0}$ may be determined from the spectrum using the formula:

$$\frac{N_v}{N_0} = \frac{I_{v'v''}\lambda_0 A_0}{I_0\lambda_{v'v''}A_{v'v''}} \tag{8.113}$$

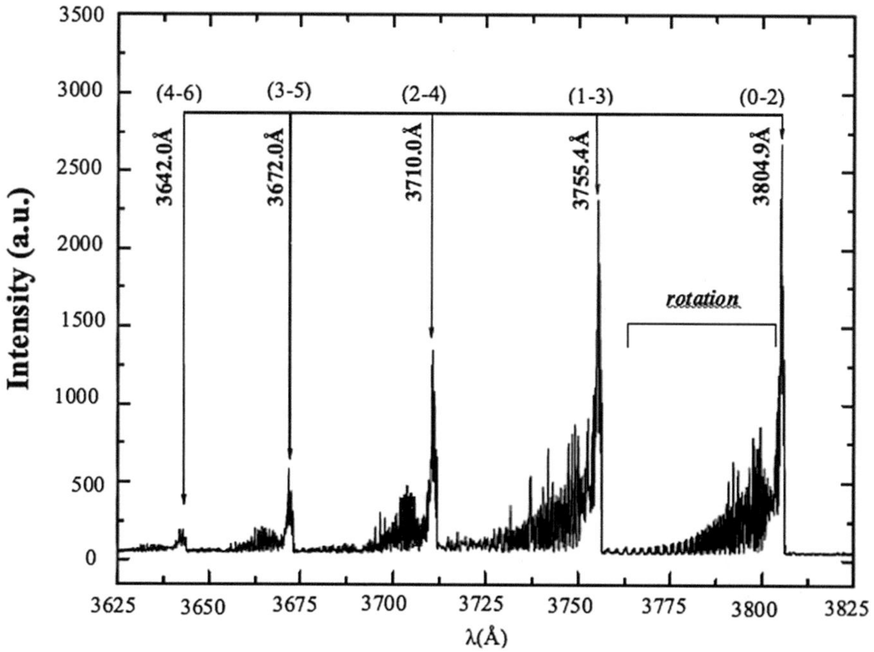

Fig. 8.19 Spectrum of the sequence $\Delta\nu = -2$ of the second positive system of nitrogen recorded in a positive column of a nitrogen discharge with pressure of 2 Torr and current of 20 mA (Nagai 2004)

8.7 Electronic Density by Line Stark Broadening

Measurement of electron density in electrical discharges is one of the most important diagnostics to characterize low-temperature plasmas once the charges regulate the physics and chemistry of the medium. To have access to the density of electrons, optical emission spectroscopy is a very interesting technique because it is non-invasive, like for example electrical probes, and offers good spatial and temporal resolutions.

In this chapter we will focus on plasmas with electron number densities greater than $5 \times 10^{13}\text{cm}^{-3}$ where the technique of measurement of electron density from line broadening may be used. This range of density is normally found in low-temperature plasmas working at atmospheric and sub-atmospheric pressure ranges. For low-pressure discharges, electrical probes are normally employed although laser techniques may eventually be used as it will be discussed in Chap. 10.

Optical emission spectroscopy technique based on line broadening measurement has the possibility to do spatially and temporally resolved electron number density measurements exploring the lineshape of the Balmer β transition (4–2) of atomic hydrogen at 486.132 nm. In some experimental conditions, in higher density plasmas, Balmer α transition (3–2) of atomic hydrogen at 656.279 nm may also be used but this line is more perturbed by resonance and van der Walls widening.

This technique requires the addition to the plasma of a small amount (typically 1 or 2 % mole fraction) of hydrogen, which may either come from dissociated water vapour in humid air or from premixing H_2 into the gas stream. For detection of this line by emission, the population of the n = 4 electronic state of atomic hydrogen must be high enough for the H_β line to be distinguishable from underlying gas plasma emission (in air plasmas mostly First and Second Positive Systems of nitrogen). This condition is usually fulfilled in equilibrium air plasmas or in non-equilibrium plasmas with sufficient excitation of hydrogen electronic states.

The collisional processes occurring in plasmas perturb the emitting atoms and/or molecules influencing the shape and width of emitted spectral lines. The main mechanisms responsible for H_α and H_β line broadenings are instrumental, natural, Doppler, Stark, resonance, and van der Waals. Among these effects resonance and natural ones are not important in most experimental conditions found in atmospheric and sub-atmospheric pressure plasmas and may be neglected. The table below presents formulas to calculate the various line broadening mechanisms that affects lines H_α and H_β emitted in plasmas.

The emission line best suited for electron density measurement is the H_β because the Stark effect is more pronounced while resonance and van der Waals are less important in line widening compared with H_α line. The figure below shows the FWHM for the main broadening mechanisms when the electron density varies from 10^{13}cm^{-3} to 10^{16}cm^{-3}, values normally found in high-pressure low-temperature plasmas.

In the Fig. 8.20 the simulation was done for a typical value of electron and gas temperatures of 0.8 eV and 2000 K respectively, which are typical of atmospheric plasmas. As can be seen for electron density above 10^{14}cm^{-3} the Stark effect is the dominant mechanism for line broadening of the H_β line.

Figure 8.21 shows the H_β line profile recorded in radio-frequency atmospheric micro-plasma jet as a function of powers for various pressures (Souza-Corrêa et al. 2010). In the case of 50 W the FWHM is 2.25 Å. Considering that the apparatus function in this case is 0.2 Å and observing the Fig. 8.20, it can be seen that the most important effect for the line broadening is the Stark effect. The electron density in this case is found to be 1.3×10^{15}cm^{-3}.

Table 8.3 Full width at half maximum (FWHM) (nanometers) of various line broadening mechanisms, for the H_α and H_β lines at 656.279 nm and 486.132 nm respectively. P is the pressure in atm and T the gas temperature in K. n_e is the electron density in cm^{-3} and X_H is the mole fraction of hydrogen atoms

Line	$\Delta\lambda_{Stark}$	$\Delta\lambda_{resonance}$	$\Delta\lambda_{van\ der\ Waals}$	$\Delta\lambda_{natural}$	$\Delta\lambda_{Doppler}$
H_α	$1.0 \times 10^{-11} n_e^{0.649}$	$119\, X_H\, P\, T^{-1}$	$3.6\, P\, T^{-0.7}$	2.02×10^{-4}	$2.35 \times 10^{-4} T^{0.5}$
H_β	$2.0 \times 10^{-11} n_e^{0.668}$	$60.4\, X_H\, P\, T^{-1}$	$3.6\, P\, T^{-0.7}$	1.8×10^{-4}	$3.48 \times 10^{-4} T^{0.5}$

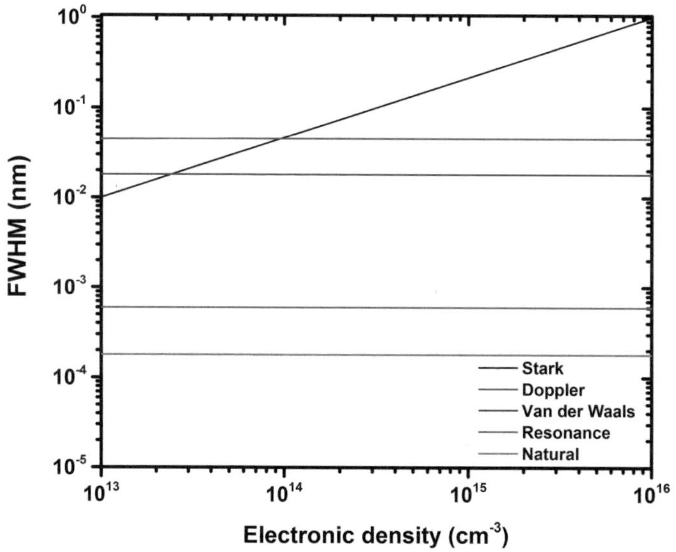

Fig. 8.20 FWHM for various broadening mechanisms of H_β line as a function of electron density in an atmospheric pressure discharge. Gas temperature of 2000 K and electron temperature of 0.8 eV (Souza-Corrêa et al. 2010)

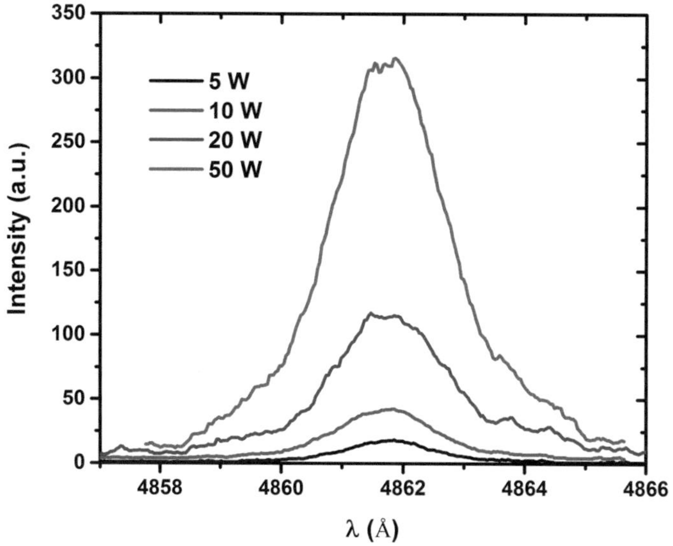

Fig. 8.21 Emission spectra of H_β line at the exit of a radio-frequency atmospheric micro-plasma jet as a function of RF power

Exercises

Exercise 8.1. Taking into account the Eq. (8.1) obtains the mean radiated power by the damped classical oscillator. Then calculates the natural line width of the emitted line emitted spontaneously from a given energy level E_i.

Resolution: The mean radiated power, from Eq. (8.1), is given by:

$$\left\langle \frac{dW}{dt} \right\rangle = \left\langle -\gamma m\dot{x}^2 \right\rangle = -\frac{\gamma}{2} m x_0^2 \omega_0^2 e^{-i\gamma t}$$

this expression shows that the mean radiated power decreases exponentially with time and the 1/*e* of its initial value is attained at $\tau_i = 1/\gamma$. In this case τ_i is the average lifetime of the level at energy E_i and is related to the Einstein coefficient A_i by $\tau_i = (A_i)^{-1}$.

Exercise 8.2. Calculate the natural, van der Waals, Resonance and Stark broadening for the Balmer H_α line.

Resolution: The energy level diagram of H_α line (transitions 3s-2p, 3p-2s and 3d-2p) is shown below (a). Others transitions that are important in the calculations of broadening mechanisms are: 3p-1s (b) and 2p-1s (c) (Fig. 8.22).

Resonant Broadening:

The transitions to be considered in the case of H_α line are n = 3 to n = 2 ($s \rightarrow i$) but also from n = 3 to n = 1 ($f \rightarrow s$) and from n = 2 to n = 1 ($f \rightarrow i$). From Eq. (8.37):

$$\Delta\lambda_{res} = \frac{3e^2}{16\pi^2\varepsilon_0 m_e c^2}\lambda_{si}{}^2\left(\lambda_{if}f_{if}\sqrt{\frac{g_f}{g_i}}n_f + \lambda_{fs}f_{fs}\sqrt{\frac{g_f}{g_s}}n_f + \lambda_{si}f_{si}\sqrt{\frac{g_i}{g_s}}n_i\right)$$

The constants for the energy levels involved in the H_α line are according to Wise et al. (1996):

$$f_{if} = 0.4162;\ f_{fs} = 0.079;\ f_{si} = 0.425$$

$$\lambda_{if} = 121.567\ \text{nm};\ \lambda_{if} = 102.5795\ \text{nm};\ \lambda_{si} = 656.279\ \text{nm}$$

As the statistical weight is $g = 2n^2$;

$$g_f = 2;\ g_i = 8;\ g_s = 18$$

Considering that the population distribution in the energy levels is given according to a Boltzmann distribution:

$$n_i = n_f\,\frac{g_i}{g_f}\,e^{-\left(\frac{E_i}{k_B T}\right)}$$

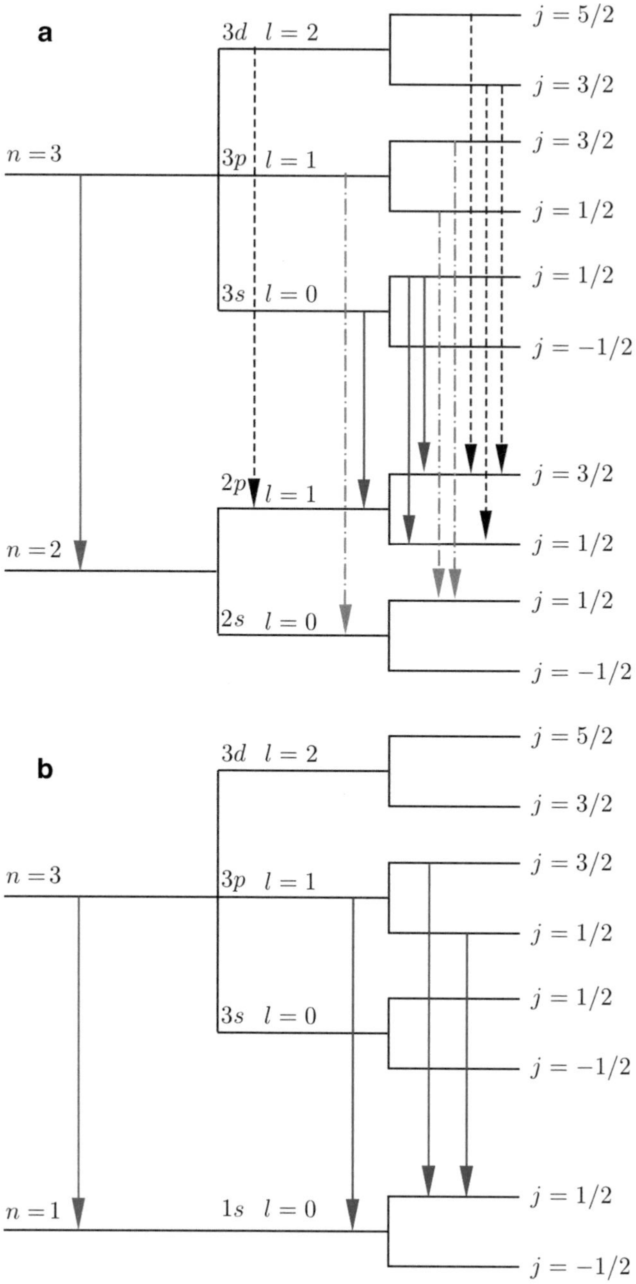

Fig. 8.22 (Continue)

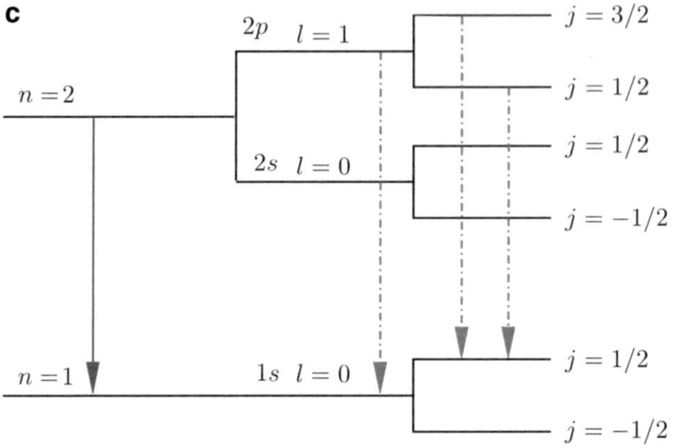

Fig. 8.22 Energy level diagram of H_α line

and that the energy of the lower level n = 2 is 1.634×10^{-18}J;

$$\Delta\lambda_{res}\,(\text{nm}) = 4.25\,\frac{P\,(atm)}{T_g(K)}\left(27.99 + 0.744\,10^3 e^{-\left(\frac{1.18\,10^5}{T_g\,(K)}\right)}\right)$$

Doppler Broadening:

$$\Delta\lambda_D = 7.17\,10^{-7}\lambda_{si}\sqrt{\frac{T(K)}{M\,\left(\frac{\text{g}}{\text{mol}}\right)}}$$

For the H_α line $\Delta\lambda_{si}$ is 656.279 nm and M = 1 g.mol^{-1} so:

$$\Delta\lambda_D\ (\text{nm}) = 4.7\,10^{-4}\,\sqrt{T\,(K)}$$

Natural Broadening:

Considering the transitions and energy levels involved in H_α line:

$$\Delta\lambda_{natural} = \frac{\lambda_{si}^2}{4\pi c}\left(\sum_{n<3} A_{3n} + \sum_{m<2} A_{2m}\right)$$

$$\begin{aligned}\sum_{n<3} A_{3n} = {} & \left(A_{32}l_{j'=1/2,j''=1/2} + A_{32}l_{j'=1/2,j''=3/2}\right)_{3s-2p} \\ & + \left(A_{32}l_{j'=1/2,j''=1/2} + A_{32}l_{j'=3/2,j''=1/2}\right)_{3p-2s} \\ & + \left(A_{32}l_{j'=3/2,j''=3/2} + A_{32}l_{j'=3/2,j''=1/2} + A_{32}l_{j'=5/2,j''=3/2}\right)_{3d-2p} \\ & + \left(A_{31}l_{j'=1/2,j''=1/2} + A_{31}l_{j'=3/2,j''=1/2}\right)_{3p-1s} = 5.149\,10^8 s^{-1}\end{aligned}$$

$$\sum_{m<2} A_{2m} = \left(A_{21} l_{j'=3/2,j''=1/2} + A_{21} l_{j'=3/2,j''=1/2}\right)_{2p-1s} = 1.253 \times 10^9 \text{ s}^{-1}$$

then $\left(\sum_{n<3} A_{3n} + \sum_{m<2} A_{2m}\right) = 1.768 \times 10^9 \text{ s}^{-1}$ and $\Delta\lambda_{natural}(nm) = 2.02\ 10^{-4}$.

Van der Waals Broadening:

$$R_j^2 = \frac{1}{2}\frac{E_H}{E_\infty - E_j}\left[5\frac{z^2 E_H}{E_\infty - E_j} + 1 - 3l_j\left(l_j + 1\right)\right]\ ;\ \begin{cases} j = s,\ upper\ level \\ j = i,\ lower\ level \end{cases}$$

$E_H = E_\infty = 13.6\ eV$ and $E_j = 12.09$ eV for the H_α transition. Z = 1 for a non-ionized particle. As the H_α line has seven multiplets, see the transitions diagram in the beginning of this solution, it can be written:

$$l_j = 0 \quad \rightarrow \quad R_j^2 l_{il=0} = 4.5033\ [5 \times 9.007 + 1\ ] \sim 207.31$$

$$\rightarrow \left(R_j^2\right)_{l_j=0}^{2/5} = 8.45$$

$$l_j = 1 \quad \rightarrow \quad R_j^2 l_{lj=1} = 4.5033\ [5 \times 9.007 + 1 - 6] \sim 180.29$$

$$\rightarrow \left(R_j^2\right)_{l_j=0}^{2/5} = 7.99$$

$$l_j = 2 \quad \rightarrow \quad R_j^2 l_{lj=2} = 4.5033\ [5\ \times\ 9.007 + 1 - 18] \sim 126.24$$

$$\rightarrow \left(R_j^2\right)_{l_j=0}^{2/5} = 6.93$$

The transition probability for H_α line in each transition is; 1.6 % for 3s-2p, 17 % for 3p-2s and 81.4 % for 3d-2p. We may now write a final expression $(R_j^2)^{2/5}$:

$$\left(R_j^2\right)^{2/5} = 0.016\left(R_j^2\right)_{l_j=0}^{2/5} + 0.17\left(R_j^2\right)_{l_j=1}^{2/5} + 0.814\left(R_j^2\right)_{l_j=2}^{2/5} = 7.135$$

$$\Delta\lambda_{waals} = \frac{1}{4c}\lambda_{si}^2\left(\frac{27h^5{\varepsilon_0}^3 R_j^2}{8\pi {m_e}^3 e^4}\right)^{2/5}\left(\frac{8k_B T}{\pi}\right)^{3/10}\frac{P}{k_B T}\sum_p \frac{\chi_p}{{\mu_{red}}^{3/10}}$$

$$\Delta\lambda_{waals} = 3.422\ 10^{-41}\lambda_{si}^2\frac{P}{k_B T^{0.7}}\sum_p \frac{\chi_p}{{\mu_{red}}^{3/10}}$$

Stark Broadening:

Using the expression given in Table 8.3 we have: $\Delta\lambda_{Stark} = 1.0 \times 10^{-11} n_e^{0.649}$, for an electron density of $n_e = 10^{12}\text{cm}^{-3}$, $\Delta\lambda_{Stark} = 2.74 \times 10^{-3}$ nm.

Exercise 8.3. Calculate the natural, van der Waals, Resonance and Stark broadening for the Balmer H_β line.

Resolution: The energy level diagram of H_β line (transitions 4s-2p, 4p-2s and 4d-2p) (a). Others transitions that are important in the calculations of broadening mechanisms are: 4p-1s (b) and 2p-1s (c) are shown below (Fig. 8.23):

Resonant Broadening:

The transitions to be considered in the case of H_β line are n = 4 to n = 2 ($s \to i$) but also from n = 4 to n = 1 ($f \to s$) and from n = 2 to n = 1 ($f \to i$). From Eq. (8.37) and taking into account the constants for the energy levels involved in the H_α line are according to Wise et al. (1996):

$$f_{if} = 0.4162;\ f_{fs} = 0.02899;\ f_{si} = 0.1193$$

$$\lambda_{if} = 121.567\ \text{nm};\ \lambda_{if} = 97.2537\ \text{nm};\ \lambda_{si} = 486.132\ \text{nm}$$

As the statistical weight is $g = 2n^2$;

$$g_f = 2;\ g_i = 8;\ g_s = 32$$

Considering that the population distribution in the energy levels is given according to a Boltzmann distribution and that the energy of the lower level n = 2 is 1.634×10^{-18}J;

$$\Delta\lambda_{res}\,(\text{nm}) = 2.34\,\frac{P\ (atm)}{T_g(K)}\left(26.0 + 0.116 \times 10^3 e^{-\left(\frac{1.18\times10^5}{T_g\ (K)}\right)}\right)$$

Natural Broadening:

Considering the transitions and energy levels involved in H_β line:

$$\Delta\lambda_{natural} = \frac{\lambda_{si}^2}{4\pi c}\left(\sum_{n<4} A_{4n} + \sum_{m<2} A_{2m}\right)$$

$$\begin{aligned}\sum_{n<4} A_{4n} = {} & \left(A_{42}l_{j'=1/2,j''=1/2} + A_{42}l_{j'=1/2,j''=3/2}\right)_{4s-2p} \\ & + \left(A_{42}l_{j'=1/2,j''=1/2} + A_{42}l_{j'=3/2,j''=1/2}\right)_{4p-2s} \\ & + \left(A_{42}l_{j'=3/2,j''=3/2} + A_{42}l_{j'=3/2,j''=1/2} + A_{42}l_{j'=5/2,j''=3/2}\right)_{4d-2p} \\ & + \left(A_{41}l_{j'=1/2,j''=1/2} + A_{41}l_{j'=3/2,j''=1/2}\right)_{4p-1s} = 1.995\ 10^8 s^{-1}\end{aligned}$$

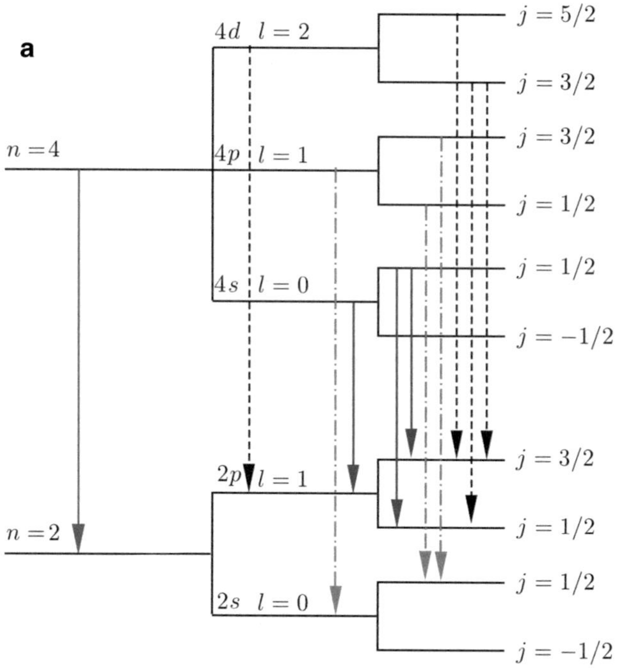

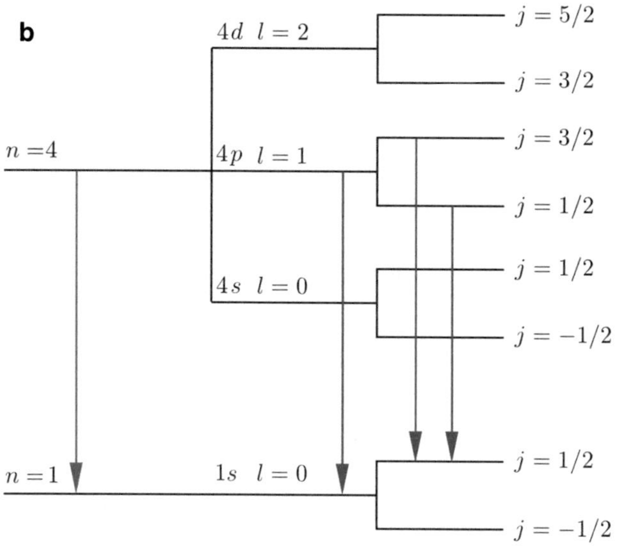

Fig. 8.23 (Continue)

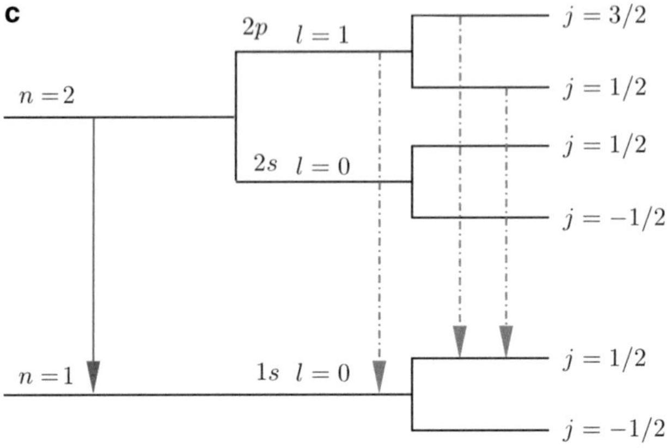

Fig. 8.23 Energy level diagram of H_β line

$$\sum_{m<2} A_{2m} = \left(A_{21} l_{j'=1/2, j''=1/2} + A_{21} l_{j'=1/2, j''=3/2}\right)_{2p-1s} = 1.253\ 10^9 s^{-1}$$

then $\left(\sum_{n<4} A_{4n} + \sum_{m<2} A_{2m}\right) = 1.543\ 10^9 s^{-1}$ and $\Delta\lambda_{natural}(nm) = 1.83\ 10^{-4}$.

Doppler broadening:

For the H_β line $\Delta\lambda_{si}$ is 486.132 nm and M = 1 g · mol^{-1} so:

$$\Delta\lambda_D\ (\text{nm}) = 3.5\ 10^{-4}\ \sqrt{T\ (K)}$$

Van der Waals broadening:

$E_H = E_\infty = 13.6$ eV *and* $E_j = 12.75$ eV for the H_β transition. Z = 1 for a non-ionized particle. As the H_β line has seven multiplets, as the H_α line, see the transitions diagram in the beginning of this solution, it can be written:

$$l_j = 0 \quad \rightarrow \quad R_j^2 l_{il=0} = 8.0\ [5 \times 16.0 + 1\,] \sim 648 \quad \rightarrow \quad \left(R_j^2\right)^{2/5}_{l_j=0} = 13.32$$

$$l_j = 1 \quad \rightarrow \quad R_j^2 l_{lj=1} = 8.0\ [5 \times 16.0 + 1 - 6] \sim 600 \quad \rightarrow \quad \left(R_j^2\right)^{2/5}_{l_j=0} = 12.92$$

$$l_j = 2 \quad \rightarrow \quad R_j^2 l_{lj=2} = 8.0\ [5 \times 16.0 + 1 - 18] \sim 504 \rightarrow \left(R_j^2\right)^{2/5}_{l_j=0} = 12.05$$

The transition probability for H_α line in each transition is; 1.9 % for 4s-2p, 21.6 % for 4p-2s and 76.5 % for 4d-2p. We may now write a final expression $(R_j^2)^{2/5}$:

$$\left(R_j^2\right)^{2/5} = 0.019\left(R_j^2\right)^{2/5}_{l_j=0} + 0.216\left(R_j^2\right)^{2/5}_{l_j=1} + 0.765\left(R_j^2\right)^{2/5}_{l_j=2} = 12.26$$

$$\Delta\lambda_{waals} = \frac{1}{4c}\lambda_{si}^2\left(\frac{27h^5{\varepsilon_0}^3R_j^2}{8\pi {m_e}^3e^4}\right)^{2/5}\left(\frac{8k_BT}{\pi}\right)^{3/10}\frac{P}{k_BT}\sum_p\frac{\chi_p}{{\mu_{red}}^{3/10}}$$

$$\Delta\lambda_{waals} = 5.88\ 10^{-41}\lambda_{si}^2\frac{P}{k_BT^{0.7}}\sum_p\frac{\chi_p}{{\mu_{red}}^{3/10}}$$

Stark broadening:

Using the expression given in Table 8.3 we have: $\Delta\lambda_{Stark} = 2.0 \times 10^{-11}n_e^{0.668}$, for an electron density of $n_e = 10^{12}$ cm^{-3}, $\Delta\lambda_{Stark} = 9.66\ 10^{-3}$ nm.

Exercise 8.4. Find the resultant FWHM of the convolution of two Gaussian profiles.
Resolution:

Convolution of two Gaussian profiles:

$$f(\Delta\nu) = \int_{-\infty}^{+\infty} f_1(\Delta\nu - \Delta\nu')f_2(\Delta\nu')\,d(\Delta\nu')$$

where $\Delta\nu = \nu - \nu_0$.

$$f(\Delta\nu) = \int_{-\infty}^{+\infty} I_1(\nu_0)\ e^{-\frac{(\Delta\nu-\Delta\nu')^2c^2}{2v^2\nu_{01}^2}}\ I_2(\nu_0)\,e^{-\frac{\Delta\nu'^2c^2}{2v^2\nu_{02}^2}}\ d(\Delta\nu')$$

But the FWHM value is:

$$\Delta\nu_{1/2} = \Delta\nu_{1,2} = \nu_{01,2}\,(v/c)\,(2\ln 2)^{1/2}$$

$$\Delta\nu_{1,2}^2 = \frac{(\nu_{01,2})^2v^2}{c^2}\,(2\ln 2)$$

$$f(\Delta\nu) = I_1(\nu_0)\ I_2(\nu_0)\int_{-\infty}^{+\infty} e^{-\frac{(\Delta\nu-\Delta\nu')^2 ln2}{\Delta\nu_1^2}}\ e^{-\frac{\Delta\nu'^2 ln2}{\Delta\nu_2^2}}\ d(\Delta\nu')$$

$$f(\Delta\nu) = I_1(\nu_0)\ I_2(\nu_0)\int_{-\infty}^{+\infty} e^{-\frac{\Delta\nu^2-2\Delta\nu\Delta\nu'+\Delta\nu'^2 ln2}{\Delta\nu_1^2}}\ e^{-\frac{\Delta\nu'^2 ln2}{\Delta\nu_2^2}}\ d(\Delta\nu')$$

$$f(\Delta\nu) = I_1(\nu_0)\ I_2(\nu_0)\int_{-\infty}^{+\infty} e^{-\left[\left(\frac{1}{\Delta\nu_1^2}+\frac{1}{\Delta\nu_2^2}\right)ln2\ \Delta\nu'^2+\frac{2\Delta\nu}{\Delta\nu_1^2}ln2\ \Delta\nu'^2+\frac{\Delta\nu^2}{\Delta\nu_1^2}ln2\right]}d(\Delta\nu')$$

Doing: $a = \left(\frac{1}{\Delta\nu_1^2} + \frac{1}{\Delta\nu_2^2}\right) ln2\ \Delta\nu'^2$; $b = \frac{2\Delta\nu}{\Delta\nu_1^2} ln2\ \Delta\nu'^2$; $c = \frac{\Delta\nu^2}{\Delta\nu_1^2} ln2$ and taking into account that:

$$\int_{-\infty}^{+\infty} e^{-\left(ax^2+bx+c\right)} dx = \sqrt{\frac{\pi}{a} e^{\left(b^2-4ac\right)/4a}}$$

$$f(\Delta\nu) = I_1 I_2 \sqrt{\frac{\pi}{\left(\frac{1}{\Delta\nu_1^2} + \frac{1}{\Delta\nu_2^2}\right) ln2}}\, e^{-\left[\frac{-\frac{\Delta\nu^2 ln2}{\Delta\nu_1^4} + \left(\frac{1}{\Delta\nu_1^2} + \frac{1}{\Delta\nu_2^2}\right)\frac{\Delta\nu^2}{\Delta\nu_1^2}}{\left(\frac{1}{\Delta\nu_1^2} + \frac{1}{\Delta\nu_2^2}\right)}\right]}$$

At FWHM, $e^{-\left[\frac{-\frac{\Delta\nu^2 ln2}{\Delta\nu_1^4} + \left(\frac{1}{\Delta\nu_1^2} + \frac{1}{\Delta\nu_2^2}\right)\frac{\Delta\nu^2}{\Delta\nu_1^2}}{\left(\frac{1}{\Delta\nu_1^2} + \frac{1}{\Delta\nu_2^2}\right)}\right]} = \frac{1}{2}$; $\frac{-\frac{\Delta\nu^2 ln2}{\Delta\nu_1^4} + \left(\frac{1}{\Delta\nu_1^2} + \frac{1}{\Delta\nu_2^2}\right)\frac{\Delta\nu^2}{\Delta\nu_1^2}}{\left(\frac{1}{\Delta\nu_1^2} + \frac{1}{\Delta\nu_2^2}\right)} = ln2$

So $\Delta\nu^2 = \Delta\nu_1^2 + \Delta\nu_2^2$

Exercise 8.5. Find the spectroscopic notation for the ground state configuration of the Sc.

Resolution: Sc has 21 electrons. The orbital configuration of the Sc ground state is: $(1s)^2(2s)^2(2p)^6(3s)^2(3p)^6(4s)^2(3d)^1$ or $[Ar](4s)^2(3d)^1$. The total angular momentum of Sc is determined by the 3d electron, once $[Ar]$ and $(4s)^2$ have $S = 0$ and $L = 0$. Since the 3d electron has $S = 1/2$ and $L = 2$, the total angular momentum may have the following values: $|2 - 1/2| \ \leq J \ \leq 2 + 1/2$, yielding $J = 3/2$ and 5/2 The possible states for the ground level are: ${}^2D_{3/2}$ or ${}^2D_{5/2}$. According the third Hund's rule, less than half of the shell is full, the total angular momentum is given by $J = |L - S| = |2 - 1/2| = 3/2$ so the lowest level is ${}^2D_{3/2}$.

Exercise 8.6. Considers the boron atom with the following orbital configuration: $(1s)^2(2s)^2(2p)^1$. Then the angular momentum of the boron is determined by the 1p electron, with $S = \frac{1}{2}$ and $L = 1$, once 1s and 2s subshells are completely filled. Determine the lowest energy level of the atom.

Resolution: The coupling of $S = \frac{1}{2}$ and $L = 1$ yields $J = 1/2$ or 3/2 resulting in two possible states: ${}^2P_{1/2}$ or ${}^2P_{3/2}$. By the third Hund's rule for atoms, the 2p shell of the boron is less than half full and the value of J corresponding to the lowest state is $J = |L - S| = 1 - 1/2 = 1/2$.; hence ${}^2P_{1/2}$ is the lower energy state.

Exercise 8.7. The hydrogen molecule has a rotational constant, in the electronic ground state, of 60.8 cm^{-1}. Calculate the moment of inertia and obtain an expression for the rotational energy knowing that the equilibrium distance is 0.74×10^{-10} m.

Resolution: The reduced mass of the hydrogen molecule is 8.35×10^{-28} kg. As the moment of inertia is given by $I = m_a r_a^2 + m_b r_b^2$ and the equilibrium distance $R_e = r_a + r_b$, the moment of inertia may be written as $I = \frac{m_a m_b}{m_a + m_b} R_e{}^2 = \mu R_e{}^2$.

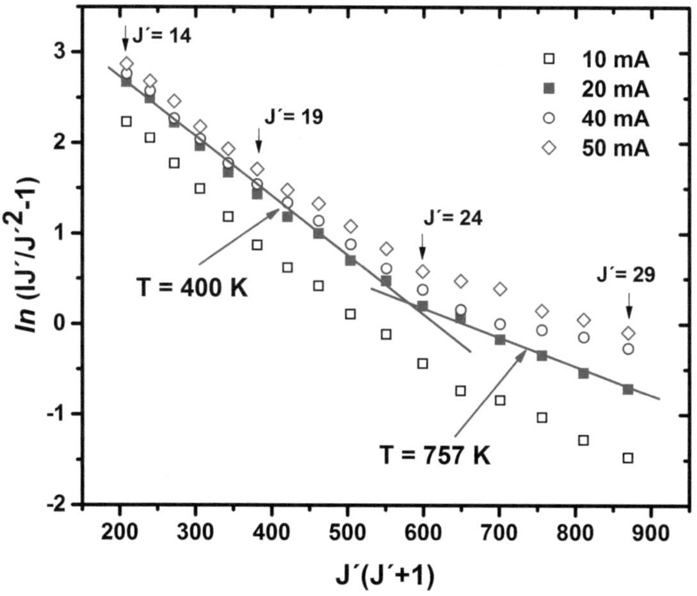

Fig. 8.24 Experimental points obtained from spectra of the second positive system of nitrogen, R branch, pressure of 2 Torr as a function of discharge currents. Two regions with different slopes is readily identified

Then $I = 4.6 \times 10^{-48}$ kg · m^2. Using formula (8.55) it can be easily verified that $E_{rot} = 1.2 \times 10^{-21} J\,(J + 1)$ *Joule.*

Exercise 8.8. From Fig. 8.17 calculate the rotational (gas) temperature, i.e. the lower temperature, and the higher temperature of the nitrogen molecules for currents of 10, 40 and 50 mA assuming that $B'_v = 1.815$ cm^{-1} for $N_2(C^3\Pi_u)$ state.

Resolution: From formula (8.111) we have:

$$T_r = \frac{B'_v J'\,(J' + 1)\,hc}{k \ln\left(\frac{IJ'}{J'^2 - 1}\right)} = \frac{2.61}{\ln\left(\frac{IJ'}{J'^2-1}\right)/J'\,(J' + 1)}$$

As an illustration for the 20 mA current (Fig. 8.24);
Using the same procedure for the other current, we have:

Discharge current (mA)	Gas temperature (K)
10	370
20	400
40	420
50	447

References

W. Demtröder, *Atoms, Molecules and Photons: An Introduction to Atomic-, Molecular- and Quantum Physics*, 2nd edn. (Springer, Heidelberg, 2010)

M.A. Gigosos, V. Cardenoso, New plasma diagnosis tables of hydrogen Stark broadening including ion dynamics. J. Phys. B Atomic Mol. Opt. Phys. **29**, 19964795–19964838 (1996)

B. Gordiets, C.M. Ferreira, M.J. Pinheiro, A. Ricard, Self-consistent kinetic model of low-pressure N_2-H_2 flowing discharges. Plasma Sources Sci. Technol. **7**(1), 363 (1998)

H.R. Griem, *Plasma Spectroscopy* (McGraw-Hill Book Company, New York, 1964)

H.R. Griem, *Principles of Plasma Spectroscopy* (Cambridge University Press, Cambridge, 1997)

G. Herzberg, *Molecular Spectra and Molecular Structure*, vol. 1 Ed. (Van Nostrand Reinhold Company, New York, 1950)

I. Kovacs, *Rotational Structure in the Spectra of Diatomic Molecules* (Akademiai Kiado, Budapest and Adam Hilger Ltd, London, 1969)

H.G. Kuhn, *Atomic Spectra* (Academic, New York, 1969)

A. Loftus, P.A. Krupenie, The spectrum of molecular nitrogen. J. Phys. Chem. Ref. Data **6**, 113–307 (1977)

J. Nagai, Ph.D. Thesis, Instituto Tecnológico de Aeronáutica, 2004 (in Portuguese)

D.K. Otorbaev, V.N. Ochkin, P.L. Rubin, S.Y. Savinov, N.N. Sobolev, S.N. Tskhai et al., Electron-impact excitation levels of the rotational levels of molecular electron states in gas discharges. Proc. Lebedev Phys. Inst. Acad. Sci. USSR **179**, 1 Nova Sci. Publ. (1989)

D. Pagnon, J. Amorim, J. Nahorny, M. Touzeau, M. Vialle, On the use of actinometry to measure the dissociation in O_2 DC glow discharges: determination of the wall recombination probability. J. Phys. D Appl. Phys. **28**, 1856–1868 (1995)

A. Ricard, *Reactive Plasmas*, 1st edn. (Societé Francaise du Vide, Paris, 1996)

J.A. Souza-Corrêa, C. Oliveira, M.P. Gomes, J. Amorim, Electric and spectroscopic properties of argon-hydrogen RF microplasma jets at atmospheric pressure. J. Phys. D Appl. Phys. **43**, 395203 (2010)

J.I. Steinfeld, *Molecules and Radiation: An Introduction to Modern Molecular Spectrsocopy*, 2nd edn. (Dover Publications Inc., New York, 1974)

C.E. Treanor, J.W. Rich, R.G. Rehm, Vibrational relaxation of anharmonic oscillators with exchange-dominated collisions. J. Chem. Phys. **48**, 1798–1807 (1968)

H.E. White, *Introduction to Atomic Spectra* (McGraw-Hill Book Company, New York, 1934)

W.L. Wise, M.W. Smith, B.M. Glennon, *Atomic Transitions Probabilities Vol. 1: Hydrogen Through Neon* (Government Printing Office, Washington, DC, 1996). NSRDS-NBS4

Chapter 9
Absorption Spectroscopy

It was shown in the preceding chapter that emission spectroscopy could be only used to determine the absolute density of species if the optical apparatus was calibrated as a whole and supposing that the emission intensity has a linear response as a function of species densities. This is a very difficult task if we are dealing with radiative species. Concerning ground state atoms, the technique could be employed but it is still more complicated to access to absolute values. Another more complicated situation, for emission spectroscopy, is to measure species densities in post-discharges, spatially or temporally resolved, where there is not excitation of radiative states. In these conditions the best solution is the absorption spectroscopy.

The principle is simple; the medium to be probed need to be illuminated with photons with wavelength corresponding to allowed transitions and related to the state and specie to be interrogated. The absorption is higher when the population of the probed specie is high and the transition has a strong oscillator strength.

The main sources of radiation employed are: continuous radiation sources like high pressure arc discharges, line sources that emit the line to be absorbed and coherent sources like dye lasers. The continuous radiation sources are normally employed to measure absorptions over many lines or bands of atoms or molecules respectively. The inconvenient is that the emitting intensity is shared in a broad region of the spectrum and the absorbable energy in a given thin line is feeble. The line sources such as spectral or hollow cathode lamps are good choices for line absorption and allow an excellent match between wavelength emitted and absorbed wavelength. Dye lasers are good to synchronize the emitted and absorbed wavelength and provide high-photon flux with high spatial, temporal and frequency resolutions but are more complicate to operate and more expensive. These lasers sources will be discussed in the next chapter.

The original version of this chapter was revised. An erratum to this chapter can be found at DOI 10.1007/ 978-3-319-09253-9_12

J.M.A.H. Loureiro, J. de Amorim Filho, *Kinetics and Spectroscopy of Low Temperature Plasmas*, Graduate Texts in Physics,
DOI 10.1007/978-3-319-09253-9_9

In this chapter we will study the basis of an incoherent absorption, i.e. absorption of light originated from line sources, that emit spectra of lines that are absorbed by the medium under study. This technique induces a very small perturbation of the plasma and its implementation is very simple.

9.1 Principles of Absorption Spectroscopy

The basis of this method consists in the measurement of the absorption coefficient of lines related to allowed optical transitions from the energy level we want to measure the density. The spectral lines are emitted by a reference source, which generates an incoherent beam of light that passes throughout the medium under study. A very good reference in this field is the book of Mitchell and Zemanski (1961).

A very simple experimental arrangement is presented in Fig. 9.1. A beam of light with intensity I_0, emitted by a reference source, i.e. the incident beam intensity, passes through the discharge tube with length L. The transmitted beam intensity I_t is analyzed with aid of a monochromator or spectrometer.

The incoherent absorption is usually employed in the density measurement of atoms, ions or molecules in their ground states or still to probe densities of metastable states that sometimes are more populated than radiative ones. In these conditions the stimulated emission, discussed in the Chap. 8, is negligible if compared to the absorption. If the concentration of absorbing species is homogeneous along the absorption length L, the transmitted intensity I_t may be written as:

$$I_t = I_0 e^{-k_\lambda L} \tag{9.1}$$

k_λ is the absorption coefficient of the line for a given value of wavelength λ. The expression for k_λ relies on the line to be absorbed, the nature and concentration of absorbing atoms or molecules and the line shape of the emitted line and of the absorbing atoms or molecules and the shape of the emitted line. If the beam, from a reference source emitting a continuous spectrum is sent through the absorption

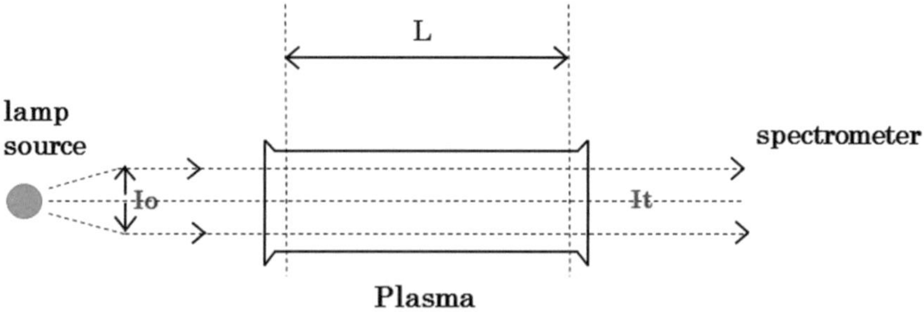

Fig. 9.1 Sketch of an experimental apparatus for optical absorption measurements

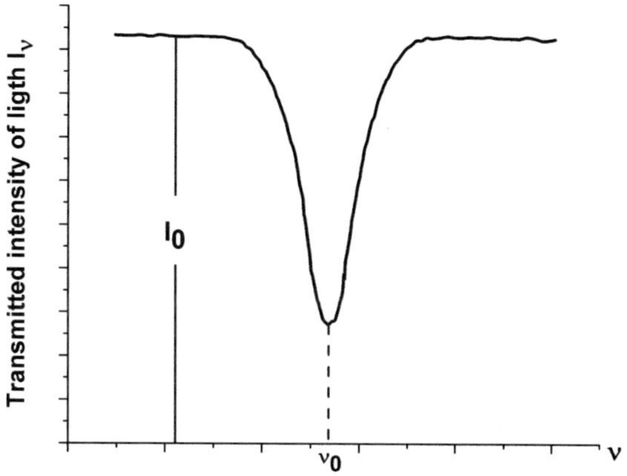

Fig. 9.2 Intensity of a transmitted light as a function of frequency in an absorption experiment of a monoatomic gas

column, the intensity of the transmitted beam I_t through a monoatomic gas shows a frequencies distribution, like the one shown in Fig. 9.2.

If the medium under study has an isotropic radiation (emitted or absorbed) with intensity I_ν of frequencies between ν and $\nu + d\nu$ and a population of atoms with the ground state 1 and an excited state 2, we define the following probability coefficients (Mitchell and Zemanski 1961):

$B_{1,2}I_\nu$ = probability per second that an atom in the state 1 submitted to the illumination of the light beam of frequencies between ν and $\nu + d\nu$ and intensity I_ν will absorb a quantum of energy $h\nu$ and be excited to the state 2.

$A_{2,1}$ = probability per second that the atom in the excited state 2 will spontaneously emits in all direction, a quantum with energy $h\nu$ and pass to the level 1.

$B_{2,1}I_\nu$ = probability per second that an atom will undergo the transition from the level 2 to the level 1 when absorbs an isotropic radiation of frequencies between ν and $\nu + d\nu$ and intensity I_ν emitting thereby a photon with energy $h\nu$ in the same direction as the stimulating photon.

If the medium is in thermodynamic equilibrium between the radiation field and the atoms, Einstein showed that;

$$\frac{A_{2,1}}{B_{1,2}} = \frac{8\pi h\nu^3}{c^3}\frac{g_1}{g_2} \tag{9.2}$$

and

$$\frac{B_{2,1}}{B_{1,2}} = \frac{g_1}{g_2} \tag{9.3}$$

here g_1 and g_2 are the statistical weights of levels 1 and 2 respectively, and c is the velocity of light. The spontaneous emission coefficient $A_{2,1}$ is related to the radiative lifetime τ of the level 2 by:

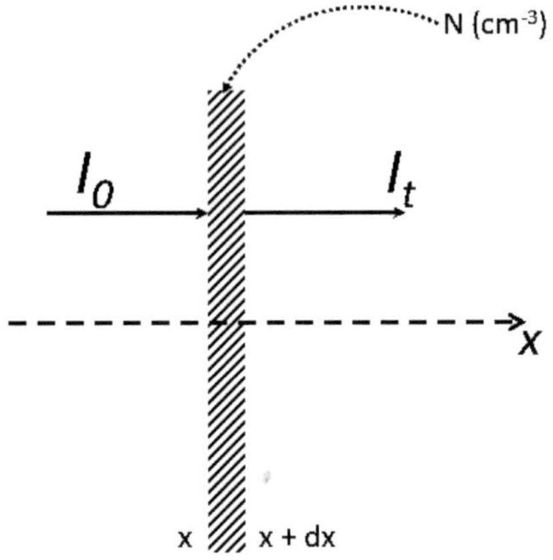

Fig. 9.3 Infinitesimal slab dx where the probe beam is absorbed

$$A_{2,1} = \frac{1}{\tau} \tag{9.4}$$

Consider a parallel beam of light with frequencies between ν and $\nu + d\nu$ and intensity I_ν passing through a layer of thickness dx of absorbing atoms. The layer is bounded by planes at x and $x + dx$, see Fig. 9.3.

Assuming that in N atoms by cm^{-3} of which δN_ν are capable of absorbing the radiation frequency between ν and $\nu + d\nu$ and in N' atoms by cm^{-3} with $\delta N'_\nu$ may emit in this frequency range. Neglecting re-emission, the amount of the light beam absorbed in the slab is:

$$d(I_\nu \delta\nu) = -\left(\delta N_\nu dxh\nu B_{1,2}\frac{I_\nu}{4\pi} - \delta N'_\nu dxh\nu B_{2,1}\frac{I_\nu}{4\pi}\right) \tag{9.5}$$

where $I_\nu/4\pi$ is the intensity of the equivalent isotropic radiation for the defined Einstein coefficient for absorption and $B_{2,1}$ is the Einstein coefficient for induced emission. Rearranging Eq. (9.5) results:

$$-\frac{1}{I_\nu}\frac{dI_\nu}{dx}\delta\nu = \frac{h\nu}{4\pi}\left(B_{1,2}\delta N_\nu - B_{2,1}\delta N'_\nu\right) \tag{9.6}$$

From Eq. (9.1) the left-hand side of Eq. (9.6) is

$$k_\nu \delta\nu = \frac{h\nu}{4\pi}\left(B_{1,2}\delta N_\nu - B_{2,1}\delta N'_\nu\right) \tag{9.7}$$

Integrating over the whole absorption line profile we obtain:

$$\int k_\nu d\nu = \frac{h\nu_0}{4\pi}\left(B_{1,2}N_\nu - B_{2,1}N'_\nu\right) \tag{9.8}$$

where ν_0 is the central frequency of the absorption line profile. Substituting (9.2), (9.3), and (9.4) into (9.8),

$$\int k_\nu d\nu = \frac{\lambda_0^2}{8\pi} \frac{g_2}{g_1} \frac{N}{\tau} \left(1 - \frac{g_1}{g_2} \frac{N'}{N}\right) \tag{9.9}$$

If the only source of creation of excited state is the absorption of the probe light beam $N' \ll N$ the Eq. (9.9) becomes:

$$\int k_\nu d\nu = \frac{\lambda_0^2}{8\pi} \frac{g_2}{g_1} \frac{N}{\tau} \tag{9.10}$$

The Eq. (9.10) is very important because it shows that the integral of the absorption coefficient is proportional to the density N and remains constant when N is constant.

In the case of a Gaussian profile due to the Doppler effect, the absorption coefficient of the line is (Mitchell and Zemanski 1961)

$$k_\lambda = k_{\lambda_0} e^{-\left[\frac{2(\lambda - \lambda_0)}{\Delta\lambda_{1/2}^D} \sqrt{\ln 2}\right]^2} \tag{9.11}$$

k_{λ_0} is the absorption coefficient at the line center λ_0, $\Delta\lambda_{1/2}^D$ is the FWHM of the absorption line. From Eq. (9.11) it may be plotted k_λ as a function of the wavelength λ as it is shown in Fig. 9.4.

The total breadth of the curve or FWHM is at the place where the maximum intensity of the absorption coefficient $k_{max.}$ falls to one-half of its value and has a

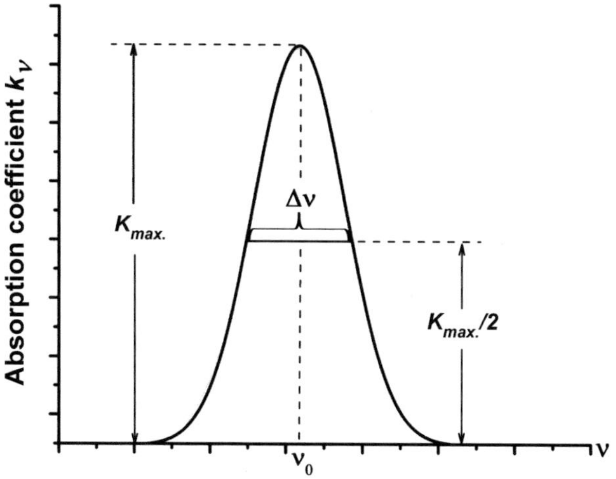

Fig. 9.4 Absorption coefficient k_λ as a function of the wavelength λ for an absorption line. Note the breadth $\Delta\nu$ at $k_{max.}/2$

breadth $\Delta\lambda$. In general, the absorption coefficient of a gas depends on the nature of the molecules or atoms gas, the particle velocity and the interaction with foreign molecules or atoms.

In the Eq. (9.11) the absorption coefficient at the line center k_{λ_0} depends on the Doppler width of the absorption line $\Delta\lambda^D_{1/2}$, the oscillator strength f of the transition and on the concentration n_M of the atoms or molecules working as absorbers. In MKS units the expression for k_{λ_0} is given by Mitchell and Zemanski (1961):

$$k_{\lambda_0} = \frac{e^2}{4\epsilon_0 mc^2}\sqrt{\frac{\ln 2}{\pi}}\frac{2}{\Delta\lambda^D_{1/2}} f n_M \tag{9.12}$$

where m is the electron mass.

The oscillator strength f of the absorbed line being known, the Eq. (9.12) shows that the measurement of k_{λ_0} leads to the determination of the concentration n_M of the atoms or molecules, once the Doppler width of the absorption line is known.

Experimentally it is not possible to measure directly k_{λ_0} because the monochromators have an apparatus function that need to be taken into account. During the experiments they are measured the integrated absorption of the line emitted by the source, the line shape of the transitions of absorbing particles in the medium and the apparatus function used to record the line. The intensity is given by:

$$I_t(\lambda, L) = C\int_{-\infty}^{+\infty} I_0(\lambda - y)\, e^{-k_\lambda(\lambda-y)L} F(y)dy \tag{9.13}$$

C is a constant that represents the absolute calibration of the system and F(y) is the apparatus function of the monochromator used to record the line.

The experimental set-up for absolute concentration measurements is depicted in Fig. 9.5. Although many configurations may be adopted, we illustrate this chapter with a system conceived to measure absolute ground state of oxygen or hydrogen atoms. The set-up is consisted of a microwave discharge lamp that generates the 125 nm or 130 nm resonance lines for absorption measurements on oxygen or hydrogen ground-state atoms respectively. The lamp in this case is a surface-wave discharge generated by a Surfatron device and operated with a flux of 2 % O_2–98 % He and microwave power of 60 W. The oxygen or hydrogen is injected in the sense indicated in the figure to reduce the cold absorption layer formed along the optical path. The buffer gas used in this lamp is helium.

The radiation from the microwave lamp is transmitted through a MgF_2 window (optically transparent for Vaccum Ultraviolet VUV absorption experiments) and absorbed in a dc positive column of 30 cm length. The detection system includes a VUV spectrometer of 50 cm focal length, a sodium salicylate scintillator, and a photomultiplier tube (PMT). In this system set-up a PMT for the UV-Vis region was used and a scintillator was employed to re-emit in the visible region when it absorbs UV radiation. Photomultiplier tubes with windows in F_2 and MgF_2 with

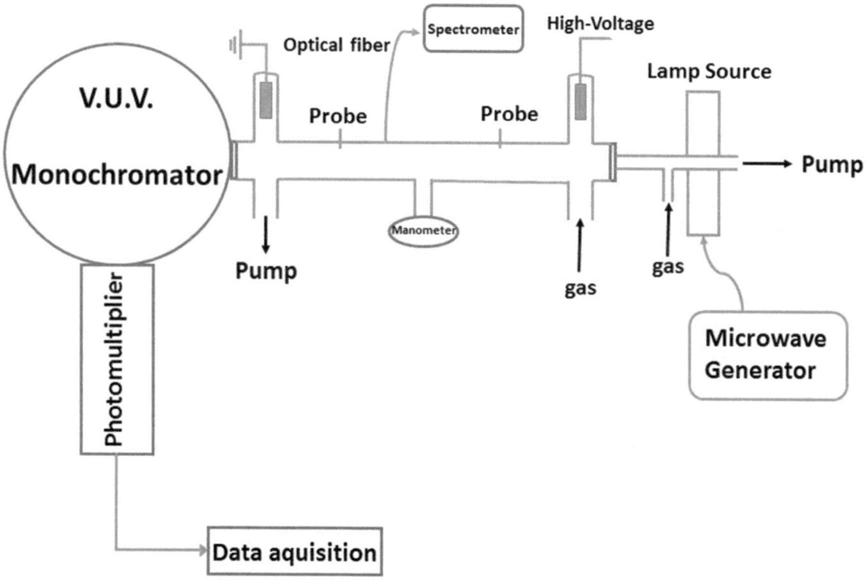

Fig. 9.5 Typical absorption experimental apparatus for probing atoms in a low-pressure positive column discharge (Pagnon et al. 1995)

high quantum efficiency, named solar blinds, may also be employed. In order to improve the signal to noise ratio, a lock-in amplifier was used.

Considering an absorption experiment to probe atoms density in a low-pressure positive column of a direct current discharge, the atoms are cold and the apparatus function is much broader than the width of the lines studied. In this case the total intensities of incident (I_0) and transmitted (I_t) beams may be written as:

$$I_0 = \int I_0(\lambda)\, d\lambda \tag{9.14}$$

$$I_t = \int I_0(\lambda)\, e^{-k_\lambda L} d\lambda \tag{9.15}$$

The integration is done along the line profile. The absorption A_L over a given length L is:

$$A_L = \frac{I_0 - I_t}{I_0} \tag{9.16}$$

To relate the absorption A_L with the density of absorbers n_M it is important to know the profile of the line emitted by the reference source. Profiles of the incident line (source) and the absorption line (absorbing medium) need to be known because they are generally different from pure Gaussian profile as is the case in low-pressure plasmas.

In order to employ the absorption method developed by Landerburg and Richie (1913), as presented above, the resonance lamp must satisfy some conditions as negligible pressure broadening of the line to be absorbed and negleting Stark effect. The main mechanisms responsible for line breadth need to be natural and Doppler broadening. As natural width of lines may be considered negligible, the only mechanism that broadens the line of an ideal lamp is Doppler. It is needed also to construct a source of radiation that avoid self-absorption of the line inside the reference lamp. This is not always feasible and very often it is necessary to work with lamps excited by electron bombardment or with resonance lamps in which either the vapor pressure or the thickness of the emitting layer or both are not accurately known. In these cases, is convenient to use for frequency distribution of the emitted radiation an empirical expression which represents roughly the line broadening resulting from the vapor pressure and temperature conditions within the lamp. We will present a procedure to take these effects into account now.

When both profiles, from the incident line and absorption line, are Gaussian the following coefficient, introduced by Mitchell and Zemanski (1961), is useful to simplify the expression for absorption.

$$\alpha = \frac{FWHM\ of\ the\ incident\ beam\ \left(\Delta\lambda_{1/2}^{S}\right)}{FWHM\ of\ the\ absorption\ line\ \left(\Delta\lambda_{1/2}^{D}\right)} = \frac{\gamma}{\Delta\lambda_{1/2}^{D}} \tag{9.17}$$

$\Delta\lambda_{1/2}^{D}$ is representative of the Doppler broadening produced by the movement of the atoms in the plasma. We can remark that a value of α equal to unity implies that the line emitted by the lamp is of the same shape (Gaussian in this case) and breadth as that one of the absorption line of the atom in the absorption discharge tube, indicating that the reference source and the absorption cell are at the same temperature. Values of α greater than one represent a line of the same shape and greater width than that of the absorption discharge tube or cell.

The intensities of incident and transmitted beams may be written as:

$$I_0(\lambda) = \int I_0(\lambda_0)\, e^{-\left[\frac{2(\lambda-\lambda_0)}{\gamma}\sqrt{\ln 2}\right]^2} d\lambda \tag{9.18}$$

$$I_t(\lambda) = \int I_0(\lambda_0)\, e^{-\left[\frac{2(\lambda-\lambda_0)}{\gamma}\sqrt{\ln 2}\right]^2} e^{-k_\lambda L} d\lambda \tag{9.19}$$

Supposing the concentration of absorbing species constant along the absorption length L of the discharge tube and assuming the lines emitted by the reference source and by the absorption with Gaussian profile, the absorption A_L defined by Eq. (9.16), taking into account the α parameter, can be written in a form of series like (Mitchell and Zemanski 1961):

$$A_L\,(\alpha,\ k_{\lambda_0}L) = \sum_{n=1}^{\infty} \frac{(-1)^{n+1}(k_{\lambda_0}L)^n}{n!\sqrt{1+n\alpha^2}} \tag{9.20}$$

The Fig. 9.6 depicts graphs, in form of abacuses, of $k_{\lambda_0}L$ as a function of A_L and α for the range $0 \le \alpha \le 5$ and $0.1 \le k_{\lambda_0}L \le 10$.

Experimentally with the value of α determined from Eq. (9.17) one can use the curves of $A_L\,(\alpha,\ k_{\lambda_0}L)$ and find the $k_{\lambda_0}L$ corresponding to experimentally measured values of absorption. Using Eq. (9.12), knowing $\Delta\lambda^D_{1/2}$ and f, one determines the concentration n_M of the atoms or molecules. To illustrate the theory presented until now we present a case study of evolution of the metastable state of argon $^3P_2$, in a positive column direct current discharge.

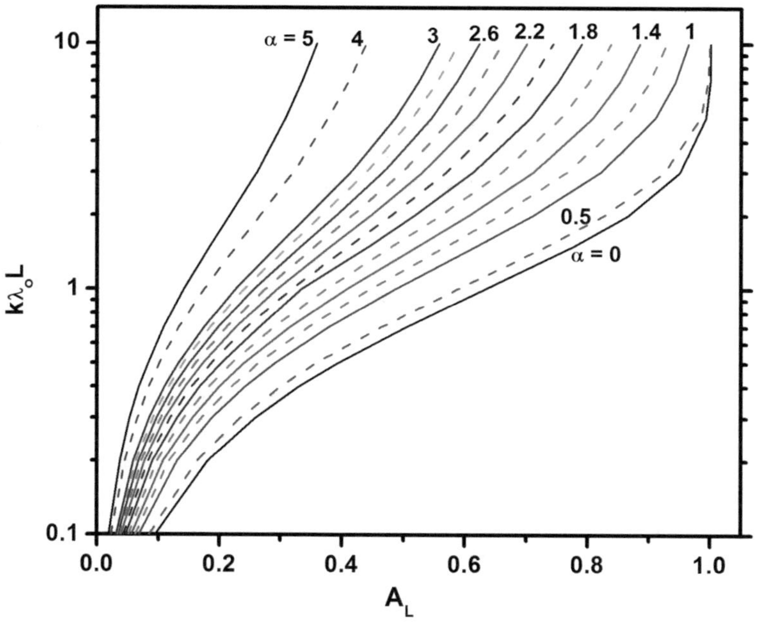

Fig. 9.6 Absorption coefficient $k_{\lambda_0}L$ as a function of the absorption A_L and parameter α

9.2 Metastables in the Positive Column of Ar Direct Current Discharge

The case proposed is the study of the evolution of the population of $^3P_2$ metastable state of argon in a low-pressure positive column of argon using the absorption apparatus similar to the one shown in Fig. 9.4. This case study is useful to use the concepts presented in this chapter and to show to the student a practical application of the absorption spectroscopy in a real experimental situation.

The Fig. 9.7 is similar to the Fig. 3.9 but is reproduced here with more details needed in the study of evolution of metastable and resonant levels of argon. The levels $^3P_0$ and $^3P_2$ are metastable state, see the selection rules in Chap. 8. The energy levels $^1P_1$ and $^3P_1$ are optically connected to the ground state level, so they are called resonant states.

9.2.1 Study of the Evolution of Metastable and Resonant Levels

To study the evolution of metastable and resonant levels in a positive column of argon as a function of discharge parameters, we need to establish the balance between the creation and destruction of the states involved. The conservation equation for the population of these states with the creation and loss mechanisms is;

$$\frac{\partial n_M}{\partial t} = D_M \nabla^2 n_M - P + S \tag{9.21}$$

where D_M is the diffusion coefficient of excited atoms, P is the loss rate of excited atoms, as a function of time and volume, and S is the creation rate of excited atoms.

Firstly, we describe the form of source term and the loss term, evaluating the possible destruction routes. After that, we write down an expression for the density of excited states in the stationary regime.

Initially, lets comment on the source term. In this case, we will take into account the main process of excited state creation, i.e. the electronic excitation from the ground state level. An excited state M is created by electronic collision with the ground state and de-excited by a radiative transition. The source term may be written as:

$$S = n_e n_0 C^M_{e-0} \tag{9.22}$$

C^M_{e-0} is the electron rate coefficient for creation of the excited state M and relies on the electron distribution function, i.e. the electron temperature T_e, n_e is the electron density and n_0 is the density of Ar atoms in the ground state.

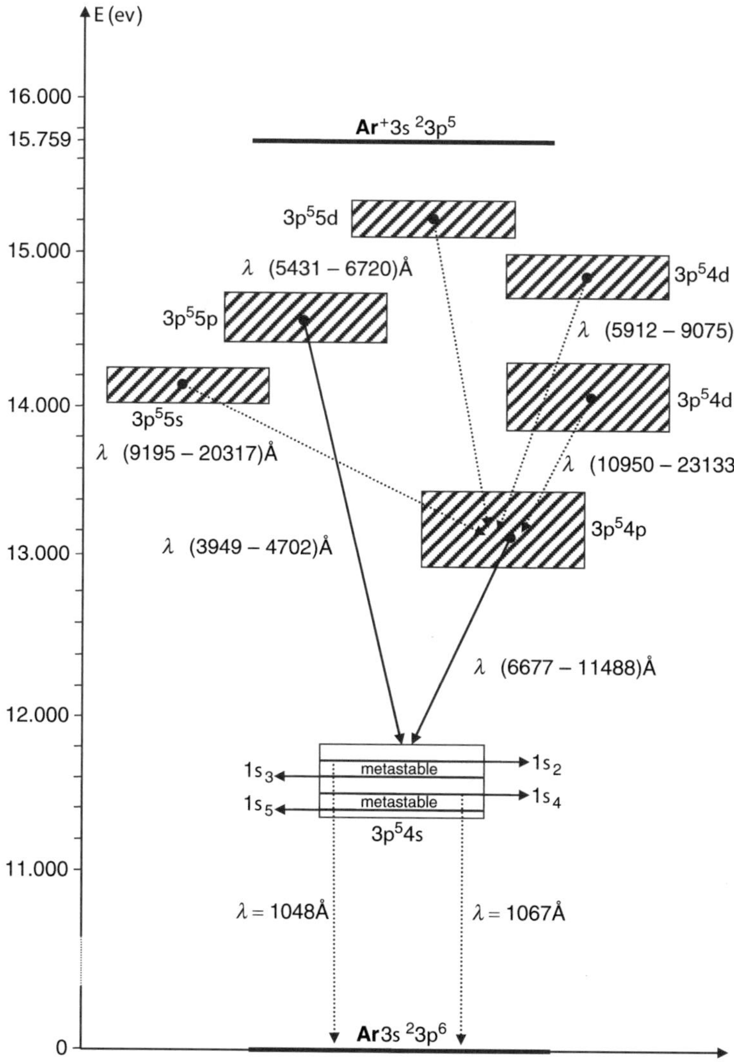

Fig. 9.7 Schematic drawing of the first excited levels of argon with the transitions from the $3p^5$ 4p levels to the $3p^5$ 4s ones

The destruction of excited states may occur through different ways. Here we will take into account the most important ones in order to have a general picture of the kinetics of excited states in a positive column of an Ar direct current discharge. The destruction pathways discussed are the radiative de-excitation, diffusion loss to the walls, electron collisions, collisions with neutrals and metastable states. The radiative lifetime of metastable states is long if compared with the time between collisions. In these conditions they are destroyed before they emit radiation. The resonant states, $^1P_1$ and $^3P_1$ of argon, that are optically coupled to the ground state levels behave differently. They emit radiation before a collision takes place. This mechanism of loss is efficient if the photon escapes from the medium. If the photon is reabsorbed along its path by an atom in the fundamental state, the net radiative loss is zero. The radiative level behaves as a pseudo-metastable. The theory that addresses this phenomenon is complex and will not be discussed here. However it can be remarked that this radiation trap increases as a function of pressure.

In the diffusion of metastable atoms to the walls of the gas discharge tube, the atoms created in volume diffuse to the walls where they are deactivated. This mechanism of diffusion creates a radial gradient of excited atoms.

The diffusion frequency may be defined by (Wieme et al. 1980):

$$\nu = \frac{D_M}{\Lambda^2} \tag{9.23}$$

Λ is the characteristic diffusion length; $\Lambda = R/2.4$ for a Bessel radial distribution, D_M is the diffusion coefficient as stated before depending on the level considered. For the argon metastable state $^3P_2$, the measured value for D_M at 1 Torr (Wieme 1980) is:

$$D_M = 3.2 \times 10^{-3} T^{1.68} \ \left(\mathrm{cm^2 s^{-1}}\right) \tag{9.24}$$

considering $T = 300$ K, $D_M = 46.42\ \mathrm{cm^2 s^{-1}}$.

The metastable losses by collisions with electrons are the most important. They may be responsible for stepwise ionization, which plays a key role in discharge plasmas. Superelastic collisions reduce de population of metastable states and increase with electron density. The coefficient for excitation of metastable levels by electron collisions $C_{e\text{-}M}$ depends on the electron energy distribution function, i.e. the electron temperature T_e. The metastable states may also be destroyed by collisions with neutrals. This de-excitation reaction is called quenching. The coefficient for this destruction is $C_{o\text{-}M}$.

Collisions between metastables atoms may ionize one of the atoms if enough energy is available; this is called Penning ionization reaction;

$$A^M + A^M \rightarrow A^+ + e + A \tag{9.25}$$

this reaction becomes important when the metastable density n_M is around $10^{12}\ \mathrm{cm^{-3}}$.

From Eq. (9.21) and considering the balance between creation and loss of metastable states in stationary regime, the density may be written as:

$$n_M = \frac{n_e n_o C_{e-o}^M}{\frac{D_M}{\Lambda^2} + n_o C_{o-M} + n_e C_{e-M}} \tag{9.26}$$

When the electron density is low the losses by electron collisions may be discarded and the expression (9.26) is reduced to:

$$n_M = \frac{n_e n_o C^M_{e-o}}{\frac{D_M}{\Lambda^2} + n_o C_{o-M}} \tag{9.27}$$

The denominator of the above expression is a destruction frequency that relies on the pressure and the gas temperature but that is not depended on electron temperature.

On the other hand, if the collision with electrons are dominant the metastable density is given by:

$$n_M = \frac{n_o C^M_{e-o}}{C_{e-M}} \tag{9.28}$$

As can be seen, in this case the density of metastable states depends only on pressure and electron temperature, the latter through the electron rate coefficients. As the electron density increases the metastable density increase linearly until a plateau is reached.

In this case study we will obtain the coefficients C^M_{e-o} and C_{e-M} from the measurements of metastable concentration n_M as a function of electron density n_e, i.e. discharge current. The experimental set-up presented in Fig. 9.5 is composed of gas discharge tube with inside diameter of 2.6 cm and length of 50 cm. The direct current discharge is started between an indirectly heated cathode and an anode that are placed 30 cm from each other. Two tungsten probes placed at a radial position $r = R/2$ and spaced 10 cm apart were provided for measurements of the axial electric field strength and electron density. The population of the $^3P_2$ metastable state of argon for various currents and pressures was determined by the resonant absorption method employing two spectral lines, 696.5 nm (transition 1 s_5 – $2p_2$) and 714.7 nm (transition 1 s_5 – $2p_4$). The metastable states may also be destroyed by collisions with neutrals. This quenching reaction has the coefficient C_{o-M} which was proposed, for the state $^3P_2$, to be $1.4 \times 10^{-10}\text{cm}^3\text{s}^{-1}$ (Ferreira et al. 1985).

9.2.2 *Determination of Metastable Density*

In order to determine the metastable density, it is necessary to know some data. First of all we will complete the table below calculating the Doppler broadening of the absorption lines, the parameter α knowing that the gas discharge temperature is 300 K and that the spectral lamp source emits a line with Gaussian profile and has the temperature indicated in the table for each line.

We have the oscillator strengths and the temperature of the source and plasma gas discharge. The reference lamp employed is a medium with high temperature and generally is optically thick. In these conditions it can be assumed for

Table 9.1 Determination of $\Delta\lambda^S_{1/2}$, $\Delta\lambda^D_{1/2}$ and α

Transition	Wavelength (nm)	f	T_{source} (K)	$\Delta\lambda^S_{1/2}\times10^{-3}$ nm	T_{plasma} (K)	$\Delta\lambda^D_{1/2}\times10^{-3}$ nm	α
$1s_5-2p_2$	696.5	2.9×10^{-2}	762	*2.18*	300	1.37	*1.59*
$1s_5-2p_4$	714.7	3.0×10^{-3}	358	*1.53*	300	1.40	*1.09*

Table 9.2 Absorption on transition $1s_5-2p_2$ (696.5 nm)

Pressure (Torr)	Id (mA)	A_L	$k_{\lambda_0}L$	$Ar(^3P_2)(cm^{-3})$
	0.2	0.26	*0.64*	2.52×10^{10}
	0.3	0.34	*0.89*	3.48×10^{10}
	0.4	0.41	*1.19*	4.68×10^{10}
	0.6	0.48	*1.56*	6.12×10^{10}
	0.8	0.54	*1.93*	7.55×10^{10}
0.75	1.0	0.58	*2.23*	8.72×10^{10}
	2.0	0.64	*2.86*	1.12×10^{11}
	4.0	0.66	*3.27*	1.28×10^{11}
	6.0	0.69	*3.62*	1.42×10^{11}
	10.0	0.70	*3.90*	1.52×10^{11}
	20.0	0.70	*3.88*	1.53×10^{11}

didactic proposes that the spectral lines have Gaussian profile with large breadths. The Doppler broadening of the absorption line is given by:

$$\Delta\lambda_D = 7.17\times10^{-7}\lambda_0\sqrt{\frac{T(K)}{M\left(\frac{g}{\text{mol}}\right)}} \tag{9.29}$$

The FWHM of the source above and then the medium were calculated using the expression above and then the parameter α was determined (results are in italics) (Table 9.1).

9.2.3 *Determination of the Absorption Coefficient $k_{\lambda_0}L$ Using the Abacuses*

The Tables 9.2 and 9.3 give the results (italics) of experiments of absorption on the lines 696.5 nm (pressure of 0.75) and 714.7 nm (pressure of 0.75 Torr) realized in a positive column of an argon direct current discharge. From the absorption A_L, for a given pressure and current, we obtain the absorption coefficient $k_{\lambda_0}L$, using the abacuses of Fig. 9.6. Using Eq. (9.12) we calculate the metastable density $[Ar(^3P_2)]\ cm^{-3}$.

Table 9.3 Absorption on transition 1 $s_5 - 2p_4$ (714.7 nm)

Pressure (Torr)	Id (mA)	A_L	$k_{\lambda_0}L$	$Ar(^3P_2)(cm^{-3})$
0.075	0.4	0.06	0.097	3.66×10^{10}
	0.6	0.09	0.132	5.01×10^{10}
	0.8	0.11	0.170	6.43×10^{10}
	1.0	0.13	0.207	7.85×10^{10}
	2.0	0.22	0.364	1.38×10^{11}
	4.0	0.29	0.525	1.99×10^{11}
	6.0	0.325	0.605	2.29×10^{11}
	10.0	0.331	0.636	2.41×10^{11}
	20.0	0.342	0.657	2.49×10^{11}

9.2.4 *Density of Ar($^3P_2$) State as a Function of Electronic Density*

Employing the continuity equation and assuming the gas temperature equal to 300 K, E/N = 100 Td (pressure = 0.75 Torr) and E/N = 10 Td (pressure = 0.075 Torr), we find the variation of Ar($^3P_2$) as a function of n_e for the two pressures chosen. From Fig. 2 of the paper (Ferreira and Ricard 1983) we may estimate the drift velocity $V_d = 8 \times 10^6$ cm/s for p = 0.075 Torr, while for p = 0.75 Torr we have $v_d = 10^6$cm/s.

It can be seen in Fig. 9.8 that the density of metastable Ar($^3P_2$) state and the electronic density n_e increase with discharge current and saturate for currents higher than 7 mA. When the pressure increases from 0.075 to 0.75 Torr the electronic density increases and the density of metastable Ar($^3P_2$) state decreases. For the Ar($^3P_2$) state the main excitation and losses processes can be depicted from the Figs. 9.9 and 9.10 at low pressure.

The principal collisional and radiative processes used to explain with aid of a simple kinetic scheme the evolution of populations in the *3p⁵4s* levels are:

$$Ar\ \left(^1S_0\right) + e\ \rightarrow Ar\ \left(s_j\right) + e \tag{9.30}$$

$$Ar\ \left(^1S_0\right) + e\ \rightarrow Ar^*\ + e \tag{9.31}$$

$$Ar^*\ \rightarrow \cdots \rightarrow\ Ar\ \left(s_j\right) + h\nu \tag{9.32}$$

$$Ar\ \left(s_j\right) + e\ \leftrightarrows\ Ar\ \left(s_i\right) + e \tag{9.33}$$

$$Ar\ \left(s_j\right) + e\ \rightarrow Ar\ \left(p_k\right) + e \tag{9.34}$$

$$Ar\ \left(p_k\right)\ \rightarrow\ Ar\ \left(s_i\right) + h\nu \tag{9.35}$$

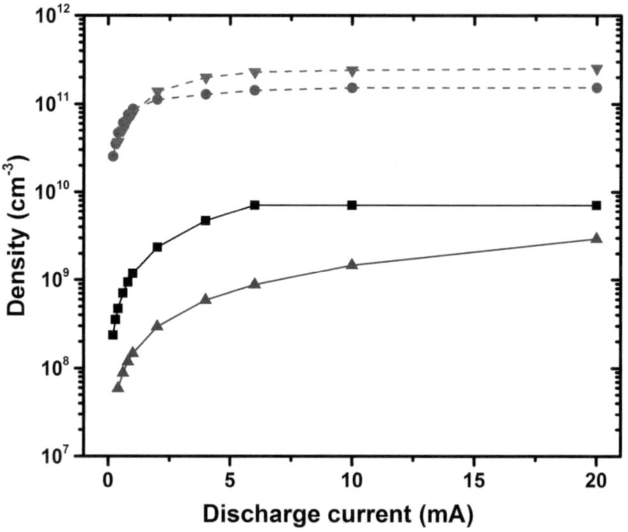

Fig. 9.8 Concentration of metastable Ar($^3P_2$) state and electronic density calculated as a function of pressure and discharge current. The *Solid curves* and *dashed curves* are for electron density and metastable concentration, respectively. (*filled circle, filled square*) 0.75 Torr and (*filled triangle, filled inverted triangle*) 0.075 Torr

Here, Ar($^1S_0$) is the ground state of the argon atom, Ar(p_k) denotes a level of the *3p54p* configuration, and Ar(s_j) denotes a level of the $3p^54p$ configuration. These latter levels often are referred to as s_j latter levels will often be are referred the correspondence between the values of *j* and the levels being the same as in Pachen notation (i.e., $s_2 - {}^1P_1$, $s_3 - {}^3P_0$; $s_4 - {}^3P_1$; $s_5 - {}^3P_2$) (Figs. 9.9 and 9.10)

As can be seen for currents lower than 10 mA the $^3P_2$ state is lost by diffusion while for higher currents the transitions to other $3p^5$ $4s$ levels by electron collisions, directly and via the $3p^5$ $4p$ levels, become more significate. For higher pressures the diffusion is less important and the losses by electron collisions become more intense.

However, at the higher currents the electron collisions establish a strong coupling between the $3p^5$ $4s$ levels and redistribute the total excited population among them. In particular, the $^3P_0$ metastable is predominantly created and destroyed by these excitation transfer processes, directly and via the 10 levels of the $3p^54p$ configuration, for discharge currents higher than 10 mA. The $^3P_2$ metastable is also principally destroyed by electron excitation transfers to the other $3p^5$ $4s$ levels at currents larger than about 10 mA. Therefore, the metastable populations tend to saturate with increasing current as experimentally observed. In spite of the partial trapping of the resonance radiation the radiative decay constitutes by far the major loss process for the $^3P_1$ and $^1P_1$ levels and their population increase steadily with current under the conditions presented in this case study. Owing to the strong collisional mixing between the levels the radiative decay of the resonance

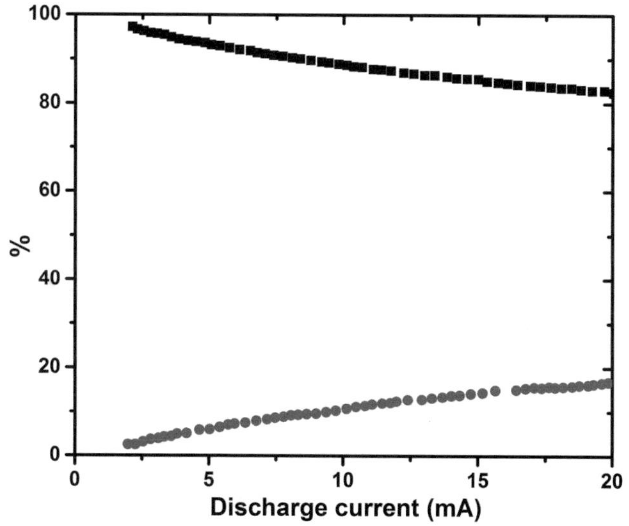

Fig. 9.9 Percentage contributions of excitation from the ground state (processes 9.30–9.32, *filled square*) and from the other $3p^54s$ levels (processes 9.33–9.35, *red filled circle*), to the total excitation of $^3P_2$ state as a function of the discharge current for $p = 0.075$ Torr (Ferreira et al. 1985)

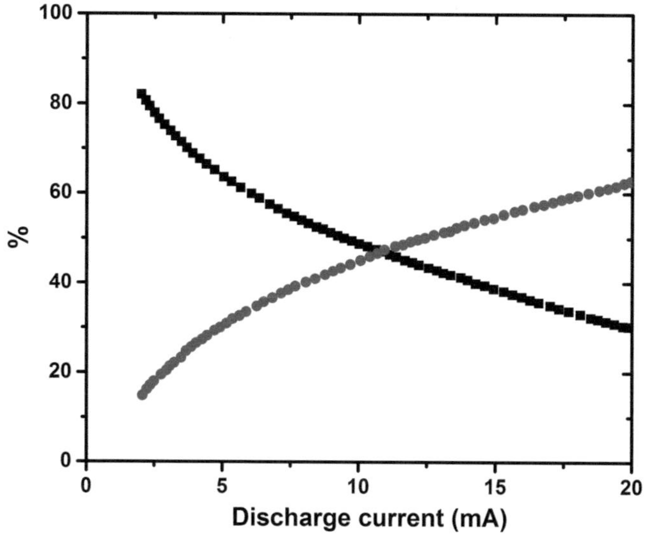

Fig. 9.10 Percentage contributions of various processes to the total loss rate of $^3P_2$ state as a function of the discharge current for $p = 0.075$ Torr: (*filled square*) diffusion; (*red filled circle*) transitions to the other $3p^54s$ levels by electron collisions, directly and via the $3p^54p$ levels (Ferreira et al. 1985)

levels constitutes at the end the principal process through which the total excited population and radiative states is lost at higher currents.

Note finally that the ionization of the $3p^5\ 4s$ levels by electron collisions and the collisions between pairs of $3p^5\ 4s$ levels have a negligible effect in determining the excited populations. These processes play, however, a fundamental role in the discharge ionization balance (Ferreira et al. 1985).

9.2.5 Electron Rate Coefficients for Creation of the Excited and Metastable States as a Function of Pressure

The balance between creation and loss of metastable states in stationary regime given by Eq. (9.26) help us to determine the C^M_{e-o} and C_{e-M} coefficients when the approximations of low and high discharge current are realized according to Eqs. (9.27) and (9.28). Considering T = 300 K, $D_M = 46.42\ \mathrm{cm^2 s^{-1}}$ and the characteristic diffusion length $\Lambda = R/2.405 = 0.54$ cm, in the region of low current and pressure values we can obtain the C^M_{e-o} coefficient (Table 9.4). It can be assumed a mean value of $C^M_{e-o} = 7.94 \times 10^{-8} cm^3\ S^{-1}$.

In the domain of high current and pressures values we can estimate the C_{e-M} coefficient, i.e. the coefficient for metastable losses through electronic collisions (Table 9.5).

It can be assumed a mean value of $C_{e-M} = 1.34 \times 10^{-2}\ \mathrm{cm^3 s^{-1}}$. Note that in this case $D_M/\Lambda^2 = 159.19\ \mathrm{s^{-1}}$, $n_o C_{o-M} = 3.39 \times 10^6\ \mathrm{s^{-1}}$ and $n_e C_{e-M} = 2.95 \times 10^8\ \mathrm{s^{-1}}$

Table 9.4 Determination of electron density n_e, metastable Ar($^3P_2$) density and C^M_{e-o} coefficient

Pressure (Torr)	Id (mA)	n_e (cm^{-3})	Ar($^3P_2$) (cm^{-3})	C^M_{e-o} (cm^{3}s^{-1})
0.075	0.4	5.89×10^7	3.66×10^{10}	8.70×10^{-8}
	0.6	8.83×10^7	5.01×10^{10}	7.95×10^{-8}
	0.8	1.18×10^8	6.43×10^{10}	7.63×10^{-8}
	1.0	1.47×10^8	7.85×10^{10}	7.48×10^{-8}

For p = of 0.075 Torr

Table 9.5 Determination of electron density n_e, metastable Ar($^3P_2$) density and C_{e-M} coefficient

Pressure (Torr)	Id (mA)	n_e (cm^{-3})	Ar($^3P_2$) (cm^{-3})	C_{e-M} (cm^{3}s^{-1})
	4.0	4.71×10^9	1.28×10^{11}	1.5×10^{-2}
	6.0	7.07×10^9	1.42×10^{11}	1.35×10^{-2}
0.75	10.0	1.18×10^{10}	1.52×10^{11}	1.26×10^{-2}
	20.0	2.36×10^{10}	1.53×10^{11}	1.25×10^{-2}

For p = 0.75 Torr

justifying the assumption of the collisions with electrons are dominant for metastable losses.

In the next chapter Laser Spectroscopy techniques will be introduced to probe atoms and molecules in low-temperature plasmas. They are based on absorption of photons produced in a coherent beam of light as a laser. This technique opens many other possibilities to interrogate reactive species if compared with classical emission and absorption spectroscopies presented in Chaps. 8 and 9.

Exercises

Exercise 9.1. A 5 Torr pressure hydrogen discharge is operating with a current of 40 mA, a gas temperature of 400 K, electron temperature of 1 eV, and dissociation degree of 3.31×10^{-4}.

(a) Find the density of atoms in the discharge.
(b) Supposing a Boltzmann distribution for the population of excited states, determine the density of atoms in the levels n = 2 and n = 3.
(c) In the detection of excited atom density, explain why atomic emission is more sensitive to discharge instabilities than atomic absorption.
(d) If during an experiment the absorption measured was 0.42 for an absorption path of 10 cm, find the absorption cross section for the $H(1s) \rightarrow H(2p)$ transition.

Resolution:

(a) The density of hydrogen atoms in the discharge may be found once the dissociation degree is known.
$p = NkT$ then $N = \frac{p}{kT} = \frac{5\times 133.32}{1.38\times 10^{-23}\times 400} = 1.21 \times 10^{23}\text{m}^{-3} = 1.21 \times 10^{17}\text{cm}^{-3}$.

$\frac{[H]}{[H_2]} = 3.31 \times 10^{-4}$, so
$[H] = 3.31 \times 10^{-4} \times [H_2] = 3.31 \times 10^{-4} \times 1.21\ 10^{17}\text{cm}^{-3} = 4.0\ 10^{13}\text{cm}^{-3}$

(b) $N_i = N_1 \exp - \left({}^{\Delta E}\!/\!_{kT} \right)$
For the n = 2 and n = 3 levels the corresponding energies are:
$E_n = -\frac{13.6\text{ eV}}{n^2}$,
for $n = 2$; $E_2 = -\frac{13.6\text{ eV}}{4} = -3.4$ eV and for $n = 3$; $E_3 = -\frac{13.6\text{ eV}}{9} = -1.51$ eV

$$N_2 = 4 \times 10^{13} \exp - \left({}^{(13.6 - 3.4)}\!/\!_{1} \right) = 1.40 \times 10^{9}\text{cm}^{-3}$$
$$N_3 = 4 \times 10^{13} \exp - \left({}^{(13.6 - 1.51)}\!/\!_{1} \right) = 2.24 \times 10^{8}\text{cm}^{-3}$$

(c) Atomic emission is severely affected by fluctuations in temperature since the intensity is dependent on the number of atoms in the excited state. However, in the case of atomic absorption, the intensity depends on the number of atoms in ground state that absorb energy. The number of atoms in ground state is very high as related to the number of atoms in the excited states as can be seen by a simple calculation;

$\frac{N_2}{N_1} = 3.5 \times 10^{-5}$ and $\frac{N_3}{N_1} = 5.6 \times 10^{-6}$

(d) The absorption experiment may be described by the Lambert-Beer law:

$I_t = I_0 exp - N\sigma l$

The absorption is given by: $A = 1 - \frac{I_t}{I_0}$, then $0.42 = 1 - \frac{I_t}{I_0}$ which results:
$0.58 = \exp\left(-4 \times 10^{13} \times \sigma \times 10\right)$
$\sigma = 1.36 \times 10^{-15} \mathrm{cm}^{-2}$

Exercise 9.2. A portable monochromator registered a current of 80 pA when detecting a radiation in an absorption experiment. The dark current response was eliminated putting a shutter in front of the entrance slit and zero current was recorded with no light striking the detector. The absorption medium was replaced and a new current of 50 pA was recorded. Determine:

(a) The transmittance of the medium
(b) The absorption coefficient of the sample.
(c) The transmittance of the sample when the absorber is replaced by one another with concentration that is one fourth of the former.

Resolution:

(a) I_0 corresponds to a relative value of 80 while I_t corresponds to a relative value of 50, therefore:

$$T = \frac{I_t}{I_0} = \frac{50}{80} \times 100\% = 62.5\%$$

(b) The absorption is given by:

$$A = 1 - \frac{I_t}{I_0} = 1 - \frac{50}{80} = 0.38$$

(c) The transmittance becomes ¼ higher when a medium with lower concentration is used:

$$A = \left(1 - \frac{50}{80}\right) \times \frac{1}{4} = 0.09$$

The transmittance may be obtained from:

$$T = \frac{I_t}{I_0} = 1 - A = 0.91$$

Exercise 9.3. In gas discharge spectroscopy, we are interested in the identification of chemical species absorbing or emitting radiation as well as in the determination of their populations as the gas discharge parameters are changed. These qualitative and quantitative analyses often require different monochromator slit widths. Explain why the regulation of the monochromator is necessary.

Resolution: For identification of chemical species absorbing or emitting radiation, i.e. in a qualitative analysis, wide slits are usually employed in order to increase light throughput, increasing the sensitivity of the measurement. Wide slits in qualitative analysis will result in overlapping of peaks losing spectral resolution. However, in quantitative analysis, i.e. the determination of species density, the use of narrower slit widths is required to improve resolution and thus to catch the small spectroscopic features in details to identify a given specie.

Exercise 9.4. A gas discharge tube in argon operates at 1 Torr with a discharge current of 30 mA and has a length of 10 cm. An absorption apparatus is employed to measure the metastable atomic density ($1s_5$) employing the line 696.5 nm originated from a lamp source with a FWHM of $\Delta\lambda^S_{1/2} = 2 \times 10^{-3}$ nm with a coefficient $\alpha = 1.33$. Find:

(a) The gas discharge temperature
(b) Knowing that the measured absorption is 0.4, $f = 2.9 \times 10^{-2}$, determine the argon metastable atomic density Ar($1s_5$).

Resolution:

(a) To determine the gas temperature we need first to obtain the FWHM of the Doppler broadened profile.

$\alpha = \frac{\Delta\lambda^S_{1/2}}{\Delta\lambda^D_{1/2}}$, thus $\Delta\lambda^D_{1/2} = \frac{2\times10^{-3}\ \text{nm}}{1.33} = 1.5 \times 10^{-3}$ nm

$\Delta\lambda_D = 7.17 \times 10^{-7}\lambda_0\sqrt{\frac{T(K)}{M(\frac{g}{\text{mol}})}}$;

$$1.5 \times 10^{-3}\ \text{nm} = 7.17 \times 10^{-7} \times 696.5\ \text{nm} \times \sqrt{\frac{T(K)}{1}}$$

$T = 360$ K

(b) Using the abacus of Fig. 9.6 and knowing that α is 1.33 and A = 0.4, we find that $k_{\lambda_0}L = 0.91$. As the absorption length is 10 cm, $k_{\lambda_0} = 9.1\ \text{m}^{-1}$.
We know that:

$$k_{\lambda_0} = \frac{e^2}{4\epsilon_0 mc^2}\sqrt{\frac{\ln 2}{\pi}}\frac{2}{\Delta\lambda^D_{1/2}} f n_M$$

Substituting the values:

$$9.1 = \frac{e^2}{4\epsilon_0 mc^2}\sqrt{\frac{\ln 2}{\pi}}\frac{2}{1.5\times 10^{-3}\ \mathrm{nm}}\times 2.9\times 10^{-2}\times n_M; n_M = 5.68\times 10^{10}\mathrm{cm}^{-3}$$

Exercise 9.5. In chemistry, it is common to use the absorbance rather than absorption, which is defined as the common logarithm of the ratio of incident to transmitted radiant power through a material.

$$A = \log\left(\frac{I_0}{I_t}\right) = -\log T = \varepsilon bC$$

where I_t is the transmitted intensity of light, I_0 is the intensity of light received and T is the transmittance of that material. ε is the molar absorptivity $\left(\mathrm{cm}^{-1}\mathrm{M}^{-1}\right)$, b is the sample thickness and C is the analyte's concentration.

Consider a line source that emits radiation at two wavelengths, λ' and λ''. The absorbance at these two wavelengths are designed as A' and A''. When the two wavelengths are measured simultaneously, the resulting absorbance is:

$$A = \log\left(\frac{I_0' + I_0''}{I_t' + I_t''}\right)$$

If the molar absorptivity at λ' and λ'' is the same, show that the absorbance is given by: $A = \varepsilon bC$.

Resolution:

$A' = \log\left(\frac{I_0'}{I_t'}\right) = \varepsilon' bC$ and $A'' = \log\left(\frac{I_0''}{I_t''}\right) = \varepsilon'' bC$ substituting in

$$A = \log\left(\frac{I_0' + I_0''}{I_t' + I_t''}\right)$$

results:

$$A = -\log\left(\frac{10^{-\varepsilon' bC} I_0' + 10^{-\varepsilon'' bC} I_0''}{I_0' + I_0''}\right)$$

As the molar absorptivity at λ' and λ'' is the same, i.e. $\varepsilon' = \varepsilon'' = \varepsilon$;

$$A = -\log\ 10^{-\varepsilon bC} - \log\left(\frac{I_0' + I_0''}{I_0' + I_0''}\right) = \varepsilon bC$$

References

C.M. Ferreira, A. Ricard, Modelling of the low pressure argon positive column. J. Appl. Phys. **54**, 2261–2271 (1983)

C.M. Ferreira, J. Loureiro, A. Ricard, Populations in the metastable and the resonance levels of argon and stepwise ionization effects in a low pressure argon positive column. J. Appl. Phys. **57**, 82–90 (1985)

A.C.G. Mitchell, M.W. Zemanski, *Resonance. Radiation and Excited Atoms* (Cambridge University Press, New York, 1961)

R. Landerburg, F. Richie, Über selektive absorption. Annalen der Physik **347**, 181–209 (1913)

D. Pagnon, J. Amorim, J. Nahorny, M. Touzeau, M. Vialle, On the use of actinometry to measure the dissociation in O_2 DC glow discharges: determination of the wall recombination probability. J. Phys. D Appl. Phys. **28**, 1856–1868 (1995)

W. Wieme, J. Lenaerts, Diffusion of metastable atoms in rare gases. Physica B+C **98**, 229–234 (1980)

Chapter 10
Laser Spectroscopy

In order to improve the existing plasma techniques and to develop new processes, physicists and engineers need predictive models based on plasma physics, chemical physics and hydrodynamics. The knowledge of physical and chemical data must be checked or determined through experiments. Moreover, the reproducibility of a process is largely related to the control of some determining plasma parameters, which can be performed through diagnostics. For all these reasons plasma diagnostic techniques are of fundamental importance. In the last 30 years, many types of plasma diagnostics have been developed (Lochte-Holtgreven 1968; Donnelly et al. 1990; Preppernau et al. 1993; Jolly 1995), each of them allowing one to access one or more plasma parameters, but no one being able to fully characterize plasmas. So, plasma physicists have to simultaneously or successively use various diagnostics depending on the objective of their study. Among the main techniques currently used to characterize plasmas we may mention electric and magnetic probes, mass spectrometry, emission and absorption spectroscopy and laser-based techniques.

At the region close to surface, at the interface surface-plasma, a region called sheath is characterized by high gradients of the main parameters that constitute the plasma. The diagnostics need to be done "in-situ" by non-perturbing methods with good spatial-temporal resolution.

The most commonly used of all these diagnostics is undoubtedly plasma-induced emission (PIE) spectroscopy. This very sensitive, in situ, non-intrusive, easy-to-use and relatively inexpensive technique, based on the spectral analysis of de-excitation radiation of the excited species, is employed in a large number of situations. Moreover, PIE allows spatial and temporal resolution for heterogeneous, time-dependent or transient plasmas. Optical emission spectroscopy is very useful in order to detect atoms, radicals or molecules present in the plasma, but it concerns only the excited species that have concentrations that are quite low compared to their parent species in ground or metastable states. The mass spectrometry diagnostic

J.M.A.H. Loureiro, J. de Amorim Filho, *Kinetics and Spectroscopy of Low Temperature Plasmas*, Graduate Texts in Physics,
DOI 10.1007/978-3-319-09253-9_10

technique is also useful for the detection of numerous molecular species, including large molecules not accessible by PIE, but this ex-situ diagnostic gives no spatial resolution.

Laser-based plasma diagnostics constitute a complementary technique allowing in situ detection of ground-state atoms, small molecules and radicals. The characteristics of the laser techniques are related to those of the lasers. The directivity of the laser beam allows high spatial resolution, the relatively short pulse duration, with pulsed lasers, is useful for temporal resolution, the narrow spectral line width permits the selectivity of the species and the high flux, with a focused laser beam, allows transitions with weak absorption cross sections. Laser-based plasma diagnostic techniques have largely contributed to advances in the understanding of active plasma kinetics. These diagnostics have achieved a tremendous evolution since the 1970s and the development of solid-state lasers changed the picture of laser-based diagnostics.

The spectroscopy based on lasers to diagnose plasmas, initially devoted to fundamental research, to probe the main species and parameters of plasmas, nowadays are used to monitor processes like deposition of thin films, advanced nanotechnologies products, solar cells, highly efficient combustion motors, and treatment of cancer cells.

Ideal diagnostics need to help to access to fundamental data that intervenes in the equations governing the plasma, like Boltzmann equation, Poisson equation, transport equation etc., it means the species densities, their distribution functions as a function of energy and the electric field that is responsible to accelerate charged particles. It is also important in the control of processes measuring for example the deposition/etching rates, end-point detector and analyzes the surface state. Nowadays we are far to have all the techniques needed to measure and control the species and processes. Moreover, a certain number of spectroscopic measures are possible aiding to determine local properties of plasmas as:

- the nature and density of certain reactive species like atoms, molecules/radicals, ions in their neutral or excited states.
- temperature or energy of these species: gas temperature, vibrational and rotational temperatures, translation energy
- electric field

With relation to the surface, the laser diagnostics may give a certain number of information about their physical properties, electric and chemical composition of the substrate.

The laser systems currently used for plasma diagnostics are composed of two or three parts; the first one is the laser source used to pump the dye, the more current sources are solid-state Nd:YAG lasers with two possible pump wavelengths (532 and 355 nm, which are, respectively, the second and third harmonics) and excimer lasers emitting in the near-ultraviolet (307–380 nm, depending on the gases used); the second part is the dye laser containing a dispersion system in the laser cavity, which ensures a continuously tunable wavelength within the spectral range of the dye. The total spectral range obtained with the various dyes available covers the near-UV and visible region (350–800 nm) (Amorim et al. 2000). It is worth to mention that accordable OPO solid state lasers are becoming a very attractive solution to tune the laser wavelength with transitions in atoms and molecules; the third component of the system is the UV generator system to extend laser emission toward the ultraviolet region ($\lambda > 200$ nm) by frequency doubling and/or mixing the dye laser in nonlinear crystals.

In this chapter we will introduce in Sect. 10.1 the principles of laser spectroscopy followed by an explanation of many gas lasers in Sect. 10.2, solid-state lasers in Sect. 10.3 and liquid lasers in Sect. 10.4. Experiments with absorption of one photon are presented in Sect. 10.5. Absolute density measurements are discussed in Sect. 10.6 while multiphoton laser induced fluorescence is studied in Sect. 10.7.

10.1 Principles of Laser Spectroscopy

The discovery of lasers revolutionized the spectroscopy given the opportunity to the experimentalist a monochromatic light source, powerful, directive and sometimes tunable. These properties enable the laser source to be an instrument that is not comparable with others used in spectroscopy, allowing by its spectral finesse the selective excitation of atoms and molecules, by its directivity diagnostics with an excellent spatial resolution and by the spectral power density the excitation of transitions with low absorption coefficient or multiphoton processes.

The atom or molecule excited by the absorption of one photon may, if the excited level is a radiative state, lose all or part of its energy emitting one photon of fluorescence. This is the principle of the laser-induced fluorescence (LIF). This technique is the most popular between the laser-based techniques and allows one to obtain information about the population of ground-state or long-lived, non-radiative, excited-state atoms (see Table 10.1), molecules or radicals (see Table 10.2). This technique has been employed for many years for diagnostics in flames (Aldén et al. 1982), combustion (Arnold et al. 1990) and plasmas (Amorim et al. 2000). The LIF

Table 10.1 Some atoms detected by one-photon LIF (Amorim et al. 2000)

Species	Plasma source	Transition	λ_{laser} (nm)	$\lambda_{fluorecene}$ (nm)	References
Al	RF	$3p\ ^2P^o - 4s\ ^2S$	394	396	Omenetto et al. (1984)
As	RF	$4p^3\ ^4S^o - 5s\ ^4P$	193	245	Selwyn (1987)
Cu	RF	$4s\ ^2S - 4p\ ^2P^o$	325	325 and 510	Leong et al. (1986)
Fe	Magnetron	$a\ ^5D - y\ ^5D^o$	302	382	Hamamoto et al. (1986)
Ge	RF	$3p^2\ ^3P_0 - 4s\ ^3P_1$	265	275	Hata et al. (1987)
Mo	RF	$a\ ^7S - {}^7P^o$	313	317	Omenetto et al. (1984)
Si	RF	$3p^2\ ^3P_0 - 4s\ ^3P_1^o$	251	253	Roth et al. (1984)
Zn	RF	$4s^2\ ^1S - 4p\ ^1P^o$	214	214	Leong et al. (1986)

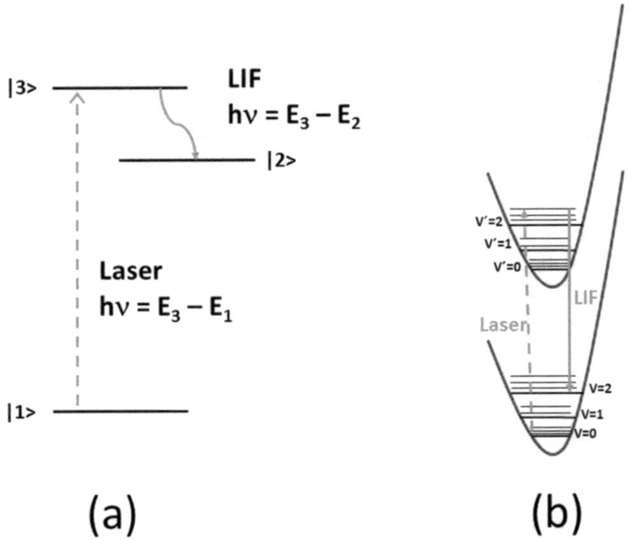

Fig. 10.1 Energy scheme of the laser induced fluorescence detection in (**a**) atoms and (**b**) molecules

scheme can be depicted as a two-step process. First, a specific atomic or molecular species in its ground state $|1\rangle$ (energy E_1) is excited by resonant absorption of laser photons ($h\nu_{laser}$) to a higher energy electronic level $|3\rangle$ of energy E_3 with $h\nu_{laser} = E_3 - E_1$. Then, the excited state relaxes by spontaneous emission of a fluorescence radiation to a lower state $|2\rangle$ of energy E_2 with $h\nu_{LIF} = E_3 - E_2$. The LIF radiation is emitted in the whole space surrounding the interaction volume between the laser photon and the target. A simplified three-level scheme of LIF in atoms and molecules is given in Fig. 10.1.

The LIF spectroscopy can be used in two different ways: as in fluorescence spectrum mode or excitation spectrum mode. In the fluorescence spectrum mode, the laser frequency is fixed at one absorption frequency of the species studied and the fluorescence spectrum, which corresponds to the allowed radiative transitions towards lower energy levels, is recorded by scanning the spectrometer wavelength. In the latter case, the laser frequency is scanned and the induced fluorescence is observed through an optical filter or a low-resolution spectrometer.

The fluorescence is detected when the laser frequency matches a resonant absorption between the levels $|1\rangle$ and $|3\rangle$. In this case, high spectral resolution can be achieved since it depends only on the line width of the laser and of the absorption line, as illustrated in Fig. 10.2.

To implement laser-based diagnostics there is a large range of lasers available, some operating in continuous mode and others pulsed. In this last case the repetition frequency may have a key role if we consider the average power delivered.

Table 10.2 Examples of molecules and radicals detected by one-photon LIF (Amorim et al. 2000)

Species	Plasma source	Transition	λ_{laser} (nm)	$\lambda_{fluorecene}$ (nm)	Reference
BCl	RF	$A\,^1\Pi \rightarrow X\,^1\Sigma^+$	272	272	Gottscho and Mandich (1985)
CCl	RF	$A\,^2\Delta \rightarrow X\,^1\Sigma^+$	278	278	Gottscho et al. (1982)
CF	Pulsed	$B\,^2\Delta \rightarrow X\,^2\Pi$	193	194	Hansen et al. (1988)
CF_2	Pulsed	$\tilde{A}\,^1A \rightarrow \tilde{X}\,^1A$	248 and 266	257 and 271	Hansen et al. (1988)
$Cl_2{}^+$	RF	$A\,^2\Pi \rightarrow X\,^2\Pi$	386	396	Donnelly et al. (1982)
CH	ECR	$B\,^2\Sigma^- \rightarrow X\,^2\Pi$	387	390	Jacob et al. (1994)
CH	ECR	$A\,^2\Delta^- \rightarrow X\,^2\Pi$	413	430	Jacob et al. (1994)
CH	Microwaves	$C\,^2\Sigma^- \rightarrow X\,^2\Pi$	314	314	Hummernbrum et al. (1992)
CN	DC	$B\,^2\Sigma \rightarrow X\,^2\Sigma$	388	421	Hayaud et al. (1997)
FeO	DC	$B_0 \rightarrow X_0$	579	609	Niemi et al. (2001)
$N_2{}^+$	DC	$B\,^2\Sigma_u^+ \rightarrow X\,^2\Sigma_g^+$	391	428	Davis and Gottscho (1983)
NH	DC	$A\,^3\Pi \rightarrow X\,^3\Sigma^-$	336	336	Amorim et al. (1995)
NO	Pulsed	$A\,^2\Sigma^- \rightarrow X\,^2\Pi$	226	248	Fresnet et al. (1999)
OH	Afterglow	$A^2\Sigma^+ \rightarrow X\,^2\Pi$	281 and 284	312	Adams et al. (1989)
SiCl	Afterglow	$B^2\Sigma \rightarrow X\,^2\Pi$	275 and 295	280 and 320	Singleton et al. (1992)
SiH	RF	$A\,^2\Delta \rightarrow X\,^2\Pi$	413	413	Tachibana et al. (1991)
SiH_2	Afterglow	$\tilde{A}\,^1B \rightarrow \tilde{X}\,^1A$	580	618	Kono et al. (1993)
SiN	DC	$\tilde{B}\,^2\Sigma \rightarrow \tilde{X}\,^2\Sigma$	396	414	Walkup et al. (1984)
SiO	PECVD	$A^1\Pi \rightarrow X^1\Sigma$	221	248	Van der Weijer and Zwerver (1989)

For a given spectroscopic diagnostic, fluorescence spectrum mode or excitation spectrum mode, the laser need to be fixed in frequency or emitting in a broad range of frequencies where it is tunable.

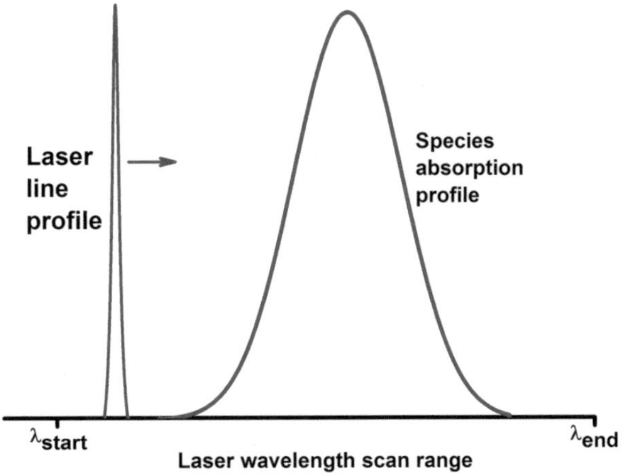

Fig. 10.2 LIF scheme when the data acquisition is done when the probing laser is scanned over the absorption line profile of the atom or molecule probed

The continuous and fixed frequency noble gas lasers most employed are He-Ne and the ionic ones Ar^+ and Kr^+. The former due to its low power, some mW, are used in few diagnostic techniques, e.g. interferometry. The ionic lasers, mainly Ar^+, containing a tube with a length of the order of 1 m, can generate 10 W or 20 W of output power in the green spectral region at 514.5 nm, using several tens of kilowatts of electric power. The voltage drop across the discharge tube is few hundred volts, whereas the current can be several tens of amperes. The total wall-plug efficiency is thus very low, usually below 0.1 % (Paschotta et al. 2008). The laser can be switched to other wavelengths such as 457.9 nm (blue), 488.0 nm (blue–green), or 351 nm (ultraviolet) by rotating the intracavity prism. The highest output power is achieved on the standard 514.5 nm line. Without an intracavity prism, argon ion lasers have a tendency for multi-line operation with simultaneous output at various wavelengths.

There are similar noble gas ion lasers based on krypton instead of argon. Krypton ion lasers typically emit at 647.1 nm, 413.1 nm, or 530.9 nm, but various other lines in the visible, ultraviolet and infrared spectral region are accessible. Multi-watt argon ion lasers can be used, e.g. for pumping titanium–sapphire lasers and dye lasers. They are in competition today with frequency-doubled diode-pumped solid-state lasers. The latter are far more power efficient and have longer lifetimes, but are more expensive. Argon tubes have a limited lifetime of the order of a few thousand hours. An argon laser may thus be preferable if it is used only during a limited number of hours, whereas a diode-pumped solid-state laser is the better solution for reliable and efficient long-term operation. The gas lasers present, in the absence of intracavity line narrowing optics, a linewidth that can be of the order of 1 GHz. On the other hand, the laser linewidth from stabilized low-power continuous-wave lasers can be very narrow and reach down to less than 1 kHz. Although these lasers have a very narrow linewidth the spectral domain is limited. They are employed essentially in absorption experiments to dose density of species when the laser must have a linewidth that is very low if compared with the Doppler broadening of the absorption line.

The pulsed lasers are generally more simple design and are more employed in plasma diagnostics. They are customarily cheaper and easier to use. In the next sections we will present some examples of these lasers.

10.2 Gas Lasers

These lasers are composed of a low-density gas medium where the broadening of the laser line is small, few GHz, due to the weak influence of the line broadening mechanisms if compared with solids. Doppler effect in this case essentially governs the broadening because the pressure of the gas in the discharge tube is on the order of few tens of Torr. The excitation is governed by the electrical current circulating in the gas but other mechanisms like chemical pumping, gas expansion or optical pumping may be found. The gas lasers addressed in this book will be the ones more employed in spectroscopy diagnostics, i.e. nitrogen, excimers, cooper vapor lasers and CO_2.

10.2.1 Nitrogen Lasers

The nitrogen lasers are the most simple and less expensive of the vibronic lasers. Its most important oscillation is at 337 nm, belonging to the category of self-terminating lasers, i.e. laser transitions cannot be continuously operated due to the accumulation of population in the lower laser level. The laser action occurs in the transition $C\,^3\Pi_u \rightarrow B\,^3\Pi_g$, the Second Positive System of N_2. The $C\,^3\Pi_u$ state is excited in a gas discharge tube and has a lifetime of 40 ns while the lifetime of the $B\,^3\Pi_g$ state is 10 μs reason why this class of laser cannot operate in continuous mode. In the gas discharge tube a high pressure mixture is usually employed with 40 mbar of nitrogen in 960 mbar of helium. With nitrogen lasers high power pulses of some MW may be obtained with short duration, around 10 ns and the repetition rate may attain 100 Hz. Energies of some mJ may be used to pump dye lasers enabling to obtain tunable laser beam with some μJ of energy.

10.2.2 Excimer Lasers

Excimer lasers are included in an important class of molecular lasers where the transitions occur between different electronic states of excimers molecules. These molecules are formed by excited dimer. The excited electronic states show a minimum in the potential curve while the ground electronic state is repulsive, meaning that stable molecules cannot exist in this state. The stable bound molecule exists only in the excited state forming a dimer. If a large fraction of molecules is excited in a given volume, the laser action can then occur on the transition between the upper-bound state and the lower-unbound (free) state, i.e. a bound-free transition. The first excimer laser was done in Ne_2^* by Basov et al. (1971) in 1971 emitting at $\lambda = 170$ nm.

These lasers present interesting properties; since the transition occurs between different electronic states the emission is in UV range of the spectrum. As the lower state of the transition is a repulsive or quasi-repulsive state the emission present a feature of a broad continuum. Actually the gas discharge medium is a mixture of rare gas (Kr, Ar, Xe)-(F, Cl) halogen gas that leads to the formation of the excimer molecules. The most important examples are ArF ($\lambda = 193$ nm), KrF ($\lambda = 248$ nm), XeF ($\lambda = 351$ nm) and XeCl ($\lambda = 309$ nm).

10.2.3 Cooper Vapor Lasers

Cooper vapor lasers are employed in applications where the high-peak power is not necessary like in high-speed photography, micromachining and also to pump dye lasers. One important application is in ^{235}U isotope separation where large facilities based on cooper-vapour pumped dye lasers exists in some countries. In this last case many modules of hundreds of watts has been built. The main transition of this laser is $^2P \rightarrow {}^2S_{1/2}$ transition, which is allowed by electric-dipole, emitting photon in the green ($\lambda = 510$ nm) and in the yellow ($\lambda = 578$ nm) colors. Commercially available cooper-vapour lasers have powers of 100 W and pulse duration of 30–50 ns with high-repetition rates up to 10 kHz.

10.2.4 CO_2 Lasers

The laser transitions occur between two vibrational levels of the CO_2 molecule. These lasers generally employ a mixture of CO_2, N_2 and He to improve the efficiency. Is one of the most powerful lasers in the market with output powers of 100 kW and efficiencies up to 20 %. The main laser transition wavelength is at $\lambda = 10.6$ μm. Nitrogen is used in the mixture to boost large population in the upper laser level while helium is responsible to remove population of the lower laser level. The laser tube works generally with total pressures of 20 Torr in a gas mixture of 1:1:8 of CO_2:N_2:He when operating in continuous mode. The CO_2 lasers may have some constructional designs like in axial flows, sealed, waveguide transversely excited atmospheric pressure (TEA) for example. Typical applications of these lasers in the industry are cutting metals, marking and ablation of metals and plastics.

10.3 Solid-State Lasers

Solid-state lasers belong to a class of high-density active media lasers well used in spectroscopy. In this subsection we will present some general concepts, engineering details with emphasis in its main performances like wavelengths, output power or energy, wavelength tunability etc. These lasers are generally produced from the introduction of an impurity, ions belonging to the transition elements of the periodic table, in a transparent dielectric host material in crystalline or glass structure. In particular, rare-earth or transition metal ions have been used as active impurities. For the host material crystals like Al_2O_3 or fluorides like $YLIF_4$ are the most used. The combination of oxides enables to form a synthetic garnet as $Y_3Al_5O_{12}$ (called YAG for the acronym for yttrium aluminum garnet), which is one of the most popular examples. Ions Al^{3+} site can accommodate transition metal ions while Y^{3+} is used in combination with rare-earth ions due to geometrical properties of these sites. The rare-earth elements used to build lasers are Nd, Er, Yb, Tm and Ho while for transition metals we have Ti, Co, Cr, and Ni.

Undoubtedly the most popular solid-state laser is the neodymium ones. The host is a crystal YAG in which some of the Y^{3+} is replaced by the Nd^{3+} ions. Some other host media may be $YLiF_4$ or YVO_4. The doping levels are some % in Nd:YAG lasers. The main pumps bands are at 730 and 800 nm. These bands are coupled by non-radiative decay to the upper level of the laser transition $^4F_{3/2}$. The laser action occurs in a transition $^4F_{3/2} \rightarrow {}^4I_{11/2}$ at 1.064 μm. The long lifetime of the upper level, around 230 μs, makes the Nd:YAG lasers very suitable for Q-switched operation and its broad emission bandwidth is suitable to operate in mode-locking to generate picosecond pulses. Nd:YAG lasers may operate in continuous or pulsed mode, being pumped either by lamps or AlGaAs semiconductor lasers (Svelto et al. 1998). Flash lamps configurations are used in when one or two lamps are used to illuminate the rod, according to the desired performance of the laser. Gases used in the lamps are Xe, with pressures of 500–1500 Torr, and Kr, pressures of 4–6 atm, when the operating mode is pulsed or continuous respectively, with output powers up to 3 kW. Continuous output powers of 15 W are obtained for longitudinally diode-pumping and 100 W for transversely diode pumping scheme. Efficiencies of 10 % was mentioned in diode pumping while 3 % is a good value when lamps are employed as pump element.

Applications such as drilling and welding demands Nd:YAG lasers with pulse energy of 5–10 J and pulse duration of 1–100 ms at repetition rates between 10 and 100 Hz. Other uses of these lasers are in the medical domain where it is useful in coagulation and tissue evaporation. Through an optical fibre and with aid of an endoscope, the beam may be sent to stomach, lungs and bladder. In spectroscopy Q-switched lasers with energies of hundreds of mJ and pulses from 0.1 to 20 ns and repetition frequencies of 10–50 Hz are used in the pupping of dye lasers. For that second-harmonic ($\lambda = 532$ nm), third-harmonic ($\lambda = 355$ nm) and even fourth-harmonic ($\lambda = 266$ nm) of the fundamental wavelength may be used in a variety of spectroscopic diagnostics like probing atoms and molecules in low-temperature plasmas or in combination with many dyes to generate a broad range of wavelength.

Tunable solid-state lasers are today a class of lasers much used in spectroscopy. The most important tunable solid-state lasers are Alexandrite ($Cr:BeAl_2O_4$), Ti:sapphire and Cr:LiSAF among others. Titanium Sapphire laser is undoubtedly the most widely used tunable solid-state laser. It operate over a broad tuning range, from 660 to 1180 nm, being the largest bandwidth laser. Ti:Sapphire lasers may be pumped by the green beam of an Ar laser. In pulsed operation mode it may be pumped by the frequency doubled Nd:YAG at 532 nm, as well as flashlamps may also be employed. Relatively large energy devices, with energies from 20 mJ to 1 J with pulse width of 20–100 fs is a sophisticated apparatus to be used in the study of fast reaction kinetics.

Another class of lasers are the ones based in semiconductors. They are usually not included in the solid-state lasers, due to the pumping mechanisms and laser action be quite different from that employed in solid-state lasers. These lasers are very important today because of the large variety of applications in spectroscopy but also as pump source of solid-state lasers.

10.4 Liquid Lasers

The lasers in liquid media are generally tunable and made of a dye laser pumped by a fixed frequency laser described above. The dye lasers have an active medium constituted of an organic solution in a solvent as ethyl or methyl alcohol, water or glycerol (Svelto et al. 1998). Organic dyes belong to the group of polyatomic molecules with long chains of carbon atoms with unsaturated bonds. Example of these groups of molecules are; (a) polymethine dyes used in lasers oscillating in the red or infrared ($\lambda = 0.7 - 1.5$ μm), (b) xanthene dyes when the laser operates in the visible region ($\lambda = 400 - 700$ μm) and (c) coumarin dyes which oscillate in the blue region ($\lambda = 400 - 500$ μm). These molecules have an extended absorption band and a wide fluorescence spectrum. This is due to the large number of ro-vibrational levels and the effective line-broadening mechanisms in liquids which result in an unresolved structure at ambient temperature. The solution of these complex molecules absorbs the radiation in a region of the spectra, according to their chemical structure, and emits at higher wavelengths;

$$\lambda_l(dye) > \lambda_l(pump) \tag{10.1}$$

With these lasers tunable radiation may be generated from near UV until near IR (350–900 nm) according to the dye employed and the wavelength of the pumping laser. An example of the region obtained with a dye laser pumped by the second harmonic of an Nd-YAG is presented in the Fig. 10.3.

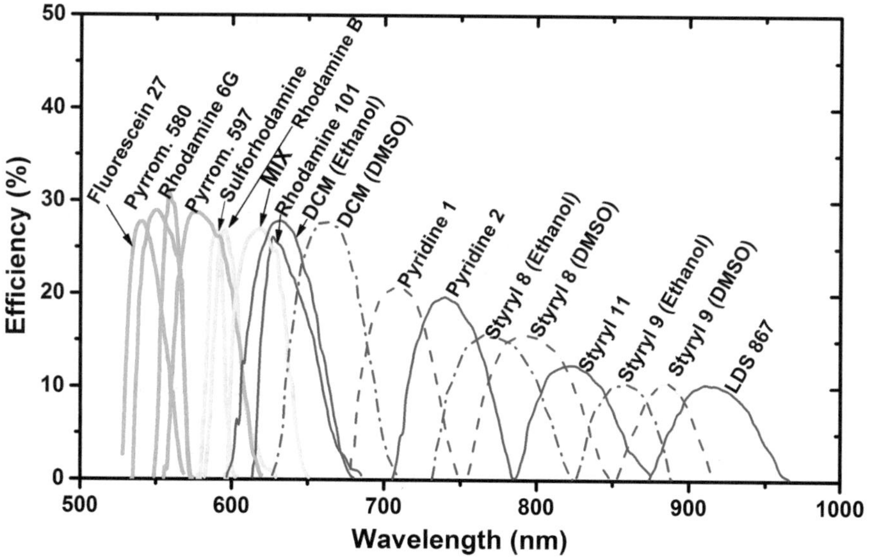

Fig. 10.3 Spectral domain obtained with a dye laser pumped by the second harmonic of a Nd:YAG laser (http://www.spectra-physics.com/products/high-energy-pulsed-lasers/cobrastretch?cat=tunable&subcat=pulsed. Accessed in November 2014)

The transitions between singlet and triplet states of the dye need to be evaluated in order to adjust the decay time and the quenching rate of these states for pulsed or continuous operation of the laser.

The dye lasers pumped by continuous lasers enable to obtain a monochromatic beam with very high finesse and stable in frequency with resolution of about 1 MHz. The power delivered in this case is lower than 1 W.

On the other hand, the pulsed lasers, of simplest conception, are characterized by lower linewidths, less than 1 GHz, but they are capable to deliver peak powers of 10 MW, which may be useful to generate harmonics with aid of non-linear crystals thus to produce an enlargement of the spectral domain of the source.

The spectral domain of dye lasers covers wavelengths from 350 to 900 nm according to the pumping laser source used. To extend the range of application of these lasers, non-linear crystals are used enabling the harmonic generation or frequency addition. Others techniques of frequency conversion may be employed like that using the Raman Anti-Stokes effect. In this case a cell with hydrogen at high pressure is employed to generate a UV laser beam.

To make use of frequency conversion techniques using crystals the dye laser power needs to attain a certain threshold value because these non-linear effects have conversion efficiencies that strongly rely on the power of the incident radiation.

For example, in the case of second harmonic generation through a KDP crystal the conversion efficiency is around 50 % for an incident beam power of 500 MW but it is only few percent in the case of powers lower than 1 MW.

The KDP (potassium dihydrogen phosphate) crystal allows to extend the wavelength domain of the dye laser to UV range until 218 nm, the limit of KDP transmission. The wavelength range of a dye laser pumped by a Nd:YAG laser may be enlarged by using a BBO crystal (β barium borate) which allows to go to 198 nm. One example of harmonic generation and frequency addition is found when the experimentalists need photons with 205 nm to detect hydrogen atoms in its ground state. This can be achieved in the following way;

Dye laser	KDP	Second harmonic	BBO	Addition of frequencies
ν_0 $\lambda = 615\ nm$	$\rightarrow$	$2\nu_0, \nu_0$ $\lambda = 307.5\ and\ 615\ nm$	$\rightarrow$	$3\nu_0 = 2\nu_0 + \nu_0$ $\lambda = 205, 307.5\ and\ 615\ nm$

The separation of the different beams is done with aid of a Pellin-Broca prism placed at the exit of the BBO crystal. For example, with a dye laser delivering 70 mJ at 615 nm, 1.5–2.0 mJ of energy can be obtained at 205 nm.

10.5 One-Photon Laser Induced Fluorescence

The one-photon LIF is the simplest case of laser diagnostics. It is desired when we are interested to probe the density of the fundamental state. The correlation between the fluorescence signal and the concentration of the ground state is easy to work out.

Considering a collection of atoms (or molecules) where a classical three level scheme can be depicted as in Fig. 10.1. The ground state $\left|1\right\rangle$ with density n_1 and energy E_1 is submitted to a laser beam in order it may absorb laser photons ($h\nu_{laser}$) and may be excited to a higher electronic level $\left|3\right\rangle$ of energy E_3 with $h\nu_{laser} = E_3 - E_1$. The excited state relaxes by spontaneous emission emitting a fluorescence radiation to a lower state $\left|2\right\rangle$ of energy E_2 with $h\nu_{LIF} = E_3 - E_2$. The LIF radiation is emitted in the whole space surrounding the interaction volume between the laser photon and the target with a signal intensity given by:

$$I_{LIF} \propto A_{3-2}\, n_3 \tag{10.2}$$

where A_{3-2} is the Einstein transition probability for spontaneous emission and n_3 is the population density of the excited level $|3\rangle$. The evolution of the density of the level $|3\rangle$, assuming a non-saturated absorption, is given by:

$$\frac{dn_3}{dt} = \left(n_1 - \frac{g_1}{g_3}n_3\right) W_{1-3} - n_3\left(Q_3 + A_3\right) \tag{10.3}$$

where n_1 is the ground-state population density, g_1 and g_3 are the statistical weights of the levels $|1\rangle$ and $|3\rangle$, respectively, W_{1-3} is the excitation rate due to the absorption of the laser photons is given by;

$$W_{1-3} = B_{1-3} I_l \tag{10.4}$$

B_{1-3} is the Einstein absorption coefficient, I_l is the laser intensity which can be expressed as the number of photons of energy $h\nu_l$ that takes place in the excitation process. The quenching rate of the level $|3\rangle$, due to the collisional depopulation, Q_3 is given by;

$$Q_3 = K_{qM}\ [M] \tag{10.5}$$

where K_{qM} is the quenching coefficient and [M] the gas density. A_3 is the sum of the spontaneous de-excitation probabilities of the level $|3\rangle$ toward all the lower levels which equals to the inverse of the mean radiative lifetime of level $|3\rangle$. For measurements made at a time scale $> t_{laser}$, being t_{laser} the laser pulse duration, two situations can be found; low-laser intensity and high-laser intensity.

In stationary regime $dn_3/dt = 0$, at low laser intensity, the LIF signal is proportional to the n_1 population and to the laser intensity according to:

$$I_{LIF} = C\, B_{1-3} I_l\, n_1 \frac{A_{3-2}}{A_3 + Q_3} \tag{10.6}$$

where C is a calibration factor that relies on the geometry and the spectral response of the optical imaging and detection system. In principle, it is possible to calculate the absolute density n_1 provided that all the constants and calibration factors are known. In practice this is rarely the case and in most experimental situations the factor C is very difficult to evaluate. The light detected comes from the gas volume formed by the interaction of the laser beam and the collected optical field of the detector. In general this volume is very small, lower than mm^3, providing measures with a very good spatial resolution. Strictly speaking one cannot talk about a stationary regime for a short laser pulse. However, if the laser pulse duration is short when compared with the characteristic time for excited atomic state relaxation, the Eq. (10.6) remains valid in a first approximation and it express the maximum intensity of fluorescence as a function of the laser intensity I_l and the concentration n_1. An estimation of typical values allows us to obtain the detection limit of one photon LIF of about $10^5 - 10^6$ particles cm^{-3}.

At high laser intensity the absorption on the $|1\rangle \rightarrow |3\rangle$ transition may saturate during the laser pulse. The density n_3 becomes so high that the level $|3\rangle$ is destroyed by stimulated emission induced by the laser beam. In this case the laser induced de-excitation $|3\rangle \rightarrow |1\rangle$ occurs equilibrating the population between the levels $|1\rangle$

and $|3\rangle$. When this occurs a further increase of the laser intensity has no more effect on the population n_3. In this condition the density n_3 is given by:

$$n_3 = \frac{B_{1-3}}{B_{3-1}} n_1 = \frac{g_3}{g_1} n_1 \tag{10.7}$$

and the LIF intensity is independent of the laser intensity I_l and proportional to the concentration n_1;

$$I_{LIF} = C \frac{g_3}{g_1} n_1 \frac{A_{3-2}}{A_3 + Q_3} \tag{10.8}$$

Actually complex non-linear phenomena related to finite spectral width of the laser might appear for high laser energies. Only density matrix equations may describe precisely the pumping of atoms and molecules by resonant laser radiation and the LIF mechanisms discussed above must be considered as merely descriptive. However, in most conditions used in plasma diagnostics, the rate equation approach is largely sufficient to describe the evolution of population density of the states and the emission intensities. A two-level system is adequate to describe the physics involved in the LIF process when the laser probe pulse duration is much shorter than the radiative lifetime of the excited state created by the laser absorption. The density of species of the excited state $n_3(\Delta t_{laser})$ at the end of the laser pulse with duration Δt_{laser} is given by (Amorim et al. 2000):

$$n_3\left(\Delta t_{laser}\right) = \frac{g_3}{g_1 + g_3} \frac{S}{S+1} n_1 \left\{1 - e^{\left[-(S+1)A_{3-2}\Delta t_{laser}\right]}\right\} \tag{10.9}$$

here S is the saturation parameter given by:

$$S = \frac{g_1 + g_3}{g_3} A_{3-1}^{-1} \int_{\nu} B_{1-3}(\nu)\, u_{laser}(\nu, r)\, d\nu \tag{10.10}$$

is result from the convolution of the spectral absorption coefficient $B_{1-3}(\nu)$ with the spectral laser energy density $u_{laser}(\nu, r)$ which relies on the radial position r of the laser beam. If the fluorescence time is long compared to the laser pulse duration, we need to consider a three-level system with all radiative and non-radiative loss terms of the level $\left|3\right\rangle$. Assuming that the collection of fluorescence photons starts just after the end of the laser pulse, the fluorescence signal is given by:

$$\begin{aligned} I_{LIF} &= \tfrac{K}{4\pi} \int_0^{l_s} \int_a \int_0^{\Delta t} n_3\left(\Delta t_{laser}\right) A_{3-2} e^{-t/\tau}\, dt\, dA\, dl \\ &= \tfrac{K}{4\pi}\, l_s\, A_{3-2}\, n_1\, \tfrac{g_3}{g_1+g_3}\, \tau \left[1 - e^{\Delta t/\tau}\right] \tfrac{S}{S+1} \left\{1 - e^{\left[-(S+1)A_{3-1}\Delta t\right]}\right\} dA \end{aligned} \tag{10.11}$$

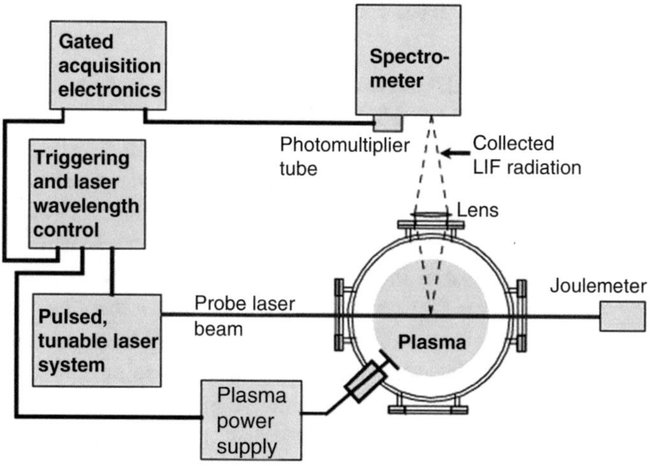

Fig. 10.4 Schematic diagram of a typical set-up employed in LIF diagnostics

where Δt is the time for signal integration, l_s is the length of the interaction volume, between the probe laser and the species, which is determined by the slit width and height of the spectrometer, and τ is the effective lifetime of the excited state of the species being studied, in the interaction volume. The detection volume the Eq. (10.11) needs to be evaluated taking into account losses due to diffusion out of the volume being probed.

The spatial dependence of the laser beam energy is considered in Eq. (10.11) by the integration $\int_a dA$ over the front end (A) of the volume under study. To determine the absolute specie densities from the LIF measurements, each term in the Eq. (10.11) needs to be evaluated in order to find a relation between I_{LIF} and n_1.

A typical set-up employed in the LIF diagnostics is shown in Fig. 10.4. As can be seen the detection of fluorescence is done perpendicularly to the laser beam.

The region probed is imaged onto the entrance slit of a spectrometer by optical lenses. The spectrometer is adjusted to transmit the fluorescence wavelength with a bandwidth of typically 2–3 nm. The LIF light may be detected by a photomultiplier tube, where the resulting electrical signal is integrated, averaged and stored in an acquisition data system triggered by the laser. The detection gate depends on the radiative and collisional lifetime of the fluorescent level. It should be as short as possible in order to limit the quenching effect, which can become important, depending on the species, for pressures greater than a few Torr. The spatial resolution is obtained by moving together the laser beam and the optical detection system. A direct spatial resolution can be obtained by imaging, through an optical filter, the interaction volume onto the photo-cathode of a gated, CCD intensified camera.

10.6 Absolute Calibration Procedures

The detection of species by LIF leads only to relative values of the densities. Absolute values can be obtained through the knowledge of the laser beam characteristics, of the spectroscopic constants involved in the process, and the careful calibration of the optical detection system. Alternative calibration techniques can be made by substituting the probed plasma medium by a source of known species concentration or by using other absolute measurement technique with the same plasma. In the latter case, the degree of confidence of the calibration is largely related to the possibility of generating plasmas as similar as possible for both the absolute concentration measurement and the laser-based diagnostic calibration if they are not performed simultaneously. The techniques usually employed in the absolute calibration method are resonant absorption spectroscopy, resonance enhanced multiphoton ionization (in afterglows only), titration and Rayleigh scattering. Another method is based on a comparative measurement at a spectrally close two-photon resonance of a noble gas (Niemi et al. 2001). In this chapter we will discuss briefly the Rayleigh scattering, readers interested in others calibration techniques may refer to (Amorim et al. 2000).

One of the main difficulties in calibrating LIF experiments is the evaluation of the factor K, see Eq. (10.11), which takes into account the geometry and the spectral response of the optical imaging and detection of the fluorescence.

The factor K can be determined by a Rayleigh scattering calibration performed with the same system (laser and detection) as in the LIF experiment. The Rayleigh signal at a laser frequency $\nu_R = \omega_R/2\pi$ is expressed as:

$$S_R = K\, l_S\, n_G\, \frac{d\sigma}{d\Omega}\, \frac{1}{\hbar\omega_R} \int_A dA \int_\nu c\, \Delta t_{laser}\, u_{laser}\,(\nu, r)\, d\nu \tag{10.12}$$

where n_G is the density of the gas used to perform the Rayleigh scattering (for example N_2) for which the Rayleigh scattering cross section $d\sigma/d\Omega$ is known. The Rayleigh signal, like the LIF signal, depends on the spatial profile of the laser energy density; this leads to the identical integration term $\int_A dA$ written in Eq. (10.11). This technique is particularly suitable in LIF experiments for which the wavelengths of the laser excitation and fluorescence detection are identical or very close. This is the case for numerous atoms, molecules and radicals (see Tables 10.1 and 10.2).

10.7 Multiphoton Laser Induced Fluorescence

Atoms belonging to the first three lines of periodical table of chemical elements like H, N, O... have important role in the physics and chemistry of low-temperature plasmas. The figure below shows a typical energy diagram of the first levels of these atoms.

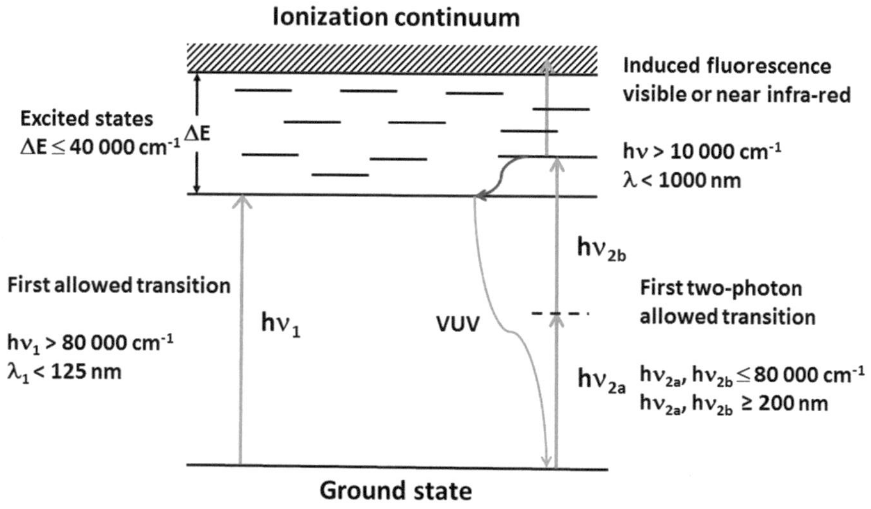

Fig. 10.5 Energy levels diagram of light atoms (H, C, N, O, Cl....). The left side of the figure corresponds to an one-photon excitation while the right part describes a two-photon excitation scheme

The first excited level is found at energies $\geq 80,000\ \mathrm{cm}^{-1}$ above the ground state energy level corresponding to transitions with wavelengths ≤ 125 nm. In this case the one-photon laser induced fluorescence is difficult to be implemented because lasers in the VUV are needed. The solution consists in exciting the atom to the first allowed two-photon transition. The wavelength of the laser for a two-photon transition is ≥ 200 nm corresponding to a spectral domain of certain dye lasers, see the right part of the Fig. 10.5. The excited atom may emit spontaneously a photon and de-excite to a lower level, radiating a photon in the visible or near-infrared. It may also absorb another photon and then be ionized. These two mechanisms of de-excitation and ionization, may have the same probability. In the first case we have a fluorescence induced by two-photon excitation while in the second case there is a creation of electric charges such that in an electrical discharge may induce a photo-galvanic effect, i.e. the modification of plasma impedance by absorption of light.

In the two-photon absorption case the equations governing the evolution of states are more complicated than the one-photon pumping scheme, see Fig. 10.6. For an ideal atom with three levels and considering the ionization we have:

$$\frac{dn_3}{dt} = [n_1 - n_3]\, W_{1,3} - n_3\, \{W_{3,i} + Q_3 + A_{3,2}\} - B_{3,2} \frac{I_{se}}{c} g_D(\nu)\, [n_3 - n_2] \tag{10.13}$$

where $W_{1,3}$ is the excitation rate of level 3 due to absorption of two photons, $W_{3,i}$ is the ionization rate, Q_3 is the quenching rate of the level 3, see Eq. (10.5), $A_{3,2}$ and $B_{3,2}$ are the Einstein coefficients for spontaneous and induced emission respectively,

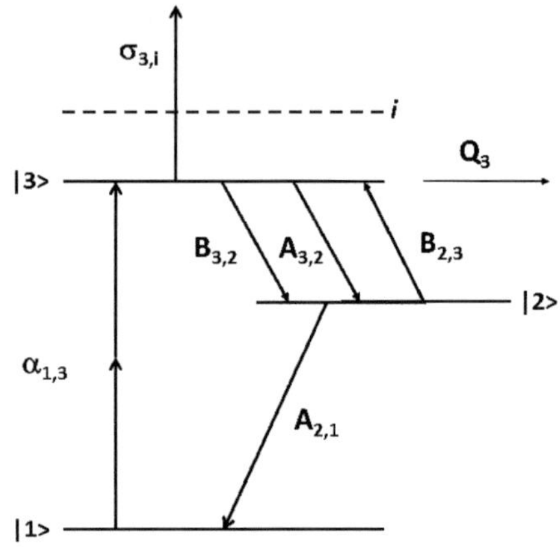

Fig. 10.6 Scheme of a four level atom, including ionization, to explain the two-photon induced fluorescence

and I_{se} is the two-photon allowed laser induced stimulated emission associated with the transition $3 \rightarrow 2$. The excitation rates for a two-photon interaction and ionization are respectively given by:

$$W_{1,3} = \frac{\alpha_{1,3}\phi_l^2}{h\nu_l} \quad \text{and} \quad W_{3,i} = \frac{\sigma_{3,i}\phi_l}{h\nu_l} \tag{10.14}$$

where ϕ_l is the laser power density, $\alpha_{1,3}$ the cross section for two-photon absorption, $\sigma_{3,i}$ the ionization cross section and ν_l is the laser frequency. As the cross section for two-photon absorption is feeble, the laser power density needs to be high in order the rate excitation of level 3 be large enough to produce a large number of induced fluorescence photons.

To pump the transition $1 \rightarrow 3$, usually one needs to use a lens to focalize the laser beam in the region where the interaction is induced. For example, using available commercial lasers, it is possible to obtain power densities higher than 100 Wcm^{-2} with a laser beam of 1 mJ of energy. The multiphoton transition occurs at the focal volume where photon density is higher.

In general, where the depopulation terms of the level 3 are comparable, a simple reasoning is not possible. If the multiphoton excitation is high a population inversion may occur between levels 2 and 3 ($n_3 > n_2$). The stimulated emission created induces a non-linearity and it is necessary to solve the Eq. (10.13) together with the differential equations that govern the evolution of states 1, 2 and of the ions formed by photon absorption from the level 3. One still needs to add to the coupled ensemble of equations regulating the levels population, a set of equations describing the intensity of stimulated emission in the focal volume, as well as an expression for the spatio-temporal distribution of the laser flux ϕ_l. The equations governing the evolution of each level are:

$$\frac{dn_1}{dt} = -\left(n_1 - \frac{g_1}{g_2}n_3\right)\frac{\alpha_{1,3}\phi_l^2}{h\nu_l} + n_2A_{2,1} \tag{10.15}$$

$$\frac{dn_2}{dt} = n_3A_{3,2} - n_2A_{2,1} + \frac{B_{3,2}I_{se}}{c} g_D(\nu)\left(n_3 - \frac{g_3}{g_2}n_2\right) \tag{10.16}$$

$$\frac{dn_3}{dt} = \left(n_1 - \frac{g_1}{g_3}n_3\right)\frac{\alpha_{1,3}\phi_l^2}{h\nu_l} - n_3\left(\frac{\sigma_{3,i}\phi_l}{h\nu_l} + Q_3 + A_{3,2}\right) - \frac{B_{3,2}I_{se}}{c} g_D(\nu)\left(n_3 - \frac{g_3}{g_2}n_2\right) \tag{10.17}$$

$$\frac{dn_i}{dt} = n_3\frac{\sigma_{3,i}\phi_l}{h\nu_l} \tag{10.18}$$

n_j $(j = 1, 2, 3, i)$ are the concentration of the appropriate level j, $h\nu_l$ is the laser photon energy, $\alpha_{1,3}$ is the two-photon excitation cross section, $\sigma_{3,i}$ is the ionization cross section, g_j is the statistical weight of the level j, $A_{i,j}$ is the spontaneous emission rate coefficient for a transition from the level i to level j, Q_3 is the quenching rate coefficient of level 3, $B_{3,2}$ is the Einstein stimulated emission coefficient for a transition form the level 3 to level 2, c is the speed of light, and $g_D(\nu)$ is the Doppler line-shape factor.

In this solution some reasonable approximations need to be made such as considering negligible the quenching of level 2, see Fig. 10.6, the redistribution of population between levels 1 and 2 due to the quenching of the level 3, and ignoring the radiation trapping in the transition 2 → 1. If the quenching of level 3 is negligible the problem is strictly limited to a four levels system and the following closure relation is imposed:

$$n_1(t = 0) = n_1 + n_2 + n_3 + n_i \tag{10.19}$$

This system of differential equations is difficult to solve analytically and in general only a numerical solution is possible. As the stimulated emission is not isotropic one needs to take into account the geometry of the convergence of the laser beam (caustic) and to integrate in space and in the focal volume.

As an example of comparison between theory and experiment, a simulation is performed for the detection of hydrogen atoms, by two-photon absorption LIF (TALIF), with the following parameters: laser beam diameter $D = 0.5$ cm, divergence $\theta = 5 \times 10^{-4}$ rad, lens focal length $f = 35$ cm. A Gaussian distribution profile is assumed for the laser power density ϕ_l, as a function of both the radius r and the time t, is given by:

$$\phi_l(x, r, t) = \phi_m\left(\frac{S_0}{S_x}\right) exp\left[-\delta\left(\frac{r}{r_0}\right)^2 - \left(\frac{t - t_0}{\tau}\right)^2\right] \tag{10.20}$$

where ϕ_m is the maximum intensity of the laser flux at $t = t_0$, $r = 0$ and $x = 0$, r_0 is the radius of the circle determined by the intersection of a plane perpendicular to the laser beam and the caustic at distance x, δ is the parameter related to the radial distribution of the laser beam, τ is the FWHM of the laser pulse duration, and x is the distance along the beam path. The intersection of surfaces at planes and the caustics at $x = 0$ and x are respectively, $S_0 = \pi b^2$ (with $b = f\,\theta/2$) and $S_x = \pi r_0^2$. The interaction volume is a stretched pseudo-hyperboloid of cylindrical symmetry [5]. In this example for detection of hydrogen atoms the geometry dimensions charactering the interaction volume are for length $2f^2\theta/D = 2.45$ cm and diameter $f\theta = 1.75 \times 10^{-2}$ cm. In this example the diffusion of detecting atoms out of the interaction volume is neglected and the laser pulse duration is ≤ 10 ns.

In order to take into account in the calculations the evolution of stimulated emission, the volume needs to be divided into small transversal slices in which $\phi_l(x)$ is assumed constant. In each slice the coupled set of rate Eqs. (10.15, 10.16, 10.17, and 10.18) are solved. To take into account the propagation of the two-photon absorption laser induced stimulated emission (TALISE), the photon transport equations in the forward and backward directions are solved simultaneously with the rate equations for each specific slice, considering the flux coming from adjacent cells, as shown in the Fig. 10.7.

The photon transport equations are:

$$\frac{dI_{se_f}}{dx} = \left(n_3 - \frac{g_3}{g_2}n_2\right)\frac{B_{3,2}I_{se_f}}{c}\,g_D(\nu)\,h\nu + A_{3,2}n_3\Delta\Omega g_D(\nu)\,h\nu \tag{10.21}$$

$$\frac{dI_{se_b}}{dx} = -\left[\left(n_3 - \frac{g_3}{g_2}n_2\right)\frac{B_{3,2}I_{se_b}}{c}\,g_D(\nu)\,h\nu + A_{3,2}n_3\Delta\Omega' g_D(\nu)\,h\nu\right] \tag{10.22}$$

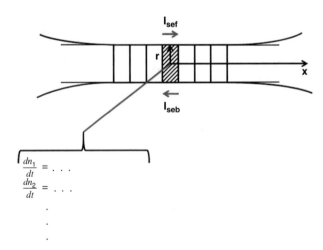

Fig. 10.7 Scheme of the geometry of focused laser beam. I_{se_f} and I_{se_b} are respectively the forward and backward intensities of TALISE induced by TALIF

Table 10.3 Constants used in the solution of differential equations (Amorim et al. 2000)

Parameter	Value	Unity
$\alpha_{1,3}$	5.0×10^{-28}	cm^4W^{-1}
$\sigma_{3,i}$	9.0×10^{-20}	cm^2
$A_{2,1}$	4.7×10^8	s^{-1}
$A_{3,2}$	4.4×10^7	s^{-1}
$B_{3,2}$	1.09×10^{27}	$cm^3J^{-1}s^{-2}$
Q_3	19.9×10^{-10}	cm^3s^{-1}

where $h\nu$ is the fluorescence photon energy, I_{se_f} and I_{se_b} are the TALISE intensities in the forward and backward directions, respectively, and $\Delta\Omega$ and $\Delta\Omega'$ are the fractions of the isotropic spontaneous emission that contribute to the TALISE in both directions, given by (Amorim et al. 2000):

$$\Delta\Omega = 2\pi\left(1 - \frac{(L-x)}{\left[(L-x)^2 + d^2\right]^2}\right) \tag{10.23}$$

where L is the length of the interaction region, x is a given position region, and d is the diameter. Considering only in x coordinate, the above system of equations was solved in one dimensional space. When the number of cells increases, rapid convergence is obtained.

In certain conditions the depopulation of the level 3 by TALISE is negligible when compared with other loss channels, e.g. spontaneous emission and quenching. As an example we may cite the low-pressure discharges when the concentration of atoms is generally weak and/or the laser power density is not high. The repopulation of the level 3 due to reabsorption of radiation is also negligible.

The solution of the rate equations coupled to the equations for photons transportation is possible if the terms that intervene in the equations, such as Einstein's coefficients, cross sections, quenching coefficient and the experimental parameters, e.g. $\phi_l(x, r, t)$, are known.

The Fig. 10.8 shows the evolution of the populations n_1, n_2, n_3 and n_i obtained by solving the differential equations taking into account the TALISE and the data shown in Table 10.3. In this condition if the laser energy is constant and the discharge pressure does not change, i.e. the density of ground-state atoms is constant. The fluorescence signal is proportional to the concentration n_1 of the ground state n_1. It can be observed a depopulation of the ground state level n_1 after the laser pulse and the creation of ions with density n_i.

The TALIF and TALISE, integrated over the laser pulse duration, are shown in Fig. 10.8. It can be noticed that the TALIF and TALISE intensities are calculated in this simulation as functions of the laser energy and compared with experimental results for a ground-state hydrogen atom density of $10^{14}cm^{-3}$.

In Fig. 10.9 the TALIF and TALISE intensities, theoretical calculations, and experimental points, are plotted as a function of the laser pulse energy. It can

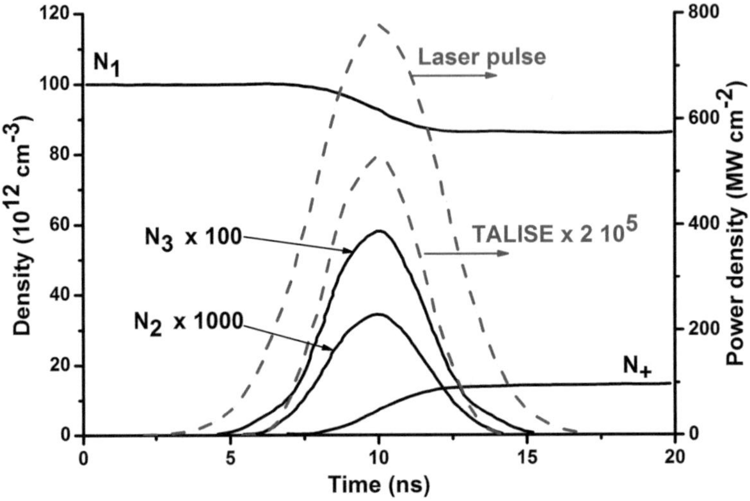

Fig. 10.8 Temporal evolution of the population n_1, n_2, n_3 and n_i obtained through the resolution of differential equations for the particular case of hydrogen atoms. The numerical resolution was realized for the following conditions: H ground-state atom density of 10^{14} cm^{-3}; laser energy (at $\lambda = 205$ nm) 1.0 mJ with a laser pulse of 5 ns duration. The coefficients used are given in Table 10.3

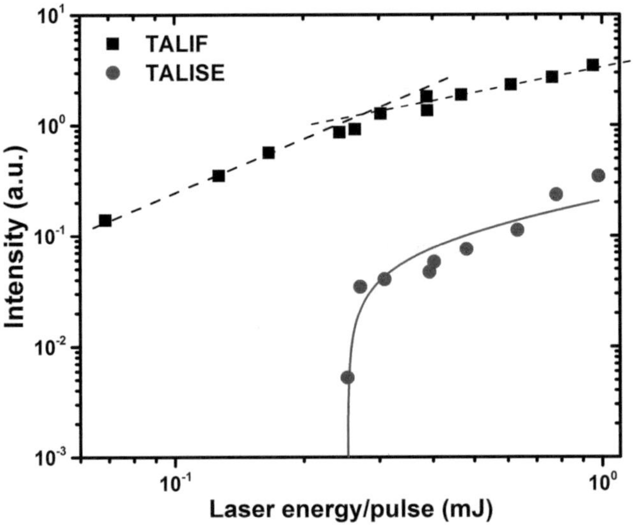

Fig. 10.9 TALIF and TALISE intensities as a function of the laser pulse energy. Calculated (*lines*) and measured TALIF (*filled square*) and the TALISE (*filled diamond*). The numerical results and measured TALISE intensity are normalized at the laser energy of 0.8 mJ

Table 10.4 Examples of atoms detected by TALIF in the ground state (Amorim et al. 2000)

Atom	Transition	λ_{laser} (nm)	$\lambda_{fluorescence}$ (nm)
C	$2p^2\ {}^3P - 3p\ {}^3D$	280	910
Cl	$3p^5\ {}^2P^o - 4p\ {}^4S^o$	233	725, 775
F	$2p^5\ {}^2P^o - 3p\ {}^2D^o$	170	776
H	$1s\ {}^2S - 3d\ {}^2D$	205	656
N	$2p^3\ {}^4S^o - 3p\ {}^4D^o$	211	869
O	$2p^4\ {}^3P - 3p\ {}^3P$	226	845
S	$3p^4\ {}^1D - 4p\ {}^1F$	288	166.7
I	$5p^5\ {}^2P^o - 2p\ D^o$	304.7	178.3
Xe	$5p^6\ {}^1S - 4p\ {}^4S^o$	250	828

be noticed a good agreement between theory and experiment. The intensity of TALIF is calculated over the whole solid angle while for TALISE the intensity is integrated along the laser emission direction. Experimental and calculated curves are normalized for a laser pulse energy of 0.8 mJ. In this figure, the TALISE signal is represented below the TALIF to give an insight in the threshold for TALISE with respect to the TALIF behavior. It can be observed that TALISE threshold is at 0.23 mJ which corresponds to the region where a TALIF begins to saturate. In this case TALISE exponential region is reached just after the threshold followed by a saturation region.

In Table 10.4 a list with some species detected by TALIF is presented. Two-photon absorption LIF is the diagnostic technique currently used for the detection of atoms in which their first excited levels have energies ≥ 6.5 eV.

Two laser photons, with wavelengths between 200 and 305 nm, are simultaneously absorbed to induce a resonant transition between the fundamental and the first excited electronic level allowed by the two-photon transition selection rules. The fluorescence signal is observed at the wavelength corresponding to the radiative de-excitation of this level towards, generally, the first excited level of the atom. In this scheme, where two-photons are absorbed, the probability for excitation is much weaker than that for the one-photon LIF. This necessitates the use of high laser fluxes ($I_{laser} \geq 10^8$ Wcm^{-2}), which are obtained by focusing the laser beam in the region probed. The experimental arrangement is typically the same as that used in one-photon LIF except the focusing of the laser beam. The detection threshold for TALIF is in the range 10^{11}–10^{12}cm^{-3}. When varying the laser energy, the TALIF intensity increases as the square of the laser energy and then saturates at larger energy. Additional depopulation pathways of level 3, see Fig. 10.6, such as the ionization by absorption of a third laser photon giving REMPI (Resonance Enhanced Multiphoton Ionization) and the TALISE, are presented. These techniques can be used as alternative diagnostics to TALIF.

In this chapter it was presented a brief overview of the main laser-induced fluorescence diagnostics used to probe atoms, molecules and radicals in plasmas. In the next chapter applications of low-temperature plasmas will be presented.

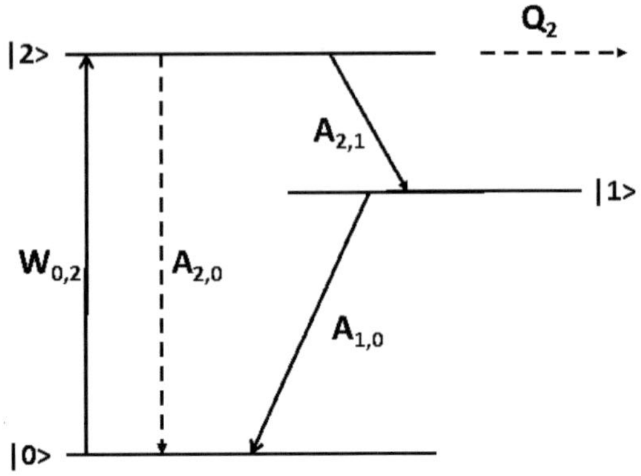

Fig. 10.10 Scheme of a three-level atom to illustrate the one-photon induced fluorescence

Exercises

To detect atoms dillluted in a molecular gas with density [M], the laser induced fluorescence may be employed. These atoms have three levels called 0,1 and 2 (Fig. 10.10). The level 0 is the fundamental state (energy $E_0 = 0$) with initial concentration n_0. The levels 1 and 2 are excited electronic levels, optically coupled between them and with the fundamental state. They have energies E_1 and E_2 with $E_1 < E_2$.

Exercise 10.1. Describe the experience that allows to detect these atoms by one photon laser induced fluorescence wavelength. Obtain the wavelength of the laser and fluorescence.

Resolution: The scheme of the energy levels are:

The laser excitation frequency may be obtained by (Fig. 10.10):

$E = h\nu$; thus $\nu_l = \frac{E_2 - E_0}{h}$. As $c = \lambda.\nu \rightarrow \lambda_l = \frac{ch}{E_2}$, considering that $E_0 = 0$.

The wavelength of the fluorescence is $\lambda_{fluo} = \frac{ch}{E_2 - E_1}$.

Exercise 10.2. Due to the laser interaction with the atomic system, the levels 1 and 2 will attain densities n_1 and n_2. Write the expression for the fluorescence emitted intensity by the excited atoms.

Resolution: The expression for the fluorescence emitted by the excited atoms is:

$$I_{fluo} = K\, n_2 A_{2,1} h\nu$$

where K is a constant related to the optical and response of the collection apparatus system.

Exercise 10.3. Write the equations for the evolution of the concentration of different energy levels in the approximation of low-power laser. Show and express the different of gain and loss terms.

Resolution: We are interested to describe the time evolution of the densities of levels involved when the laser power is low.

$$\frac{dn_0}{dt} = -n_0 W_{0,2} + n_2 A_{2,0} + n_1 A_{1,0}$$

$$\frac{dn_1}{dt} = n_2 A_{2,1} - n_1 A_{1,0}$$

$$\frac{dn_2}{dt} = n_0 W_{0,2} - n_2 Q_2 - n_2 A_{2,0} - n_2 A_{2,1}$$

For each level the gain terms are positive in the expression and the negative ones are the loss terms.

Exercise 10.4. If a continuum laser is used as probe, calculate the laser induced fluorescence intensity as a function of atoms concentration.

Resolution: If a continuum laser is employed the fluorescence intensity as a function of atoms concentration is given by:

$$I_{fluo} = K\, n_2 A_{2,1} h\nu$$

The time dependence of population of the level $|2\rangle$ is given by:

$$\frac{dn_2}{dt} = n_0 W_{0,2} - n_2 Q_2 - n_2 A_{2,0} - n_2 A_{2,1}$$

In steady state: $\frac{dn_2}{dt} = 0$, so.

$$n_0 W_{0,2} = n_2 Q_2 + n_2 A_{2,0} + n_2 A_{2,1}$$

$$n_2 = \frac{n_0 W_{0,2}}{Q_2 + A_{2,0} + A_{2,1}}$$

Considering that $W_{0,2} = B_{0,2} I_l$, the intensity of fluorescence may be written as:

$$I_{fluo} = K\, n_0\, B_{0,2} I_l \frac{A_{2,1}}{Q_2 + A_{2,0} + A_{2,1}} h\nu$$

Exercise 10.5. What happens if the laser probe becomes intense? What are the consequences for the populations of different energy levels? Consider g_0, g_1 and g_2 as the statistical weights of the levels.

Resolution: When a very intense beam is used to probe the atoms, the transitions become saturated. In this case, we have:

$$\frac{dn_2}{dt} = n_0 W_{0,2} - n_2 W_{2,0} - n_2 \left(Q_2 + A_{2,0} + A_{2,1}\right)$$

The saturation has the tendency to equilibrate the populations of levels 0 and 2. The expression for saturation is

$$n_0\, W_{0,2} = n_2 W_{2,0}$$

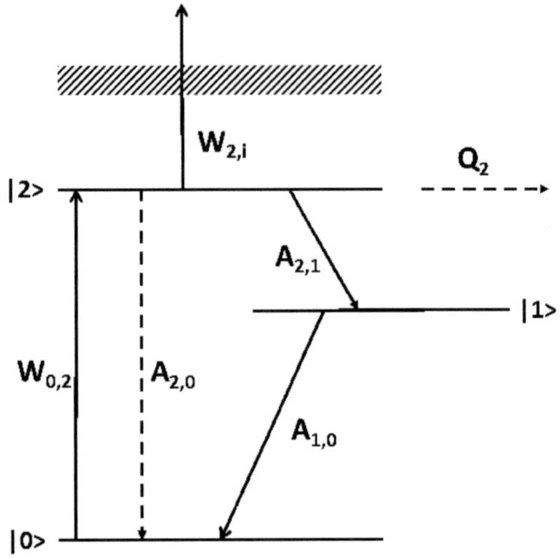

Fig. 10.11 Scheme of a three-level atom to illustrate the one-photon induced fluorescence, including ionization

$$n_0\, B_{0,2}I_l = n_2 B_{2,0} I_l$$

as $g_0 B_{0,2} = g_2 B_{2,0}$;

$$n_{2\,sat} = n_0 \frac{g_2}{g_0}$$

Exercise 10.6. If the ionization energy of the atoms is such that, $h\nu < E_i < 2h\nu$, where ν is the laser frequency employed in the induced fluorescence experiment, how can be modified the rate equations, obtained in question 3, to take into account this effect?

Resolution: As the ionization energy is $h\nu < E_i < 2h\nu$, where ν is the laser frequency employed in the induced fluorescence experiment the equations given in question 3 should take into account the absorption of one photon that ionizes the atom (Fig. 10.11):

and the equations become:

$$\frac{dn_0}{dt} = -n_0 W_{0,2} + n_2 A_{2,0} + n_1 A_{1,0}$$

$$\frac{dn_1}{dt} = n_2 A_{2,1} - n_1 A_{1,0}$$

$$\frac{dn_2}{dt} = n_0 W_{0,2} - \boldsymbol{n_2 W_{2,i}} - n_2 Q_2 - n_2 A_{2,0} - n_2 A_{2,1}$$

Exercise 10.7. Consider the fundamental energy state of density $n_0 = 10^{15}\text{cm}^{-3}$ and energies of the first and second excited levels equal to: $E_1 = 16,667\ \text{cm}^{-1}$, $E_2 = 28,571\ \text{cm}^{-1}$. The Einstein emission coefficients of the upper and intermediary levels are $5 \times 10^7\text{s}^{-1}$ and $2.5 \times 10^8\text{s}^{-1}$. Calculate the absorption coefficient for

the transition from the ground state to the upper level. Considering $g_0 = 1, g_1 = 3$ and $g_2 = 5$.
Resolution: Given:

$$A_{2,0} = 2.5 \times 10^8 \mathrm{s}^{-1}$$

$$A_{2,1} = 5 \times 10^7 \mathrm{s}^{-1}$$

$$\lambda = \frac{1}{E_2 - E_1} = \frac{1}{(28,571 - 16,667)\,\mathrm{cm}^{-1}} = 8.4 \times 10^{-5}\mathrm{cm} = 840\ \mathrm{nm}$$

$$\frac{A_{2,0}}{B_{2,0}} = \frac{8\pi h\nu^3}{c^3};\ \frac{A_{2,0}}{\frac{g_0}{g_2} B_{0,2}} = \frac{8\pi h\nu^3}{c^3}$$

$$\frac{A_{2,0}}{B_{0,2}} = \frac{g_0}{g_2}\,\frac{8\pi h\nu^3}{c^3} = \frac{g_0}{g_2}\,\frac{8\pi h}{\lambda^3}$$

$$B_{0,2} = \frac{g_2}{g_0}\,\frac{A_{2,0}\lambda^3}{8\pi hc^3} = \frac{5}{1}\,\frac{2.5 \times 10^8 \mathrm{s}^{-1} \times (840\,\mathrm{nm})^3}{8\pi \times 6.62 \times 10^{-34}\mathrm{Js}} = 4.45 \times 10^{22}\mathrm{m}^3\mathrm{J}^{-1}\mathrm{s}^{-2}$$

Exercise 10.8. In the approximation of low-power probe laser and considering negligible quenching and ionization, what is the laser energy in a spectral interval unit needed to obtain the concentration $n_2 = 10^{-6} n_0$? Calculate the photon density that corresponds to a laser spectral interval of $\Delta\nu = 2.5$ GHz.
Resolution: Considering $Q_2 = 0$ and $W_{2,i} = 0$, the laser energy density in a spectral interval to obtain the concentration $n_2 = 10^{-6} n_0$ is:

$$\frac{dn_2}{dt} = 0 = n_0 W_{0,2} - n_2 A_{2,0} - n_2 A_{2,1},$$

$$n_0 W_{0,2} = 10^{-6} n_0 \left(A_{2,0} + A_{2,1}\right),$$

$$W_{0,2} = B_{0,2} I_l g\left(\omega\right)$$

$$I_l g\left(\omega\right) = \frac{10^{-6}\left(A_{2,0}+A_{2,1}\right)}{B_{0,2}} = \frac{10^{-6}\left(2.5 \times 10^8 + 5\times 10^7\right)\mathrm{s}^{-1}}{4.45\times 10^{22}\mathrm{m}^3\mathrm{J}^{-1}\mathrm{s}^{-2}} = 6.74\times 10^{-21}\mathrm{Jm}^{-3}\mathrm{s}$$

The density of photons for the spectral interval $\Delta\nu = 2.5$ GHz is:

$$n\left(\nu\right) = \frac{I_l g\left(\omega\right)}{h\nu} = \frac{I_l g\left(\omega\right)\lambda}{hc} = \frac{6.74\times 10^{-21}\mathrm{Jm}^{-3}\mathrm{s} \times 2.5\times 10^9 \mathrm{s}^{-1} \times 840\times 10^{-9}\mathrm{m}}{6.62\times 10^{-34}\ Js \times 310^8 \mathrm{ms}^{-1}}$$
$$= 7.13 \times 10^7\ \mathrm{photons/m}^3$$

Exercise 10.9. The photo-ionization cross-section of the upper state is $\sigma = 2 \times 10^{19}\text{cm}^2$, the quenching coefficient is $k_{qM} = 2 \times 10^{-10}\text{cm}^3\text{s}^{-1}$ and $[M] = 2 \times 10^{16}\text{m}^{-3}$. Calculate the different terms of gain and loss terms and discuss the approximations done in exercise 10.8.

Resolution: Let us evaluate the approximations committed in exercise 10.8

We have: $\sigma_i = 2 \times 10^{-19}\text{cm}^2$, $k_{qM} = 2 \times 10^{-10}\text{cm}^3\text{s}^{-1}$ and $[M] = 2 \times 10^{16}\text{m}^{-3}$.

In the equation for the temporal evolution of the state with n_2 density:

$\frac{dn_2}{dt} = n_0 W_{0,2} - n_2 W_{2,i} - n_2 Q_2 - n_2 A_{2,0} - n_2 A_{2,1}$ the gain and loss terms are:

Gain: $n_0 W_{0,2}$

Loss: $n_2 W_{2,i}$, $n_2 Q_2$, $n_2 A_{2,0}$, $n_2 A_{2,1}$

$$W_{0,2} = B_{0,2} I_l g(\omega) = 4.45 \times 10^{22}\ \text{m}^3\text{J}^{-1}\text{s}^{-2} \times 6.74 \times 10^{-21}\text{Jm}^{-3}\text{s} = 300\ \text{s}^{-1}$$

$$I_l = 6.74 \times 10^{-21}\text{Jm}^{-3}s \times 2.5 \times 10^9\text{s}^{-1} = 1.69 \times 10^{-11}\text{Jm}^{-3} = 1.69 \times 10^{-17}\text{Jcm}^{-3}$$

$$W_{2,i} = \sigma_i c \phi_l = \sigma_i c \frac{I_l}{h\nu} = \sigma_i c \frac{I_l \lambda}{hc} = 2 \times 10^{-19}\text{cm}^2 \times \frac{1.69 \times 10^{-17}\text{Jcm}^{-3} \times 840 \times 10^{-7}\text{cm}}{6.62 \times 10^{-34}\ \text{Js}}$$

$$W_{2,i} = 4.29 \times 10^{-7}\text{s}^{-1}$$

$$Q_2 = k_{qM}[M] = 2 \times 10^{-10}\text{cm}^3\text{s}^{-1} \times 2 \times 10^{10}\text{cm}^{-3} = 4\ \text{s}^{-1}$$

$$A_{2,0} = 2.5 \times 10^8\text{s}^{-1}$$

$$A_{2,1} = 5 \times 10^7\text{s}^{-1}$$

Calculating the gain and loss terms:

$$n_0 W_{0,2} = 2 \times 10^{16}\text{cm}^{-3} \times 300\ \text{s}^{-1} = 6\ 10^{18}\text{cm}^{-3}\text{s}^{-1}$$

$$n_2 W_{2,i} = 2 \times 10^{10}\text{cm}^{-3} \times 4.29 \times 10^{-7}\text{s}^{-1} = 8.58 \times 10^3\text{cm}^{-3}\text{s}^{-1}$$

$$n_2 Q_2 = 2 \times 10^{10}\text{cm}^{-3} \times 4\ \text{s}^{-1} = 8 \times 10^{10}\text{cm}^{-3}\text{s}^{-1}$$

$$n_2 A_{2,0} = 2 \times 10^{10}\text{cm}^{-3} \times 2.5 \times 10^8 s^{-1} = 5 \times 10^{18}\text{cm}^{-3}\text{s}^{-1}$$

$$n_2 A_{2,1} = 2 \times 10^{10}\text{cm}^{-3} \times 5 \times 10^7\text{s}^{-1} = 10^{18}\text{cm}^{-3}\text{s}^{-1}$$

By the results, it can be concluded that: $n_0 W_{0,2}$, $n_2 A_{2,0}$, $n_2 A_{2,1} \gg n_2 W_{2,i}$, $n_2 Q_2$ justifying the approximations assumed in exercise 10.8.

References

N.G. Adams, C.R. Herd, D. Smith, Development of the flowing afterglow Langmuir probe technique for studying the neutral products of dissociative recombination using spectroscopic techniques - OH production in the HCO-2(+)+E reaction. J. Chem. Phys. **91**, 963–973 (1989)

M. Aldén, H. Edner, P. Grafström, S. Svanberg, Two-photon excitation of atomic oxygen in a flame. Opt. Commun. **42**, 244–246 (1982)

J. Amorim, G. Baravian, A. Ricard, Production of N, H and NH active species in N_2–H_2 dc flowing discharges. Plasma Chem. Plasma Process. **15**, 721–731 (1995)

J. Amorim, G. Baravian, J. Jolly, Laser-induced resonance fluorescence as a diagnostic technique in non-thermal equilibrium plasmas. J. Phys. D Appl. Phys. **33**, R51–R65 (2000)

A. Arnold, H. Becker, R. Suntz, P. Monkhouse, J. Wolfrum, R. Maly, W. Pfister, Flame front imaging in an internal-combustion engine simulator by laser-induced fluorescence of acetaldehyde. Opt. Lett. **15**, 831–833 (1990)

N.G. Basov, V.A. Danilychev, Y.M. Popov, Stimulated emission in the vacuum ultraviolet region. Sov. J. Quantum Electron. **1**, 18–22 (1971)

G.P. Davis, R.A. Gottscho, Measurement of spatially resolved gas-phase plasma temperatures by optical-emission and laser-induced fluorescence spectroscopy. J. Appl. Phys. **54**, 3080–3086 (1983)

V.M. Donnelly, D.L. Flamm, G. Collins, Laser diagnostics of plasma-etching – measurement of $Cl_2{}^+$ in a chlorine discharge. J. Vac. Sci. Technol. **21**, 817–823 (1982)

V.M. Donnelly, *Plasma–Surface Interactions and Processing of Materials* (NATO ASI Series E, vol. 176) (Kluwer, Dordrecht, 1990)

F. Fresnet, G. Baravian, S. Pasquiers, C. Postel, V. Puech, A. Rousseau, M. Rozoy, *Proc. 14th Int. Symp. Plasma Chemistry* (Prague, Czech Republic, 1999), p. 2661

R.A. Gottscho, R.H. Burton, G.P. Davis, Radiative lifetime and collisional quenching of carbon monochloride (A 2-Delta) in an alternating-current glow-discharge. J. Chem. Phys. **77**, 5298–5301 (1982)

R.A. Gottscho, M.L. Mandich, Time-resolved optical diagnostics of radio-frequency plasmas. J. Vac. Sci. Technol. A **3**, 617–624 (1985)

M. Hamamoto, T. Ohgo, K. Kondo, T. Oda, A. Iiyoshi, K. Uo, Coaxial laser-induced fluorescense spectroscopic system for impurity diagnostics in plasmas. Jap. J. Appl. Phys. **25**, 99–102 (1986)

S.G. Hansen, G. Luckman, S.D. Colson, Measurements of F-star, CF and CF_2 formation and decay in pulsed fluorocarbon discharges. Appl. Phys. Lett. **53**, 1588–1590 (1988)

N. Hata, A. Matsuda, K. Tanaka, Spectroscopic diagnostics of plasma-chemical vapor deposition from silane and germane. J. Appl. Phys. **61**, 3055–3060 (1987)

C. Hayaud, PhD Thesis, Université Paris-Sud, Orsay, France, 1997 (in French)

F. Hummernbrum, H. Kempkens, A. Ruzicka, H.D. Sauren, C. Schiffer, J. Uhlenbusch, J. Winter, Laser-induced fluorescence measurements on the C(2) Sigma(+) -X(2)Pi(r) transition of the CH radical produced by a microwave excited process plasma. Plasma Sources Sci. Technol. **1**, 221–231 (1992)

W. Jacob, M. Engelhard, W. Moller, A. Koch, Absolute density determination of CH radicals in a methane plasma. Appl. Phys. Lett. **64**, 971–973 (1994)

J. Jolly, Diagnostics laser dans les plasmas froids. J. Phys. III **5**, 1089–1113 (1995)

A. Kono, N. Koike, K. Okuda, T. Goto, Laser-induced fluorescence detection of SiH_2 radicals in a radiofrequency silane plasma. Jap. J. Appl. Phys. **32**, L543–L546 (1993)

M.B. Leong, A.P. D'Silva, V.A. Fassel, Evaluation of stimulated raman-scattering of tunable dye-laser radiation as a primary excitation source for exciting atomic fluorescence in an inductively coupled plasma. Anal. Chem. **58**, 2594–2598 (1986)

W. Lochte-Holtgreven, *Plasma Diagnostics* (North-Holland, Amsterdam, 1968)

K. Niemi, V. Schulz-von der Gathen, H.F. Doebele, Absolute calibration of atomic density measurements by laser-induced fluorescence spectroscopy with two-photon excitation. J. Phys. D Appl. Phys. **34**, 2330–2335 (2001)

N. Omenetto, H.G.C. Human, P. Cavalli, G. Rossi, Laser excited atomic and ionic non-resonance fluorescence detection limits for several elements in an argon inductively coupled plasma. Spectrochim. Acta Part B At. Spectrosc. **39**, 115–117 (1984)

R. Paschotta, *Article on 'Argon Ion Lasers' in the Encyclopedia of Laser Physics and Technology*, 1st edn. (Wiley-VCH, 2008) ISBN 978-3-527-40828-3

B.L. Preppernau, T.A. Miller, *Glow Discharge Spectroscopies* ed. by R. Kenneth Marcus (Plenum, New York, 1993)

R.M. Roth, K.G. Spears, G. Wong, Spatial concentrations of silicon atoms by laser-induced fluorescence in a silane glow-discharge. Appl. Phys. Lett. **45**, 28–30 (1984)

G.S. Selwyn, Atomic arsenic detection by ArF laser-induced fluorescence. Appl. Phys. Lett. **51**, 167–168 (1987)

S. Singleton, K.G. McKendrick, R.A. Copeland, J.B. Jeffries, Vibrational transition-probabilities in the B-X and B'-X systems of the SiCl radical. J. Phys. Chem. **96**, 9703–9709 (1992)

O. Svelto, *Principles of Lasers*, 4th edn. (Plenum Press, New York, 1998)

K. Tachibana, T. Mukai, H. Harima, Measurement of absolute densities and spatial distributions of Si and SiH in an rf-discharge silane plasma for the chemical vapor- deposition of A-Si-H films. Jap. J. Appl. Phys. **30**, L1208–L1211 (1991)

P. Van der Weijer, B.H. Zwerver, Laser-induced fluorescence of OH and SiO molecules during thermal chemical vapor-deposition of SiO_2 from silane oxygen mixtures. Chem. Phys. Lett. **163**, 48–54 (1989)

R. Walkup, P. Avouris, R.W. Dreyfus, J.M. Jasinski, G.S. Selwyn, Laser detection of diatomic products of plasma sputtering and etching. Appl. Phys. Lett. **45**, 372–374 (1984)

Chapter 11
Applications of Low-Temperature Plasmas

Low-temperature plasmas are present in a broad field of technological applications and experimented an enormous growth after the end of the twentieth century. The most important examples of applications are lamps, pretreatment of polymer materials, packaging materials, treatment of surfaces, waste, air pollution mitigation, microelectronics, and flat panels display. Recently new applications in medicine, pharmacy, foods, biology, biomass and biofuel processing, attracted the interest of the industry and the scientific community.

The purpose of this chapter is not to be exhaustive but illustrates some important industrial and technological applications. The fields where low-temperature plasmas are being employed today is vast, rapidly growing, and cannot be described in a single book chapter, so clearly well beyond this textbook. Of course, much will be left out, however a special emphasize will be stressed here in these new breakthrough applications of plasmas in health science, production of biofuels and agriculture.

11.1 Plasmas for Materials Processing

Plasma technologies are being used in industry due to their ability to offer a wide spectrum of possible treatments of materials. These plasmas are in fact source of heat and reactive species that have unique physical and chemical properties induced by charged particles. The main applications of these discharges are (Bonizzoni and Vassallo 2002): destruction of toxic/harmful materials, superficial modification of materials and creation of new materials.

The original version of this chapter was revised. An erratum to this chapter can be found at DOI 10.1007/ 978-3-319-09253-9_12

J.M.A.H. Loureiro, J. de Amorim Filho, *Kinetics and Spectroscopy of Low Temperature Plasmas*, Graduate Texts in Physics,
DOI 10.1007/978-3-319-09253-9_11

In industry basically two kind of plasmas are employed; the plasma torches which are thermal plasmas produced at high pressure by direct current, alternating electric fields like RF or microwaves. These plasmas have electron temperatures around 1 eV and low-ionization degrees. One important application of these plasmas is the mitigation of toxic-harmful substances or, as in the case of the plasma spray, to produce coatings of thick films.

Low-temperature discharges a kind of plasma characterized by the electron temperature higher than the ion temperature. It may be excited in low-pressure, with low power with help of DC, RF or microwave sources. Nowadays cold atmospheric plasmas are another option to create active species to treat materials. The interactions of particles with the materials produce the modification of the surfaces adding different functional properties with respect to the bulk of the material.

One example is the change of functional characteristics of any organic materials, since the plasma gas is at room temperature, and the organic samples cannot withstand high-temperatures. Some properties of the surfaces can be modified with help of electrical discharges. The plasma treatment processes produce low environmental impact, competitive costs and particularly the possibility to modify the surface properties of any materials. The reactive particles of the plasma, through collisions with the material surface, break the chemical bonds producing free radicals on the surface. These are subjected to additional reactions that depend on the type of gas plasma used. This results in generation of layers on the surfaces that have very different properties with respect to the bulk. The type of modification depends on the pretreatment and composition of the substrate, on the type and the quantity of reactive gas, the total reactor pressure, on the type of excitation power source and the process time. Now we will present some plasma applications in fields like modification of polymers, aeronautical, chemical, biomedical, automotive, microelectronic and textile industries.

The polymers are a class of materials with excellent optical, structural and chemical properties, however, they are characterized by low surface energy. This property directly reflects the ability of adhesion of other materials on its surface, for example, glues, inks, primers, adhesives, printing, among others (Vurzel and Polak 1970). Plasmas may be employed to functionalize the surface creating functional groups increasing the surface energy. Many industries make extensive use of polymers in packing and health care for example.

In a recent article Fuerst et al. (2015) deposited coatings by plasma-enhanced chemical vapor deposition at low temperature onto a variety of substrates including flexible polyethylene terephthalate (PET) and CdTe solar cells. The authors did a design, fabrication, and evaluation of flexible, multilayer optical films using hybrid nano-laminates consisting of TiO_2 and silicone ($SiO_xC_yH_z$) as high and low refractive index materials. They obtained an increase in the absolute transmission of about 3 % over 410–850 nm wavelengths. An infrared reflector was designed and applied to PET which was found to provide 70 % reflectance in the near-IR while maintaining >80 % transmittance for visible light. The optical performance of these flexible coatings on PET remained unchanged after automated bend testing, and were shown to be robust with respect to humidity and thermal shock tests.

Another important application is the production of polycrystalline diamond coatings by microwave, RF or DC plasmas. Diamond films have many potential applications as: wear-protection, coatings on tools, low-friction coatings, optical components coatings, high-temperature electronics and heat sinks. Diamond has the highest thermal conductivity and excellent electrical insulation. The dissipation of heat with elements coated with diamond film reduces the temperature in transistors by more than 100 °C (Suchentrunk et al. 1997). Diamond-like carbon (DLC) films present an extremely low-coefficient of friction which is useful for lubricant-free bearings.

Modern aircraft industry uses various materials in pursuit of high performances. Among them, aluminum, carbon fibers, glass fibers, resins, polymers, steel, titanium, and composites. Neverthless, there is an intrinsic difficulty in joining these materials. As most often the glue is made with water-based materials, plasma technology significantly increases the power of accession allowing the use of materials that were previously not used due to superficial difficulties.

For maintenance reasons aircrafts have to be stripped from time to time. Today the removal of thick paint films is performed worldwide by using toxic chemicals. Alternative processes using plasma stripping is especially desired for small and medium-sized aircraft components (Suchentrunk et al. 1997).

In the industry, various chemical processes using powders with lower solubility are employed. With the aid of plasma technology one can let them water soluble, for example, reducing production costs and time needed for the preparation of these powders.

Many processes in non-equilibrium low-temperature plasmas which are of commercial interest are (Vurzel and Polak 1970): oxidation of nitrogen contained in air in microwave plasma, production of C_2F_4 in glow discharge, decomposition of nitrous oxide in silent discharge, synthesis of hydrocyanic acid in a nitrogen-methane mixture, reduction of tetrachloride titanium by means of hydrogen, reduction of zirconium halides, production of boranes (B_2H_6, $B_{10}H_{16}$, etc.), and hydrazine.

In the biomedical industry, plasmas may be employed for various applications whose results provide very high performance products like:

- Bonding gaskets in gastric tubes;
- Bonding of plastic parts and injected joints;
- Increased adhesion between metal-adhesive-polymer parts

Low-temperature non-equilibrium plasmas have been extensively used in the modification of surfaces in order to manage the interactions between materials and biological systems, rendering the treated surfaces biocompatible. Plasma processing in the health-care by transformation of biomaterials experimented spectacular growth in the last years. New materials and modified surfaces have different functionalities, for example, to enhance the adhesion of living cells; non-fouling coatings tailored to inhibit completely the adhesion of biomolecules, cells and bacteria "*in vivo*". These plasmas are also employed to grow primer layers for the immobilization of peptides, enzymes, antibodies, and other types of biomolecules. Another interesting application is in the design of "smart" drug-release systems.

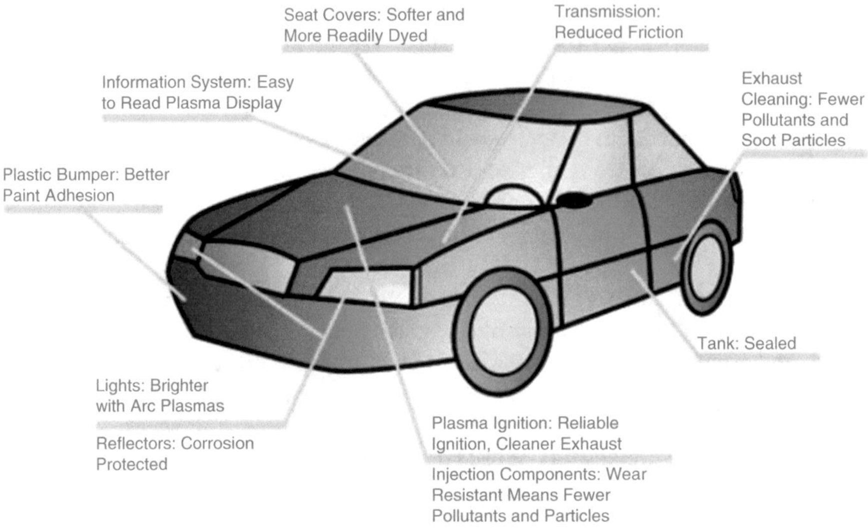

Fig. 11.1 Plasmas in automotive industry (Suchentrunk et al. 1997)

The automotive industry requires high performance materials (d'Agostino et al. 2005), mechanical properties with superior feedback and efficient solutions and low-cost. Poor compliance or inadequate wettability is one of the problems commonly encountered when using materials such as glass, polymers in general, composites and metals. Operations requiring adhesion and bonding with advanced adhesives, paint, varnish and coatings can be solved with the application of plasmas (Fig. 11.1).

The microelectronic industry is no doubt the more successful application of plasmas in industry. The 1970s have seen a rapidly growing of plasma technologies in microelectronics, like in dry etching processes for large scale integrated (LSI) circuits, and plasma diagnostic methods to monitor the processes of etching. Emission spectroscopy started to be applied extensively in attempts to understand etching, deposition mechanisms and kinetics, coupled with the surface analysis techniques to characterize the etched silicon and correlate with the radicals and ions produced in gas phase. The current dimension, down to 90 nm, has only become possible through massive research and investment in plasma like etching, PECVD, and surface modification or cleaning. Today, two-thirds of all process steps in the fabrication of semiconductor devices involve plasmas, and this trend is steadily increasing. The annual growth rate in last years in this sector has been roughly 20 % (Suchentrunk et al. 1997). Plasmas processing is also playing an ever-increasing role in the production and processing of solar cells, and in fabricating amorphous or microcrystalline thin layers of silicon by PECVD.

Plasma technology can be used in pre-treatment of surfaces to increase adhesion of solder, printing and marking, silicon wafer bonding, and removal of surface metal oxides, for example. Of course, newly manufactured fabrics have low interaction with pigments. The fabric treatment implies a hydrophilizing of the surface. With this treatment, the absorption of the pigment is made much more efficiently, *i.e.* with the same amount of dye, pigment can be a larger amount of tissue.

11.2 Biomass Processing and Biofuels Production

The development of methodologies to satisfy needs of the present life without compromising those of future generations is an urgent issue for the modern society affecting the developed and developing countries in order to promote advances in new energy sources, materials, food and the design of new cities.

In the context of energy, despite of significant growth in proven and predicted fossil fuel reserves over the next two decades, notably heavy crude oil, deep water wells, and gas, present great uncertainties in the economics of their exploitation *via* current extraction methodologies. More crucial is that an increasing proportion of such carbon resources cannot be burned without breaching the United Nations Framework Convention on Climate Change (UNFCC) targets for a 2 °C increase in mean global temperature relative to the preindustrial level. There is clearly a rising energy demands, predicted to climb 50 % globally by 2040 and the requirement to mitigate current CO_2 emissions and hence climate change.

Similar considerations apply to ensuring a continued supply of organic materials for applications including polymers, plastics, pharmaceuticals, optoelectronics and pesticides, which underpin modern society, and for which significant future growth is anticipated. The quest for sustainable resources to meet the demands of a rapidly rising world population represents one of this century's grand challenges. Plasma technologies show significant potential to address the aforementioned challenges.

The processing of biological materials using plasmas is quite wide and it seems that few possibilities have been explored to date. The polymerization of the surface of biological materials has been reported frequently (Songa et al. 2013). In reference (Maksimov and Nikiforov 2007), the authors report a study based on simulations to test the feasibility of using plasma in gas or liquid phase to bleaching and/or delignification of cotton or cellulose. In these applications, there appear to be a wide range of chemical possibilities to manipulate biological materials using plasma for promoting the conversion of organic molecules or macromolecules of little value in high value-added substances.

The gasification of biomaterials for gas synthesis production is a technology already implemented, for example by Westinghouse Plasma Corporation in China for biomass gasification and Mepl, Pune, India, for hazardous waste. However, research in this field is still active, with a view to improving energy efficiency, the quality of the gas synthesis and the type of biomass used. One innovation is the possibility of generating gas synthesis in gaseous fuels, using plasmas in liquids (Zhang and Cha 2015). An interesting direction for research in this area is the development of compact burners with high energy density and more efficient, clean ways to generate gas synthesis for applications in urban transport, waste recycling, generators, etc.

Use of ozone-generated plasmas to delignification of biomass was tested in laboratory scale by various groups for different types of biomass as sugarcane bagasse (Souza-Corrêa et al. 2013a), wheat straw (Schultz-Jensen et al. 2011) and Japanese cedar (Miura et al. 2012). The technology is effective superior to conventional pre-treatment methods in terms of conversion efficiency of enzymatic hydrolysis, but more studies about the energy efficiency and economic point of view are needed. In this sense, the research to obtain more effective ozonizers could enable economically viable processes. Ozone interaction mechanism with the biomass is relatively well known (Souza-Corrêa et al. 2013b), but more research is needed to unravel the influence and the role of other radicals (e.g. singlet states of atomic and molecular oxygen, OH, H_2O_2) in biomass degradation processes. Employing mass spectrometry authors could monitor the neutral chemical species from sugar cane bagasse that could volatilize during the bagasse ozonation process. Lignin fragments and some radicals liberated by direct ozone reaction with the biomass structure were detected as can be seen in the figure below.

In Fig. 11.2 the yields of radicals as a function of time can be seen. The results, are expressed in yields of specific masses (relative contribution to the total signal) instead of counts per second obtained directly from the Molecular Beam Mass Spectrometry (MBMS) to show the relative importance of each species during the treatment. It should be mentioned that the species appearing in Fig. 11.2 are the most abundant. Others species with less count rates, *i.e.* faint yields, may be important due to its reactivity but it is not shown here.

From Fig. 11.2 it can be also noted that CO is the first radical formed when the treatment is launched. The production almost instantaneously is in good agreement with the reaction scheme proposed by Criegee (Souza-Corrêa et al. 2013a). It disappears after around 1 h and 15 min of treatment. The breaking of C=C bonds is the origin of high rate of CO production in the first instant of the treatment. This result clearly shows the accuracy of Criegee's model to explain the ozonation of lignin and corroborates the interpretation of oxidation of β-O-4 bonds. Important intermediary states such as O_2, CH_3OH, OH, and H_2O formed as a result of the cleavage of aromatic rings, which is a slow reaction, can be seen in the second phase of the treatment. Further experiments are needed to discriminate O_2 and CH_3OH since they have the same mass ($m/z = 32$). Moreover, by using the MBMS technique it is not possible to discriminate which amount of H_2O detected by the equipment was directly related to the Criegee's reactions and which came from the moisture in the bagasse samples. In this sense, other experiments should be carried out in order to clarify this aspect as well. In the third phase, HCOOH, CO_2 and H_3O radicals are formed. The formation of these radicals is due to oxidation of the cathecol, which is a secondary process and explains why these species appear lately in the treatment. These are important results to validate the reaction paths during the interaction of ozone with the biomass.

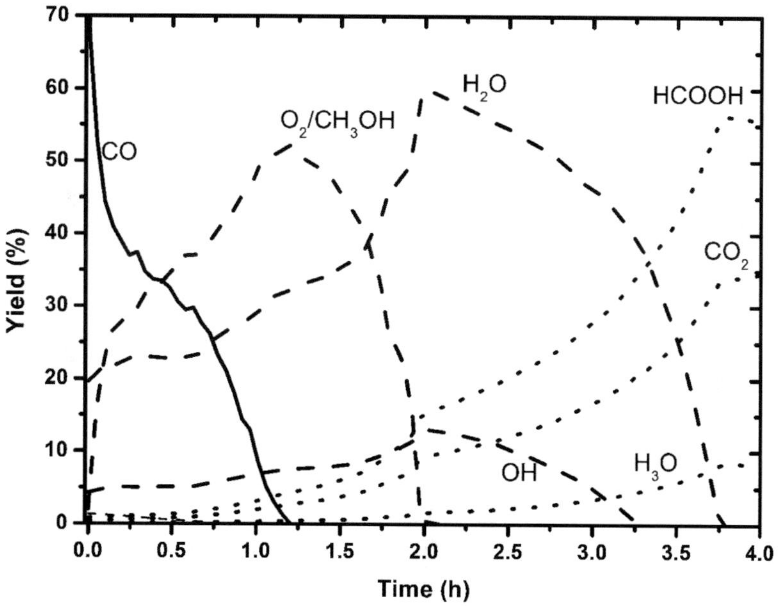

Fig. 11.2 Species yield (%) as a function of bagasse ozonation treatment time (moisture of 50 %) (Souza-Corrêa et al. 2013a)

Figure 11.3a–d illustrates SEM images of a bagasse surface before and after the ozonation. From Fig. 11.2a, it can be observed that raw (or in natura) bagasse parenchyma shows a bundle and homogeneous structure, while the bagasse parenchyma tissue treated for 4 h has started to disrupt its surface (see Fig. 11.3b). Moreover, in Fig. 11.3c, one can see the epidermis of the raw material, which is composed by stomata, which are responsible for controlling the moisture content in a plant. From Fig. 11.3c it is possible to observe some rough structures. After 4 h of ozonation, that rugosity was reduced in comparison with the raw bagasse, smoothing the epidermis surface (see Fig. 11.3b). All these results were an indication that the ozone has really attacked the bagasse surface promoting its oxidation.

In this study the ozone concentration was monitored during the ozonation by optical absorption spectroscopy. The optical results indicated that the ozone interaction with the bagasse material was better for bagasse particle sizes less than or equal to 0.5 mm. Both techniques have shown that the best condition for the ozone diffusion in the bagasse was at 50 % of its moisture content. Fourier transform infrared spectroscopy (FTIR) was employed to analyze the lignin bond disruptions and morphology changes of the bagasse surface that occurred due to the ozonolysis reactions as well. Appropriate chemical characterization of the lignin content in bagasse before and after its ozonation was also carried out and the reader interested can see in this paper for more details (Souza-Corrêa et al. 2013b, 2014).

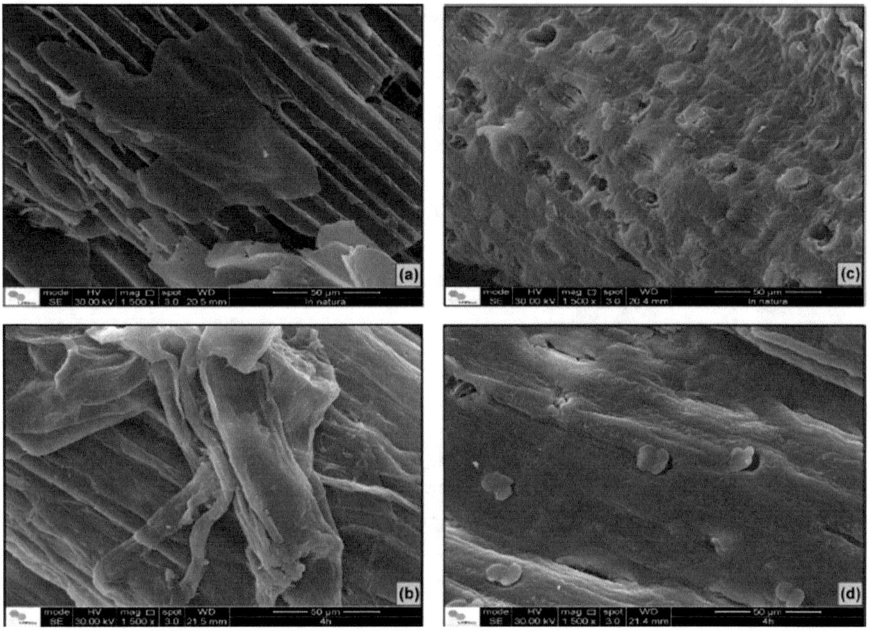

Fig. 11.3 SEM images for (**a**) raw bagasse parenchyma (1500× magnification); (**b**) bagasse parenchyma treated by ozone for 4 h (1500× magnification); (**c**) raw bagasse epidermis (1500× magnification); (**d**) bagasse epidermis treated by ozone for 4 h (1500× magnification) (Souza-Corrêa et al. 2013a)

Sugarcane bagasse samples pretreated with ozone in a downstream of an atmospheric-pressure O_2 DBD plasma showed delignification efficiency of approximately 80 % in 6 h of treatment. The efficiency of the ozonolysis process for different pretreatment conditions was evaluated by chemical composition analyses and enzymatic hydrolysis. The quantity of moisture in the bagasse had a large influence on delignification and saccharification of the bagasse, where 50 % moisture content was found to be best for delignification and 65 % of the cellulose was converted into glucose (Souza-Corrêa et al. 2013a).

In another experiment, bagasse moisture was fixed in 50 %. The delignification efficiency had small improvement as a function of particle size, varying from 75 up to 80 % when the particle size varied from 2.0 to 0.08 mm, respectively (Souza-Corrêa et al. 2014).

Another strategy is to treat the biomass in direct contact with the plasma, both in gas phase or in liquids. The cell wall is responsible for the architecture, defense and perception of plant environment. Furthermore, these walls are rich in sugars, such as glucose, essential for the production of ethanol, widely used biofuel today. The study of cell wall degradation, which is done by enzymes produced by the plant itself, is promising in the production area of cellulosic ethanol. Low-temperature plasmas produce reactive oxygen and nitrogen species, discussed in Sects. 11.3 and 11.4, that interact with constituents of the cell wall degrading by redox reactions and open access to the cellulose.

Amorim et al. (2013) employed an atmospheric-pressure argon RF microplasma jet to treat lignin. The treatment time was carried out from 30 min to 4 h resulting in a strong degradation of the lignin structure. The most important degradation was observed in functional groups having asymmetric in-phase ring stretching as well as in C=C and C=O stretching vibrations. Simple calculations done put in evidence that OH bond breakings by electron collisions inside the microplasma occured. In the afterglow, it was not observed any change on the lignin spectra.

Diffuse Reflectance Fourier Transform Infrared Spectrometry (DRIFT) was used to characterize the lignin powder pellets and the results are shown in Fig. 11.4. As can be seen many peaks changed as the treatment time increased. After 2 h of treatment, the bands located at 3800–3100 cm^{-1} wavenumber range suffered strong depletion. For example, the area under the 3374 cm^{-1} band had a reduction of 90 % showing an excellent efficiency by the microplasma jet in the breaking OH bonds of phenolic rings. Other important information from Fig. 11.4 was the strong structure reduction of bands around 2800–2900 cm^{-1}, after 1 h of treatment, which was due to CH radicals of side-chain and metoxyl groups. It can be seen in the same figure that the 1220 cm^{-1} band suffered a drastic reduction when the treatment time increased. These reductions in the areas under the bands indicated that C—C + C—O + C=O stretching in guaiacyl group and asymmetric in-phase ring stretching and C=C were destroyed by the microplasma jet particles. Another band severely reduced was the 830 cm^{-1} due to the breaking of bounds responsible for vibrations of C—H out-of-plane bonds on the aromatic ring.

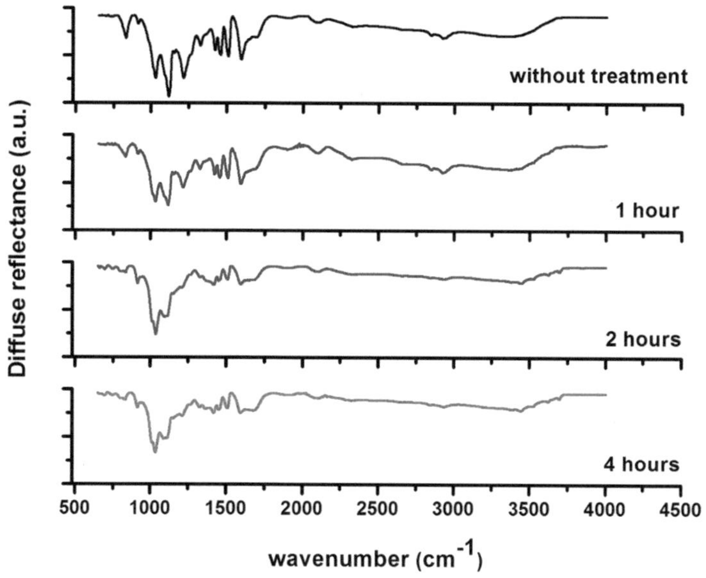

Fig. 11.4 DRIFT spectra for lignin treated in glow region of the atmospheric pressure argon microplasma jet as a function of time. Argon flow of 4.0 SLM and rf power of 20.0 W

Further studies are needed, which could make an important contribution in the field of biomass engineering processes, such as the discovery of new methods for obtaining high-added value products. The experimental investigation of plasma interaction mechanisms with biomass is a field to be explored, and experimental techniques of monitoring in real time the effect of plasma on biomass need to be developed.

The work described in (Song et al. 2009) reports the conversion of cellulose to simple sugars by plasmas. This result has not been confirmed by other studies yet, but the concept appears to be feasible. Even with this confirmation result, simple sugars are products with a low value for the energy cost of the plasma production so that the most cost effective alternative in this field appears to be the demand for biomass conversion methods into higher-added value products.

While many alternative sources of renewable energy have the potential to meet future demands for stationary power generation, biomass offers the most readily implemented, low cost solution to a drop-in transportation fuel for blending with and/or replacing conventional diesel via the bio-refinery concept, illustrated by carbohydrate pyrolysis/hydrodeoxygenation or lipid transesterification. Heterogeneous catalysis has a rich history of facilitating energy efficient selective molecular transformations and contributes to 90 % of chemical manufacturing processes and to more than 20 % of all industrial products (Shuit et al. 2013). In a post-petroleum era, catalysis will be central to overcoming the engineering and scientific barriers to economically feasible routes to alternative source of both energy and chemicals, notably bio-derived and solar-mediated via artificial photosynthesis.

The conventional heterogeneous catalysts involved in biodiesel production include mixed metal oxides, alkaline metal oxides, ion-exchange resins, sulfated oxides and immobilized enzymes. Heterogeneous catalysis has emerged as the preferred alternative for biodiesel production because the products are easy to separate, the catalysts are reusable, and the process is environmentally friendly. However, this method suffers from limitations, such as mass transfer problems, high cost and low catalyst stability, that diminish its economic feasibility and low environmental impact on the entire biodiesel process. Carbon nanotubes (CNTs) appear to be a promising catalyst support for biodiesel production due to their ability to overcome the limitations faced by conventional heterogeneous catalysts such as high specific area (Souza-Corrêa et al. 2013b). Thus, important application is the use of functionalized CNTs as catalyst support in biodiesel production, overcoming issues such as the limitations encountered by conventional heterogeneous catalysts, the advantages offered by functionalized CNTs and possible methods to functionalize CNTs to serve as catalyst support in biodiesel production. Another promising catalyst is the graphene sheets (Tatarova et al. 2014). Based on the recent research findings by the authors, functionalized CNTs and graphene sheets (Lambert et al. 2010) can easily be produced by plasma routes and also hold great potential to be a breakthrough technology in the biodiesel industry.

11.3 Plasmas in Medicine and Pharmacy

Plasma applied to medicine is a new emerging technology that is in the intercrossing of some fields of science like plasma physics, life science and clinical medicine. Plasma medical research is only a few years old, with experiments done in a broad range of different experimental conditions, e.g. various atmospheric pressure plasma sources in different in vitro/in vivo tissues, which become the comparison among them very limited and difficult. Only a systematic study of the different plasma sources with their respective reactive component composition will allow to evaluate the importance of plasma induced biological effects. In this section, we will describe the main experiments done until the present and possible explanations. The energetic electrons produced in the plasma suffer elastic and inelastic collisions with neutrals heating the gas or the liquid that are transported outside. It is also worth to mention that plasmas produce radiation in a vast range of the electromagnetic spectrum that can be important in a given process and reactive radicals.

In general, plasma medicine can be subdivided into two principal fields (von Woedtke et al. 2013b):

- Plasma-based or plasma supplemented techniques to treat bio-relevant surfaces, materials and devices to specific medical applications. These include the changes of surfaces morphology and wettability, functionalization to favor the adhesion of living cells or inhibit adhesion of proteins, bacteria or cells. Also important is the plasma treatment of surfaces for sterilization or bio-decontamination.
- Direct application of low-temperature plasmas in human body for therapeutic treatment in direct interaction with living tissue. This is the main activity of plasma medicine that constitutes the use of plasmas in dermatology, plastic surgery/dentistry, tissue regeneration, infected wounds and inflamed skin diseases.

Non-equilibrium plasmas are a particular medium because it contains neutral and ions at near room temperature and energetic electrons with temperatures at 20,000 K and higher. The main problem to use low-pressure plasmas in processing is the need of vacuum reactors. As the vacuum system is inconvenient to operate due to costs, low flexibility, materials need to be introduced in the chambers by batches during which contamination can occur. Furthermore, the reaction rates in this case are low becoming the treatment time relatively long. Additionally some material cannot withstand the low pressure conditions, as for example, organic materials, liquids, foods, living animal and plants tissues.

To circumvent the limitations of low-pressure plasmas in medical applications, atmospheric plasma sources were developed. Its exceptional chemical activity allows in principle to treat any surface while its low-background temperature makes them suitable to treat heat-sensitive surfaces. At atmospheric pressure the densities of reactants are orders of magnitude higher than at low-pressure which lead to reactions much faster.

Non-thermal plasma sources at atmospheric pressure can be tailored in such way to operate at human body temperature emitting little or no harmful radiation. With these sources a controllable amount of active species can be produced and addressed to the area to be treated without spreading over the whole target body. This property is important in order to limit to a minimum damage the organism in a non-destructive intervention. This is the most important motivation in introducing new plasma techniques in health care.

Plasma-induced physical and chemical modifications of the living tissue leading to possible therapeutic applications had been recently reviewed by researchers (Souza-Corrêa et al. 2014; Amorim et al. 2013; Song et al. 2009; Tatarova et al. 2014; Lambert et al. 2010; von Woedtke et al. 2013b). Here we present a brief discussion about each review article.

Kong et al. (2009) presented an introductory review on plasma health-care to provide the community with an overview of the current status of this emerging field, its scope, and its broad interdisciplinary approach, ranging from plasma physics, chemistry and technology, to microbiology, biochemistry, biophysics, medicine and hygiene. The article focuses on plasma interaction with prokaryotic cells, eukaryotic cells, cell membranes, DNA etc., taking care that the unfamiliar terminology to the physicists is covered and explained. The authors address the delivery of active substances, at the molecular or ionic level responsible for cell walls permeabilization, cell excitation (paracrine action) and the role of reactive species into cell cytoplasm. Electric fields, charging of surfaces, current flows etc., are also discussed once they can also affect the tissue.

Fridman et al. (2008) present the discharge plasmas applied to medicine as a non-equilibrium medium that is able to initiate, promote, control, and catalyze various complex behaviors and responses in biological systems. Plasmas can be tuned to achieve the desired medical effect, especially in medical sterilization and treatment of different kind of skin diseases. Wound healing and tissue regeneration can be achieved following various types of plasma treatment in a multitude of wound pathologies. Non-equilibrium plasmas promote non-destructive treatment of tissue, in safe, and effective matter in order to inactivate the various parasites organisms that attack the skin.

Laroussi (2005) did a survey on the recent progress on reduced-pressure plasma-based sterilization and/or decontamination. He also presented an overview on the atmospheric pressure non-equilibrium plasmas employed to treat cells of bacteria *i.e.* prokaryotic cells. In that paper he discussed the inactivation kinetics and the roles of the various plasma agents in the inactivation process. Plasma temperature, the UV emission, and concentrations of various reactive species measurements in air plasmas were also presented.

Stoffels (2007) reviewed applications of gas plasmas in medicine and introduced a device called plasma needle that operates in atmospheric pressure to be used in controlled tissue coagulation by an atmospheric plasma jet, bacterial disinfection using cold micro-plasmas and its relation to tissue-saving treatment of dental caries and non-inflammatory tissue removal based on sublethal plasma stimulation of living cells. Stoffels also discussed typical diagnostic methods for assaying the influence of cold plasmas on cells and bacteria.

Mofill et al. (2009) addressed an important issue that is the infection in hospital environment. In the paper the authors propose a new atmospheric plasma dispenser which has large area, robust electrode design in order to create atmospheric pressure discharges in air to be used in hospital disinfection presenting advantages, like higher bactericidal and fungicidal efficiency over current fluid disinfection systems. This device can be employed in future fight against the alarming and growing threat posed by bacterial infections in hospitals and community associated.

von Woedtke et al. (2013b) in a review article did a good analysis about the emerging field of plasma medicine. After a survey on the history of applications of plasma in the health care, the authors discussed present applications with an established basis of modern plasma medicine with focus on the design, development, characterization and challenges to be faced in the applications of plasma sources for therapeutic applications with an outlook of future projects.

In another timely and good topical review Graves (2009) stressed the emerging role of reactive oxygen (ROS) and reactive nitrogen species (RNS) in redox biology and presented some applications in medicine and biology. The purpose of this paper was to do a non-exhaustive review to identify some aspects concerning the actions of RONS, and to suggest that plasma biomedicine researchers consider these species as the probable active agents in related biomedical observations. Graves identified some of the most useful and most relevant works in the field of plasma biomedicine.

von Woedtke et al. (2013a) and Kong et al. (2009) from a general picture of medical applications of discharges proposed a classification of plasmas with relevance for medicine as:

- Direct plasmas—the plasma interacts directly with the living tissue. The skin or other tissue work as electrode in a manner that the electric current passes through it, e.g. DBD discharges.
- Indirect plasmas—the surface under treatment is exposed to the gas reactive flow that is rich in radicals and charges. The discharge is burned into two electrodes and the living tissue to be treated is located downstream, e.g. plasma needle or torche.
- Hybrid plasmas—is a combination of direct and indirect cases like a corona discharge.

As aforementioned, the atmospheric plasmas are more adapted to be used in medicine. Differently to the most common homogenous low-temperature discharges, the atmospheric pressure plasma devices are strictly localized. It poses difficulties to treat large surface areas but some solutions have been proposed (Isbary et al. 2010). The most interesting property of strong localization of atmospheric pressure discharges is the possibility of using the plasma reactivity to treat a well-defined target area. This characteristic can be exploited to access hard-to-reach regions such as the interior of organs through endoscopes (Polak et al. 2012) or catheters (Sato et al. 2008). Here, we will emphasize two innovative plasma devices used in medical applications. Others propositions exist and the reader may search in the review articles cited above in order to have a broad view about what had been published.

The first one device designed to be treat cells, bacterias, tissues and teeth is the plasma needle developed by Stofells' group (Stoffels et al. 2004; Stoffels 2007; Graves 2009). It is a He gas discharge jet that flows in a small diameter, 1 cm, passing in a thin needle electrode working at 13.56 MHz. The plasma is produced in a tip of the sharp needle that sometimes may attain length of some millimeters. If a small amount of oxygen is added to the helium flow ROS radicals are created. Without O_2 addition, RONS are created due to the inflow of surrounding air into the plasma. The figure below shows model predictions of various RNOS created along the plasma column (Graves 2009) (Fig. 11.5).

Another innovative device, DBD-based, capable of generating a low-temperature atmospheric pressure plasma inside a 5 m long flexible tube with 2 mm inner diameter is shown in Fig. 11.6.

The electrode arrangement consists of a double-walled PTFE tube with an inner and an outer tube concentrically aligned with two equidistant twisted electrodes in between. The distance between the electrodes is in the range of mm. The inner diameter of the double-walled tube is 2 mm, whereas the complete wall thickness is about 1 mm. The discharge operates in an argon gas flow of 1–2 slm (with possible oxygen and/or nitrogen admixture). The high-voltage source of 10 kHz with amplitude up to 11 kV is used. Homogeneous plasma can be ignited inside of several meters long catheters. This electrode configuration was primarily developed as an alternative construction of biopsy channels of endoscopes to guarantee a better decontamination and reprocessing. Due to a jet-like plasma expanding outside the end of the tube it can be used as an endoscope for cancer treatment. In harder-to-reach regions inside the body such as lungs, pancreas and duodenum, a 15 μm sized microplasma jet based on a hollow-core optical fiber was proposed (Polak et al. 2012). With this very localized plasma a single-cellular level treatment might be possible. It could be demonstrated that the generated plasma jets are sufficient to induce apoptosis but not necrosis in different in vitro cultivated tumor cell types (Kim et al. 2010).

Apoptosis is a form of programmed cell death, forming vesicles from the outer surface of the plasma membrane and the nuclear envelope, chromatin condensation, cleavage of DNA and formation of apoptotic bodies, see Fig. 11.7. Apoptosis is a highly regulated, energy-dependent processes while necrosis results from a trauma. Cells that die by apoptosis don't experiment rupture and therefore don't cause inflammation.

In necrosis, the cells are physically damaged, which results in rupture of membranes and release of cellular contents causing inflammation of tissue. Necrosis is a catastrophic injury that breaks down the mechanisms that maintain cell integrity. The membrane is damaged and the cytoplasm leaks out, releasing enzymes, that are recognized as antigens for which the organism responds with an alarm reaction known as inflammation. It works as a natural reaction that helps to remove

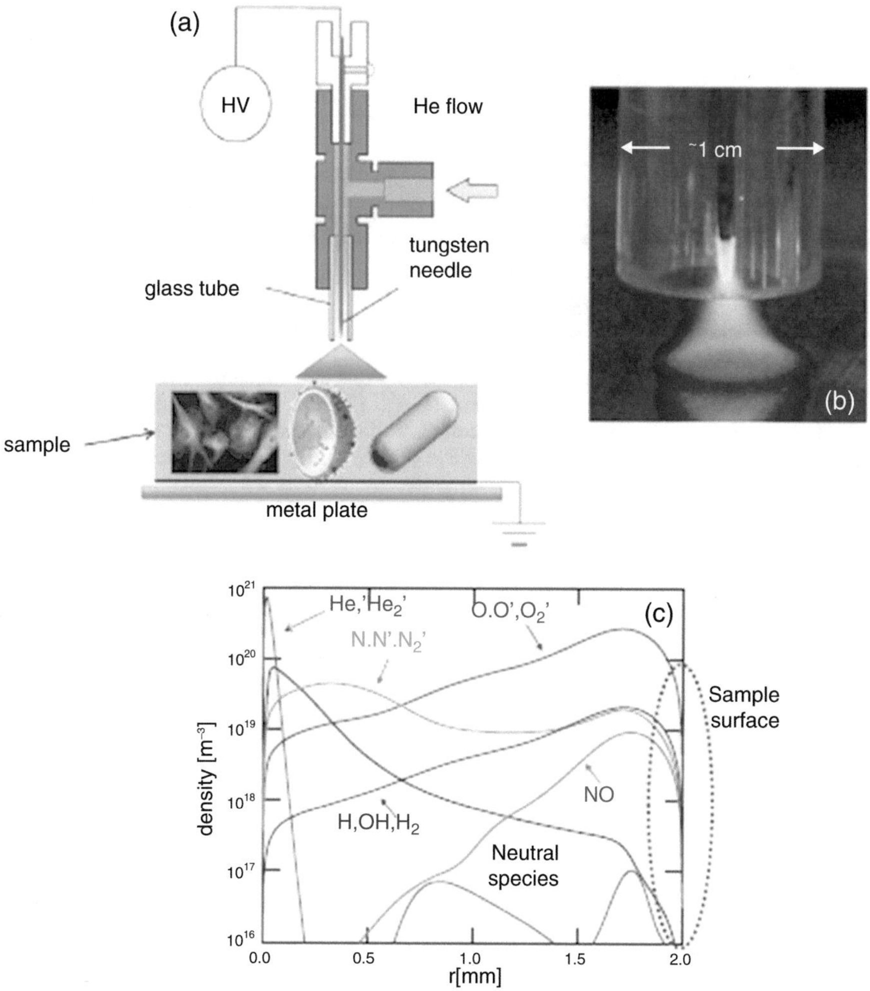

Fig. 11.5 (**a**) Sample been treat with plasma needle configuration in helium; (**b**) photograph of typical discharge. (**c**) Plot of neutral reactive species densities, following plasma model of He plasma needle near plasma-air boundary, with sample boundary at $r = 2.0$ mm (Graves 2009)

dangerous substance and dispose the necrotic tissue. One of undesired consequence of inflammation is fibrosis or connective tissue formation, which leads to scars and stenosis.

Beyond cancer treatment, further applications in gastroenterology is possible with endoscopic plasma due to a combination of antibacterial effects with healing and anti-inflammatory effects these sources could be useful for several indications.

There is no doubt today that low-temperature plasmas used in medical therapy create large amounts of reactive oxygen and nitrogen species, ROS and RNS respectively (Graves 2009) and have crucial role in biology and medicine (Halliwell

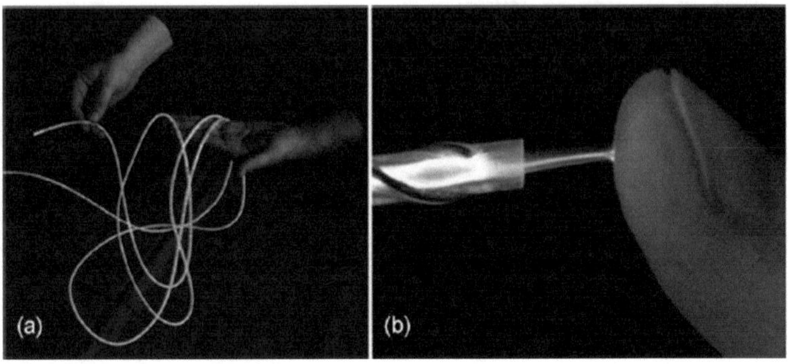

Fig. 11.6 Dielectric barrier discharge inside a long flexible tube with an inner diameter of 2 mm. The gas mixture was 1.5 slm Ar + 20 sccm N_2. (**a**) Commercially manufactured fluorinated ethylene propylene tube with flat wire electrodes, wall thickness: 0.75 mm. (**b**) Jet-like plasma at the end of a modified polytetrafluoroethylene (PTFE) tube with round electrode wires, wall thickness: 1 mm (Polak et al. 2012)

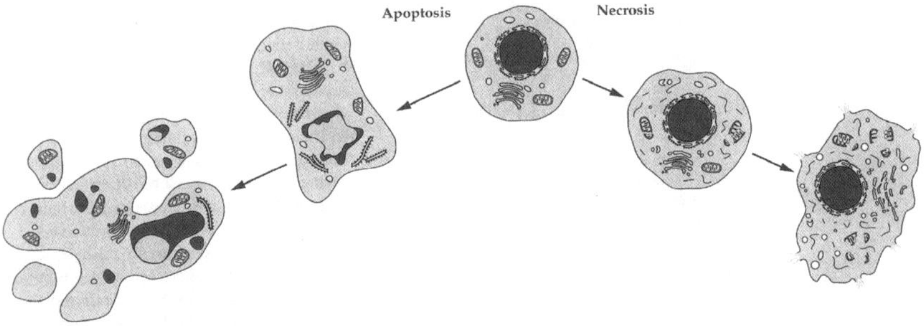

Fig. 11.7 Apoptosis and necrosis (Dangl et al. 2000)

and Gutteridge 2007). As short-living radicals, produced from air or water during the plasma treatment, RONS delivered to the cells can dissolve in physiological fluids and react with the living cells. However, the complete picture of this interaction is not well understood nowadays.

The ROS are molecular species that received electron(s) and are thus reduced forms of oxygen. The ROS includes two free radical species, the superoxide anion O_2^- and its protonated form the perhydroxyl radical HO_2, the uncharged and non-radical species like hydrogen peroxide H_2O_2, the highly reactive hydroxyl radical OH, and singlet oxygen $O_2(a^1\Delta_g)$. They are toxic molecules in which destructiveness relies on their reactivity. High concentrations may result in non-controlled oxidation of cell structures including DNA, proteins and cell membranes.

Due to their high reactivity ROS reacts with the cells' components. Atomic oxygen O is one of the most important ROS that is readily formed by electron impact dissociation and also by dissociative recombination;

$$e^{-} + O_2 \rightarrow O + O + e \quad (11.1)$$

$$e^{-} + O_2 \rightarrow O^{+} + O + 2e \quad (11.2)$$

$$e^{-} + O_2 \rightarrow O^{-} + O \quad (11.3)$$

$$e^{-} + O_2^{+} \rightarrow O + O \quad (11.4)$$

Atomic oxygen is a highly oxidizing agent that reacts with hydrogen compound and participates in the etching and oxidation of proteins.

Hydroxyl radical OH, are generated by the presence of water molecules in the plasma due for example the mixing of the gas with the surrounding air. They are formed by electron impact dissociation;

$$H_2O + e^{-} \rightarrow H + OH + e^{-} \quad (11.5)$$

and by dissociative electron recombination of H_3O^{+} that are rapidly formed in the gas;

$$H_3O^{+} + e^{-} \rightarrow OH + H_2 \quad (11.6)$$

Metastable states $O\,(^1D)$, $N_2\left(A\,^3\Sigma_u^+\right)$ and $O_2(a\,^1\Delta_g)$ formed in large concentrations in the gas phase may lead to the production of OH;

$$H_2O + N_2\left(A\,^3\Sigma_u^+\right) \rightarrow \mathrm{OH} + \mathrm{H} + N_2\left(X\,^1\Sigma_g^+\right) \quad (11.7)$$

$$H_2O + O\left(^1D\right) \rightarrow 2\mathrm{OH} \quad (11.8)$$

Another important ROS is ozone which may be formed by three-body reactions:

$$O + O_2 + O_2 \rightarrow O_3 + O_2 \quad (11.9)$$

$$O + O_2 + N_2 \rightarrow O_3 + N_2 \quad (11.10)$$

Others radicals may also be formed in large concentrations like H_2O_2, O_2^{-}, OOH, $ONOO^{-}$.

Reactive nitrogen species RNS formed in discharges has also great importance. One important radical is NO formed, for example, by:

$$N + O_3 \rightarrow NO + O_2 \quad (11.11)$$

$$N + NO_2 \rightarrow NO + NO \quad (11.12)$$

$$O + NO_2 \rightarrow NO + O_2 \quad (11.13)$$

Apart the DNA breaks discussed above, the RNOS' may lead to modification of unsaturated fatty acid, amino acid oxidation, peptide bind cleavage, sugar modifications and protein-protein crosslinks,

Besides reactive species, plasmas may generate ultraviolet radiation in a wavelength range from 100–380 nm. The Vacuum UV is in the range of 100–200 nm, UV-C from 200 to 280 nm, UV-B in the range 280–315 nm and UV-A from 315 to 380 nm. These UV photons have antimicrobial effect due to strand breaks that destruct and modify the DNA preventing the multiplication of bacterial cells.

As an example of ROS action, plasma interacts with water producing the peroxide anion O_2^{2-} which can react with organic molecules of the tissue. The O_2^{2-} is one radical that participates in the natural defense of the organism to inactivate the invading fungi or bacteria and to dispose damaged cells. This later process is very important in the body to renewal damaged or dangerous cells, like cancerous, by apoptosis.

The importance of ROS and RNS in medicine and biology was pointed out recently in many articles (Graves 2009, Novo and Parola 2008). These radicals are important in oxidative stress and work as mediators playing outstanding role in major human diseases, including atherosclerosis, diabetes, cardiovascular diseases, cancer, neurodegenerative disorders, chronic liver and lung diseases. Halliwell (2006) discussed the role of these radicals as antioxidant agents and their function to keep down the levels of free radicals enabling them to perform their functions without causing too much damage.

Graves (2009) in his review identified some aspects related to the effect of ROS in organism lifetime. Today is well accepted by the scientific community that ROS are associated with ageing and its related diseases. For example Ristow and Zarse (2010) put in evidence that ROS participating in mitochondrial metabolism may be associated with increasing in lifespan. The researchers concluded that ROS are essential signaling molecules, which are required to promote health and longevity, due to calorie restriction and specifically reduced glucose metabolism in mitochondria resulting in extend life span.

Cancer is recognized as the "emperor of all maladies" (Graves 2009). Cancer is a tumor or malignant neoplasm that can be presented in different diseases in many organs of the body. Today, cancer remains one of the most important diseases being the second cause of death after cardiovascular ones. Cancer leads to deformation of the abnormal cells and loss of cell differentiation. More than seven million people worldwide lose their lives due to cancer and this number may reach 10–15 million by 2020 (Nokhandani et al. 2015).

Recent works have shown that RONS species may be generated in plasmas that can help in the killing of cancer cells (Kong et al. 2009; von Woedtke 2013; Graves 2009). Recently Ralph et al. (2010), revisited the causes of cancer in a review that focuses on the evidence of the role of mitochondria as drivers of elevated ROS production during malignant transformation and hence, their potential as targets for cancer therapy. In fact, the chemistry of cancer starts with the creation of oxidants in cells which leads to DNA damage, mutation and modification in gene expression.

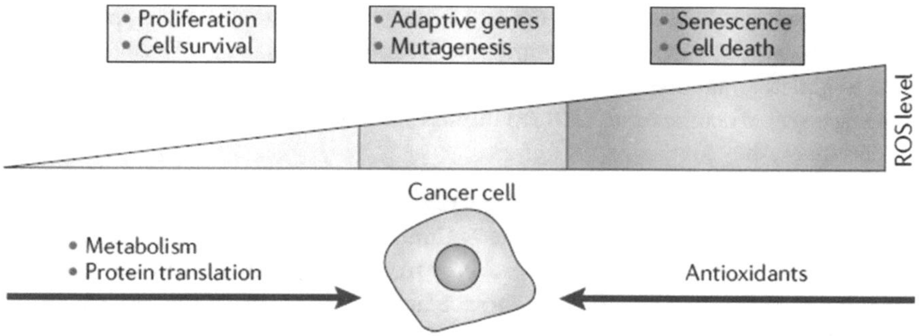

Fig. 11.8 ROS and cancer cell (Cairns et al. 2011). The level of ROS present in cell is decisive to determine the proliferation or cell death. Low levels of ROS (*left*) provide a beneficial effect, supporting cell proliferation and survival pathways, while excessively high ROS dose (*right*), may cause detrimental oxidative stress that can lead to cell death. In a cancer cell, highly metabolic and with protein translation generate abnormally high levels of antioxidants to counterbalance the ROS density to safe levels. Mutagenesis may occur even with moderate concentration of ROS

The importance of ROS and antioxidants aging on cancer cells is depicted in Fig. 11.8 according to Cairns et al. (2011) in a review article about the cancer cell regulation and metabolism. Higher levels of ROS may lead to cell death but relatively low-levels can lead to cell proliferation and pathways.

Some radicals are recognized to have strong reactivity in the biological medium. For example, NO plays a signaling hole in vascular smooth muscle relaxation (Marletta et al. 1988) and is a key specie in macrophage product responsible for cytostatic and respiratory inhibition in tumor target cells (Stuehr and Nathan 1989). It works as a signaling molecule in the cardiovascular system, discovery that was honored with a Nobel Prize in Medicine in 1998 (Graves 2009).

Hydrogen peroxide H_2O_2 is another ROS that is important as signaling molecule. Inside the cell the NO creates O_2^- which is converted to H_2O_2 and reacts in the intracellular fluid, cytosol, in order to oxidize enzymes. By a complex intracellular chemistry, the hydrogen peroxide reacts with enzymes inactivating them.

In a recent work Yagi et al. (2015) treated mouse melanoma cells using a nanosecond pulsed streamer discharge and determined the correlation between the rate of cell death, mainly induced by apoptosis, and the densities of OH, O, and NO, measured using laser-induced fluorescence (LIF). It was shown that reactive species derived from water vapor such as OH are responsible for the melanoma cell death, whereas those from O_2, such as O and NO, are less likely responsible. The correlation between OH density and cell death does not necessarily mean that only the flux of OH radicals is responsible for the cell death. Other reactive species, such as HO_2, H_2O_2 and HNO_x that are derived from water vapor, may also be responsible for the cell death because their fluxes are expected to depend on the gas flow rate similar to OH flux. They also indicate the importance of water evaporation from the culture medium surface in cell treatment. Others radicals are important in the cell signaling and the reader interested can see in that paper a more complete discussion (Graves 2009).

In the last years, a new field derived from plasma medicine has appeared and grown fast which is the plasma pharmacy. In reality, it is the use of discharge plasmas to generate, modify and stabilize pharmaceutical preparations in development of drugs. von Woedtke et al. (2013a) discuss the possibility of plasma applications in pharmacy, they are:

Plasma-Based Generation of Biologically Active Liquids Simple liquids like water or physiological saline become antimicrobially active after treatment with atmospheric pressure plasmas. These properties are attributable to the transient on-site generation of different low-molecular reactive species. The persistence of antiseptic activity in liquids and consequently the long-term stability of antimicrobially active substances generated in the liquid need further detailed research.

Plasma-Based Preparation, Optimization, or Stabilization of Pharmaceutical Preparations A very interesting new field is the use of atmospheric pressure plasmas to induce or catalyze chemical reactions in liquids. Plasmas opens the door to solubilize and/or stabilize poorly soluble or non-soluble substances in aqueous media using micelles without the detrimental side effect of high surfactant concentrations of the respective liquids.

Nanotechnology processes induced by plasmas contributes to the synthesis, functionalization and processing of nanomaterials and nanoparticles used for the fabrication and optimization of innovative nanoparticulate drug-delivery systems. Plasma-based activation of drugs in liquids may be used to activate substances which are not stable in liquids by first-pass metabolic processes inside the body after application and absorption.

Discharge plasmas can contribute in the preparation or modification of controlled drug release systems as multilayered tablets or functionalized composite powders for matrix systems by specific plasma-surface interactions.

Support of Drug Transport Across Biological Barriers Reversible manipulation of transfer characteristics of cell membranes as well as biological barriers like skin is called "plasma poration" that is modifying the permeability of cell membrane. Plasma source can be helpful to bring active agents into living cells or to enhance transdermal drug permeation. Future plasma treatments could be used to support intracellular effectivity of pharmacological substances in cancer therapy, for example, or to permit or enhance transcutaneous drug delivery.

Plasma-Based Stimulation of Biotechnological Processes Plasmas emit radiation in a broad range domain and specially in UV region and -are sources of highly reactive radicals which can promote the generation and improvement of strains production. These radiation and species may also be used to induce desired changes of the DNA. Low-pressure plasmas may treat spores and improve the fermentation efficiency for production of bacteria used commercially in animal health and agriculture.

Biotechnology plasma has potential to become an important tool to breed high-yielding strains for production of drugs and to promote the transfer of DNA into suitable expression systems. This field of plasma application is at the beginning and much efforts should be done to develop the potentialities.

To conclude, the application of plasmas in medicine and pharmacy will need to understand and control the hole of the radicals delivered to the right place with right dose in a given time. These exogenous sources of species and photons must be studied in order to establish methods and protocols to control the most important species in a given treatment of cells and tissues composed of proteins, carbohydrates and lipids. In the next section we will see how low-temperature plasmas may be used to treat plants, seeds and foods.

11.4 Plasmas in Biology and Agriculture

Plasmas applied in biology is another recent field of plasmas applications due to their ability to generate reactive species and photons. As discussed in Sect. 11.3 the discharge plasmas may generate RONS' that participate in the redox biology. RONS' may be generated inside the cells through many different pathways where enzymes have an important role, this is called endogenous source of RONS'. The plasmas as an external source of RONS may induce a series of different reactions than the ones produced inside the cells. In a recent review about the role of reactive oxygen and nitrogen species in redox biology and implications of plasmas on medicine and biology, Graves (2009) writes:

> One can imagine many possible future applications, including the use of plasma to protect plants from fungus or other microbial or even parasite attack, but this will require much additional work. In the meantime, it will be helpful to investigate more deeply the roles played by RONS in conventional plant physiology and disease.

Plasmas may work as exogenous sources of RNOS' modifying the redox chemistry of plants in order to promote processes in the cells like signaling, metabolism and immunity in plants. One example, may be the unique today, is the work published by Puač et al. (2006) where a plasma needle set-up was employed to induce effects in plant tissue. The authors analyzed the interaction of the plasma needle with gametophytes and calli as representatives of small multicellular plant organisms, two month old prothalli of *Polypodium vulgare L.* and parts of the calli *Fritillaria imperialis*. Cell death (necrosis) of the *Polypodium prothalium* occurred after high doses (power treatment time) of plasma. In these experiments the treatment time is in the range of 120 s at a power of 100 W of the RF source but the effective power at the plasma needle is 1 W, which is a condition that reduces the thermal transfer to the surrounding gas and the living tissue. Figure bellow shows the effect of the plasma in the plant tissue (Fig. 11.9).

In the treatment of fresh weight of compact calli the plasma needle treatment doubled all treated calli fresh weight compared with the untreated sample, even for longer exposure times. Increase in the fresh weight is an obvious implication of calli growth, which is an irreversible increase in size, accomplished by a combination

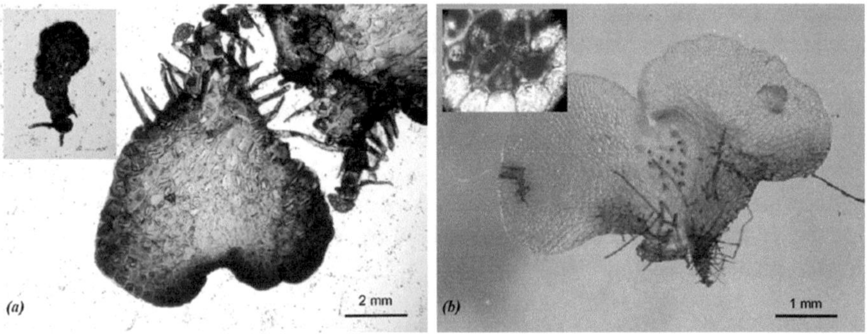

Fig. 11.9 Plasma treatment of sweet fern gametophyte. (**a**) The treatment time is 120 s, and RF source power 100 W. Insert: 1 month old gametophyte; (**b**) the treatment time is 30 s and RF source power 300 W. Insert: necrotic cells (Puač et al. 2006)

of cell division and cell enlargement. Authors concluded that RONS' formed in plasmas are responsible for the effects observed in plants.

Plant cells are continuously producing ROS as products of normal aerobic metabolism and it is not surprisingly that plants have enzymatic and non-enzymatic antioxidant mechanisms to counter balance the growth of ROS which may leads to reach destructive levels, regulating the cellular redox balance. ROS are not only toxic by-products acting as a cell damage agent but have other important functions in plants. If their concentration is low, they may have a function of signaling molecules that act at the interface between abiotic and biotic stress. They have important role as coordinators of cell biology responses to numerous environment stimuli that include pathogen attack, drought, irradiation, temperature and salt stresses.

Several cellular compartments participate in the generation of ROS (Desikan et al. 2005). Non-enzymatic mechanisms of ROS generation, such as electron transfer to molecular oxygen during photosynthesis and respiration, are done in chloroplasts and mitochondria, respectively, and may also be created as by-products of various enzymes such as photo-respiratory glycolate oxidase in peroxisomes for example. In fact, exogenous ROS may cause a response in cells not normally exposed to it as in pharmacological manipulation of endogenous ROS via application of drugs that interact with enzymes that produce or inhibit ROS.

One illustration is given to show how plants through ROS react to attacks of bacteria, fungi, viruses or other aggression (Torres 2010). Figure 11.10 shows that ROS created inside and outside of the plant cell react to a pathogen and as a consequence activated mechanisms called "hypersensitive response" induce a programmed cell death at the site of infection.

Surowsky et al. (2015) in a review article summarized the research done in the interactions of low-temperature plasmas at atmospheric pressure in contact with solid and liquid food systems. Quality and safe are major concerns that attract much attention today to improve the efficiency and sustainability of foods.

Atmospheric pressure plasmas are a recent technology that can be decisive in antimicrobial efficacy mainly in temperature-delicate foods. In last 20-years the

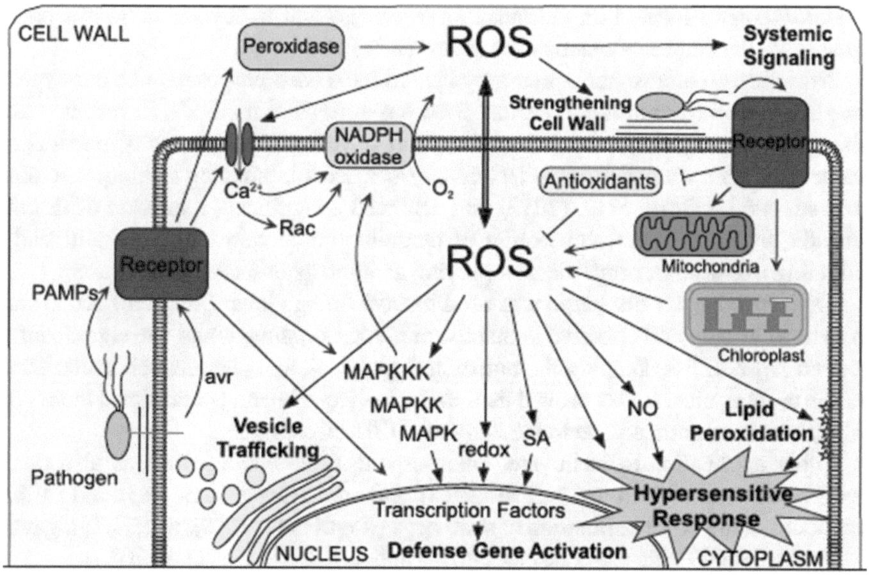

Fig. 11.10 Pathogen recognition leads to ROS production. Thin arrows depict signaling events that point to ROS production both in the apoplast and inside the plant cell. Double-head *arrow* indicates the cross talk between ROS in these compartments. Thick *arrows point* to the functions of these ROS in relation to activation of plant defenses (Torres 2010)

number of publications dealing with microbial inactivation in foods has grown strongly attaining around 800 papers in 2013 (Surowsky et al. 2015).

The first paper with direct treatments of foods was done by Montenegro et al. (2002) where non-thermal plasma discharges were applied directly into reducing the number of *Escherichia coli* O157:H7 cells in apple juice, by several orders of magnitude. This is a good example of application of plasmas in liquids, a field that recently attracted the interest of plasmas physicists (Bruggeman et al. 2009). A plasma discharge in a liquid is medium of complex kinetics that includes a combination of various physical-chemical processes that rely on a large number of factors such as electric conductivity of the liquids, pH, polarity, electrodes' geometry, and other liquid properties. Non-thermal plasmas can be created by an electrical discharge between two electrodes placed within a liquid food. Electrons in non-thermal plasmas are hot whereas the other larger species generated in the solution, such as ions, excited species, and free radicals are at ambient temperature.

Cold atmospheric gas plasma has been applied in the surface treatment of lettuce, potato and strawberry in order to inactivate microorganisms, e.g. Salmonella enterica serovar Typhimurium, located in the food surfaces (Fernández et al. 2013).

Employing a plasma jet with helium and oxygen Perni et al. (2008) studied the impact of low-temperature plasma on E. coli and Saccharomyces cerevisiae in the pericarp of mango and honeydew melon. Inactivation were achieved after 5 s for E. coli on both surfaces and after 10 s for S. cerevisiae on mango and 30 s on honeydew melon. Longer treatment times are necessary for the same microorganisms located on cut surfaces of mango and honeydew melon in order to achieve the degree of inactivation observed before.

Non-thermal atmospheric pressure plasmas have been proposed to be a prospective alternative to traditional thermal food pasteurization methods. In recent years the effect of plasmas in the formation of off-flavors and the losses of nutritional value have been studied. Plasma technology as a decontaminating technique in milk was studied by Gurol et al. (2012) that achieved a significant reduction of E. coli in milk, by more than a three order of magnitude reduction without significantly affecting pH or color properties, employing an atmospheric corona discharge.

Dielectric barrier discharge was used to treat orange juice (Shi et al. 2011) and proved to be safe and effective in inactivating microorganisms and can significantly extend its shelf life, keeping the nutritional, physical, and chemical characteristics of the orange juice. It was proved that, under the experimental conditions, there is a pH decrease of orange juice induced by the DBD discharge.

Other application to inactivated microorganisms in apple juice was also done with good results (Surowsky et al. 2014). Others examples can be found in the literature in the decontamination of shell eggs (Ragni et al. 2010), water (Kamgang-Youbi et al. 2009), cheese (Lee et al. 2012) and tomato (Pankaj et al. 2013).

An interesting application of plasmas in agriculture is the treatment of seeds to improve germination. Ling et al. (2014) showed that cold plasma had an active effect on soybean seed germination and in the vigor indices, which is an index that reveals the ability of a seed to withstand a variety of different stress factors, were increased by cold plasma treatments. Another seeds like tomato (Zhou et al. 2011) and wheat (Sera et al. 2010) treat by plasmas significantly improved the germination rate.

An emerging and powerful method for surface decontamination of both foodstuffs and food packaging materials is a new trend to decontaminate in-package, offering non-thermal treatment of foods post-packaging (Pankaj et al. 2014). In a recent study, non-thermal plasma was used to degrade pesticide residues on strawberries. The use of pesticides in agriculture is important to the stabilization of crop production. However, it can be a source of environmental and health problems associated with and cannot be overlooked. Fungicides, such as azoxystrobin, cyprodinil, fludioxonil and pyriproxyfen are relatively new pesticides that have been introduced into the marketplace. Misra et al. (2014) employed a DBD discharge, in atmospheric-pressure in air, to degrade pesticides residues in an in-packed strawberries, see Fig. 11.11.

The authors quantified the degradation of fungicides, namely azoxystrobin, cyprodinil, fludioxonil, and pyriproxyfen on strawberry surface by GC-MS/MS analysis, under the influence of DBD discharge treatment. They explained the observed effects doing at the same time of the treatment the electrical and optical characterization of the plasma. They concluded that DBD discharge may ensure chemical food safety, in addition to its proven microbicidal effects. They had shown that plasma successfully degraded pesticide residues on strawberries. Operating in the filamentary regime, the RONS´ were responsible to reduce the levels of azoxystrobin, cyprodinil, fludioxonil and pyriproxyfen by 69, 45, 71 and 46 % respectively after 5 min of treatment at 80 kV (RMS).

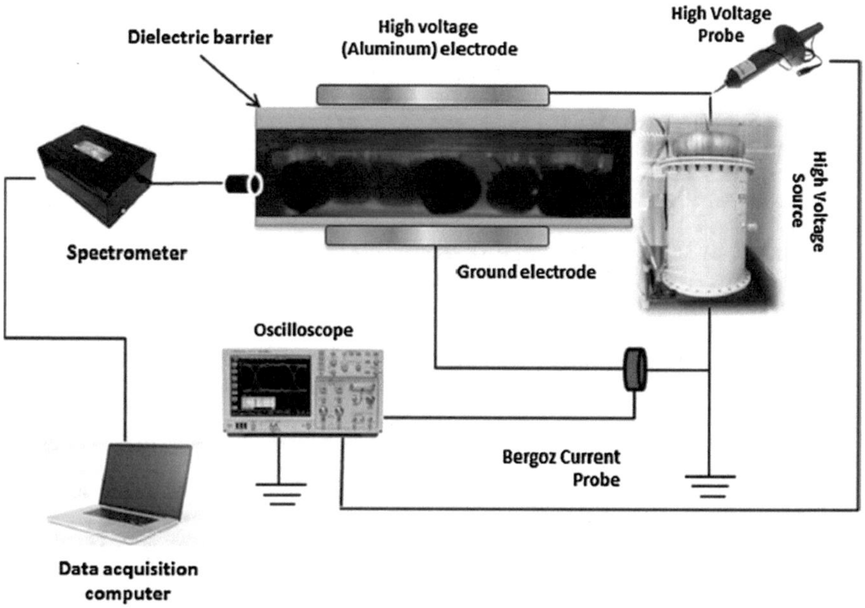

Fig. 11.11 Schematic of the DBD experimental set-up for in-package treatment. The package shown in the figure is filled with helium for demonstration purpose (Misra et al. 2014)

The aim of this chapter was to present some important applications of low-temperature plasmas (LTP). LTP discussed in this book, are partially ionized gases or liquids with a broad use in many technological applications such as microelectronics, light sources, lasers, biology and medicine. LTPs lead to the production of atomic and molecular excited states, chemically reactive radicals, photons, and activated surface sites, which are in the origin of the deposition of thin films, advanced nanotechnologies products, solar cells, highly efficient combustion motors, and treatment of cancer cells. The field where low-temperature plasmas are being used today is large, rapidly growing, and cannot be described in just one chapter. Of course, much is not presented but some new applications such as the use plasmas in health care, production of biofuels and agriculture were addressed.

References

J. Amorim, C. Oliveira, J.A. Souza-Corrêa, M.A. Ridenti, Treatment of sugarcane bagasse lignin employing atmospheric pressure microplasma jet in argon plasma. Processes Polym. **10**, 670–678 (2013)

G. Bonizzoni, E. Vassallo, Plasma physics and technology; industrial applications. Vacuum **64**, 327–336 (2002)

P. Bruggeman, C. Leys, Non-thermal plasmas in and in contact with liquids. J. Phys. D Appl. Phys. **42**, 053001 (2009)

R.A. Cairns, I.S. Harris, T.W. Mak, Regulation of cancer cell metabolism. Nat. Rev. Cancer **11**, 85–95 (2011)

C.-L. Song, Z.-T. Zhang, W.-Y. Chen, L. Cheng, Converting cornstalk into simple sugars with high pressure nonequilibrium plasma. IEEE Trans. Plasma Sci. **37**, 1817–1824 (2009)

J.L. Dangl, R.A. Dietrich, H. Thomas, *Biochemistry & Molecular Biology of Plants*, eds. by B. Buchanan, W. Gruissem, R. Jones. American Society of Plant Physiologists. (Rockville, MD, 2000)

R. d'Agostino, P. Favia, C. Oehr, M.R. Wertheimer, Low-temperature plasma processing of materials: past, present, and future. Plasma Processes Polym. **2**, 7–15 (2005)

R. Desikan, J. Hancock, S. Neil, in *Antioxidants and Reactive Oxygen Species in Plants*. ed. by N. Smirnoff, 2005, 169

A. Fernández, E. Noriega, A. Thompson, Inactivation of Salmonella enterica serovar typhimurium on fresh produce by cold atmospheric gas plasma technology. Food Microbiol. **33**, 24–29 (2013)

G. Fridman, G. Friedman, A. Gutsol, A.B. Shekhter, V.N. Vasilets, A. Fridman, Applied plasma medicine. Plasma Processes Polym. **5**, 503–533 (2008)

T.F. Fuerst, M.O. Reese, C.A. Wolden, PECVD synthesis of flexible optical coatings for renewable energy applications. Plasma Process. Polym. 201500114 (2015)

D.B. Graves, The emerging role of reactive oxygen and nitrogen species in redox biology and some implications for plasma applications to medicine and biology. J. Phys. D Appl. Phys. **45**, 263001 (2009)

C. Gurol, F.Y. Ekinci, N. Aslan, M. Korachi, Low temperature plasma for decontamination of E. coli in milk. Int. J. Food Microbiol. **157**, 1–5 (2012)

B. Halliwell, Reactive species and antioxidants. Redox biology is a fundamental theme of aerobic life. Plant Physiol. **141**, 312–22 (2006)

B. Halliwell, J.M.C. Gutteridge, *Free Radicals in Biology and Medicine* (Oxford University Press, New York, 2007)

G. Isbary, G. Morfill, H.U. Schmidt, M. Georgi, K. Ramrath, J. Heinlin, S. Karrer, M. Landthaler, T. Shimizu, B. Steffes, W. Bunk, R. Monetti, J.L. Zimmermann, R. Pompl, W. Stolz, A first prospective randomized controlled trial to decrease bacterial load using cold atmospheric argon plasma on chronic wounds in patients. Br. J. Dermatol. **163**, 78–82 (2010)

G. Kamgang-Youbi, J.-M. Herry, T. Meylheuc, J.-L. Brisset, M.-N. Bellon-Fontaine, A. Doubla, M. Naïtali, Microbial inactivation using plasma-activated water obtained by gliding electric discharges. Lett. Appl. Microbiol. **48**, 13–18 (2009)

J.Y. Kim, J. Ballato, P. Foy, T. Hawkins, Y. Wei, J. Li, S.O. Kim, Single-cell-level cancer therapy using a hollow optical fiber-based microplasma. Small **6**, 1474–1478 (2010)

M.G. Kong, G. Kroesen, G. Morfill, T. Nosenko, T. Shimizu, J. van Dijk, J.L. Zimmermann, Plasma medicine: an introductory review. New J. Phys. **11**, 115012 (2009)

T.N. Lambert, C.C. Luhrs, C.A. Chavez, S. Wakeland, M.T. Brumbach, Graphite oxide as a precursor for the synthesis of disordered graphenes using the aerosol-through-plasma method. Carbon **48**, 4081–4089 (2010)

M. Laroussi, Low temperature plasma-based sterilization: overview and state-of-the-art. Plasma Processes Polym. **2**, 391–400 (2005)

H.J. Lee, S. Jung, H. Jung, S. Park, W. Choe, J.S. Ham, C. Jo, Evaluation of a dielectric barrier discharge plasma system for inactivating pathogens on cheese slices. J. Anim. Sci. Technol. **54**, 191–198 (2012)

L. Ling, J. Jiafeng, L. Jiangang, S. Minchong, H. Xin, S. Hanliang, D. Yuanhua, Effects of cold plasma treatment on seed germination and seedling growth of soybean. Sci. Rep. **4**(5859), 1–7 (2014)

A.I. Maksimov, A.Y. Nikiforov, Comparison of plasma and plasma–solution modifications of polymer materials in the liquid phase. High Energy Chem. **41**, 454–459 (2007)

M.A. Marletta, P.S. Yoon, R. Iyengar, C. Leaf, J.S. Wishnok, Macrophage oxidation of L-arginine to nitrite and nitrate-nitric-oxide is an intermediate. Biochemistry **27**, 8706–8711 (1988)
N.N. Misra, S.K. Pankaj, T. Walsh, F. O'Regan, P. Bourke, P.J. Cullen, In-package nonthermal plasma degradation of pesticides on fresh produce. J. Hazard. Mater. **271**, 33–40 (2014)
T. Miura, S.-H. Lee, S. Inoue, T. Endo, Combined pretreatment using ozonolysis and wet-disk milling to improve enzymatic saccharification of Japanese cedar. Bioresour. Technol. **126**, 182–186 (2012)
J. Montenegro, R. Ruan, H. Ma, P. Chen, Inactivation of E-coli O157: H7 using a pulsed nonthermal plasma system. J. Food Sci. **67**, 646–648 (2002)
G.E. Morfill1, T. Shimizu, B. Steffes, H.-U. Schmidt, Nosocomial infections-a new approach towards preventive medicine using plasmas. New J. Phys. **11**, 115019 (2009)
A.M. Nokhandani, S.M.T. Otaghsara, M.K. Abolfazli, M. Karimi, F. Adel, H. Babapour, G. Atae, A review of new method of cold plasma in cancer treatment. Scholars Acad. J. Biosci. **3**, 222–230 (2015)
E. Novo, M. Parola, Redox mechanisms in hepatic chronic wound healing and fibrogenesis. Fibrogenesis Tissue Repair **1**, 1–58 (2008)
S.K. Pankaj, N.N. Misra, P.J. Cullen, Kinetics of tomato peroxidase inactivation by atmospheric pressure cold plasma based on dielectric barrier discharge. Innovative Food Sci. Emerg. Technol. **19**, 153–157 (2013)
S.K. Pankaj, C. Bueno-Ferrer, N.N. Misra, V. Milosavljević, C.P. O'Donnell, P. Bourke, K.M. Keener, P.J. Cullen, Applications of cold plasma technology in food packaging. Trends Food Sci. Technol. **35**, 5–17 (2014)
S. Perni, D.W. Liu, G. Shama, M.G. Kong, Cold atmospheric plasma decontamination of the pericarps of fruit. J. Food Prot. **71**, 302–308 (2008)
M. Polak, J. Winter, U. Schnabel, J. Ehlbeck, K.-D. Weltmann, Innovative plasma generation in flexible biopsy channels for inner-tube decontamination and medical applications. Plasma Processes Polym. **9**, 67–76 (2012)
N. Puač, Z.L. Petrović, G. Malović, A. Đordević, S. Živković, Z. Giba, D. Grubišić, Measurements of voltage–current characteristics of a plasma needle and its effect on plant cells. J. Phys. D Appl. Phys. **39**, 3514–3519 (2006)
L. Ragni, A. Berardinelli, L. Vannini, C. Montanari, F. Sirri, M.E. Guerzoni, A. Guarnieri, Non-thermal atmospheric gas plasma device for surface decontamination of shell eggs. J. Food Eng. **100**, 125–132 (2010)
S.J. Ralph, S. Rodríguez-Enríquez, J. Neuzil, E. Saavedra, R. Moreno-Sánchez, The causes of cancer revisited: Mitochondrial malignancy and ROS-induced oncogenic transformation – why mitochondria are targets for cancer therapy. Mol. Aspects Med. **31**, 145–70 (2010)
M. Ristow, K. Zarse, How increased oxidative stress promotes longevity and metabolic health: the concept of mitochondrial hormesis (mitohormesis). Exp. Gerontol. **45**, 410–418 (2010)
T. Sato, O. Furuya, K. Ikeda, T. Nakatani, Generation and transportation mechanisms of chemically active species by dielectric barrier discharge in a tube for catheter sterilization. Plasma Processes Polym. **5**, 606–614 (2008)
B. Sera, P. Spatenka, M. Sery, N. Vrchotova, I. Hruskova, Influence of plasma treatment on wheat and oat germination and early growth. IEEE Trans. Plasma Sci. **38**, 2963–2968 (2010)
J.A. Souza-Corrêa, M.A. Ridenti, C. Oliveira, S.R. Araújo, J. Amorim, Decomposition of lignin from sugar cane bagasse during ozonation process monitored by optical and mass spectrometries. J. Phys. Chem. B **117**, 3110–3119 (2013a)
J.A. Souza-Corrêa, C. Oliveira, L.D. Wolf, V.M. Nascimento, G.J.M. Rocha, J. Amorim, Atmospheric pressure plasma pretreatment of sugarcane bagasse: the influence of moisture in the ozonation process. Appl. Biochem. Biotechnol. **171**, 104–116 (2013b)
J.A. Souza-Corrêa, C. Oliveira, L.D. Wolf, V.M. Nascimento, G.J.M. Rocha, J. Amorim, Atmospheric pressure plasma pretreatment of sugarcane bagasse: the influence of biomass particle size in the ozonation process. Appl. Biochem. Biotechnol. **172**, 1663–1672 (2014)
R. Suchentrunk, G. Staudigl, D. Jonke, H.J. Fuesser, Industrial applications for plasma processes - examples and trends. Surf. Coat. Technol. **97**, 1–9 (1997)

N. Schultz-Jensen, F. Leipold, H. Bindslev, A.B. Thomsen, Plasma-assisted pretreatment of wheat straw. Appl. Biochem. Biotechnol. **163**, 558–572 (2011)
X. Shi, G. Zhang, X. Wu, Y. Li, Y. Ma, X. Shao, Effect of low-temperature plasma on microorganism inactivation and quality of freshly squeezed orange juice. IEEE Trans. Plasma Sci. **39**, 1591–1597 (2011)
S.H. Shuit, K.F. Yee, K.T. Lee, B. Subhash, S.H. Tan, Evolution towards the utilization of functionalized carbon nanotubes as a new generation catalyst support in biodiesel production: an overview. RSC Adv. **3**, 9070–9094 (2013)
Z. Songa, J. Tang, J. Li, H. Xiao, Plasma-induced polymerization for enhancing paper hydrophobicity. Carbohydr. Polym. **92**, 928–933 (2013)
E. Stoffels, R.E.J. Sladek, I.E. Kieft, H. Kersten, R. Wiese, Power outflux from the plasma: an important parameter in surface processing. Plasma Phys. Controlled Fusion **46**, B167–B177 (2004)
E. Stoffels, in *Applications of Gas Plasmas in Medicine*, ed. by J. Amorim. Gas Discharges Fundamentals & Applications (Transworld Research Network, Kerala, 2007), pp. 83–112
D.J. Stuehr, C.F. Nathan, Nitric-oxide- a macrophage product responsible for cytostasis and respiratory inhibition in tumor target-cells. J. Exp. Med. **169**, 1543–1555 (1989)
B. Surowsky, A. Fröhling, N. Gottschalk, O. Schlüter, D. Knorr, Impact of cold plasma on Citrobacter freundii in apple juice: inactivation kinetics and mechanisms. Int. J. Food Microbiol. **174**, 63–71 (2014)
B. Surowsky, O. Schlüter, D. Knorr, Interactions of non-thermal atmospheric pressure plasma with solid and liquid food systems: a review. Food Eng. Rev. **7**, 82–108 (2015)
E. Tatarova, A. Dias, J. Henriques, A.M. Botelho do Rego, A.M. Ferraria, M.V. Abrashev, C.C. Luhrs, J. Phillips, F.M. Dias, C.M. Ferreira, Microwave plasmas applied for the synthesis of free standing graphene sheets. J. Phys. D Appl. Phys. **47**, 385501 (2014)
M.A. Torres, ROS in biotic interactions. Physiol. Plant. **138**, 414–429 (2010)
F.B. Vurzel, L.S. Polak, Plasma chemical technology-future of chemical industry. Ind. Eng. Chem. **62**, 8–22 (1970)
T. von Woedtke, B. Haertel, K.-D. Weltmann, U. Lindequist, Plasma pharmacy – physical plasma in pharmaceutical applications. Pharmazie **68**, 492–498 (2013a)
T. von Woedtke, S. Reuter, K. Masur, K.-D. Wektmann, Plasmas for medicine. Phys. Rep. **530**, 291–320 (2013b)
I. Yagi, Y. Shirakawa, K. Hirakata, T. Akiyama, S. Yonemori, K. Mizuno, R. Ono, T. Oda, Measurement of OH, O, and NO densities and their correlations with mouse melanoma cell death rate treated by a nanosecond pulsed streamer discharge. J. Phys. D Appl. Phys. **48**, 424006 (2015)
X. Zhang, M.S. Cha, The reformation of liquid hydrocarbons in an aqueous discharge reactor. J. Phys. D Appl. Phys. **48**, 215201 (2015)
Z.W. Zhou, Y.F. Huang, S.Z. Yang, W. Chen, Introduction of a new atmospheric pressure plasma device and application on tomato seeds. Agric. Sci. **2**, 23–27 (2011)

Erratum

Kinetics and Spectroscopy of Low Temperature Plasmas

Jorge Loureiro and Jayr Amorim

J.M.A.H. Loureiro, J. de Amorim Filho, *Kinetics and Spectroscopy of Low Temperature Plasmas*, Graduate Texts in Physics,
DOI 10.1007/978-3-319-09253-9

DOI 10.1007/978-3-319-09253-9_12

The publisher regrets the errors in the copyright page, preface, and chapters 3, 5, 8, 9 and 11 in the print and online versions of this book. The corrected versions are given on the following pages:

The online version of the updated original book can be found at
http://dx.doi.org/10.1007/978-3-319-09253-9

J.M.A.H. Loureiro, J. de Amorim Filho, *Kinetics and Spectroscopy of Low Temperature Plasmas*, Graduate Texts in Physics,
DOI 10.1007/978-3-319-09253-9_12

The affiliation of the author “Henriques Loureiro” has been incorrectly captured in page iv and the correct affiliation is as follows:

“Instituto Superior Técnico”

The sentence “low-temperature plasmas in Instituto Superior Técnico” in page vii (Preface) has been incorrectly captured. The correct sentence is as follows:

“low-temperature plasmas at Instituto Superior Técnico”

The equation 3.12 (page 90, Chapter 3) is followed by an alphabet “*r*” which is incorrect, “*r*” should be deleted.

The sentence in first paragraph of Section 3.3.1 (page 110, Chapter 3) is incorrectly captured as “density case, in which”. The correct sentence is as follows:

“density, case in which”

The sentence in first paragraph (page 209, Chapter 5) is incorrectly captured as “so that, the effective ionization”. The correct sentence is as follows:

“so that the effective ionization”

The last display equation (page 350, Chapter 8) is incorrectly captured. The correct equation is as follows:

$$\sum_{m<2} A_{2m} = \left(A_{21} l_{j'=3/2, j''=1/2} + A_{21} l_{j'=3/2, j''=1/2}\right)_{2p-1s} = 1.253 \times 10^{9}\ \mathrm{s}^{-1}$$

The word in second paragraph (page 361, Chapter 9) is incorrectly captured as “frequency”. The correct word is as follows:

“frequencies”

The words in second paragraph (page 374, Chapter 9) is incorrectly captured as “becomes more”. The correct usage is as follows:

“become more”

The sentence in third paragraph (page 376, Chapter 9) is incorrectly captured as “of metastable losses”. The correct sentence is as follows:

“for metastable losses”

The coefficient in Table 9.4 caption (page 376, Chapter 9) is incorrectly captured as “C_{e-M}”. The correct coefficient is as follows:

“C_{e-o}^{M}”

The sentence in second paragraph (page 426, Chapter 11) is incorrectly captured as "figure bellow show". The correct sentence is as follows:

"The figure below shows"

The sentence in third paragraph (page 430, Chapter 11) is incorrectly captured as "one radical that participate". The correct sentence is as follows:

"one radical that participates"

The word in second paragraph (page 434, Chapter 11) is incorrectly captured as "interacts". The correct usage is as follows:

"interact"

Index

J.M.A.H. Loureiro, J. de Amorim Filho, *Kinetics and Spectroscopy of Low Temperature Plasmas*, Graduate Texts in Physics,
DOI 10.1007/978-3-319-09253-9

C

V

W